ASTEROIDS, COMETS, METEORS

IAU SYMPOSIUM No. 229

COVER ILLUSTRATION: IAUS229-ACM2005 Logo

This is the logo of the IAUS229-ACM2005 developed by Thiago Moeda Santanna, staff member of the Observatório Nacional, host institution of the meeting. The logo was designed to represent the hosting city and the scientific interest of the meeting. The city of Búzios (Rio de Janeiro, Brazil) is symbolized by a typical fisherman while small asteroids, comets and meteors are trapped in its nest. The following citation best describes the logo:

"Like the fishermen at Búzios we cast our net to the sky to catch the Asteroids, Comets, Meteors and Trans-Neptunian Objects"
The Local Organizing Committee

INTERNATIONAL ASTRONOMICAL UNION

UNION ASTRONOMIQUE INTERNATIONALE

ASTEROIDS, COMETS, METEORS

PROCEEDINGS OF THE 229th SYMPOSIUM OF THE
INTERNATIONAL ASTRONOMICAL UNION
HELD IN BÚZIOS, RIO DE JANEIRO, BRASIL
AUGUST 7–12, 2005

Edited by

DANIELA LAZZARO
Observatório Nacional, Rio de Janeiro, Brazil

SYLVIO FERRAZ MELLO
IAG, São Paulo University, São Paulo, Brazil

and

JULIO ANGEL FERNÁNDEZ
Astronomy Department, Faculty of Sciences, Montevideo, Uruguay

CAMBRIDGE UNIVERSITY PRESS
The Edinburgh Building, Cambridge CB2 2RU, UK
40 West 20th Street, New York, NY 10011–4211, USA
477 Williamstown Road, Port Melbourne, VIC 3207, Australia
Ruiz de Alarcón 13, 28014 Madrid, Spain
Dock House, The Waterfront, Cape Town 8001, South Africa

First published 2006

Printed in the United Kingdom at the University Press, Cambridge

Typeset in System LaTeX 2_ε

A catalogue record for this book is available from the British Library

Library of Congress Cataloguing in Publication data

ISBN-13 978 0521 85200 5 hardback
ISBN-10 0521 85200 5 hardback
ISSN 1743-9213

Table of Contents

Preface

Since the realization of the first meeting, more than 20 years ago, in Uppsala (Sweden), ACM has grown to become one of the most important conferences in Planetary Sciences gathering, this year, 263 participants from 31 different countries. As in previous occasions, ACM was dedicated not only to the study of these three categories of objects themselves, but special attention was given to their interrelations. In addition, one window on the small natural satellites was kept open, as these objects may have an important relation with the small bodies wandering in our Solar System.

It was shown in the successive conferences the enormous progress of the area thanks to several space missions aimed at the study of small bodies, but also to the introduction of new observational techniques using large telescopes, and also to the increasing number of observation facilities dedicated to the search and discovery of new objects that may approach the Earth in the course of their motion.

We are glad that this time ACM convened in Brazil. This country has firmly supported investigations in Astronomy and Astrophysics being a partner in several international projects such as Gemini, SOAR, CoRoT, Auger, etc. The association with ESO in the recent past allowed the development of an extensive mineralogical survey of asteroids. The Brazilian astronomers are glad every time they can host an international conference as ACM and increase their contacts with the international community. Furthermore, the location favoured the participation of a large number of scientist from South American countries. The organization of the ACM 2005 was supported by many national agencies and institutions and special thanks are addressed to Observatório Nacional, the Conselho Nacional de Desenvolvimento Científico e Tecnológico (CNPq), the Coordenação de Aperfeiçoamento de Pessoal de Nível Superior (CAPES), the Financiadora de Estudos e Projetos (FINEP), and the Research Foundations of the states of São Paulo (FAPESP) and Rio de Janeiro (FAPERJ). We also have to thank the support of the International Astronomical Union, which accepted to include ACM2005 among its official Symposia. Thanks are also due to the Centro LatinoAmericano de Física (CLAF), the European Space Agency (ESA) and the Organizing Committee of ACM 1996 for their financial support.

This volume includes, with just one exception, the invited lectures presented at ACM 2005 meeting following the order of their presentation at the meeting. The exception is one of the included papers that the SOC programmed not as a lecture but as a series of concatenated oral communications. We thank the colleagues that acted as referees of these papers for their fast response and for the many suggestions helping to improve the contents and the completeness of the papers. It would have been impossible to include in one volume all the nearly 400 contributions presented at ACM 2005, without reducing them to extended abstracts of small interest. We thank Giovanni Valsecchi and the journal *Earth, Moon and Planets* for accepting to publish a special issue with contributed papers presented at ACM 2005.

Last, but not least, we have to thank the members of the organizing committees: the members of the LOC for the large amount of work they have done in order to have this meeting organized properly and the members of the SOC and the Advisory Board for their active participation in the organization of the scientific program.

Sylvio Ferraz Mello and Julio Angel Fernández, co-chairs SOC,
Daniela Lazzaro, chair LOC
São Paulo, Montevideo, Rio de Janeiro, November 15, 2005

THE ORGANIZING COMMITTEE

Scientific

R.P. Binzel (USA)
A. Coradini (Italy)
S. Ferraz-Mello (co-chair, Brazil)
J.A. Fernández (co-chair, Uruguay)
G. Hahn (Germany)

D. Lupishko (Ukarine)
A. Morbidelli (France)
J. Ticha (Czech Rep.)
J. Watanabe (Japan)
I.P. Williams (UK)

Advisory Board

M.F. A' Hearn (USA)
M.A. Barucci (France)
H. Boehnhardt (Germany)
E. Bowell (USA)
H. Campins (USA)
A.W. Harris (Germany)

A.W. Harris (USA)
D. Jewitt (USA)
H.U. Keller (Germany)
A. Milani (Italy)
H. Rickman (Sweden)
V. Zappalá (Italy)

Local

C. Beaugé (Argentina)
M. Florczak (Brazil)
R. Gil-Hutton (Argentina)
D. Lazzaro (chair, Brazil)
T. Michtchenko (Brazil)

F. Roig (Brazil)
R. Scorzelli (Brazil)
G. Tancredi (Uruguay)
W. Vilas-Boas (Brazil)
T. Yokoyama (Brazil)

Acknowledgements

The symposium is sponsored and supported by the IAU Division III (Planetary Systems Sciences) and by the IAU Commissions No. 15 (Physical Studies of Comets and Minor Planets), No. 20 (Positions and Motions of Minor Planets, Comets and Satellites), No. 22 (Meteors, Meteorites and Interplanetary Dust) and No. 7 (Celestial Mechanics and Dynamical Astronomy).

The Local Organizing Committee operated under the auspices of the
Observatório Nacional, Rio de Janeiro.

Funding by the
International Astronomical Union – IAU,
Conselho Nacional de Desenvolvimento Científico e Tecnológico – CNPq,
Financiadora de Estudos e Projetos – FINEP,
Coordenação de Aperfeiçoamento de Pessoal de Nível Superior – CAPES,
European Space Agency – ESA,
Organizing Committee of ACM 96,
Fundação de Apoio à Pesquisa do Estado de São Paulo – FAPESP,
Fundação de Apoio à Pesquisa do Estado do Rio de Janeiro – FAPERJ,
Centro Latino-Americano de Física – CLAF,
and
Observatório Nacional,
is gratefully acknowledged.

Support by the
Prefeitura de Búzios
is also acknowledged.

Participants

Paul **Abell**, NASA Johnson Space Center, Huston, USA — paul.a.abell1@jsc.nasa.gov
Michael F. **A'Hearn**, University of Maryland, College Park, USA — ma@astro.umd.edu
Alvaro A. **Alvarez-Candal**, Observatório Nacional, Rio de Janeiro, Brazil — alvarez@on.br
Diana **Andrade-Pilling**, Instituto de Química, UFRJ, Rio de Janeiro, Brazil — dianaufrj@gmail.com
Oscar **Arratia**, University of Valladolid, Valladolid, Spain — oscarr@eis.uva.es
Alexander V. **Bagrov**, Institute of Astronomy of Russian Acad. of Sci., Moscow, Russia — abagrov@inasan.ru
Kiran S. **Baliyan**, Physical Research Laboratory, Ahmedabad, India — baliyan@prl.res.in
Maria A. **Barucci**, LESIA, Observatoire de Paris, Meudon, France — antonella.barucci@obspm.fr
Irina **Belskaya**, Astronomical Observatory of Kharkiv University, Kharkov, Ukraine — irina@astron.kharkov.ua
Paula G. **Benavidez**, Universidad de Alicante, Alicante, Spain — paula@dfists.ua.es
Fabrizio **Bernardi**, Institute for Astronomy, University of Hawaii, Honolulu, USA — bernardi@ifa.hawaii.edu
Bidushi **Bhattacharya**, Spitzer Science Center, Caltech, Pasadena, USA — bhattach@ipac.caltech.edu
Jens **Biele**, DLR RS/MUSC, Köln, Germany — jens.biele@dlr.de
Richard P. **Binzel**, Massachusetts Institute of Technology, Cambridge, USA — rpb@mit.edu
Nicolas **Biver**, LESIA, Observatoire de Paris, Meudon, France — nicolas.biver@obspm.fr
Carlo **Blanco**, Dip. di Fisica e Astronomia, Università di Catania, Catania, Italia — cblanco@ct.astro.it
Andrea **Boattini**, INAF, Osservatorio Astronomico di Roma, Roma, Italia — boattini@mporzio.astro.it
Dominique **Bockelee-Morvan**, LESIA, Observatoire de Paris, Meudon, France — dominique.bockelee@obspm.fr
Heloisa **Boechat-Roberty**, Observatório do Valongo, UFRJ, Rio de Janeiro, Brazil — heloisa@ov.ufrj.br
Herman **Boehnhardt**, MPI for Solar System Research, Katlenburg-Lindau, Germany — boehnhardt@linmpi.mpg.de
Daniel **Boice**, Southwest Research Institute, Boulder, USA — dboice@swri.edu
Jeremie **Boissier**, LESIA, Observatoire de Paris, Meudon, France — jeremie.boissier@obspm.fr
Jiří **Borovička**, Astronomical Institute of the Academy of Sciences, Ondrejov, Czech Rep. — borovic@asu.cas.cz
William **Bottke**, Southwest Research Institute, Boulder, USA — bottke@boulder.swri.edu
Edward **Bowell**, Lowell Observatory, Flagstaff, USA — ebowell@lowell.edu
Ramon **Brasser**, Queen's University, Kingston, Ontario, Canada — brasser_astro@yahoo.com
Noah **Brosh**, Wise Observatory, Tel Aviv University, Tel Aviv, Israel — noah@wise.tau.ac.il
Miroslav **Brož**, Astronomical Institute, Charles University, Prague, Czech Rep. — mira@sirrah.troja.mff.cuni.cz
Rosario **Brunetto**, Dipartimento di Fisica, Università di Lecce, Lecce, Italy — rbrunetto@ct.astro.it
Schelte J. **Bus**, University of Hawaii, Hilo, USA — sjb@ifa.hawaii.edu
Renato N. **Cabral**, Observatório Nacional, Rio de Janeiro, Brazil — renato@on.br
Fernando **Cachucho**, UNIFEV, Votuporanga, Brasil — cachucho@fev.edu.br
Nelson **Callegari Jr.**, DEMAC, Universidade de São Paulo, Rio Claro, Brazil — calleg@rc.unesp.br
Humberto **Campins**, University of Central Florida, Orlando, USA — campins@physics.ucf.edu
Adriano **Campo Bagatin**, Universidad de Alicante, Alicante, Spain — adriano@dfists.ua.es
David **Čapek**, Astronomical Institute, Charles University, Prague, Czech Rep. — capek@sirrah.troja.mff.cuni.cz
Valerio **Carruba**, IAG, Universidade de São Paulo, São Paulo, Brazil — valerio@astro.iag.usp.br
Jorge Márcio F. **Carvano**, LESIA, Observatoire de Paris, Meudon, France — jorge.carvano@obspm.fr
Alberto **Cellino**, INAF, Osservatorio Astronomico di Torino, Pino Torinese, Italy — cellino@to.astro.it
Clark R. **Chapman**, Southwest Research Institute, Boulder, USA — cchapman@boulder.swri.edu
Carolina A. **Chavero**, Observatório Nacional, Rio de Janeiro, Brazil — carolina@on.br
Andrew **Cheng**, Johns Hopkins University Applied Physics Laboratory, Laurel, USA — andrew.cheng@jhuapl.edu
Steven **Chesley**, Jet Propulsion Laboratory, Pasadena, USA — steven.r.chesley@jpl.nasa.gov
Young-Jun **Choi**, Jet Propulsion Laboratory, Pasadena, USA — young-jun.choi@jpl.nasa.gov
François **Colas**, IMCCE / Observatoire de Paris, Paris, France — colas@imcce.fr
Maria **Colín-Garcia**, Instituto de Ciencias Nucleares, Mexico — mcolin@nucleares.unam.mx
Angioletta **Coradini**, INAF–IASF, Tor Vergata, Roma, Italy — angioletta.coradini@rm.iasf.cnr.it
Marcello **Coradini**, European Space Agency, Paris, France — marcello.coradini@esa.int
Ricardo Reis **Cordeiro**, Departamento de Física, UFV, Viçosa, Brazil — cordeiro@ufv.br
Cristiano **Cosmovici**, Istituto di Fisica dello Spazio Interplanetario, Roma, Italy — cosmo@ifsi.rm.cnr.it
Jacques **Crovisier**, Observatoire de Paris, Meudon, France — jacques.crovisier@obspm.fr
Enio **da Silveira**, Pontificia Universidade Catolica, Rio de Janeiro, Brazil — enio@fis.puc-rio.br
John **Davies**, UK Astronomy Technology Centre, Edimburgh, UK — jkd@roe.ac.uk
Donald **Davis**, Planetary Space Institute, Tucson, USA — drd@psi.edu
Francisco X. **de Araújo**, Observatório Nacional, Rio de Janeiro, Brazil — araujo@on.br
Ramiro J. **de La Reza** , Observatório Nacional, Rio de Janeiro, Brazil — delareza@on.br
Julia **de León**, Instituto de Astrofisica de Canárias, La Laguna, Spain — jmlc@iac.es
Francesca **De Meo**, Massachusetts Institute of Technology, Cambridge, USA — fdemeo@mit.edu
Cristina M. **de Sanctis**, INAF, Roma, Italy — mariacristina.desanctis@rm.iasf.cnr.it
Alan **Delamere**, Delamare Support Services, Boulder, USA — alan@delamere.biz
Marco **Delbó**, INAF, Osservatorio Astronomico di Torino, Torino, Italy — delbo@to.astro.it
Andrey **Delsanti**, Institute of Astronomy, Hawaii, USA — delsanti@ifa.hawaii.edu
Gabriel **Denicol**, Instituto de Física, UFRJ, Rio de Janeiro, Brazil — gsdenicol@if.ufrj.br
Romina P. **di Sisto**, Fac. de Ciencias Astron. e Geofisicas, UNLP, La Plata, Argentine — romina@fcaglp.unlp.edu.ar
Pablo X.**do Prado Cocher**, UNESP, Rio Claro, Brazil — paprado@rc.unesp.br
Henry **Dones**, Southwest Research Institute, Boulder, USA — luke@boulder.swri.edu
Michael **Drake**, Lunar and Planetary Laboratory, University of Arizona, Tucson, USA — drake@lpl.arizona.edu
René **Duffard**, Observatório Nacional, Rio de Janeiro, Brazil — duffard@on.br
Daniel D. **Durda**, Southwest Research Institute, Boulder, USA — durda@boulder.swri.edu
Piotr A. **Dybczynski**, Astronomical Observatory, A. Mickiewicz University, Poznam, Poland — dybol@amu.edu.pl
Beata **Dziak-Jankoska**, Institute of Geophysics, Warsaw University, Warsaw, Poland — bdziak@igf.fu.edu.pl
Vacheslav **Emel'yanenko**, South Ural University, Chelyabinsk, Russia — vvemel@math.susu.ac.ru
Gustavo S. **Faraudo**, PUC-Rio, Rio de Janeiro, Brazil — gfaraudo@yahoo.com
Lucio **Farenzena**, PUC-Rio, Rio de Janeiro, Brazil — lucio@vdg.fis.puc-rio.br
Gislaine **Felipe**, UNESP, Guaratinguetá, Brazil — xlaine@uol.com.br
Julio A. **Fernández**, Depto. de Astronomia, Faculdad de Ciencias, Montevideo, Uruguay — julio@fisica.edu.uy
Yan R. **Fernández**, University of Central Florida, Orlando, USA — yan@physics.ucf.edu
Sylvio **Ferraz-Mello**, IAG-USP, São Paulo, Brazil — sylvio@astro.iag.usp.br
George **Flynn**, SUNY, Plattsburgh, USA — flynngj@plattsburgh.edu
Marcello **Fulchignoni**, LESIA, Observatoire de Paris, Meudon, France — marcello.fulchignoni@obspm.fr
Ryszard **Gabryszewski**, Space Research Center, Polish Acad. of Sci., Warsaw, Poland — kacper@cbk.waw.pl
Adrián **Galád**, Comenius University, FMPI, Bratislava, Slovak Rep. — galad@fmph.cas.cz
Tabare **Gallardo**, Faculdad de Ciencias, Montevideo, Uruguay — gallardo@fisica.edu.uy
Ricardo **Gil-Hutton**, Félix Aguilar Observatory, San Juan, Argentina — rgilhutton@educ.ar
Silvia M. **Giuliatti Winter**, UNESP, Guaratinguetá, Brazil — silvia@feg.unesp.br
Brett **Gladman**, University of British Columbia, Vancouver, Canada — gladman@astro.ubc.ca
Fred **Goesmann**, MPI for Solar System Research, Katlenburg-Lindau, Germany — goesmann@mps.mpg.de
Raymond **Golstein**, Southwest Research Institute — rgoldstein@swri.edu
Larissa **Golubeva**, Shemakha Astrophysical Observatory, Baku, Azerbaijan — land@azdata.net

Rodney S. **Gomes**, Observatório Nacional, Rio de Janeiro, Brazil — rodney@on.br
Michael **Granvik**, Observatory, University of Helsinki, Helsinki, Finland — michael.granvik@helsinki.fi
Simon **Green**, PSSRI, The Open University, Milton Keynes, UK — s.f.green@open.ac.uk
Giovanni-Federico **Gronchi**, Dept. of Mathematics, University of Pisa, Pisa, Italy — gronchi@dm.unipi.it
Olivier **Groussin**, University of Maryland, College Park, USA — groussin@astro.umd.edu
Pedro **Gutierrez**, Instituto de Astrofísica de Andalucía, Granada, Spain — pedroj@iaa.es
Edith **Hadamcik**, Service d' Aéronomie, Verrières le Buisson, France — edith.hadamcik@aerov.jussieu.fr
Therese **Häggström**, Dept. of Earth Sciences, Goteborg University, Göteborg, Sweden — therese@gvc.gu.se
Alan W. **Harris**, Space Science Institute, Boulder, USA — awharris@spacescience.org
Alan W. **Harris**, DLR Institute of Planetary Research, Berlin, Germany — alan.harris@dlr.de
Walter **Harris**, University of Washington, Seattle, USA — wmharris@u.washington.edu
Pedro H. **Hasselman**, UFRJ e Observatório Nacional, Rio de Janeiro, Brazil — pedrohasselmann@yahoo.com.br
Daniel **Hestoffer**, IMCCE, Observatoire de Paris, Paris, France — hesto@imcce.fr
Martin **Hilchenbach**, MPI für Sonnen System for Schung, Katlenburg-Lindau, Germany — hilchenbach@linmpi.mpg.de
Jonathan **Hillier**, The Open University, Milton Keynes, UK — j.k.hillier@open.ac.uk
James **Hilton**, U.S. Naval Observatory, Washington, USA — jhilton@aa.usno.navy.mil
Keith **Holsapple**, University of Washington, Seattle, USA — holsapple@aa.washington.edu
Paulo **Holvorecem**, Holvorcem Consultoria e Comércio de Software Ltda., Campinas, Brazil — holvorecem@mpc.com.br
Kevin **Housen**, The Boeing Co., Seattle, USA — kevin.r.housen@boeing.com
Ellen **Howell**, Arecibo Observatory, Arecibo, USA — ehowell@naic.edu
Henry H. **Hsieh**, Institute for Astronomy, University of Hawaii, Honolulu, USA — hsieh@ifa.hawaii.edu
Hauke **Hussmann**, IAG-USP, São Paulo, Brazil — hauke@astro.iag.usp.br
Serguei **Ipatov**, University of Maryland, College Park, USA — sipatov@hotmail.com
Takashi **Ito**, National Astronomical Observatory of Japan, Tokyo, Japan — tito@cc.noa.ac.jp
Violeta **Ivanova**, Institute of Astronomy, Bulgarian Academy, Sofi, Bulgaria — ivanova@astro.bas.bg
Peter **Iza**, PUC, Rio de Janeiro, Brazil — izapeter@vdg.fis.puc-rio.br
Kandy **Jarvis**, Jacobs Sverdrup / ESCG, USA — kandy.s.jarvis@jsc.nasa.gov
Robert **Jedicke**, Institute for Astronomy, University of Hawaii, Honolulu, USA — jedicke@ifa.hawaii.edu
Klaus **Jockers**, MPI för Sonnen System Schung, Germany — jockers@mps.mpg.de
Daniel **Jones**, Queen Mary, University of London, London, UK — d.c.jones@qmul.ac.uk
Hans U. **Kaeufl**, European Southern Observatory, Garching, Germany — hukaufl@eso.org
J.J. **Kavelaars**, Herzberg Institute of Astrophysics, Victoria, Canada — jjk@astro.ubc.ca
Horst Uwe **Keller**, MPI fur Sonnen System for Schung, Katlenburg-Lindau, Germany — keller@linmpi.mpg.de
Nicolai **Kiselev**, Institute of Astronomy of Kharkiv University, Kharkiv, Ukraine — kiselev@kharkov.ukrtel.net
Boris **Klumov**, MPI fuer Extraterrestrische Physics, Garching, Germany — klumov@mpe.mpg.de
Zoran **Knežević**, Astronomical Observatory, Belgrade, Serbia & Montenegro — zoran@aob.bg.ac.yu
Matthew **Knight**, University of Maryland, College Park, USA — mmk8a@astro.umd.edu
Ludmilla **Kolokolova**, University of Maryland, College Park, USA — ludmilla@astro.umd.edu
Pavel **Koten**, Astronomical Institute, Academy of Sciences, Odrejov, Czech Rep. — koten@asu.cas.cz
Malgorzata **Królikowska**, Space Research Center, Polish Acad. of Sci., Warsaw, Poland — mkr@cbk.waw.pl
Agnieszka **Kryszczynska**, Astronomical Observatory, A. Mickiewicz Univ., Poznan, Poland — agn@amu.edu.pl
Michael **Küppers**, MPI für Sonnensystem for Schung, Katlenburg-Lindau, Germany — kuppers@mps.mpg.de
Luisa M. **Lara**, IAA – CSIC, Granada, Spain — lara@iaa.es
Jérémie **Lasue**, Service d'Aéronomie/Univ. Paris 6, Verrières le Buisson, France — jeremie.lasue@aerov.jussieu.fr
Conor **Laver**, University of California, Berkeley, USA — conor@astro.berkeley.edu
Daniela **Lazzaro**, Observatório Nacional, Rio de Janeiro, Brazil — lazzaro@on.br
Susan **Lederer**, California State Univ. SB, San Bernardino, USA — slederer@csusb.edu
Jacek **Leliwa-Kopystynski**, Warsaw University, Warsaw, Poland — jkopyst@mimuw.edu.pl
A.-C. **Levasseur-Regourd**, Paris VI/Serv. Aéronomie, Verrières le Buisson, France — chantal.levasseur@aerov.jussieu.fr
Javier **Licandro**, ING & IAC, Santa Cruz de La Palma, Spain — licandro@ing.iac.es
Carey **Lisse**, John Hopkins University, Applied Physics Laboratory, USA — carey.lisse@jhuapl.edu
Silvia **Lorenz-Martins**, Observatório do Valongo, UFRJ, Rio de Janeiro, Brazil — slorenz@ov.ufrj.br
Amy **Lovell**, Agnes Scott College, Decatur, USA — alovell@agnesscott.edu
Stephen **Lowry**, Queen's University, Belfast, UK — s.c.lowry@qub.ac.uk
Dimitrij F. **Lupishko**, Inst. of Astronomy, Kharkiv University, Kharkiv, Ukraine — lupishko@astron.kharkov.ua
Patryk S. **Lykawka**, Kobe University, Kobe, Japan — patryk@kobe-u.ac.jp
Teemu **Mäkinen**, Finish Meteorological Institute, Helsinki, Finland — teemu.makinen@fmi.fi
Ingrid **Mann**, Institut für Planetologie, Müenster Univ., Müenster, Germany — imann@uni-muenster.de
Franck **Marchis**, University of California, Berkeley, USA — fmarchis@berkeley.edu
Rafael **Martinez**, PUC-Rio, Rio de Janeiro, Brazil — rodrigues@vdg.fis.puc-rio.br
Paulo **Martini**, INPE, São José dos Campos, Brazil — martini@dsr.inpe.br
John **Matese**, University of Louisiana at Lafayette, Lafayette, USA — matese@louisiana.edu
Neil **Mc Bride**, The Open University, Milton Keynes, UK — n.m.mcbride@open.ac.uk
Lucy **McFadden**, University of Maryland, College Park, USA — mcfadden@astro.umd.edu
William **McKinnon**, Washington University, Saint Louis, USA — mckinnon@wustl.edu
Karen **Meech**, Institute for Astronomy, University of Hawaii, Honolulu, Hawaii — meech@ifa.hawaii.edu
Mario **Melita**, IAFE, UBA, Buenos Aires, Argentine — melita@iafe.uba.ar
Frederic **Merlin**, Observatoire de Paris, Paris, France — frederic.merlin@obspm.fr
William **Merline**, Southwest Research Institute, Boulder, USA — merline@boulder.swri.edu
Tadeusz **Michalowski**, Astronomical Observatory, A. Mickiewicz Univ., Poznan, Poland — tmich@amu.edu.pl
Claude **Michaux**, Jet Propulsion Laboratory, Pasadena, USA — claude.michaux@jpl.nasa.gov
Tatiana **Michtchenko**, IAG-USP, São Paulo, Brazil — tatiana@astro.iag.usp.br
Andrea **Milani**, Dept. Mathematics, University of Pisa, Pisa, Italy — milani@dm.unipi.it
David W. **Mittlefehldt**, NASA Johnson Space Center, Huston, USA — david.w.mittlefehdt@nasa.gov
Alessandro **Morbidelli**, CNRS – Observatoire Côte D' Azur, Nice, France — alessandro.morbidelli@obs-nice.fr
Thais **Mothé Diniz**, LESIA, Observatoire de Paris, Meudon, France — thais.mothe@obspm.fr
Stefano **Mottola**, DLR – German Aerospace Center, Köln, Germany — stefano.mottola@dlr.de
Olivier **Mousis**, Observaoire de Besançon, Besançon, France — olivier.mousis@obs-besancon.fr
Thomas **Mueller**, MPI Fuer Extraterrestrisohe Physik, Garching, Germany — tmueller@mpe.mpg.de
Michael **Müller**, DLR Berlin, Inst. of Planetary Research, Berlin, Germany — michael.muller@dlr.de
Eirik **Mysen**, Inst. Of Theoretical Astrophysics, Oslo, Norway — emysen@astro.uio.no
Andreas **Nathues**, MPI für Sonnen System for Schung, Katlenburg-Lindau, Germany — nathues@linmpi.mpg.de
David **Nesvorný**, Southwest Research Institute, Boulder, USA — davidn@boulder.swri.edu
Erica C. **Nogueira**, Inst. Física, UFRJ, Rio de Janeiro, Brazil — erica.nogueira@ov.ufrj.br
Michael **Nolan**, Arecibo Observatory, Arecibo, USA — nolan@naic.edu
Keith S. **Noll**, Space Telescope Science Institute, Baltimore, USA — noll@stsci.edu
David **O'Brien**, Observatoire de La Côte D'Azur, Nice, France — obrien@obs-nice.fr
José L. **Ortiz**, Instituto de Astrofisica de Andalucía – CSIC, Granada, Spain — ortiz@iaa.es
David **Osip**, Las Campanas Observatory, La Serena, Chile — dosip@lco.cl
Eric **Pantin**, Service d'Astrophysique SACLAY, Gif-sur-Yvette, France — epantin@cea.fr
John **Papaloizou**, DAMTP Cambridge, Cambridge, UK — j.c.b.papaloizou@maths.qmw.ac.uk
Gabriela M. **Parisi**, Dep. de Astronomia, Universidad de Chile, Santiago, Chile — gparisi@das.uchile.cl

Neil **Parley**, The Open University, Milton Keynes, UK — n.r.parley@open.ac.uk
Thierry **Pauwels**, Koninklijke Sterrenwacht van Belgie, Bussel, Belgium — thierry.pauwels@oma.be
Rade **Pavlovic**, Astronomical Observatory Belgrade, Belgrade, Serbia & Montenegro — rpavlovic@aob.bg.ac.yu
Nuno **Peixinho**, CAAUL, Observatório Astronómico de Lisboa, Lisboa, Portugal — peixinho@oal.ul.pt
Elena **Petrova**, Space Research Inst. of Russia Academy of Sciences, Moscow, Russia — epetrova@iki.rssi.ru
Carlé **Pieters**, Dep. of Geological Sciences, Brown University, Providence, USA — carle_pieters@brown.edu
Sérgio **Pilling**, Inst. Química, UFRJ, Rio de Janeiro, Brazil — sergiopilling@yahoo.com.br
Noemi **Pinilla-Alonso**, IAC, Fundacion Galileo Galilei, Santa Cruz de La Palma, Spain — npinilla@tng.iac.es
Jana **Pittichová**, Institute for Astronomy, University of Hawaii, Honolulu, USA — jana@ifa.hawaii.edu
David **Polishook**, Tel-Aviv University, Tel-Aviv, Israel — david@wise.tau.ac.il
Vladimir **Porubcan**, Fac. of Math., Phys. & Inform., Comenius Univ., Bratislava, Slovak Rep. — porubcan@fmph.uniba.sk
Petr **Pravec**, Astronomical Inst., Acad. of Sci. of the Czech Rep., Ondřejov, Checz Rep. — ppravec@asu.cas.cz
Thomas **Prettyman**, LANL, Los Alamos, USA — thp@lanl.gov
Dina **Prialnik**, Tel-Aviv University, Tel-Aviv, Israel — dina@planet.tau.ac.il
Carol **Raymond**, Jet Propulsion Laboratory / Caltech, Pasadena, USA — carol.raymond@jpl.nas.gov
Douglas **Revelle**, Los Alamos National Laboratory, Los Alamos, USA — revelle@lanl.gov
Derek **Richardson**, University of Maryland, College Park, USA — dcr@astro.umd.edu
Hans **Rickman**, Uppsala Astronomical Observatory, Uppsala, Sweden — hans@astro.uu.se
Andrew **Rivkin**, The Johns Hopkins Univ. Applied Physics Lab., Laurel, USA — asrivkin@alumni.mit.edu
Ana Monica **Rodrigues**, Inst. Química, UFRJ, Rio de Janeiro, Brazil — anamfsnn@yahoo.com.br
Fernando V. **Roig**, Observatório Nacional, Rio de Janeiro, Brazil — froig@on.br
Reinhard **Roll**, MPS for Solar System Research,, Katlenburg-Lindau, Germany — roll@linmpi.mpg.de
Françoise **Roques**, LESIA, Observatoire de Paris, Meudon, France — francoise.roques@obspm.fr
Philippe **Rousselot**, Observatoire de Besançon, Besançon, France — rousselot@obs-besancon.fr
Christopher **Russel**, University of California, Los Angeles, USA — ctrussell@igpp.ucla.edu
Galina **Ryabova**, Research Inst. of Applied Mathematics and Mechanics, Tomsk, Russia — ryabova@niipmm.tsu.ru
Nalin **Samarasinha**, NOAO & PSI, Tucson, USA — nalin@noao.edu
Maria Eugenia **Sansaturio**, University of Valladoid / NEODYS, Valladolid, Spain — genny@pisces.eis.uva.es
Pablo **Santos Sanz**, Instituto de Astrofísica de Andalucía, CSIC, Granada, Spain — psantos@iaa.es
Gal **Sarid**, Tel-Aviv University, Tel-Aviv, Israel — galahead@post.tau.ac.il
Sho **Sasaki**, National Astronomical Observatory of Japan, Tokyo, Japan — sho@miz.nao.ac.jp
Peter Scheirich, Astronomical Institute ASCR, Ondrejov, Czech Rep. — petr.scheirich@centrum.cz
Peter **Schultz**, Brown University, Providence, USA — peter.schultz@brown.edu
Rita **Schulz**, ESA Research & Scientific Suppor Dep., ESTEC, Noordwijk, The Netherlands — rita.schulz@rssd.esa.int
Gerhard **Schwehm**, European Space Agency, ESTEC, Noordwijk, The Netherlands — gerhard.schwehm@esa.int
Eduardo **Seperuelo**, CEFET–Nilopolis/PUC-Rio, Rio de Janeiro, Brazil — esduarte@gmail.com
Scott **Sheppard**, Carnegie Institution of Washington, Washington, USA — sheppard@dtm.ciw.edu
Yuriy **Shkuratov**, Astronomical Institute of Kharkov National Univ., Kharkiv, Ukraine — shkuratov@vk.kh.ua
Bruno **Sicardy**, LESIA, Observatoire de Paris, Meudon, France — bruno.sicardy@obspm.fr
Grzegorz **Sitarski**, Space Research Center, Warsaw, Poland — sitarski@alpha.uwb.edu.pl
Colin **Snodgrass**, Queen's University Belfast, Belfast, UK — c.snodgrass@qub.ac.uk
Andrea **Sosa**, Depto. de Astronomia, Faculdad de Ciencias, Montevideo, Uruguay — andsosa@fisica.edu.uy
Nancy **Sosa**, Depto. de Astronomia, Faculdad de Ciencias, Montevideo, Uruguay — nsosa@fisica.edu.uy
Pavel **Spurny**, Astronomical Inst. of the Czech Academy of Sciences, Ondrejov, Slovak Rep. — spurny@asu.cas.cz
Raphael **Steinitz**, Ben-Gurion University, Beer-Sheva, Israel — raphael@bgu.ac.il
Giovanni **Strazzulla**, INAF–Osservatorio Astrofisico, Catania, Italy — gianni@ct.astro.it
Jessica M. **Sunshine**, SAIC, Chantilly, USA — sunshinej@saic.com
Ján **Svoreň**, Astronomical Inst., Slovak Academy of Sciences, Tatranska Lomnica, Slovak Rep. — astrsven@ta3.sk
Mark **Sykes**, Planetary Science Institute, Tucson, USA — sykes@psi.edu
Slawomira **Szutowicz**, Space Research Center, Polish Acad. of Sci., Warsaw, Poland — slawka@cbk.waw.pl
Gonzalo **Tancredi**, Depto. Astronomia, Faculdad de Ciencias, Montevideo, Uruguay — gonzalo@fisica.edu.uy
David **Tholen**, University of Hawaii, Honolulu, USA — tholen@ifa.hawaii.edu
Joanna **Thomas Osip**, Las Campanas Observatory, La Serena, Chile — jet@lco.cl
Cristina **Thomas**, Massachusetts Institute of Technology, Cambridge, USA — cathomas@mit.edu
Jana **Tichá**, Klet Obsertory, Ceske Budejovice, Czech Rep. — jticha@klet.cz
Milos **Tichy**, Klet Observatory, Ceske Budejovice, Czech Rep. — mtichy@klet.cz
Imre **Toth**, Konkoly Observatory, Budapest, Hungary — tothi@konkoly.hu
Juraj **Tóth**, Fac. of Math., Phys. & Inform., Comenius Univ., Bratislava, Slovak Rep. — toth@fmph.uniba.sk
Cleofas **Uchoa**, Osservatório Astronómico de Búzios, Búzios, Brazil — cleofas.uchoa@terra.com.br
Giovanni B. **Valsecchi**, INAF – IASF, Roma, Italy — giovanni@rm.iasf.cnr.it
Jeremie **Vaubaillon**, University of Western Ontario, Ontario, Canada — jvaubail@uwo.ca
Pierre **Vernazza**, LESIA, Observatoire de Paris, Paris, France — pierre.vernazza@obspm.fr
Maria Jesus **Vidal**, Instituto de Astrofísica de Andalucía – CSIC, Granada, Spain — mjvn@iaa.es
Ernesto **Vieira Neto**, UNESP, Guaratinguetá, Brazil — ernesto@feg.unesp.br
José W. **Vilas-Boas**, INPE, São José dos Campos, Brazil — jboas@das.inpe.br
Marcos R. **Voelzke**, Universidade Cruzeiro do Sul, São Paulo, Brazil — mrvoelzke@zipmail.com.br
Kevin **Walsh**, University of Maryland, College Park, USA — kwalsh@astro.umd.edu
Xiao-Bin **Wang**, Yunnan Observatory, Kunming, Peoples Republic of China — wangxb@ynao.ac.cn
Jun-ichi **Watanabe**, NAO, Tokyo, Japan — jun.watanabe@nao.ac.jp
Michael **Weiler**, Institute of Planetary Research, DLR, Berlin, Germany — michael.weiler@dlr.de
Paul **Weissman**, Jet Propulsion Laboratory, Pasadena, USA — paul.r.weissman@jpl.nasa.gov
Dennis **Wellnitz**, University of Maryland, College Park, USA — wellnitz@astro.umd.edu
Iwan **Williams**, Queen Mary, London University, London, UK — i.p.williams@qmul.ac.uk
Mark **Willman**, Institute for Astronomy, University of Hawaii, Honolulu, USA — willman@ifa.hawaii.edu
Othon **Winter**, UNESP, Guaratinguetá, Brazil — ocwinter@feg.unesp.br
Kristin R. **Wirth**, European Space Agency, ESTEC, Noordwijk, The Netherlands — kristin.wirth@esa.int
Diane **Wooden**, NASA Ames Research Center, Moffet Field, USA — wooden@delphinus.arc.nasa.go
Laura **Woodney**, University of Central Florida, Orlando, USA — woodney@physics.ucf.edu
Masahisa **Yanagisawa**, Univ. of Electro-Comunications, Tokyo, Japan — yanagi@ice.uec.ac.jp
Tadashi **Yokoyama**, UNESP, Rio Claro, Brazil — tadashi@ms.rc.unesp.br
Fumi **Yoshida**, National Astronomical Observatory of Japan, Tokyo, Japan — yoshdafm@cc.nao.ac.jp
Michael E. **Zolensky**, NASA Johnson Space Center, Houston, USA — michael.e.zolensky@nasa.gov

Small Bodies of the Solar System: Asteroids, Comets, Meteors, TNOs and Small Planetary Satellites

Asteroids, Comets, Meteors
Proceedings IAU Symposium No. 229, 2005
D. Lazzaro, S. Ferraz-Mello & J.A. Fernández, eds.
© 2006 International Astronomical Union
doi:10.1017/S1743921305006642

Physical properties of small bodies from Atens to TNOs

C. R. Chapman

Southwest Research Institute, 1050 Walnut St., Suite 400, Boulder, CO 80302, USA
email: cchapman@boulder.swri.edu

Abstract. Properties of small, heliocentric bodies in the solar system share many attributes because of their small sizes, yet vary in other ways because of their different locations of formation and the diverse subsequent evolutionary processes that have affected them. Our insights concerning their properties range from highly detailed knowledge of a few specific bodies (like Eros), to rich knowledge about unspecific bodies (meteorite parent bodies), to no knowledge at all (other than existence and rough limits on size) concerning much smaller and/or more distant bodies. Today's state of learning about physical properties of TNOs is analogous to that for main-belt asteroids 35 years ago. This invited review attempts to elucidate linkages and differences concerning these populations from the highly heterogeneous data sets, emphasizing basic properties (size, shape, spin, density, metal/rock/ice, major mineralogy, presence of satellites) rather than the highly detailed knowledge we have of a few bodies or their dynamical properties. The conclusion is that there are vital interrelationships among these bodies that reinforce the precept that guided the original ACM meetings, namely that we should all think about small bodies in an integrated way, not just about subsets of them, whether divided by size, composition, or location.

Keywords. comets: general, Kuiper Belt, minor planets: asteroids, planets and satellites: general, infrared: solar system

1. Introduction

In some ways, the physical properties of small bodies in the solar system is the largest topic in solar system research, if it can be considered to be a single topic at all. Besides being nearly infinite in numbers, small bodies have an enormous variety in physical properties, ranging from the nickel-iron alloys of metallic meteorites and asteroids to the underdense, volatile-rich bodies of the outer solar system, some with transient atmospheres. Their locations in the solar system range from perhaps inside the orbit of Mercury, although none of the hypothetical "vulcanoid" have yet been found, to the outermost reaches of the solar system; some would include the recently discovered tenth planet, 2003 UB313 ("Xena") at 97 AU distance from the Sun to be a "small body" (certainly its moon is). Diverse techniques are being utilized to divine the physical properties of small bodies, ranging from state-of-the-art laboratory examination of meteorites and interplanetary dust particles (IDPs), to groundbased astronomy (both passive [UV/optical/IR/radio] and active [radar]) utilizing the largest and most advanced facilities in the world, to orbital, *in situ*, and sample return studies of representative small bodies by spacecraft.

The breadth of this "meta-topic" under review was determined by the Organizing Committee of the ACM-2005 meeting, who also assigned it as the first talk of the first session. Inasmuch as it is impossible for me to prepare a definitive review of such an unwieldy topic and because many other invited reviews in this volume cover, at appropriate depth, subsets of this topic, this review should be considered as a "meta-review", summarizing some general themes to establish a context for many of the other contributions to this volume. Indeed, I must further delimit the scope of this review and treat some topics in only the briefest way. For example, a major branch of planetary science – meteoritics and cosmochemistry – is focused chiefly, though not exclusively, on measurement and interpretation of the physical properties of samples of small bodies, the meteorites. In view of the general astronomical orientation of ACM, however, I purposefully don't do justice to meteoritics in this review.

Let me begin by defining the various classes of "small bodies" whose physical properties are being researched. Although one might classify small bodies by composition, the two most useful classifications are (1) by distance from the Sun or, more specifically, by type of orbit, and (2) by size. By orbital type, one could list them roughly by increasing distance from the Sun: the still hypothetical vulcanoids, inner-Earth objects (IEOs, or Apoheles, of which three are currently known), Near Earth Asteroids (NEAs, including their sub-classes the Atens, Apollos, and Amors), main-belt asteroids (including the Hungarias, Cybeles, Hildas, and others separated by large gaps from the densely populated main belt torus), Trojans (chiefly of Jupiter, but also of Mars and Neptune), Centaurs, Scattered Disk Objects (SDOs), Kuiper Belt Objects (KBOs, including Plutinos and classical Cubewanos), more distant objects that might be considered to be in the inner Oort cloud, comets (including Jupiter Family Comets [JFCs] and longer period comets, including those newly arrived from the Oort cloud; one might also include Damocloids, presumed dead comets), and planetary satellites. The term Trans-Neptunian Object (TNO) is often applied to the ensemble of outer solar system small bodies, sometimes including those that do not strictly adhere to the definition of having semi-major axes larger than that of Neptune. A final type of small body is a moon orbiting one of the other types. Rapidly increasing numbers of such moons have been discovered in the last few years orbiting NEAs, main-belt and Trojan asteroids, and TNOs. In principle, moons may orbit around small-body moons, although the first triple asteroid discovered (87 Sylvia) has two small moons orbiting around the main asteroid, not around each other. In this review, I concentrate on NEAs, main-belt asteroids, and TNOS, and largely ignore planetary satellites, even as the Cassini spacecraft is revealing a wealth of new information about the latter, including the fascinating interrelationships between small moons and Saturn's ring particles (which, of course, are small bodies – or conglomerations of small bodies – themselves).

One may also classify small bodies by size. In order of ascending size, there are IDPs, meteorites, and meteoroids at the small end; the middle range of diverse, astronomically observable small bodies roughly 10 m to 1000 km in diameter; and those larger than 1000 km, such as Pluto, 2003 UB313, Sedna, and the larger planetary satellites. In this review, I concentrate on bodies in the middle size range.

The kinds of information about physical properties that we seek to learn range from the very basic properties of size, density, and spin to the most highly detailed characterization of small-scale features (geology) and composition (chemistry, mineralogy). Generally, of course, we can determine or constrain some of the most basic properties for thousands or even tens of thousands of small bodies by simple telescopic observation from Earth whereas the most detailed physical characterization by close-up spacecraft studies can be done for only a few bodies. Disjoint from this spectrum of knowledge is the case of meteoritical studies, which measure in exquisite detail the isotopic, chemical, and mineralogical properties of small fragments of presumably hundreds of different small bodies; but almost none of this knowledge can be assigned to a specific small body, and assignment even to a class of small body (e.g. properties of ordinary chondrites to S-type asteroids) is fraught with uncertainty. The one likely exception is that there is a good case for believing that most eucrites, howardites, and diogenites (the HED achondrites) are derived, at least indirectly, from the asteroid Vesta.

Associated with the extreme range in specificity of knowledge of the physical properties of small bodies is the issue of observational biases and lack of representation. While it is elementary that objects that are bigger, closer to Earth, and have higher albedos are over-represented compared with small, black, distant bodies, there remains an unconscious bias towards "what-you-see-is-what-is-there" or, in the case of vulcanoids (notoriously difficult to find because of proximity to the brilliant Sun), "if you haven't found them we must presume that they don't exist." It is still not fully appreciated that the Jupiter Trojans are nearly as populous as main-belt asteroids. One must expect observational biases to be especially applicable to the TNOs and other outer solar system small bodies; indeed, apart from Pluto and Chiron, none of these bodies were known until 1992. Although more than 1,000 such bodies have been found in the subsequent 13 years, that places their statistics where the asteroids were in the 1920s, shortly after asteroid families were first recognized by Hirayama. Of course, the kind of detailed physical and compositional characterization of asteroids that can be done with modern astronomical techniques such as spectroscopy has so far been applied, to various degrees of precision, to only about 1% of the nearly 300,000 asteroids with reasonably well known orbits; although colors for tens of thousands of asteroids are being released by the Sloan Survey, the vast majority of asteroids still are characterized only by their orbital properties and rough apparent brightnesses. Because of the extreme faintness of most TNOs, considerable time on the largest telescopes is required to obtain physical data on even the brightest TNOs that is comparable to what is routinely obtained for asteroids. Thus TNO researchers must be particularly aware of the dangers of over generalizing results from a few well-observed bodies, over-interpreting noisy data, or ignoring potential observational biases.

The goals of astronomical observation of small body physical properties are to characterize size, three-dimensional shape, mass, spin rate and pole direction, albedo, spectral reflectance properties (from the UV into the infrared, perhaps revealing minerals or ices), thermal emission spectrum (mid-IR to radio), photometric and polarimetric properties, and temporal variations in many of the above (that might indicate spatial variations revealed by rotation or actual outbursts of dust or volatiles). Eventually, using radar on the closest NEAs but generally requiring spacecraft exploration, the goal is to observe bodies with sufficient spatial resolution (and even measure properties *in situ* or from returned samples) so that they are transformed from astronomical objects into geological/geophysical/geochemical worlds the way Mars is currently being studied by the numerous spacecraft orbiting or roving around on that planet. Because of their vast numbers, however, it will always remain the case that we will be able to study only a tiny percentage up close. So we will have to develop reliable ways to extrapolate our specific knowledge of the few to the general statistical population observable only from afar.

2. Colors and Spectral Properties

My first theme concerns "colors", by which I mean the approximate characterization of the spectral reflectance properties of the surfaces of small bodies. A highly precise and accurate reflection spectrum throughout the Sun's spectral range offers specific insight into the presence of some particular minerals and ices, but not of others. Even where useful absorption bands exist, the interpretation is sometimes ambiguous, and quantitative estimates of proportions of constituent materials are rendered difficult by uncertainties in surface particle sizes and other factors. In practice, however, spectral data are subject to further limitations due to signal-to-noise, variability in sky conditions, spectral bands where the sky is opaque, etc. There are further complications due to the fact that the optical surfaces of small bodies are subject to modification and damage due to "space weathering," caused by impacts of solar wind particles and micrometeorites. Thus the optical layers may not be representative of the bulk regolith on a body, let alone the material of which the body is predominantly composed. Still, such spectral reflectance data represent the best evidence we have about the composition of a distant body. For most small bodies, however, observational limitations restrict us to much lower photometric precision and much coarser spectral resolution than is required for determining specific mineralogy. Such "color" data nevertheless permit the development of a colorimetric taxonomy, and one may assume that members of a taxonomic group have the more specific characteristics of certain members of that group, which – because they are brighter or just happen to have been studied much more thoroughly – have available high-precision spectra.

By 1970, several dozen asteroids had been observed for UBV colors. Several researchers proposed that there were between 2 and 4 color groups, although hoped for correlations with meteorite colors were not apparent. By 1975, spectral reflectance data (for wavelengths shortwards of 1 micron) were available for a couple hundred asteroids, and UBV colors for many more. Some clustering was visible in color-color plots, and there were other statistically significant differences in color, whether or not there was an actual bifurcation into separate groups. Together with a statistically significant number of asteroids measured by 10-micron thermal infrared radiometry (yielding albedos), the database on spectra and colors permitted the development of the "C, S, M . . ." taxonomy, which has now consumed most of the letters of the alphabet. Such a taxonomy has been very useful for organizing the massive database on asteroidal physical properties. It has been augmented in recent years both by extension to beyond 2 microns as well as augmentation in sample size to several thousand asteroids. In 1975, the statistics were already sufficient to de-bias the data and start researching such statistical properties of asteroids as variation of taxonomic type with orbital elements, size distributions of taxonomic types, etc. The last comprehensive de-biased study of asteroid physical properties was in the late 1980s. Given the massive augmentation in the asteroid database since then, another comprehensive study is long overdue.

A currently exciting frontier is replicating for TNOs the kinds of studies that were being done three decades ago for main-belt asteroids. Beginning in 1998, it was proposed that colors of KBOs fell into two distinct groups, and debates ensued. The current situation is shown in Fig. 1, where clearly Centaur colors are bimodal, Plutinos are possibly bimodal, and SDOs and Cubewanos appear monomodal. The mean colors of the latter two groups differ from each other, however; it is not clear if their different colors are related to one or the other modes of the bimodal Centaurs. Weak correlations of colors of TNOs with different orbital properties are getting stronger as the sample increases. For example, it appears that there is a broad range of B-R colors for those with perihelion distance $q < 38$ AU whereas lower values of B-R are missing at larger q's, except for a group with inclination $i > 25°$. It is interesting that comet nuclei tend to have colors dissimilar from objects in their presumed source regions, which implies some kind of processing. Moderate resolution infrared spectra, obtained for a few of the largest TNOs, are revealing absorption bands characteristic of several types of ices (e.g. water, methane, nitrogen); already there appears to be considerable variety in surface compositions of TNOs. There is much discussion in the recent literature about the degree to which these studies reveal the inherent attributes of primitive bodies, or instead various kinds of processing, including ongoing space weathering.

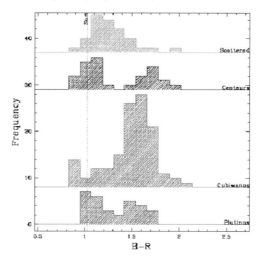

Figure 1. Histograms of B-R colors for Scattered Disk Objects, Centaurs, Cubewanos, and Plutinos, showing degrees of bimodality and monomodality. (From the database of Hainaut and Delsanti 2005.)

Another arena of recent progress in spectral reflectance studies is NEAs. It is clear that larger NEAs share the diversity of spectral properties seen in the inner and middle parts of the main asteroid belt, consistent with recent dynamical research suggesting that NEAs should representatively sample large volumes of the inner and middle belt. However, at diameters smaller than 5 km (and especially < 2 km), the range in colors of the higher albedo NEAs (of S and Q taxonomic types) spreads to include shallower spectral slopes like those characteristic of ordinary chondritic meteorites. This appears to be consistent with the hypothesis that most common S-type asteroids are inherently of ordinary chondritic composition, but that the colors of larger bodies are often modified by space weathering processes. Larger asteroids tend to have two characteristics that would favor their surfaces being space-weathered: (a) larger asteroids have longer lifetimes against collisional disruption, thus would have a greater chance for space weathering to reach maturity, and (b) larger asteroids have greater gravity, thus facilitating the retention and processing of regolith, also enabling maturation of space weathering. The least reddened (space-weathered) NEAs have colors and spectral reflectances very similar to spectra of ordinary chondrites. While such spectra have not yet been seen in the main belt, few main-belt asteroids significantly smaller than 5 km diameter have been observed so far. It is plausible that some of the extreme colors and spectra seen among the smallest NEAs will eventually be found in the main belt, when objects of similar small sizes can be observed. Significantly, as many as 15% of NEAs exhibit D-type colors, common only in the outer main belt and beyond (e.g. among Trojans). However, the outer main belt remains understandably somewhat under-represented among NEAs, even after correcting for observational biases against low-albedos.

One theme of small-body research has been the search for heterogeneity in compositional properties. For example, it was once thought that many asteroid families contained members of several different taxonomic groups; that might imply processes of compositional differentiation (e.g. core formation due to heating and segregation of a mantle and crust). Recent, more comprehensive studies reveal, however, little evidence of such heterogeneity. On the contrary, the precursor bodies of most families appear to have been compositionally homogeneous. The erroneous earlier results mainly resulted from the presence of interlopers and less accurate approaches for calculating proper elements and family membership. Another approach to identifying heterogeneity is to watch for color or spectral changes as a body rotates. The very first asteroid whose reflectance spectrum was measured was Vesta, in 1929; indeed its rotation period was determined from temporal variations in its color, much more recently ascribed to an olivine-rich region in its otherwise basaltic crust. But most other reports of rotational color variations have been marginal and are doubtful. A famous instance was publication of confident conclusions that the NEA Eros had slightly different spectral properties on opposite sides; these were shown to be erroneous by the NEAR Shoemaker mission to Eros, which found an extremely high degree of spectral uniformity around the body. A recent interesting case involves a report of very different colors on Karin, the largest body in a sub-family within the Koronis family that was formed very recently, 5.8 Myr ago, in a catastrophic disruption; this report awaits confirmation. Initial results of searches for rotational color variations among TNOs reveals some showing no variations, but a couple of others hinting at variations.

3. Size Distributions

Sizes of small bodies can be estimated approximately from their apparent brightnesses. Combinations of visible and thermal-IR photometry, as well as other techniques, can yield reasonably accurate sizes for small bodies; their often irregular shapes limit ultimate precision. An individual size (or volume), when combined with a measurement of mass (e.g. from perturbations of nearby spacecraft, other nearby small bodies, or even planets like Mars...and, more recently, from the orbits of moons of some bodies), yields an important constraint on composition: density. In the aggregate, however, the statistics of sizes – when properly assessed from debiased observational data – provide fundamental information about collisional processes, either low-velocity accretional processes or subsequent catastrophic disruptions. Although the size distribution of NEAs was first inferred indirectly from the size distribution of lunar craters, the more reliable approach is to measure small-body sizes more directly.

While early theoretical work predicted a single equilibrium power-law size distribution for collisionally evolved systems, early studies of debiased main-belt asteroid sizes revealed a wavy pattern (i.e. the power-law exponent varies with size). The census of main-belt asteroids is now complete down to diameters of a couple tens of km, and debiased statistics of samples of smaller asteroids are valid down to about 3 km (Fig. 2). Relative to a single power law, there is an excess of asteroids about 100 km diameter. This is widely believed, as first proposed four decades ago, to be the collisionally un-evolved remnant of the primordial population of asteroids. The relatively steep-sloping "tail" of the distribution for asteroids <30 km diameter is due to collisional evolution; these are the products of catastrophic collisions during the last 4 Gyr. (One issue that has not been revisited recently in any comprehensive way is late-1970s indications that different taxonomic types, and different dynamical groups of asteroids in and beyond the main belt, have different size distributions.)

Studies of the NEA size distribution incorporate not only astronomical data on NEAs but also statistics of fireballs and meteors as well as inferences from impact craters on the Earth and the Moon. The NEA size distribution differs noticeably from the main-belt case, probably due to size-dependent processes (like the Yarkovsky Effect) that extract NEAs from the main belt. It is slightly wavy, but the data are closely approximated by a single power-law exponent (the straight dashed blue line in Fig. 3); the data are inconsistent with attempts to match the lunar crater size distribution (red dashed and

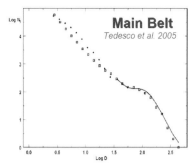

Figure 2. Incremental size distribution for main-belt asteroids larger than 3 km diameter, from the "Standard Asteroid Model" of Tedesco *et al.* (2005), shown as open squares. The dots are from an older model.

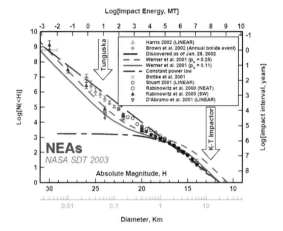

Figure 3. Cumulative size distribution for numbers of NEAs brighter than absolute magnitude H (Near-Earth Object Science Definition Team 2003). Equivalent axes for NEA diameter and Earth impact energy (in megatons) are shown. The data points are from astronomical observations; those based on LINEAR are more recent and reliable. The long-dashed blue line is a power-law that approximately fits the good data. The red curves represent unsuccessful attempts to fit a standard crater curve derived mainly from lunar craters, assuming two different albedos for NEAs.

solid curves), probably because most small lunar craters are produced by secondary ejecta from larger primary craters rather than by direct impacts by small NEAs.

The frontier of research on size distributions is in the outer solar system. It has been notoriously difficult to measure directly the sizes of comet nuclei, partly because of their activity. The latest results indicate that a power-law-like size distribution starts to become truncated at sizes <4 km diameter with very few comets smaller than 0.5 km. This is consistent with evidence for a paucity of small craters (except secondaries) on the surfaces of young satellite surfaces, like Europa, which are cratered predominantly by comets rather than asteroids. A recent discussion of the size distributions of various classes of TNOs is summarized in Fig. 4. There may be different size distributions for Cubewanos ("classical disk") compared to other TNOs. In any case, the slope of the power-law is relatively shallow below about 25 km diameter and steep at large sizes, crudely mimicking the excess of bodies ∼100 km diameter exhibited by main-belt asteroids. One can speculate that the 100 km "hump" reflects a primordial accretionary size distribution and that the different size distributions of comets and asteroids below 25 km (shallow versus

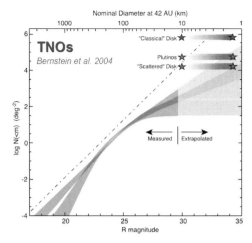

Figure 4. Limits on the size distributions of classical KBOs, shown in red, and "excited" (high inclination or resonant) TNOs, shown in green, fitting a double power-law to the data and extrapolating to smaller sizes. An earlier fit of a single power-law is shown by the dot-dash line. The horizontal bands represent theoretical estimates for three types of TNOs if they are sources for JFCs. Diagram from Bernstein *et al.* 2004.

comparatively steep, respectively) reflects different responses of the two types of bodies to collisions and other disaggregational processes.

4. Shapes, Satellites, and Geophysics

The mere fact that most small bodies exhibit double-peaked lightcurves implies that they are not spherical. In fact, some are highly irregular in shape, generally because the strengths of their constituent materials exceed the modest gravitational forces that would otherwise compress them into spheres, or into equilibrium figures for spinning bodies. Inversion of lightcurves ("photometric geodesy") yields fairly coarse constraints on three-dimensional shapes and requires time-consuming observations over many years to obtain diverse observing and illumination geometries. More recently, various additional techniques (high-resolution imaging by adaptive optics [AO] or from HST, radar delay-doppler mapping, stellar occultations, and close-up imaging from spacecraft) have greatly augmented our knowledge of the shapes of NEAs, other asteroids, and a few comet nuclei. Lightcurves of TNOs are beginning to suggest that they are commonly less spherical than comparably sized asteroids. This is especially true for one of the largest TNOs, 2003 EL61, which appears to be a highly elongated quasi-equilibrium figure (Jacobi ellipsoid) due to its very rapid rate of spin (just 3.9 h); this body's length may exceed Pluto's diameter, although it has only one-third Pluto's mass.

Small bodies have a wide diversity of shapes and configurations (Fig. 5). One of the most profound changes in our gestalt of small bodies in the last dozen years has been the transformation from a perspective that few or none of them had satellites or were other than single bodies to the recognition that satellites or double configurations are extremely common. Despite unconfirmed earlier reports, not until 1994 was the first satellite of a small body discovered – Dactyl, orbiting around the main-belt asteroid Ida, found in images taken during the Galileo spacecraft flyby. Since then, satellites and/or double configurations have been discovered among most classes of small bodies from NEAs to TNOs. Numerous observational techniques are being used to address this issue, including AO and HST imaging, delay-doppler radar, and analysis of lightcurves (e.g.

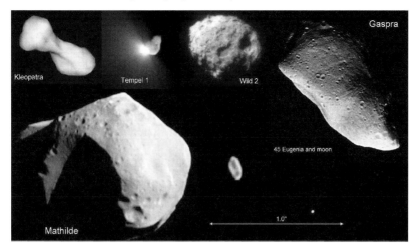

Figure 5. Pictures of diverse small bodies (not to scale) show a range of shapes, including the main-asteroid-with-moon configuration of Eugenia (lower right; from AO imaging with 1 arcsecond scale bar shown). Kleopatra is a double-lobed model based on radar delay-doppler imaging. Remaining images are from spacecraft.

"eclipsing binary" phenomena, dual-period lightcurves, etc.). At least 15% of NEAs have satellites or double configurations and widely separated satellites appear to be common among TNOs. There is every indication that percentages of bodies with satellites will increase as observational barriers are overcome (e.g. ability to detect smaller satellites near bright objects, or ability to detect closer satellites orbiting distant bodies). The common presence of satellites has the potential to dramatically improve our knowledge of the bulk densities of small bodies, from Kepler's third law, provided the volumes of the primaries can be determined fairly well.

There is great theoretical, and even *practical*, interest in the internal configurations of small bodies. Theoretical considerations of the efficiency of converting collisional kinetic energy into the kinetic energy that disperses fragments have long predicted that many small bodies are "rubble piles." The original use of this term envisioned that a collision would fragment a body into a size distribution of fragments; if the largest fragment has less than half the mass of the original body, and most of the fragments are lofted at less than escape velocity and thus reaccumulate into a conglomerate body, then the resulting rubble pile is dominated in mass by a few comparatively large bodies while the body also contains innumerable smaller fragments. Perhaps, after numerous sub-dispersive collisions, the largest components are themselves fragmented, although no physical process has been envisioned that yields a multi-component body in which all components are the same size, which has been convenient to model in computer simulations. The rubble pile concept has also been adapted to modelling comet nuclei, in which case the components of a rubble-pile nucleus might be original planetesimals gently accreted onto the growing nucleus or, alternatively, might be analogous to an asteroid rubble pile if comets are collisionally evolved.

There have been observational interpretations, for instance of large-scale geological features on Eros, suggesting a different kind of morphology, called the "shattered shard," in which it is envisioned that large-scale impacts have shattered the body but the remaining larger pieces have remained more-or-less in place. Computer hydrocode simulations of impacts, and of tidal deformation during close passages to a planet, have suggested that a variety of possible internal structures for small bodies may be produced. One clue about internal structure of NEAs has been the fact that, as of a few years ago, all NEAs >200 m diameter rotate with a period longer than 2.2 h, a period at which a cohesionless rubble pile would barely fly apart by centrifugal force; smaller rapid spinners would then have to be monoliths (bodies with inherent tensile strength). Since then, exceptions have been observed. Also, it has been argued that large, natural bodies are inherently weak, even if they have not been physically broken by collisions. So it remains for future geophysical measurements by spacecraft missions to address the internal structures of small bodies, beyond the non-specific results of bulk density measurements (a low density may imply a large fraction of voids, but it doesn't specify whether the voids are microscopic or macroscopic in scale, and there is a wide range of inherent densities of materials of which small bodies are composed, at least spanning the range from ice to nickel-iron).

5. Geology

The first small bodies to be closely examined by spacecraft were the two moons of Mars. The distributions of craters, smooth areas, and cracks differ between the two. But it is difficult to generalize from such bodies buried deeply within the gravity well of Mars to the dominant populations of small bodies in heliocentric orbits. For example, ejecta from impacts on Phobos and Deimos enter "dust belts" encircling Mars that tend to reaccrete rapidly onto those satellites. Many satellites of the outer planets were imaged, generally at rather coarse resolution for the smaller ones, by the two Voyagers; a few, like Miranda, which presents an odd appearance, were seen at fairly high resolutions. Additional images have been obtained more recently by Galileo and Cassini. Ignoring the larger, planet-sized moons, one can nevertheless say that a reasonable geologic diversity is evident among many of these bodies, although impact craters are ubiquitous except on a few exceptional portions of a few bodies like Enceladus.

The first two heliocentric bodies to be seen close-up were the S-type, main-belt asteroids Gaspra and Ida (and Ida's moon). Geologically, they appear rather different, with Gaspra displaying an angular shape, perhaps configured by remnant "facets" of older large impact scars, and an under-saturated population of small craters while Ida's craters resemble much more closely the familiar saturation-cratered terrains of the Moon. Subsequent close-up imagery of other kinds of bodies, including C-type asteroid Mathilde, NEA Eros, and several comets, continues to reveal surprising diversity among them. Mathilde is dominated by huge craters on the scale of Mathilde's own radius. Global high-resolution imaging of Eros revealed a surprising dearth of small craters, an abundance of boulders, and wholly unexpected smooth regions colloquially termed "ponds". It is noteworthy that the surface of the much smaller NEA Itokawa, imaged by the Hayabusa spacecraft, strongly resembles Eros (smooth areas, numerous boulders, nearly craterless) at the same resolution (Fig. 6).

Although images of comets Halley and Borrelly were of too coarse resolution for detailed geological analysis, the higher resolution Stardust images of comet Wild 2 reveal a jagged, pock-marked surface apparently unlike the other comets. Deep Impact's images of comet Tempel 1 shows geological features, including a couple of smooth plains, very different from Wild 2; preliminary mosaics of images taken from the D.I. impactor itself, just before it struck, are shown in Fig. 7.

Additional spacecraft missions, either contemplated or already underway, may extend our examination of small bodies by focusing on *in situ* and sample return science. These approaches will surely begin to connect the "hand-sample" science of meteoritics to the "field geology and geophysics" of asteroids being revealed by fly-by and orbital missions, both of which may ultimately be extrapolated to the countless members of these

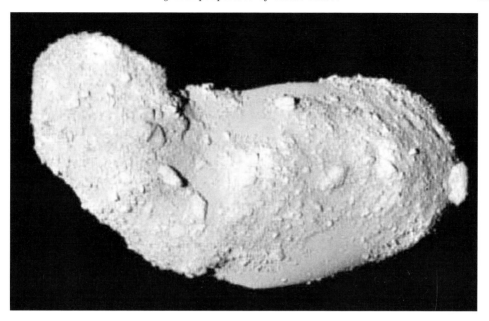

Figure 6. NEA 25143 Itokawa, as imaged by the Hayabusa spacecraft in early October 2005. The 0.5 km diameter asteroid exhibits rocks and boulders, as well as smooth areas, but essentially no impact craters. Courtesy of JAXA.

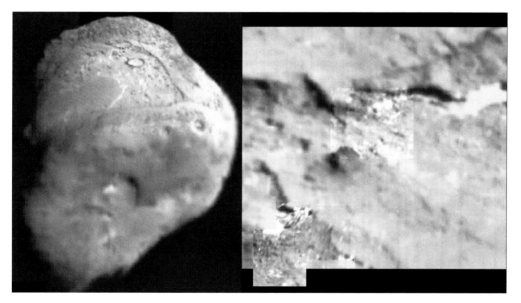

Figure 7. Preliminary mosaics of images taken by the Deep Impact impactor before impact. The nucleus of comet Tempel 1 is shown on the left. A mosaic of the final images is on the right; it is about the size of the uppermost of the two prominent craters near the top of the full-nucleus image, and is centered just below the lower part of that crater's rim.

populations observable only from a distance by astronomical techniques. Our appetite for the next phase of small body studies has been whetted by the NEAR Shoemaker mission, which actually landed successfully on Eros, although it was not designed to do so. Analysis, during the five years since the NEAR mission ended, of its comprehensive orbital remote-sensing data (especially from the X-ray spectrometer) has conclusively demonstrated that Eros is an L or LL type of ordinary chondrite. Of course, Eros is not the original parent body of these meteorites, since it is in a very transient orbit in the inner solar system, but the linkage appears largely to resolve the long-standing controversy about the nature of most S-type asteroids. Of course, S-type asteroids volumetrically constitute a tiny fraction of the main belt, which is dominated by low-albedo objects (C-, P-, and D-types). And it is presumed that we do not yet have terrestrial samples, other than IDPs, of the more volatile rich and presumably generally less-altered bodies from the colder, more distant reaches of the solar system. So the geology of small bodies is a topic in its infancy.

6. Relationship of Physical Properties to Other Issues

While this review of the meta-topic of small-body properties has attempted to summarize major elements of the study of physical properties of small bodies, it is appropriate to view physical properties in the context of small-body science as a whole. The other major meta-topic of this ACM meeting concerns the dynamics of small bodies. There is an intricate linkage between the two. I briefly consider four kinds of ways in which the topics are related.

First, dynamical processes *cause physical properties* to be the way they are. For example, it is becoming increasingly clear that effects related to the Yarkovsky Effect play a fundamental role in determining the spins and axis orientations of asteroids. Tidal interactions of small bodies with planets and with the Sun cause distortions, disruptions, and even disintegrations. Collisions and catastrophic disruptions, and the dynamics of ejecta, create families, rubble-pile structures, and satellites (determining initial spins and sizes). Second, physical properties help *elucidate dynamics*. For example, colors have helped define dynamical families of asteroids. Yarkovsky/YORP effects depend on albedo, shape, thermal inertia, spin, density, etc. Third, dynamical analysis can help *determine physical properties*. Mass, hence density, is determined by analysis of gravitational perturbations or from the orbits of satellites. I have already described how spins may, or may not, define whether a body is a rubble pile or a monolith. Non-gravitational forces constrain attributes of the physical properties of cometary nuclei. Fourth and finally, dynamical analysis helps us *study physical processes*. Precise, specific ages for asteroid families derived from dynamical analysis help us determine the rates of physical processes such as space weathering. The ways that small body perihelia evolve permit us to better understand volatilization of surface materials.

There are actually practical implications of physical properties of small bodies. Astronomy is a prime arena of ivory-tower science, solar physics being the chief exception (e.g. manifestations in electrical grids on Earth). About the only other topic in astronomy with potential practical effects are asteroids and comets. They present both a hazard, from their rare but potentially devastating impacts, and the most accessible sources of resources for use in space. Both in terms of "handling" a dangerous NEA in order to divert it from Earth impact and in terms of mining materials for use in space, the physical properties – especially surface and near-surface properties – are fundamental. How can one anchor onto the surface of a nearly gravitationless body . . . if it is made of solid metal, if it has a regolith, or if it has the consistency of "talcum powder," a term used in early descriptions of the character of comet Tempel 1 inferred from Deep Impact?

Consider the case of 320 m diameter 99942 Apophis, which at this writing has a 1-in-6000 chance of striking the Earth on 13 April 2036 by passing through a "key-hole" during its exceptionally close pass by Earth on 13 April 2029. Whether it passes through the keyhole depends, in part, on physical properties that determine how the Yarkovsky Effect will modify its trajectory. Its surficial properties would constrain our ability to attach a device (e.g. low-thrust ion engine) to its surface. Internal properties may affect how it responds to accelerations, explosions, or other approaches to averting the impact. Indeed, its physical properties will determine how it responds to Earth's strong tidal forces as it passes by below synchronous satellite altitude in 2029; calculations suggest that there may be dramatic changes to its spin state and perhaps internal readjustments, especially if it is a rubble pile. Even the consequences of an Earth impact are affected (although in a secondary way) by its density and structure; a tsunami, threatening the west coast of North America, roughly equivalent in magnitude to the South Asian tsunami of 2004, is envisioned if it were to strike in the Pacific Ocean, but detailed consequences might depend on Apophis' physical properties.

7. Concluding Themes

As I discussed at the outset, small bodies are difficult to study. They are small, distant, and extremely numerous – and their evident diversity means that seeing one is *not* like seeing them all. Many of them are dark and/or dimly illuminated. Some of them, like Apophis, are in resonant orbits that render them invisible much of the time. The past decade has seen revolutionary improvements in search techniques and in instrumentation that reveals physical properties even from our distant location on or orbiting Earth. But we are only scratching the surface. Whole new populations of small bodies may yet be discovered and studies of physical properties of outer solar system small bodies will continue to be exceptionally challenging.

There are subtler issues, however. Our remote-sensing observations, whether obtained from telescopes or spacecraft, almost all pertain to the very uppermost surfaces of small bodies. Virtually the entire volumes of these bodies remain hidden from our view. The presumption may be valid in many cases that surfaces are made of roughly the same materials as the interiors, but it remains a presumption. Beyond that, especially for these airless bodies, the surfaces that we remotely sense are the very same surfaces struck by solar wind particles, micrometeorites, ultraviolet and higher energy radiation, etc. that damage or modify grains on the immediate surfaces. So what you see is often *not* what you "get". Either we understand these space weathering effects and develop reliable theories of regolith processes, for example, or we extrapolate from surfaces to internal depths at our peril.

Another aspect of small bodies in which our intuition may fail us concerns their nearly gravity-free environments. Geologists, in particular, often interpret spacecraft data from other bodies in a "comparative planetological" approach, in which analogies from familiar terrestrial experience play a large role. Geology at almost zero-g can be very different. Transitioning from astronomical to geological perspectives of small bodies may be more difficult than it has been in the cases of the terrestrial planets.

I expect more surprises as we really learn about the structures of small bodies: their porosities, densities, strengths, etc. Why do so many comets disintegrate and vanish? Are they like "dust bunnies"? What are appropriate analogs for materials that accreted slowly and have never been heated or compacted? Styrofoam? Talcum powder? (And how does talcum powder behave at near-zero g? What are the roles of electrostatic or magnetic forces?) Are M-types stripped metallic cores? Spectral evidence suggests that many are not . . . then what are they? One issue we have to contend with is the enormous bias we have inherited from the physical characteristics of meteorites in our collections. The vast majority of even asteroidal materials, let alone cometary fragments (which have the added problem of higher velocity impacts), may not ever make it through Earth's atmosphere for collection. If most small body materials are granular, weak, or underdense, then we would have no direct evidence that they exist.

Yet another question about small bodies is "what are we missing?" Despite considerable efforts to locate additional Plutos, years went by before 2003 UB313 was discovered. It is in an unusually inclined orbit. To what degree are search programs biased, for example by competitive pressures to find "the most" objects, or to name them, etc.? It is difficult to suppress our preconceptions. Witness the radically different structures of extra-solar planetary systems, with their intra-Mercurian Jovian planets. Let us not prematurely rule out vulcanoids, or vast clouds of Trojans around other planets, or satellites of small bodies in places we haven't found them yet, just because they haven't yet been found. Louis A. Frank's mini-comets do not exist, but other populations of hypothetical small bodies (e.g. interstellar comets), or even populations never imagined, may yet exist and await discovery.

Finally, the science of extra-solar small bodies already is underway, even if we can never imagine studying an individual body at such an enormous distance. Asteroid belts, Oort clouds, and planetesimal swarms have already been interpreted to exist around some other stars, primarily from infrared astronomical detection of disks. The statistical properties of such disks may prove to be powerful counterpoints against which to consider small bodies in our own solar system.

NOTE: Due to the enormous range of topics covered by this meta-review, there is not space to provide a comprehensive list of citations. Rather than listing some arbitrary references incompletely, I choose to list none, except sources for illustrations.

References

Bernstein, G.M., Trilling, D.E., Allen, R.L., Brown, M.E., Holman, M. & Malhotra, R.2004 *Astron. J.* 128, 1364

Hainaut, O. & Delsanti, A.2005, http://www.sc.eso.org/~ohainaut/MBOSS/tnoCentaursKS_hist_B-R.gif (accessed Aug. 2005)

Near-Earth Object Science Definition Team 2003, Study to Determine the Feasibility of Extending the Search for Near-Earth Objects to Smaller Limiting Diameters, NASA Office of Space Science, Solar System Exploration Div., Washington D.C., 154 pp., http://neo.jpl.nasa.gov/neo/neoreport030825.pdf

Tedesco, E.F., Cellino, C. & Zappalá, V.2005, *Astron. J.* 129, 2869

Asteroids, Comets, Meteors
Proceedings IAU Symposium No. 229, 2005
D. Lazzaro, S. Ferraz-Mello & J.A. Fernández, eds.

© 2006 International Astronomical Union
doi:10.1017/S1743921305006654

Orbital migration in protoplanetary disks

J. C. B. Papaloizou[1,2] and E. Szuszkiewicz[3]

[1]Astronomy Unit, Queen Mary, University of London, Mile End Rd, London E1 4NS, England

[2]Department of applied Mathematics and Theoretical Physics
Centre for Mathematical Sciences, Wilberforce Road,
Cambridge CB3 0WA, United Kingdom

[3]Institute of Physics, University of Szczecin, Wielkopolska 15, 70-451 Szczecin, Poland

Abstract. We review processes leading to the orbital migration of bodies ranging from dust particles up to protoplanets in the earth mass range in protoplanetary disks. The objects most at risk of being lost from the disk have dimensions of the order of metres. To retain these it may be necessary to invoke either strong turbulence or trapping due to special flow features such as vortices. Migration also becomes important for protoplanets with mass exceeding $0.1 M_\oplus$. In this case it can also lead to the attainment of commensurabilities for pairs of protoplanets. Such pairs could be left behind after disk dispersal. We review some recent work on the attainment of commensurabilities for protoplanets in the earth mass range.

Keywords. planetary systems, planetary systems: formation, planetary systems: protoplanetary disks

1. Introduction

The discovery of extrasolar giant planets orbiting close to their host stars (Mayor & Queloz 1995; Marcy & Butler 1995, 1998) with periods of a few days has led to an appreciation of the importance of orbital migration during and post formation.

Here we review the processes that lead to orbital migration for objects ranging from meter size up to the several earth mass range. For these masses the migration is entirely determined by the disk in which they are embedded. This is because they are too small to affect the disk density profile sufficiently to provide feedback such as occurs with gap formation in the Jovian mas range. We shall not consider such phenomena here, referring the reader to eg. Lin & Papaloizou (1993), Kley (2000), Kley, Pietz & Bryden (2004), Papaloizou (2005) and references therein.

In a non turbulent disk with typically expected parameters, objects exceeding ~ 1 km. in size, normally considered to be planetesimals, and less than on the order of a lunar mass, orbital migration is unlikely to be important during an expected protostaellar disk lifetime $\sim 10^7$ y. For small masses below this lower limit hydrodynamic processes such as gas drag dominate, while for masses exceeding the upper limit gravitational perturbations leading to the excitation of density waves are more important. For objects of about one metre in dimension the migration is potentially so short as to threaten survival but this could be stalled at special locations in the disk. In a turbulent disk this picture is blurred by the effects of potentially strong density fluctuations that can lead to stochastic gravitational forces which in turn may give rise to to a stochastic component to the orbital migration (Nelson & Papaloizou 2004).

One of the important features of observed exoplanetary systems is the occurrence of mean motion resonances, which are generally supportive of post formation orbital migration. Convergent migration of two objects is believed to be a generic mechanism leading to the attainment of commensurabilities in satellite systems within the solar system (see e.g. Goldreich 1965; Gomes 1988). Exoplanetary systems exhibiting low order commensurabilities among giant planets are Gliese 876 (Marcy *et al.* 2001), HD 82943 (Mayor *et al.* 2001) and 55 Cancri (McArthur *et al.* 2001). A natural explanation is that the disk planet interaction produces orbital migration through the action of tidal torques (Goldreich & Tremaine 1980; Lin & Papaloizou 1986) which in turn may lead to orbital resonances in a two giant planet system. Several studies have verified this (Snellgrove Papaloizou & Nelson 2001; Lee & Peale 2002; Papaloizou 2003; Kley Pietz & Bryden 2004; Kley *et al.* 2005).

However, it is likely that planetary systems around other stars may harbor planets with masses in the Earth mass range as well. These should be revealed by future space-based missions, such as Darwin, COROT, Kepler, SIM and TPF.

Accordingly, it is of interest to consider what comensurabilities could be established when two embedded low mass planets migrate in a gaseous disk. We review some recent results of Papaloizou & Szuszkiewicz (2005) on this problem that indicate that two planets with near equal masses could produce stable low order commensurabilites (e.g. 3:2), more disparate masses could evolve into a stochastic regime in which resonances (e.g. 8:7) are less stable but nonetheless still potentially observable in some cases.

2. Migration of small bodies

Small particles are well coupled to the gas and may be modeled as a fluid separate from the gas. For such a two fluid system the equations of motion are (see e.g. Garaud & Lin 2004).

$$\frac{\partial \mathbf{v}}{\partial t} + \mathbf{v} \cdot \nabla \mathbf{v} = -\frac{\nabla P}{\rho} + \mathbf{f} - \nabla \Phi - \frac{\rho_d (\mathbf{v} - \mathbf{v}_d)}{\rho \tau_c} \tag{2.1}$$

and

$$\frac{\partial \mathbf{v}_d}{\partial t} + \mathbf{v}_d \cdot \nabla \mathbf{v}_d = -\nabla \Phi - \frac{(\mathbf{v}_d - \mathbf{v})}{\tau_c}. \tag{2.2}$$

Here $\mathbf{v}$ and $\mathbf{v}_d$ denote the velocities of the gas and dust respectively. The gas density, dust density and a gas pressure are ρ, ρ_d, and P respectively. The gravitational potential, normaly due to the central mass is Φ. The last term on the right hand side in the equations of motion gives the acceleration due to drag between the gas and solid components. The stopping time τ_c measures the time for a solid particle to relax to the gas velocity. Any additional accelerations acting on the gas, as a result of for example magnetic fields are denoted by $\mathbf{f}$.

For particles that are smaller than a mean free path, that can be estimated to be about $1m$ for conditions appropriate to a minimum mass solar nebula, we have $\tau_c \sim (4a\rho_\alpha)P_{orb}/(3\pi\Sigma)$. Here a is the particle radius and ρ_α its density. The gas surface density is Σ and the orbital period is $2\pi/\Omega = P_{orb}$. From this it follows that for bodies of radius $1m$ the coupling to the gas is good or $\tau_c < P_{orb}/(2\pi)$, provided $\Sigma \gtrsim 3000$ g cm^{-2} at $1AU$, a condition expected to be satisfied.

For the simple case of a steady axisymmetric disk with $\mathbf{f} = 0$ and the gas is in a state of pure rotation with $\mathbf{v} = (0, v_\varphi(r), 0)$ in cylindrical coordinates, one has, neglecting radial drifts, for the rotation of the solid component assumed to be at a rate close to that of the gas

$$v_{d,\varphi} - v_\phi = \frac{-1}{2\rho\Omega}\frac{dP}{dr}. \tag{2.3}$$

Drag between the solid and gaseous components then produces aradial drift velocity which can be found from the azimuthal component of the equation of motion of the solids in the form

$$v_{d,r} = \frac{1}{\rho \Omega^2 \tau_c} \frac{dP}{dr}. \tag{2.4}$$

Note that this depends on the pressure gradient in this case with $\mathbf{f} = 0$, but the discussion can be simply generalized to include additional forces.

The normal expectation is that the pressure increases inwards and so when $\mathbf{f} = 0$, the expected drift is inwards. Asuming the scale of variation of the pressure to be comparable to the radius, the characteristic magnitude of the drift rate is

$$v_{d,r} \sim \frac{c^2}{r \Omega^2 \tau_c}, \tag{2.5}$$

where c is the isothermal sound speed.

For bodies marginally coupled to the gas with $\Omega \tau_c = 1$, being such that $a = (3\Sigma)/(8\rho_\alpha)$, the radial drift time is $\sim (1/(2\pi))(r\Omega/c)^2 P_{orb}$. This is very short approaching ~ 70 orbits for aspect ratio $H/r = c/(r\Omega) = 0.05$.

2.1. *Gas drag*

The above discussion holds for particles smaller than the mean free path but which can reach sizes up to around one metre. For much larger objects fluid drag operates. We may take this into account by replacing $1/\tau_c$ by $3C_D \rho |v_{d,\varphi} - v_\phi|/(4\rho_\alpha a)$ in equation (2.2). Here C_D is the drag coefficient. For large objects such that the relative gas flow has a Reynold's number greater than of order unity C_D is of order unity.

In that case the radial drift rate is modified such that

$$v_{d,r} \sim \frac{C_D \rho c^4}{r^2 \Omega^3 \rho_\alpha a} \sim c \frac{C_D \Sigma c^2}{2r^2 \Omega^2 \rho_\alpha a}. \tag{2.6}$$

For $C_D = 0.2$, $\rho_\alpha = 1$ g cm^{-3}, $a = 10$ m and $\Sigma = 10^3$ g cm^{-2}, $v_{d,r} = 0.1c[c/(r\Omega)]^2$. The radial flow time scale is then $(5/\pi)(r\Omega/c_s)^3 P_{orb}$. For $c/(r\Omega) = 0.05$, this is ~ 15000 orbits.

The above discussion indicates potentially serious survival problems for objects about one metre in size. A possibility is that the lifetime in this size range is short, a situation that might arise if the sedimenting solids undergo gravitational instability (e.g. Goldreich & Ward 1973). However, it is now believed that a sedimenting disk, even though initially laminar, is subject to sufficient turbulence (e.g. Weidenschilling 1980, 2003) to prevent enough sedimentation to allow gravitational instability to occur.

An alternative possibility is that special regions of the flow are significant. In particular particles may accumulate at pressure extrema where the relative drift between gas and dust is expected to be small. In fact one generally expects attraction to pressure maxima for particles tightly coupled to the gas. If gas elements remain near such a maximum, then because the equations of motion for solid particles and gas coincide at an extremum in pressure, even particles not very tightly coupled to the gas can remain there also. Note that for the case when pressure varies only with radius, because the radial drift velocity is proportional to the pressure gradient, a pressure maximum is a stable attractor, while a minimum is unstable. This discussion emphasizes the potential significance of special locations in the disk for accumulating solids. The importance of the role of vortices in this respect has also been considered (e.g. Barge & Sommeria 1995).

For larger objects the drift time is $\propto 1/a$, and this becomes comparable or longer than expected gas disk lifetimes for objects larger than a few kilometres independently of flow features provided they are sufficiently smooth. However, as objects approach

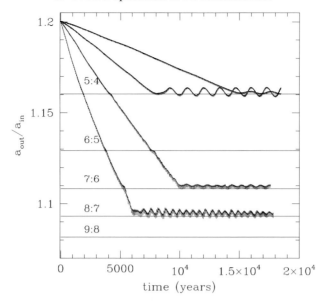

Figure 1. The evolution of the ratio of semi-major axes of two planets with masses $m_1 = 4M_\oplus$ and $m_2 = 1M_\oplus$. Starting from the lower curve and going upwards the curves correspond to the initial surface density scalings $\Sigma_0 = 4\Sigma_1$, $\Sigma_0 = 2\Sigma_1$, $\Sigma_0 = \Sigma_1$, and $\Sigma_0 = 0.5\Sigma_1$ respectively.

$\sim 0.1 M_\oplus$, direct gravitational perturbation of the disk becomes more important than local hydrodynamic processes such as drag in determining radial drift. This we consider below.

2.2. *Migration of low mass planets*

For low mass planets the disk undergoes small linear perturbations that induce density waves that propagate away from the planet. The angular momentum these waves transport away results in a potentially rapid orbital migration called type I migration (Ward 1997). In this type of migration, when the disk is laminar and inviscid, the planet is fully embedded and the surface density profile of the disk remains approximately unchanged. This, as well as the well known density wakes associated with the disk planet interaction, is illustrated in calculations such as those illustrated in Figures 1 and 2 below. The timescale of inward migration on a circular orbit can be derived from a linear response calculation and estimated for a disk with constant surface density to be given by (see Tanaka Takeuchi & Ward 2002)

$$\tau_r = \left| \frac{r}{\dot{r}} \right| = W_m \frac{M_*}{m_p} \frac{M_*}{\Sigma r^2} \left(\frac{c}{r\Omega} \right)^2 \Omega^{-1} \qquad (2.7)$$

Here M_* is the mass of the central star, m_p is the mass of the planet orbiting at distance r and the numerical coefficient $W_m = 0.3788$.

It is important to note that type I migration appropriate to a laminar disk may lead to short migration times in standard model disks, that may threaten the survival of protoplanetray cores (Ward 1997). For $m_p = M_\oplus$, and $\Sigma = 200$ gm cm^{-3}, the inward migration time at 5.2 AU is $\sim 10^6$ y, being inversely proportional to mass. Thus bodies smaller than 1 km. and larger than about $\sim 0.1 M_\oplus$ could be lost through migration within potential gas disk lifetimes of up to $\sim 10^7$ y (see e.g. Papaloizou & Larwood 2000).

However, in a disk with turbulence driven by the magnetorotational instability, the migration may be stochastic and accordingly less effective (Nelson & Papaloizou 2004).

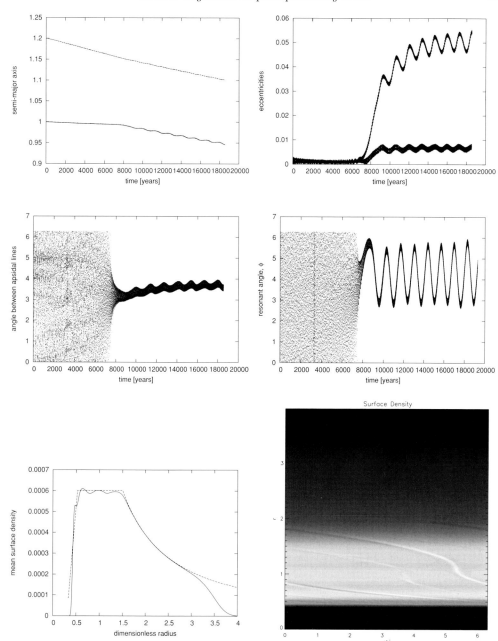

Figure 2. The evolution of the semi-major axes, eccentricities, angle between apsidal lines and the resonant angle for planets with masses, $m_1 = 4M_\oplus$ and $m_2 = M_\oplus$ migrating towards a central star, embedded in a disk with initial surface density scaling $\Sigma_0 = \Sigma_1$ (four upper panels). In this and similar plots, the resonant angle is appropriate to the commensurability attained at the end of the simulation. The mean surface density profile of the disk near the end of the simulation (solid line) and the initial surface density profile (dashed line) and a surface density contour plot are given in two lower panels.

This effect could be important for stalling migration during some evolutionary phases. Nonetheless there is considerable uncertainty as to the extent of turbulent regions in the disk resulting from uncertainties in the degree of ionization (e.g. Fromang, Terquem & Balbus, 2002) so that type I migration appropriate to a laminar disk may operate in some regions. We also note that because it is inversely proportional to the disk surface density, the migration time becomes long in low surface density regions. Hence type I migration in a laminar disk may be significant in some regions of protoplanetary disks. Hence the recent study by Papaloizou & Szuszkiewicz (2005) of the commensurabilites that could be realized by a pair of interacting planets migrating in a laminar disk with masses in the $M \oplus$ range. The density waves excited by a low mass planet with small eccentricity also lead to orbital circularization (Artymowicz 1993; Papaloizou & Larwood 2000) at a rate that can be estimated to be given by (Tanaka & Ward 2004)

$$t_c = \frac{\tau_r}{W_c} \left(\frac{c}{r\Omega}\right)^2. \tag{2.8}$$

Here, the numerical coefficient $W_c = 0.289$.

2.3. Eccentricity damping due to gas drag

We consider a body of mass m_α, density ρ_α and radius a moving in an eccentric orbit relative to the disk gas with speed v. The equation of motion for v is

$$m_\alpha \frac{dv}{dt} = -\pi a^2 C_D \rho v^2, \tag{2.9}$$

This leads to a relative velocity damping time, $t_{gd} = |v/(dv/dt)|$, equivalent to

$$\frac{1}{\Omega t_{gd}} = \frac{(36\pi)^{1/3} \Sigma C_D}{8 \rho_\alpha^{2/3} m_\alpha^{1/3}} \frac{er}{H}, \tag{2.10}$$

where we have set $v = er\Omega$ with e corresponding to the orbital eccentricity. This gives in cgs units

$$t_{gd} = 5.27 \times 10^3 \left(\frac{m_\alpha}{10^{18}}\right)^{1/3} \frac{H}{er} \left(\frac{50 \rho_\alpha^{2/3}}{\Sigma C_D}\right) \left(\frac{r}{1 \text{ AU}}\right)^{3/2} \text{ yr.} \tag{2.11}$$

Note that this time can be small for small masses, which is related to the need to evolve fast through the size range where decoupling from the gas first occurs.

2.4. Relative importance of gas drag and disk tides

Because of the dependence on protoplanet mass, gas drag is more effective for smaller masses while disk tides take over for larger masses. The transition mass, which is also the mass for which the damping time is a maximum, can be estimated by equating the gas drag time given by equation (2.11) with the eccentricity decay time resulting from disk tides given by equation (2.8). This gives

$$\frac{m_\alpha}{M_\oplus} = 5.2 \times 10^{-4} \frac{f_s^{15/8} C_D^{3/4}}{\rho_\alpha^{1/2}} \left(\frac{H/r}{0.05}\right)^3 \left(\frac{H/r}{e}\right)^{-3/4} \left(\frac{r}{5.2\text{AU}}\right)^{-3/2}. \tag{2.12}$$

Although there is significant sensitivity to disk location and aspect ratio, the crossover mass is characteristically on the order of a few percent of a lunar mass. Thus gas drag will be the dominant eccentricity damping process when planetesimals first form.

However, if many larger mass cores form and then interact during the later stages of core accumulation, disk tides will dominate.

3. Convergent migration of planets in the earth mass range and their attainment of commensurabilities

It is expected from equation (2.7) that two planets with different masses will migrate at different rates. This has the consequence that their period ratio will evolve with time and may accordingly attain and, in the situation where the migration is such that the orbits converge, subsequently become locked in a mean motion resonance (Nelson & Papaloizou 2002; Kley *et al.* 2004).

In the simplest case of nearly circular and coplanar orbits the strongest resonances are the first-order resonances which occur at locations where the ratio of the two orbital periods can be expressed as the ratio of two consecutive integers, $(p+1)/p$, with p being an integer. As p increases, the two orbits approach each other and the strength of the resonance increases. In addition, the distance between successive resonances decreases as p increases. The combination of these effects ultimately causes successive resonances to overlap and so, in the absence of gas, leads to the onset of chaotic motion.

Resonance overlap occurs when the difference of the semi-major axes of the two planets is below a limit with half-width given, in the case of two equal mass planets, by Gladman (1993) as

$$\frac{\Delta a}{a} \sim \frac{2}{3p} \approx 2 \left(\frac{m_{planet}}{M_*} \right)^{2/7}, \tag{3.1}$$

with a and m_{planet} being the mass and semi-major axis of either one of them respectively. Thus for a system consisting a two equal four earth mass planets orbiting a central solar mass we expect resonance overlap for $p \gtrsim 8$. Conversely we might expect isolated resonances in which systems of planets can be locked and migrate together if $p \lesssim 8$.

However, note that the above discussion does not incorporate the torques producing convergent migration or eccentricity damping and thus may not give a complete account of the forms of chaos that might be expected. Kary, Lissauer & Greenwig (1993) have considered the case of small particles migrating towards a much more massive planet and indeed conclude that chaotic behaviour is more extensive in the non conservative case.

In a stochastic regime, long term stable residence in commensurabilities is not possible. However, a system can remain in one for a long time before moving into another higher order commensurability. Furthermore detailed outcomes are very sensitive to input parameters. Slight changes can alter the sequence of commensurabilities a system resides in making the issue of their attainment acquire a probabilistic character. Although there may be islands of apparent stability, a system in this regime may ultimately undergo a scattering and exchange of the orbits of the two planets so that we should focus on stable lower degree commensurabilities as being physically possible planetary configurations. The arguments given above suggest that these must have $p \lesssim 8$. Our calculations indicate that the limit is even smaller.

Hydrodynamic simulations of disk planet interactions, in which the disks are modelled as flat two dimensional objects with laminar flow governed by the Navier Stokes equations and which incorporate a migrating two giant planet system which evolves into a 2:1 commensurability have been performed by Kley (2000) and Snellgrove *et al.* (2001) and successfully applied to the GJ876 system. In addition to performing additional simulations, Papaloizou (2003) has developed an analytic model describing two planets migrating in resonance with arbitrary eccentricity. In this model the eccentricities are

determined as a result of the balance between migration and orbital circularization. However, disk tides were considered to act on the outer planet only. Capture of giant planets
into resonance has also been recently investigated numerically by Kley *et al.* (2004).

4. Numerical simulations of migrating planets in resonance

Papaloizou & Szuszkiewicz (2005) performed simulations of two interacting planets
embedded in and interacting with an accretion disk. These simulations are similar in
concept to those of Snellgrove *et al.* (2001) and references therein, where the resonant
coupling in the GJ876 system induced by orbital migration was studied.

One can a system of units in which the unit of mass is the central mass M_*, the unit
of distance is the initial semi-major axis of the inner planet, r_2, and the unit of time is
$2\pi(GM_*/r_2^3)^{-1/2}$, being the initial orbital period of the inner planet.

In these dimensionless units the inner boundary of the computational domain was at
$r = r_{min} = 0.33$, and the outer boundary at $r = r_{max} = 4$. The equation of state is
locally isothermal with aspect ratio $H/r = 0.05$ and the explicit viscosity is set to zero.
The radial boundaries were open and $n_r = 384$, and $n_\varphi = 512$ equally spaced grid points
were adopted in the radial and azimuthal directions respectively.

Two planets with masses m_1 and m_2 orbiting a central star with mass M_* were
considered. The two embedded planets began on circular orbits at radii r_1 and r_2 respectively. The gravitational potential due to the planets was softened with softening
parameter $b = 0.8H$. This gives type I migration rates in reasonable agreement with
Tanaka,Takeuchi & Ward (2002) (Nelson & Papaloizou 2004). The initial disk surface
density was taken to be given by

$$\Sigma(r) = \begin{cases} 0.1\Sigma_0(15(r - r_{min})/r_{min} + 1) \text{ if } r \geqslant r_{min} \text{ and} \\ \quad r \leqslant 8r_{min}/5 \\ \Sigma_0 \text{ if } r > 8r_{min}/5 \text{ and } r < 4.5r_{min} \\ \Sigma_0(4.5r_{min}/r)^{1.5} \text{ if } r \geqslant 4.5r_{min} \end{cases}$$

where r_{min} is the inner edge of the computational domain. The planets begin in the flat
part of this profile.

The use of dimensionless units enables the results to be scaled so that they apply to
different radii and corresponding initial surface densities.

However, in order to make things definite r_2 was taken to be 5.2AU, but nonetheless
the time measured in years.

We adopt a fiducial value of the surface density scaling $\Sigma_0 = \Sigma_1 = 2 \cdot 10^3 (5.2AU/r_2)^2$
kg/m^2. Thus when $r_2 = 5.2AU$, this corresponds to a typical value attributed to the
minimum mass solar nebula at 5.2AU.

4.1. *Attainment of Commensurability*

The interaction of the planets with the disk leads to spiral wave excitation, energy and
angular momentum exchange between with the disk, orbital migration and eccentricity
damping. When the migration of the two planets is convergent, by which we mean that the
ratio of their semi-major axes increases, first order commenurabilities where the period
ratio is $(p + 1)/p$ for some integer p will inevitably be approached. As p increases, the
planets become closer together. As a result of this it is possible that the planets become
locked in mean motion resonances (e.g. Goldreich 1965, Lee & Peale 2002; Ferraz-Mello,
Beaugé & Mitchchenko 2003; Beaugé Ferraz-Mello & Mitchchenko 2003) such as 5:4,

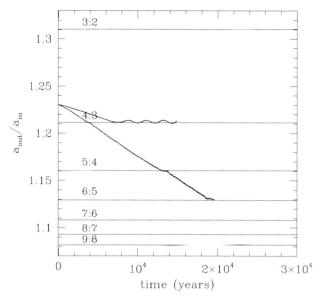

Figure 3. The semi-major axis ratio for $m_1 = 4M_\oplus$ and $m_2 = 1M_\oplus$ with surface density scaling parameter $\Sigma_0 = 0.5\Sigma_1$ – upper curve and $\Sigma_0 = \Sigma_1$ – lower curve).

6:5, 7:6, 8:7, ..., $(p + 1) : p$, ... and subsequently migrate together maintaining the commensurability for a considerable period of time or perhaps indefinitely, depending on how close the system is to an unstable stochastic regime.

Papaloizou (2003) has given a simple approximate analytic solution for two migrating planets locked in a 2:1 ($p = 1$) commensurability and Papaloizou & Szuszkiewicz (2005) have generalized this so it is applicable to the commensurabilities with larger p discussed here. Mutual resonant interaction between the planets tends to cause the eccentricities of both planets to grow with time, an effect which is counterbalanced by the effects of interaction with the disk leading to an equilibrium condition.

Close to a balanced $p + 1 : p$ resonance, the resonant angles $\phi = (p + 1)\lambda_1 - p\lambda_2 - \varpi_1$, $\psi = (p + 1)\lambda_1 - p\lambda_2 - \varpi_2$ and $\varpi_1 - \varpi_2$, with λ_i, and ϖ_i denoting the mean longitude and longitude of periapse of planet i, librate about equilibrium values which can be near to 0 or π mod 2π. Simulations for planets in the earth mass range indicated libration about a value closer to π than 0.

Which resonance (or value of p) is established depends on the rate of convergent migration. The resonances become stronger as p increases, so that increasing the relative migration rate tends to cause the attainment of larger values. However, if p becomes too large ($p \gtrsim 8$ for the planet masses considered) the motion of the planets becomes chaotic (see equation (3.1)) making commensurabilities unstable in the long term (although a system may remain in the vicinity of one for a considerable time) while introducing sensitivity of the orbital evolution to the initial conditions. In such situations a scattering may ultimately occur that interchanges the positions of the planets. Thus very high surface densities that are associated with very rapid migration rates are not conducive to the attainment of stable commensurabilities.

We performed simulations to investigate the dependence of attained commensurabilities on on the rate of convergent migration and initial conditions. To do this a pair of planets with masses $m_1 = 4M_\oplus$ and $m_2 = 1M_\oplus$ were set up in circular orbits with $r_1/r_2 = 1.2$ in disks with $\Sigma_0 = 0.5\Sigma_1$, $\Sigma_0 = \Sigma_1$, $\Sigma_0 = 2\Sigma_1$ and $\Sigma_0 = 4\Sigma_1$.

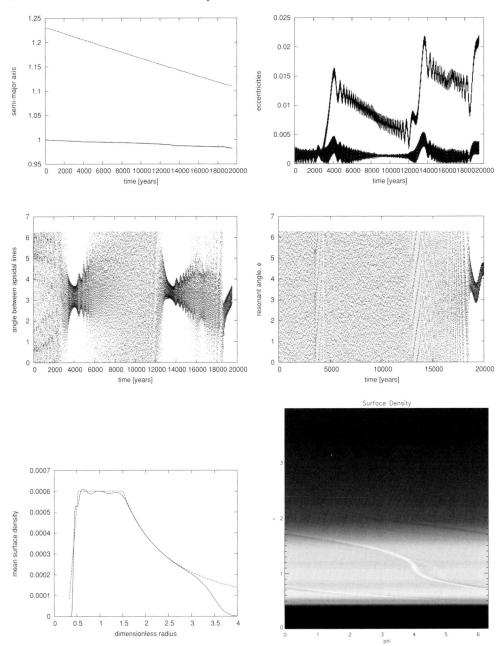

Figure 4. As for Figure 2 but when the planets are initiated with $a_1/a_2 = 1.23$.

The subsequent evolution shown in Figures 1 and 2 indicates that a resonance is established such that p increases with the value of Σ_0 or equivalently the rate of migration as expected. The planets in the disk with $\Sigma_0 = 4\Sigma_1$ become trapped in 8:7 resonance. For $\Sigma_0 = 2\Sigma_1$ a 7:6 resonance is attained. For $\Sigma_0 = \Sigma_1$ and $\Sigma_0 = 0.5\Sigma_1$ a 5:4 commensurability results.

The evolution for the simulation that started with $\Sigma_0 = \Sigma_1$ leads to attainment of a 5:4 commensurability at around 8000 years. Subsequently the inner planet semi-major

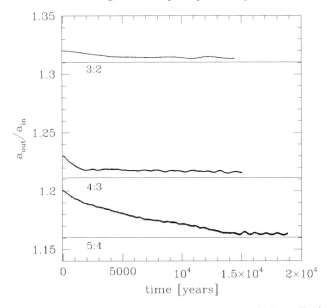

Figure 5. The semi-major axis ratio when $m_1 = m_2 = 4M_\oplus$ with $\Sigma_0 = \Sigma_1$ (uppermost curve) and with $\Sigma_0 = 4\Sigma_1$ (two lower curves).

axis changes in a characteristic oscillatory manner. These oscillations are clearly visible in Figure 2.

The eccentricity ratio at the end of the simulation is $e_1/e_2 = 0.18$. The eccentricities are still growing. The angle between the apsidal lines starts to librate at time 8000 years around $180°$ and at time 18000 years oscillates around $212°$. The resonant angle excursions have large amplitude and seem to grow with time, which might indicate that the planets will not remain in this resonance.

4.2. *Dependence on the initial separation of the planets*

The dependence of the behaviour at a commensurability on initial conditions and the general lack of stability at some resonances is indicated by the fact that when the ratio of the initial semi-major axes is changed from 1.2 to 1.23, the case with $\Sigma_0 = \Sigma_1$ passes through the 4:3 and 5:4 resonances and becomes trapped in the 6:5 resonance (see Figure 3).

The passage of the planets through the 4:3 resonance and the temporary trapping in the 5:4 resonance can be seen in Figure 4. At time 4000 years there is a rapid increase in the eccentricities of both planets followed by a slower decrease. The angle between the apsidal lines also shows the expected behaviour during resonance crossing. When the system starts with a smaller initial planet separation as illustrated in Figure 2 the effect of a prior resonance passage before approaching and becoming relatively stably trapped in the 5:4 resonance is not present. So it is likely that the mutual interaction of planets during resonance crossing influences their subsequent evolution and that these planets are either close to or in the chaotic regime.

4.3. *Equal mass planets and lower p commensurabilities*

As described above, in equilibrium the resonant interaction must balance the tendency towards convergent migration of the two planets. As indicated by use of equation (2.7),

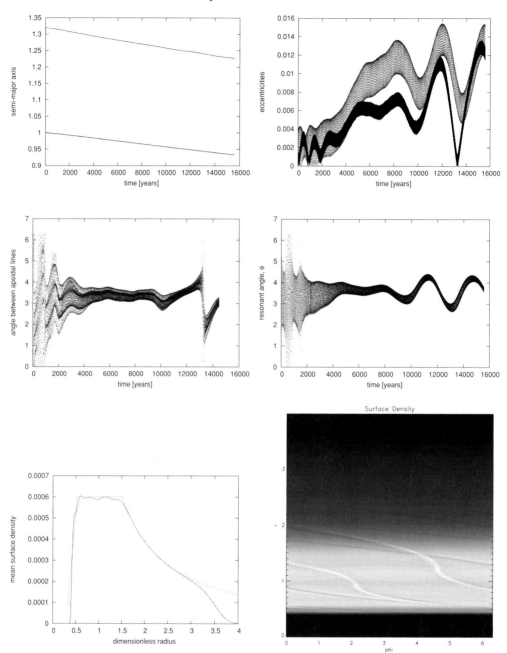

Figure 6. The evolution of semi-major axes, eccentricities, angle between apsidal lines and resonant angle for two planets with equal mass, $m_1 = m_2 = 4M_\oplus$, migrating towards a central star and embedded in a disk with $\Sigma_0 = \Sigma_1$. (four upper panels). The mean surface density profile of the disk near the end of the simulations (solid line), together with the initial surface density profile (dashed line) and a surface density contour plot near the end of the simulation are given in the lowest left and right panels respectively.

this is smaller when the planets have the same mass. In fact the ratio of the migration rates for equal masses at a commensurability is found to be $((p+1)/p)^{1/3}$. Thus the rate of convergence is about ten percent of the absolute migration rate even for $p = 3$. Accordingly, lower degree commensurabilities and less tendency to be driven into a chaotic state might be expected in this case when compared to the situation where the inner planet has significantly smaller mass.

In a simulation study, two planets of mass $4M_\oplus$ were initiated close to 3:2 resonance with $r_1/r_2 = 1.32$ in a disk with $\Sigma_0 = \Sigma_1$. The evolution of the semi-major axis ratio for the planet orbits is shown in Figure 5. We also performed simulations for the same masses starting with $r_1/r_2 = 1.23$ in a disk with $\Sigma_0 = 4\Sigma_1$ and $r_1/r_2 = 1.2$ in a disk with $\Sigma_0 = 4\Sigma_1$.

In each of these cases, the planets become trapped in the nearest available resonance, itself a good indication of stability. The evolution of the equal mass pair of planets is shown in Figure 6. The planets attained 3:2 resonance after about 4000 years, had increasing eccentricities until after about 8000 years $e_1 \sim 0.013$ and $e_2 \sim .007$. Subsequently the eccentricities oscillated with the resonant angle oscillating around $200°$.

4.4. *N-body integrations*

In order to study migrating planets with a wider range of masses and disk surface densities, Papaloizou & Szuszkiewicz (2005) studied the evolution of migrating pairs of planets using an N-body code. Migration and circularization at rates governed by equations (2.7 – 2.8) are incorporated by including appropriate non conservative terms in the equations of motion. For details of the procedure see Snellgrove *et al.* (2001), Nelson & Papaloizou (2002), or Kley *et al.* (2004). The outcome of these simulations could be well matched to those of the hydrodynamic simulations over their relatively short run times, for appropriate choices of W_m and W_c which were quite close to those suggested by Tanaka,Takeuchi & Ward (2002) and Tanaka & Ward (2004). However, there was a strong sensitivity of outcomes such as the duration for which a commensurability with specified p was maintained to these. This is symptomatic of chaotic motion as has been described by for example (Kary, Greenwig & Lissauer, 1993). However, in view of the fact that some of the sensitivity to initial conditions could be a manifestation of a sequence of passages close to hyperbolic fixed points, the extent of significant chaotic motion is unclear.

5. Discussion

We have reviewed processes that can lead to orbital migration in a protostellar disk. Masses ranging from those just decoupling from the gas, typically of radius 1 m. for standard protoplanetary disk parameters migrate quickly unless they are at special disk locations such as pressure extrema.

Objects in ranging from $\sim$1 km. to about a lunar mass are relatively safe from migration in a laminar disk over expected protoplanetary disk lifetimes.

Recent work of Papaloizou & Szuszkiewicz (2005) has shown that masses in the earth mass range can undergo convergent type I migration. When masses are disparate this normally leads to resonances such as 8:7 or 7:6 for which the motion might be stochastic. However for more equal masses comensurabilities such as 3:2 can be obtained which are much more stable.

In a well known example of such a system, the two largest mass planets orbiting a millisecond radio pulsar PSR B1257+12, are close to a 3:2 commensurability (Konacki & Wolszczan 2003). Such a resonance could be formed and maintained through the convergent migration discussed here, with the planets later moving slightly out of resonance. Studies of commensurabilities have already been applied successfully to analyse the motion of both commensurable pairs of giant planets and the pulsar planets, they proved to be a powerful tool in that context. They have the potential to be similarly useful in our search for Earth-like planets in other systems. Detection of such resonances should yield useful information about orbital migration occuring during and post planet formation.

References

Artymowicz, P. 1993, *ApJ*, 419, 155

Barge, P. & Sommeria, J. 1995, *A&A*, 295

Beaugé, C., Ferraz-Mello, S., & Mitchchenko, T.A. 2003, *ApJ*, 593, 1124

Ferraz-Mello, S., Beaugé, C., & Mitchchenko, T.A. 2003, *Cel. Mech. and Dynam. Astron.* 87, 99

Fromang, S., Terquem, C., & Balbus, S.A. 2002, *MNRAS*, 329, 18

Gladman, B. 1993, *Icarus*, 106, 247

Garaud, P. & Lin, D.N.C. 2004, *ApJ*, 608, 1050

Goldreich, P. 1965, *MNRAS*, 130, 159

Goldreich, P. & Tremaine, S. 1980, *ApJ*, 241, 425

Goldreich, P. & Ward, W.R. 1973, *ApJ*, 183, 1051

Gomes, R.S. 1998, *AJ*, 116, 997

Kary, D.M., Lissauer, J.J., & Greenzweig, Y. 1993, *Icarus*, 106, 288

Kley, W. 2000, *MNRAS*, 313, L47

Kley, W., Peitz, J., & Bryden, G. 2004, *A&A*, 414, 735

Kley, W., Lee, M.H., Murray, N., & Peale, S.J. 2005, *A&A*, 437, 727

Konacki, M. & Wolszczan, A. 2003, *ApJ*, 591, L147

Lee, M.H. & Peale, S.J. 2002, *ApJ*, 567, 596

Lin, D.N.C. & Papaloizou, J.C.B. 1986, *ApJ*, 309, 846

Lin, D.N.C. & Papaloizou, J.C.B. 1993, in: E.H. Levy & J.I. Lunine (eds.), *Protostars and Planets III* (Tucson: University of Arizona Press), p. 749

Marcy, G.W. & Butler, R.P. 1995, *187th AAS Meeting baas*, vol. 27, p. 1379

Marcy, G.W. & Butler, R.P. 1998, *Annu. Rev. Astron. Astr.*, 36, 57

Marcy, G.W. Butler, R.P., Fischer, D., Vogt, S.S., Lissauer, J.J., & Rivera, E.J. 2001, *ApJ*, 556, 296

Mayor, M. & Queloz, D. 1995, *Nature*, 378, 355

Mayor, M., Udry, S., Naef, D., Pepe, F., Queloz, D., Santos, N.C., & Burnet, M. 2001, *A&A*, 415, 319

McArthur, B.E., Endl, M., Cochran, W.D., Benedict, G.F., Fischer, D.A., Marcy, G.W., Butler, R.P., Naef, D., Mayor, M., & Queloz, D. 2001, *ApJ*, 614, L296

Nelson, R.P., Papaloizou, J.C.B., Masset, F., & Kley, W. 2000, *MNRAS*, 318, 18

Nelson, R.P. & Papaloizou, J.C.B. 2002, *MNRAS*, 333, 26

Nelson, R.P. & Papaloizou, J.C.B. 2004, *MNRAS*, 350, 849

Papaloizou, J.C.B. 2003, *Cel. Mech. and Dynam. Astron.*, 87, 53

Papaloizou, J.C.B. 2005, *Cel. Mech. and Dynam. Astron.*, 97, 33

Papaloizou, J.C.B. & Larwood, J.D. 2000, *MNRAS*, 315, 823

Papaloizou, J.C.B. & Szuszkiewicz, E. 2005, MNRAS, In press

Snellgrove, M., Papaloizou, J.C.B. & Nelson, R.P. 2001, *A&A*, 374, 1092

Tanaka, H., Takeuchi, T., & Ward, W.R. 2002, *ApJ*, 565, 1257

Tanaka, H. & Ward, W.R. 2004, *ApJ*, 602, 388

Ward, W.R. 1997, *Icarus*, 126, 261

Weidenschilling, S.J. 1980, *Icarus*, 44, 172

Weidenschilling, S.J. 2003, *Icarus*, 165, 438

Asteroids, Comets, Meteors
Proceedings IAU Symposium No. 229, 2005
D. Lazzaro, S. Ferraz-Mello & J.A. Fernández, eds.

Deep Impact:
excavating comet 9P/Tempel 1

M. F. A'Hearn[1] and The Deep Impact Team

[1]Department of Astronomy, University of Maryland, College Park MD 20742, USA
email: ma@astro.umd.edu

Abstract. The Deep Impact mission delivered 19 gigajoules of kinetic energy to the nucleus of comet 9P/Tempel 1 on 4 July 2005. Intensive observations, both from the two spacecraft and from Earth and Earth-orbit, while approaching the comet led to numerous new findings about comets. Observations of the impact event, from the flyby spacecraft have led us to a determination of numerous physical properties of the nucleus. Analysis of the near-IR spectra is still very preliminary.

Keywords. comets: general, comets: 9P/Tempel 1

1. Introduction

Deep Impact was conceived to investigate the relationship between the volatiles observed remotely in the ambient coma and ices, presumably primordial, in the bulk of the nucleus. It was also intended to investigate the physical structure of the nucleus itself. The true planning for the mission began in 1995 and the project took 5 years of selling the mission (both to the scientific review panels and to NASA itself) to reach our confirmation to proceed with construction and then an additional 5 years of building the spacecraft and flying the mission. The development of the mission design, the hardware, and the scientific background are described in a series of papers in Space Science Reviews (A'Hearn *et al.* 2005 and many additional papers cited therein). This IAU Colloquium coincides with the initial submission of the first results to Science (A'Hearn *et al.* 2005b) and thus this paper, which summarizes content both from the invited review and numerous contributed paeprs, includes only preliminary results corresponding to what is reported there (A'Hearn *et al.* 2005b). More specifically, we expect that some of the values reported here will change with more analysis. However, we have considered the likely range of errors in drawing conclusions to ensure that all our statements about processes will not change.

2. The Impact and the Nuclear Properties

The key goal of the mission was to impact the nucleus with enough energy to excavate to a depth of tens of meters. Whether this would be possible or not depended on the strength of the nuclear material in those outer tens of meters. Experts active in studying cratering had substantial disagreements over whether the formation of the crater would be controlled by the comet's gravity, by the strength of the outer layers, or by compression of porous material with moderate compressive strength, although our own position had been that the formation would likely be controlled by the comet's gravity as discussed by A'Hearn *et al.* (2005a), by Schultz *et al.* (2005) and by Richardson *et al.* (2005). In retrospect, gravitational control is precisely what happened.

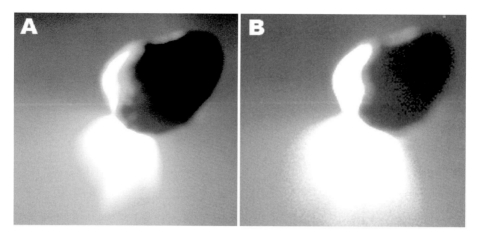

Figure 1. Highly stretched images of comet Tempel 1 taken roughly 45 and 90 minutes respectively after the impact. The ejecta cone has remained attached to the surface and is still expanding at its base as ballistic ejecta fall back to the surface outside the crater.

Our lookback imaging showed that the ejecta cone remained attached to the surface for hours, although strictly speaking we can only say that it remained within a few hundred meters of the surface since the crater itself is behind the limb as seen during lookback imaging. A pair of lookback images, taken at roughly 45 and 90 minutes after impact, is shown in figure 1 and one can easily see the cone still expanding at its lowest level visible around the limb. This is due to ballistic material from the cone falling back to the surface outside the crater and this allows us to estimate both the strength of the material near the rim of the (unseen) crater and the local gravity at the crater. With worst case assumptions, we find that the shear strength near the rim must be no more than 65 Pa, implying remarkably weak material at the rim. For the acceleration of gravity at the impact site we find 0.05 cm s^{-2}. Combining this with the volume that we have deduced we find that the bulk density of a uniform nucleus with this gravity must be 0.6 g cm^{-3}. For a dust-to-volatile ratio of 2, assuming material densities of 2.5 (some denser silicates and some less dense organics) and 1 respectively, this implies a very high porosity of 70% for the bulk properties of the nucleus.

Prior studies of comet Shoemaker-Levy 9 (SL9) had shown us that the tensile strength at km-scale was less than 100 Pa (Sekanina 1996). The success of strengthless models in showing reaccretion to the size of the observed fragments makes it plausible that the strength is that low even at the 100-m scale, but does not prove it. On Earth, strength is normally higher at smaller spatial scales but our experiment suggests that this is not the case on comets since our strength is relevant to scales from 1-m to 100-m, the size of the ejecta to the size of the crater (we will argue below that the ejecta are primarily <10 μm in size). The deduced strength is even lower than that predicted by Greenberg (1995) on the basis of his model for accreting interstellar grains.

The high porosity is not surprising since studies of the non-gravitational acceleration of comets starting with the work by Rickman *et al.* (1987) have typically deduced bulk densities between 0.1 and 1.0 g cm^{-3}, although the results are very model dependent as shown by the range of results for comet Wild 2 deduced by Farnham & Cochran (2002) and by Davidson & Gutiérrez (2004). While our own determination of the bulk density is not totally free of modelling assumptions, we think that it is much more tightly constrained than previous determinations. There is, thus far, no way to determine whether the porosity is primarily fine scale porosity or primarily large voids. The result assumes a uniform density, i.e. a uniform porosity but is not sensitive to that assumption unless the voids are comparable in size to our crater.

3. The Impact Site and the Surface Layer Properties

The impact was oblique, at 20 deg −35 deg to the horizontal. The impact site, shown in figure 2, is between two crater-like features that are roughly 300m across. The impactor approached from the top in these images, probably at about 35 deg above the horizontal. In the last figure shown, taken 13 sec before impact, the white spots are a few m in diameter. It is this terrain to which the strength measurements above apply. Whether the strength is similar in the other very different terrains that are clearly visible in the pictures of the nucleus is unknown. However, given the strength at larger scales inferred from nuclear breakup of SL9 (above) and other comets, it seems unlikely that the strength elsewhere would be dramatically different from the values deduced here. The impactor landed much closer to the lower (more southern in ecliptic coordinates, of the two crater-like features. While this area is obscured by ejecta in all post-impact images, the more northerly of the two craters is still visible after the impact with no obvious disruption. The distance from the ipact site to the southerly wall of the northerly crater thus sets an upper limit on the size of our crater at a radius of about 150 m.

The obliquity of the impact also shows up in the successive stages of brightness as illustrated in figure 3, where three successive brightenings can be seen, each displaced from the one before it. These images are contour maps of the brightness in 50-msec exposures, taken with center-to-center intervals of 62 msec. The first flash is faint, visible in only frames 64 and 65. This may correspond to the very early hot ejecta traveling backwards up the entry path or it may correspond to the subsurface flash of impactor destruction being scattered through the overlying layers. The second brightening, which appears after the first flash has disappeared, has the highest peak brightness, saturating the CCD in frames 67 and 68, which leads to a spatial broadening that is an artifact of charge bleeding along the column of the CCD. This brightening may correspond to the first material from the impact breaking through the surface. The final brightening, although not reaching the peak brightness of the second, is of much longer duration, beginning while the second flash is fading, and is the source of the bulk of the ejecta. The motion downrange between the successive brightenings is clear.

A sequence of images, from which the brightness profiles in figure 3 were derived, is shown in figure 4. This sequence, numbered horizontally and then vertically, corresponds to frames 59 through 70 of figure 3, i.e., it starts earlier than and ends in the middle of the sequence in figure 3, ending while the third brightening is still increasing. In figure 4 one can see a puff of material originating before the peak of the second brightening and rapidly moving downrange such that it is completely beyond the field of view by the end of the sequence (corresponding to frame 70). As discussed below, this is hot material driven hydronamically from the excavation which supports our interpretation of the second brightening as the first breakthrough of material at the surface of the nucleus. The horizontal row of dots in this figure shows the position of the slit of our near-infrared spectrometer during the sequence, with the bright dot in that row corresponding to the pixel that will be analyzed below. At the end of the sequence (corresponding to frame 70), the slower moving ejecta from the mechanical excavation stage have just reached the slit of the spectrograph.

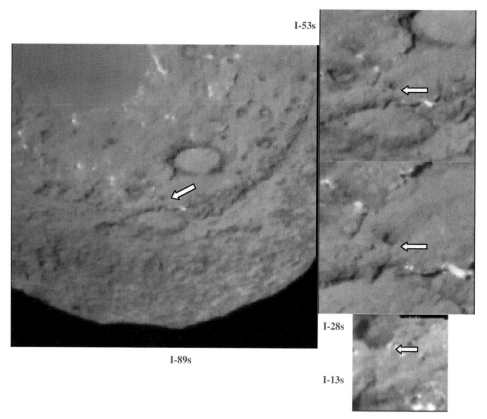

Figure 2. Successively closer images of the impact site taken by the impactor at the times shown in the figure. The ellipticity of the two craters, assuming they are both truly round, implies an approach angle of 20 deg and 35 deg respectively for the two craters, while the overall shape model agrees with the latter number.

4. Ejected particles

The particles ejected in the initial puff are very hot as measured by our near-infrared spectrometer. We obtained spectra continuously with the slit positioned as shown in figure 4, using initially an exposure time of 0.72 sec. The first of the two spectra shown in figure 5 was exposed at the single pixel shown in the slit position and was exposed completely before the impact. It shows purely a continuous spectrum with reflection from the nucleus dominating shortward of 2.5 μm and thermal emission from the nucleus dominating longward of 3.5 μm. At 0.5 sec into the exposure of the second spectrum, the expanding, lateral edge of the hot puff of material reaches the chosen pixel in frame 66 (pixels closer to the centerline of the ejecta become saturated at this point) and material from the puff remains in the slit until about the end of the exposure. In order to model the second spectrum, we require not only the thermal emission from the nucleus but also a much stronger emission from material at 850 K. Subsequently excavated material is cold and relatively unprocessed. The model for the gasous emission will be discussed below.

The emission from the hot puff has been modeled as molten, silicate droplets, although this is certainly not a unique model. The model has the droplets starting at 3500 K when they are ejected and cooling faster than they vaporize. The model predicts that the hot puff will be optically thin within 0.03 sec. The brightness of the puff drops rapidly,

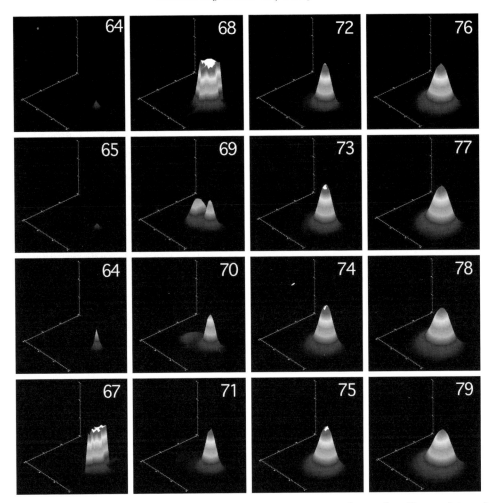

Figure 3. Brightness contours of the three successive brightenings observed at 62-msec intervals. The 3-D perspective figures have the downrange direction, approximately north to south in ecliptic coordinates, going from upper left to lower right. Individual exposures are 50-msec each.

much too fast to be explained solely by geometric dilution. However, the combination of expansion and cooling reproduces the observed peak brightnesses very well as a function of time and is also consistent with the temperatures measured at the lateral edge of the puff. After 0.42 sec, the puff has cooled such that the brightness is dominated by reflected sunlight and the fading follows an inverse square law with time solely from geometric expansion of the cloud.

Subsequent solid ejecta, measured both in front of the nucleus very soon after impact and beyond the southern limb of the nucleus at later times, are cool with temperatures near 300 to 350 K. This is warmer than the black-body temperature but relatively constant, suggesting that we are seeing the superheat expected from particles smaller than the wavelength of peak thermal emission. Although Keller *et al.* (2005) finds large particles in the ejecta when he models the radiation pressure effects for data from Rosetta, Schleicher *et al.* (2005) finds from a similar study of ground-based data that the particles are all small, of order 2 μm or less. The resolution of the discrepancy between these two results is beyond the scope of this paper. We also note that analysis of our own

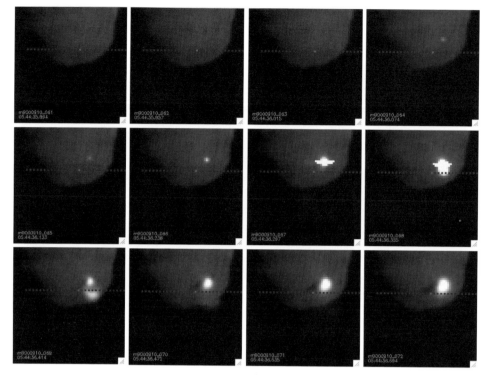

Figure 4. Sequence of images with the Medium Resolution Instrument (MRI) taken at 62-msec centers with 50-msec exposure times. Images correspond to the brightness countours of figure 3 frames 59 to 62 in the top row, 63 to 66 in the middle row, and 67 to 70 in the bottom row. A puff of rapidly moving, hot ejecta is seen in frames 65 through 69.

data is still in progress and should cast further light on this issue. If the particles are all small, *i.e.*, if the size distribution does not have the tail of large-mass particles common in ambient outgassing from comets, then we must conclude that the material prior to excavation was already in these small particles or in weak aggregates thereof. The latter makes more sense, i.e. that particles seen as large in remote sensing data are actually very weak aggregates of smaller particles. Presumably the mechanical excavation disrupted these aggregates more than does the gas drag of ambient outgassing.

The cone of ejected particles is optically thick at all times from impact until the flyby spacecraft flew past the nucleus some 14 minutes after the impact. The cone casts a prominent shadow on the nucleus as soon as 2 seconds after impact and the depth of the shadow requires an optical depth of at least a few. Meausrements of the contrast at the limb more than 10 minutes after impact yield an optical depth between 2 and 3. These results all point to the ejected mass being confined to very small particles.

5. Volatiles Released

Figure 5 shows gaseous emissions in addition to continua from reflected sunlight and thermal emission. A model has been fitted to the observed spectrum using water, hydrogen cyanide, and carbon dioxide. This is shown as the gray line in the figure. The model did not include organics, which are also obvious in the observed spectrum via the C-H stretch feature that is common to many organics over a range of wavelengths centered at 3.4 μm. The three gaseous species were all at 1400 K, but any temperature between

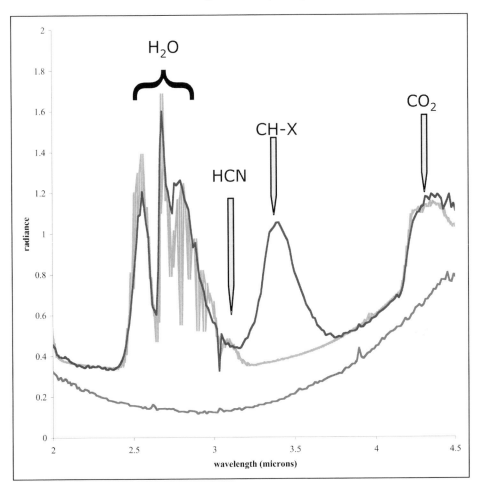

Figure 5. Two spectra, the lowest curve being the observed spectrum immediately before impact and the smooth upper curve being obtained as the puff of hot material passed in front of the slit. The upper spectrum has been fit with a model described in the text.

1000 and 2000 K provides a satisfactory fit. The details of the fit are not valid because the individual lines appear to be optically thick but the general trend of hot material in the puff followed by cooler material in later spectra is in excellent agreement with the changes seen in the continuum.

Spectra of the coma beyond the southern limb before and after impact are shown in figure 6. The two spectra are at different scales of intensity by a factor 6 in order to adequately display the structure in both. The increase in all emission features is dramatic. The biggest change in relative abundances is a large increase in the ratio of organics to water. In addition, the organic feature has become much broader and contains much more structure after impact than before. This indicates many new species that were below the detection limit prior to impact. We suggest that one of these species may be methyl cyanide (CH_3CN), for which at least two peaks are coincident with expected peaks, although we realize that this may be controversial since it would imply rather high abundances. Detailed analysis of the gaseous spectra will take a very long time.

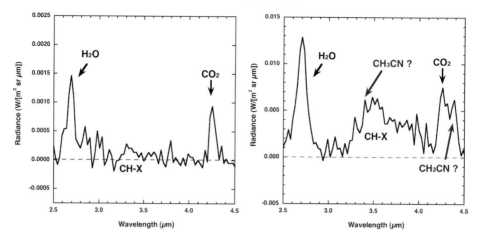

Figure 6. Spectra of the coma beyond the southern limb before (left) and after (right) impact.

6. Summary

The key conclusions from our impact are that the cometary material is extremely weak in the outer layers and that it consists of very weak aggregates of particles small compared to thermal wavelengths. The bulk gravity of the nucleus suggests a porosity well in excess of 50%. The initial excavation led to a huge increase in organics relative to H_2O and a smaller increase in CO_2 relative to H_2O. The existence of new species is clear but the identification is not.

Acknowledgements

This project was funded by NASA, as the eighth mission in the Discovery Program, through a contract to the University of Maryland and a task order to the Jet Propulsion Laboratory.

References

A'Hearn, M.F., Belton, M.J.S., Delamere, A. & Blume, W.H. 2005, *Space Sci Rev* 117, 1–21

A'Hearn, M.F., Belton, M.J.S., Delamere, W.A., *et al.* 30, 2005, *Science* 310, 265–269

Davidsson, B.J.R. & Gutiérrez, P. J. 2004, *Icarus* 168, 392

Greenberg, R. 1995, *Icarus* 10, 20

Farnham, T.L. & Cochran, A.L. 2002, *Icarus* 160, 398

Keller, H.U., Jorda, L. Küppers, M., *et al.* 2005 *Nature* 310, 281–283

Richardson, J.E., Melosh, H.J., Artemeiva, N.A. & Pierazzo, E. 2005, *Space Sci Rev* 117, 241–267

Rickman, H., Kamel, L. Festou, M.C. & Froeschle, C. 1987 in: E.J. Rolfe & B. Battrick (eds.), *Diversity and Similarity of Comets*, ESA SP278, (Noordwijk: European Space Agency), p. 471–481

Schleicher, D.G., Barnes, K.L. & Baugh, N.F. 2005, *Astron. J.* submitted

Schultz, P.H., Ernst, C.M. & Anderson, J.L.B. 2005, *Space Sci Rev* 117, 207–239

Sekanina, Z.Z. 1996, in: K.S. Noll, H.A. Weaver & P.D. Feldman (eds.), *The Collision of Comet Shoemaker-Levy 9 with Jupiter* (Cambridge: Cambridge Univ. Press), p. 55.

Asteroids, Comets, Meteors
Proceedings IAU Symposium No. 229, 2005
D. Lazzaro, S. Ferraz-Mello & J.A. Fernández, eds.

Physical properties of the dust in the Solar System and its interrelation with small bodies

I. Mann[1], A. Czechowski[2], H. Kimura[3], M. Köhler[1], T. Minato[1] and T. Yamamoto[3]

[1]Institut für Planetologie, Wilhelm-Klemm-Str. 10, 49149 Münster, Germany
email: imann@uni-muenster.de
[2]Space Research Center, Polish Academy of Sciences, Warsaw, Poland
email: ace@cbk.waw.pl
[3]Institute of Low Temperature Science, Hokkaido University, Sapporo 060-819, Japan

Abstract. Dust particles in the solar system are produced from the small bodies: asteroids, comets, meteoroids and Kuiper belt objects. A further source of dust is provided by the warm interstellar medium that the Sun is currently embedded in and that streams into the solar system. We review the physical properties of solar system dust and trace back its interrelation with the small solar system bodies. Comets contain relatively pristine material that they transport to the inner solar system. The alteration of dust in the vicinity of comets is complex and connected to the gas evolution, but a significant part of the organic dust material survives these alterations. The optical properties of cometary dust are best described with a mixture of silicate and carbon bearing materials. As far as the darkness of the cometary material is concerned, according to recent models, this is not a result of the porosity, but rather of the darkness of the carbon bearing component. This does not contradict the observation of silicate features in the thermal emission brightness of cometary dust, since porous mixtures of silicate and carbon bearing dust can produce the observed polarization and albedo characteristics, as well as the silicate features. The carbon-bearing component is most likely an organic refractory component. The relative contributions of different sources change within the solar system dust cloud and depend as well on the measurement technique considered. In particular, the dust from asteroids, which provides a large component of the dust near Earth orbit, is also preferably seen with most of the detection methods. The majority of dust inward from 1 AU is produced from cometary dust and meteoroids. Dust material evaporation induced by collisions inward from 1 AU produces a minor heavy ion component in the solar wind plasma known as inner source pick-up ions.

Keywords. comets: general, interplanetary medium, meteors, meteoroids

1. Introduction

The sources of dust in the solar system are the direct supply from asteroids by their collision fragmentation, the activity of comets and the collision fragmentation of meteoroids generated by these parent bodies. Dust particles are also produced by collision and erosion processes in the Kuiper belt and enter the solar system directly from interstellar space (see Figure 1). The activity of comets is driven by the heating and sublimation of volatiles of the cometary ice, which generates the dust seen in comae and tails of comets. Part of this dust is directly ejected in unbound orbits. The generation of larger cometary meteoroids is evident from the observation of cometary dust trails (see Figure 2). As we explain later in the text, we can assume that the majority of the solar system dust cloud near 1 AU consists of material that has undergone collisional processes.

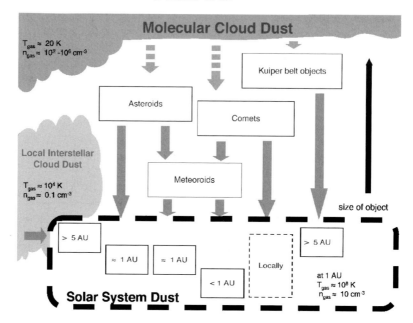

Figure 1. Dust sources in the solar system: the parent bodies of dust in the solar system are asteroids, comets and Kuiper belt objects, and interstellar dust particles directly enter the solar system from the local interstellar medium. The majority of dust outward from 5 AU is interstellar dust and dust produced in the Kuiper belt. Dust near 1 AU comes from comets and asteroids. The majority of dust from asteroids and comets is not directly released from the parent bodies but generated from collisional fragmentation of meteoroids that come from these sources. The inner solar system dust cloud is replenished from collisions of asteroidal meteoroids.

Asteroids, comets and Kuiper Belt objects were formed from molecular cloud dust during solar system formation: The majority of the heavy elements beyond He existing in the interstellar medium (ISM) are condensed into solid dust particles. Dust particles are injected into the ISM after condensation in red giant stars, supernovae and novae. They grow by condensation, collisional accretion and coagulation while at the same time, the stellar UV radiation causes chemical reactions and in particular the formation of organic refractory material in icy condensates. Icy materials form in molecular cloud regions that also provide the conditions for solar system formation.

The properties of meteroids and dust particles reveal information about their parent bodies and about their different paths of evolution from the protoplanetary dust. The following discussion of the physical properties of dust in the solar system will concentrate on the processing of the cometary dust material. This provides a relatively pristine sample of material that can be studied by various methods. We will first discuss the dust properties derived from observations of the solar system dust cloud. The next section describes observational results about the dust processing in the vicinity of comets, followed by a discussion of the cometary dust properties. We then discuss the contribution of different parent bodies to the dust cloud, and finally discuss the demise of dust in the vicinity of the Sun and interactions with the solar wind.

2. Dust properties derived from different observations

Observational data about dust in the solar system are obtained from the observation of the Zodiacal ligh and the F-corona, the part of the corona that is produced by

Figure 2. The dust trail of Comet Encke at 2 AU observed from Spitzer Space Observatory, image NASA/JPL-Caltech/M. Kelley (Univ. of Minnesota).

interplanetary dust, and from meteor observations of particles entering Earth's atmosphere. Laboratory studies are possible for collected samples of dust and micrometeorites. In-situ measurements of dust in the interplanetary medium are carried out from spacecraft. Data about cometary dust are obtained from observations of cometary comae, dust tails and dust trails as well as from spacecraft measurements during encounters with a comet.

2.1. *Zodiacal Light and F-corona*

The Zodiacal light (see Figure 3) and Zodiacal emission brightness data predominantly describe particles in the 1 to 100 μm size range at distances from about 0.7 to 1.3 AU close to the ecliptic plane (Levasseur-Regourd, Mann, Dumont *et al.* 2001). The overall distribution of dust number densities in the Zodiacal cloud can be explained with particles in low eccentricity and low inclination orbits that drift toward the Sun due to deceleration by the Poynting-Robertson effect (Mann 1998). The Zodiacal light smoothly continues into the solar F-corona brightness, but derived average optical properties are inconsistent with a gradual change of particles properties at distances smaller than about 0.5 AU from the Sun: Explaining the data requires a change of the size distribution of dust, a change of the dust cloud composition with distance from the Sun or a combination of both (Mann 1998).

Figure 3. The Zodiacal light observed in the visible light from Manua Kea, Hawaii with an analog (35 mm-film) camera with wide angle lens (focal length = 28 mm, Mukai and Ishiguru, personal communication). The brightness is produced by light scattering at dust of sizes 1 to 100 μm distributed near 1 AU.

2.2. *Meteor Observations*

High velocity impacts of meteoroids cause the meteor phenomenon in planetary atmospheres (cf. Ceplecha, Borovi, Elford *et al.* (1998)): A meteoroid that enters the Earth atmosphere and atoms ablated from the meteoroid collide with atmospheric constituents. Meteoroids, atmospheric atoms and molecules undergo dissociation and ionization and form an expanding column of partially ionized plasma along the trajectory of the meteoroid. This generated plasma cloud in the atmosphere is commonly called meteor. The meteor trail is the extended radiation observed behind the meteoroid body. The meteor head is the plasma in the immediate surrounding of the meteoroid body that moves with the speed of the meteoroid.

The sizes of entering bodies for the different detection techniques range from about 10 μm and beyond, depending on the mass, material composition, structure and entry speed. Some meteors are observed in meteor showers but the majority is observed as sporadic meteors. The detection of interstellar meteors with radar techniques was also reported (Taylor, Baggaley & Steel 1996; Baggaley & Neslusan 2002 and references therein; Meisel, Janches & Mathews 2002). The classification of the meteors as interstellar is based on their derived heliocentric velocity. The orbital speeds derived from some of the radar techniques, however, are still subject to debate among researchers (Hajduk 2001), as are the selection effects of this technique. Head echo observations were recently used to derive the velocity vectors of in-falling particles, but the analyses are still limited by statistics (Meisel, Janches & Mathews 2002; Janches, Pellinen-Wannberg, Wannberg *et al.* 2002). Head echoes are radar reflections that occur at the head when the produced ion cloud reaches a critical density (Pellinen-Wannberg & Wannberg 1994).

Recent spectroscopic observations brought progress in understanding the composition of the entering bodies (see review by Borovicka, this issue). The observations indicate the diversity in the material composition of the meteoroid particles on scales of millimeters. Material properties for meteors on near-ecliptic orbits also differ from those of meteors in orbits with higher inclination. This possibly points to different properties of cometary and asteroidal meteoroids since cometary meteoroids can be expected to be more abundant at higher latitudes.

2.3. *Collected Samples*

While meteors are observed as a result of the heating and melting of the in-falling cosmic meteoroids, smaller particles are only moderately heated during entry into the atmosphere. The smaller particles are collected in the stratosphere (Brownlee 1985) or extracted from ice samples where concentration processes that occur in melt zones allow their collection in large quantities (Maurette, Hammer, Reeh *et al.* 1986). The exact size limits for particles to survive atmospheric entry are not clearly determined since survival depends on a variety of different parameters connected to the entry velocity and to the conditions of re-radiation of the entry heat. The size limits of the collected and analyzed particles are mainly determined by collection and handling methods. Cosmic dust particles collected in the stratosphere by high flying aircraft cover the size ranges from 5 to 50 μm; these particles are often denoted as interplanetary dust particles (IDPs) (Brownlee 1985). Cosmic dust particles that are collected from Antarctic ice and Greenland ice samples as well as from the ocean floor have typically sizes of 20 μm to 1 mm (Maurette, Olinger, Michel-Levy *et al.* 1991; Kurat, G., Koeberl, C., Presper *et al.* 1994).

The mineralogical and morphological properties of these collected particles have been studied by laboratory analyses and the results have been recently reviewed by several authors in different scientific contexts (Rietmeijer 1999; Messenger 2000; Jessberger, Stephan, Rost *et al.* 2001; Rietmeijer 2002). Contamination from atmospheric constituents was detected in IDPs (see for instance Rietmeijer 1993) and some of the larger particles, extracted from ice samples are partially melted (Maurette, Olinger, Michel-Levy *et al.* 1991; Kurat, G., Koeberl, C., Presper *et al.* 1994).

The presence of solar-wind noble gases (Hudson, Flynn, Fraundorf *et al.* 1981) confirms the extraterrestrial nature of the IDPs. Also nuclear tracks, in majority generated by solar energetic particles ('solar flare tracks'), have been identified in collected stratospheric cosmic dust (Bradley, Brownlee & Fraundorf 1983). Exposure ages within the inner solar system derived from the tracks are approximately 10000 years (Bradley, Brownlee & Fraundorf 1983). Compositional characteristics, such as isotope composition (Jessberger, Stephan, Rost, *et al.* 2001) ascertain the cosmic origin of the particles. Particles extracted from Antarctica ice samples showed elemental compositions similar to chondrites, and for a fraction of particles the extraterrestrial origin was confirmed by isotopic analysis of trapped neon (Maurette, Olinger, Michel-Levy *et al.* 1991).

The thermal history of the dust particles can be used to estimate their possible parent bodies. Flynn (1989) estimated from initial orbital parameters and entry processes the heating of the dust particles: The atmospheric-entry conditions inferred for the major fraction of the collected stratospheric cosmic dust is consistent with parent bodies in the main asteroid belt. Flynn (1989) further derived that cometary dust from parent bodies with perihelia greater than 1.2 AU is heated in atmospheric entry to temperatures of approximately 900–1100 K, and dust from comets with smaller perihelia is heated to temperatures beyond 1100 K. The heating history does, however, not allow for a unambiguous determination of the parent body: Depending on the specific orbital history asteroidal dust may cross Earth orbit with a high velocity typical for dust released from a comet and vice versa. So cometary dust may have small velocities and therefore resemble from entry thermal processing an asteroidal dust particle. It is also quite possible that some collected stratospheric particles originate from the Kuiper belt. For the case of two specific particles, the density of solar flare tracks clearly exceeded the values that are typical for dust from comets or asteroids, which led Flynn (1996) to suggest they originate from the Kuiper belt. Also the capture of interstellar dust is possible, especially since they are focused by solar gravitational attraction (Mann & Kimura 2000); their sizes, however, are typically below 5 μm, wich is currently the limit for the analyzed collected IDPs.

Results of the laboratory analyses provide evidence that some of the materials in the collected IDP samples are very pristine. The so-called chondritic IDPs are thought to be among the most primitive samples (Bradley 1994). For some of them there is evidence they originate from comets. By measuring the He release pattern upon laboratory heating Joswiak, Brownlee, Bradley *et al.* (1996) could infer the entry speeds of the thus analysed particles: Those particles for which they established cometary origin were in all cases porous chondritic IDPs with unique glass/metal compounds (i.e. GEMS = glass with embedded metal and sulfides). Among them the so-called cluster IDPs are thought to be cometary dust, since their enhanced D/H ratio suggests a pristine nature (Messenger 2000). These cluster IDPs contain high abundances of GEMS. It is suggested that GEMS are either interstellar silicate dust particles or be the oldest known solar nebula solids (Bradley 1994). Studies of anhydrous chondritic porous IDPs of probable cometary origin also show much of the C to be in the form of fine grained amorphous carbon (Rietmeijer 2000; Wooden 2002), plus domains of aliphatic and aromatic carbon (Flynn, Keller, Feser *et al.* 2003; Keller, Messenger & Flynn *et al.* 2002). Also GEMS in anhydrous chondritic porous IDPs contain nano-phase C, Fe, and FeS that make these 0.1 μm substructures optically dark.

2.4. *In-situ Measurements in the Interplanetary Medium*

In-situ measurements of interplanetary dust carried out from Earth orbiting satellites detect particles of sizes typically below several μm. In-situ measurements are well fitted by a combined model of the mass distribution of dust flux density near 1 AU (Grün, Zook, Fechtig *et al.* 1985). The model also fits to meteor observations and analysis of microcraters on lunar samples. In addition, measurements onboard Pioneer, Voyager, Ulysses and Galileo determined dust fluxes in interplanetary space beyond 1 AU. Inside Earth's orbit, Helios measurements were carried out between 0.3 and 1 AU (Grün, Pailer, Fechtig *et al.* 1980). All measurements were carried out close to the ecliptic plane with the exception of Ulysses, which moved into orbit almost perpendicular to the ecliptic after a Jupiter flyby. Most experiments measure the mass and impact speed of

Figure 4. The dust coma and the faint dust trail of comet Churyumov-Gerasimenko taken by the Kiso 105 cm Schmidt telescope with R-band filters using a 2 K CCD camera (Ishiguro, Watanabe, Tanigawa *et al.* 2002). Observations of dust trails in the visible wavelength range allowed to derive the albedo of particles, which turned out to be very low.

particles. Deriving dust fluxes from the impact rates imposes some uncertainty, since derived numbers depend on assumptions for the orbital distributions of particles. Helios measurements indicate the presence of two distinct dust components: the major component is in low to medium eccentricity orbits near the ecliptic and a second component of dust is in orbit with presumably higher eccentricities and consists of dust of lower material strength (Fechtig 1982). The second component was interpreted to be made of cometary dust (Fechtig 1982).

The in-situ detection of interstellar dust entering the solar system is important for further understanding of the dust evolution. It was first identified in dust fluxes onto Earth orbiting satellite where the gravitational focusing of the interstellar dust particles caused a flux variation along Earth's orbit (Bertaux & Blamont 1976). Clear identification of interstellar dust in the solar system was obtained from measurements aboard Ulysses since impacts speeds and directions are different from those of interplanetary dust (Grün, Gustafson, Mann *et al.* 1994). Ulysses measurements allow a comparison to astrophysical models of dust properties in the local interstellar medium (Landgraf 2000; Frisch, Dorschner, Geiss *et al.* 1999; Mann & Kimura 2000; Mann & Kimura 2001). The measured mass distribution of the interstellar dust in the solar system is influenced by different forces from which particle properties can be inferred (cf. Mann 1996). Namely, the repulsion by radiation pressure force changes the mass distribution of the measured interstellar dust with distance from the Sun along the Ulysses orbit (Landgraf 2000; Mann & Kimura 2000; Mann & Kimura 2001). Kimura, Mann & Jessberger (2003) showed that the radiation pressure force infered from the data agrees with a model of core-mantle dust particles that form larger agglomerates. The core-mantle dust particles were simulated with sizes of 100 nm, while bare silicates and bare carbonaceous materials may be present as dust particles smaller than 10^{-14} g.

2.5. *Observations of Cometary Dust*

Comets are visible when they release gas and dust during that parts of their orbits which are sufficiently close to the Sun such that melting of volatiles in the nucleus causes this activity. Dust in the vicinity of comets is observed in their comae and tails and as a result of the formation history of comets the dust is expected to be more pristine than dust from other parent bodies (see Figure 5). Observations over a wide range of wavelengths allow today studying the gas and dust surrounding the cometary nucleus. A discussion of the dust and gas observations will be given in the following section 3.

Refined observations allow to detect the faint brightness of the larger dust particles ejected from comets (see Figure 4). Narrow trails of dust coincident with the orbits of periodic comets have been discovered in the data of the Infrared Astronomical Satellite (IRAS) (Sykes, Lebofsky, Hunten *et al.* 1986) and are detected with the Infrared Satellite Observatory (ISO) and with Spitzer Space Observatory. Within the IRAS data trails were studied in detail for 8 comets; alltogether more than 100 faint dust trails are suggested

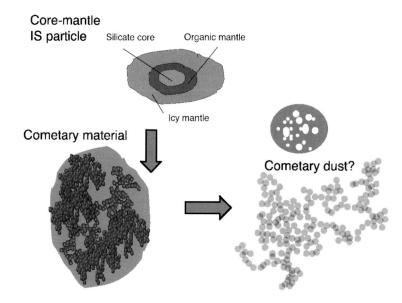

Figure 5. Formation of cometary matter: a plausiblescenario of the formation of cometary matter is that interstellarcore-mantle particles are (partially altered and) incorporated intothe cometary nucleus. The icy materialforms the matrix of the cometary matter while silicate cores- and organic refractory mantles may partially survive and provide thebasic constituents of the cometary dust observed in the solar system.

by the IRAS data (Sykes & Russell 1992). The particles are in orbits close to that of the parent comet and seen both ahead and behind the comet. Some trails exist even without an observable associated parent comet. Recently, Ishiguro, Watanabe, Tanigawa *et al.* (2002) have found a visible dust trail along the orbit of comet 22P/Kopff and a survey (see Mukai, Ishiguro & Usui 2002) showed that a large fraction of comets are also associated with visible trails. The trails consist of large (a few cm) and dark (albedo of 0.01) dust particles (see Figure 4). As Mukai, Ishiguro & Usui (2002) pointed out, the existence of dust trails along the orbit of parent comets seems to be general feature: Observations of comet trails stimulate re-evaluation of several cometary phenomena such as (1) the gas-dust interaction when the dust particles are ejected from the cometary nuclei, (2) the mass-loss rate and aging of comets, and (3) the mass distribution of small solid bodies in the solar system.

Direct in-situ detection of cometary dust, though limited by experimental conditions, was possible with the space missions to comet Halley in the 1980's (Jessberger, Christoforidis & Kissel 1988; Kissel, Brownlee, Büchler *et al.* 1986; Kissel, Sagdeev, Bertaux *et al.* 1986; Schulz, Kissel, & Jessberger 1997). Data were acquired by Giotto to a minimum flyby distance of 600 km, by the Vega spacecraft to a minimum distance of 8000 km.

Recent measurements of dust fluxes were made during the encounter of DS1 at comet Borrelli (Tsurutani 2004) and during the Stardust flyby at comet Wild2 (Tuzzolino, Economou, Thanasis *et al.* 2004). The latter Stardust measurements show strong evidence of dust fragmentation taking place in the coma, as will be discussed below.

3. Cometary Dust Evolution in the Coma

Dust is steadily lifted from the surface of the nucleus by the out-flowing gas or released when sublimation of ices below the surface of the nucleus breaks up the crust that is formed on the surface. The velocity of dust is initially determined by the flow of the surrounding gas, and at larger distances from the nucleus, by solar gravity and radiation pressure force. Local variations of the surface activity of the nucleus, chemical reactions in the gas phase as well as interactions between the gas and the dust phase influence the distribution and composition of coma gas. Extended sources of gas are the sublimation of ices and semi-refractory components in the dust, and chemical reactions in the gas phase. The highly refractory component of cometary organics remains within the dust particles that feed the solar system dust cloud.

3.1. *Dust and Coma Gas Observations*

Observations and in-situ detection of coma gases

Radio and sub-millimeter spectroscopic remote observations of comets show numerous lines from molecules, radicals and ions including the major carbohydrates as well as complex organic molecules (see Crovisier, this issue). The majority is directly produced by the outgassing of the nucleus. The presence of an extended source of coma gas is clearly seen in neutral gas measurements carried out on Giotto at comet Halley. When comparing the measured H_2O and CO radial density profiles after correction for gas kinematics and geometry effects, they show completely different behaviors: The H_2O profile outward from approximately 5000 km from the nucleus follows the radial decrease expected from the high photo destruction rate of the molecule in the solar UV radiation. The CO profile, in contrast, increases with distance from the nucleus out to 25000 km. Further out the CO profile follows a flat slope expected from the low rate of photo destruction of CO compared to the flowing time in the coma. The CO profile between the nucleus and a distance of 25000 km is explained with an extended source being present in the coma (Eberhardt 1999). Remote observations of the spatial distributions of different species in many cases show evidence for an extended source. The coma composition also changes with heliocentric distance of the comets. Disanti, Mumma, dello Russo *et al.* (1999) derive from observations of Hale-Bopp that the extended CO source may only be effective for small heliocentric distances, while at larger distances the coma gas is predominantly produced by the nucleus.

Observations of changing dust properties

Observational data suggest that also dust properties and size are changing within the coma (see Figure 6). The generally elongated shapes of isophotes seen in ground-based observations, for instance, can be explained if particle sizes change with distance from the nucleus (Combi 1994). From analyzing comae radial brightness profiles, Baum, Kreidl & Schleicher (1992) concluded that for 10 out of 14 comets the brightness decrease is steeper than would be expected from dust motion influenced by radiation pressure, meaning that either the albedo or the size of particles changes with distance from the nucleus.

Further results were obtained by recent high spatial resolution observations of brightness, polarization and colour of comet 2P/Encke. The polarization of the coma was observed to increase with projected distance from the nucleus, suggesting a change in the mean scattering properties of the dust particles on a time-of-flight timescale of 1 hr (Jewitt 2004). Jewitt (2004) suggests that disaggregation of composite, porous dust particles is a likely cause of the fragmentation. Subsequent observations confirm this result and interpretation (Jockers, Kiselev, Bonev *et al.* 2005).

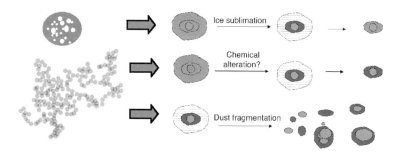

Figure 6. A possible path of dust processing in the coma: the pristine cometary material is altered by fragmentation, sublimation and chemical reactions in an unknown way. This illustration follows the Greenberg model describing the cometary dust as agglomerates of core-mantle particles.

In addition to changes in dust particle sizes and dust composition, variations of the activity over the surface of the nucleus influence the observed coma profiles. Imaging the inner coma of comet P/Halley over a 10^4 km square region showed that the near-infrared colors between 1 to $5\,\mu$m were not constant as a function of nucleocentric distance and moreover the radial decrease in surface brightness significantly differed from the jet-side of the nucleus to the opposite, tail side (Woodward, Shure, Forrest *et al.* 1996). Observations of the coma of comet Tabur prior to the perihelion passage are also interpreted with either the existence of different particle population or with dust destruction: The scattered continuum light from dust detected with visible spectroscopic observations was blue within 5300 km around the projected position of the photocenter and reddish further Sunward Turner & Smith (1999). Broadband (B, V, R) CCD imaging of the comet proves that the reflectance of the dust behaves differently in the Sun and in the tail directions (Lara, Schulz, Stüwe *et al.* 2001).

Aside from the change of dust properties within the coma, properties vary with the heliocentric distance. Narrowband filter photometry observations of comet Hyakutake indicate a change in dust color from significantly reddened at large heliocentric distances to close to the solar spectrum at small heliocentric distance (Schleicher & Osip 2002). The authors suggest the change of color with heliocentric distance implies a significant change in dust particle sizes or a changing proportion of different dust populations. For comet Hale-Bopp the observed 3 to $20\,\mu$m spectral energy distributions derived from observations were used to fit thermal emission models (Harker, Wooden, Woodward *et al.* 2002, 2004): Near perihelion, Hale-Bopp displayed higher jet activity than at larger heliocentric distance and the coma dust had a steeper size distribution, was more porous and had more crystalline silicates. The authors suggest, that these changes indicate either the fragmentation of aggregate dust particles, or that the dust particles released from highly active areas were different, possibly more pristine.

Variation of the properties from comet to comet is possibly more pronounced than this variations with heliocentric distance. Baum, Kreidl & Schleicher (1992) infer from their study that large intrinsic differences exist in the nature of the dust particle populations of the 14 considered comets, while they could not find a systematic connection of the porperties to the heliocentric distances of the comets.

In-situ dust flux measurements

The Dust Flux Monitor Instrument (DFMI) aboard Stardust measures particles in the 10^{-14} to 10^{-7} kg mass range. During the flyby at Comet 81P/Wild 2 it encountered regions of intense swarms of particles. The clouds of particles were only a few hundred meters across, which is explained by particle fragmentation (Tuzzolino, Economou, Thanasis *et al.* 2004; Green, McDonnell, McBride *et al.* 2004). Imaging also reveals a large numbers of jets projected nearly around the entire nucleus (Tsou, Brownlee, Anderson *et al.* 2004). Sekanina, Brownlee, Economou *et al.* (2004) interpret the dust flux measurements with the release of sheets of high dust density extending from small sources on the rotating nucleus. They suggest that the dust is accelerated by the expanding gas in the jets and that large cometary fragments may travel with the comet for a long time and then possibly fragment as a result of sublimation. Clark, Green, Economou *et al.* (2004) suggest that fragmentation causes these streams and that fragmentation itself could be caused by enhanced heating as well as depressurization, phase transitions, exothermic chemical reactions, centrifugal forces, and electrostatic repulsion. Green *et al.* (presented at IAU Symposium No. 229) pointed out that the mass distribution measured with DFMI changed significantly during the encounter, which they also interpret as due to fragmentation processes.

In-situ dust composition measurements

In-situ mass spectrometer measurements were made as close as 600 km from the nucleus of comet Halley. The masses and densities of the dust particles measured with the mass spectrometers aboard the Giotto and Vega spacecraft range from 10^{-19} to 10^{-14} kg (approximate diameter range 0.02 to 2 μm) with densities from 0.3 to 3 kg/m^3. The interpretation in terms of the elemental compositions of the impacting dust particles requires a detailed understanding of the impact ionization process. Though the understanding of the impact ionization process is limited, the published results (Jessberger, Christoforidis & Kissel 1988; Kissel, Brownlee, Büchler *et al.* 1986; Kissel, Sagdeev, Bertaux *et al.* 1986) provide interesting information about the dust in the coma. The detected particles are mixtures of two end-member components, often called CHON (rich in the elements H, C, N, and O) and ROCK (rich in rock-forming elements as Si, Mg, Fe). The CHON component is assumed to be refractory organic material while the ROCK component is assumed to consist of silicates, metals, and oxides. Further analysis by Schulz, Kissel & Jessberger (1997) showed the ROCK dominated particles are composed of primarily Mg-rich pyroxenes, some Mg-Fe pyroxenes and olivines and relatively rare Fe and FeS; about 70% in Fe and FeS particles.

The silicate and the carbonaceous component are mixed down to the finest scale, possibly (but not necessarily) in the form of a core-mantle structure. CHON- and ROCK-dominated particles make up each about 25% of the measured dust and most of the small particles (below 10^{-17} kg) are rich in the light elements H, C, N, and O. Lawler & Brownlee (1992) found from their analysis of the mass spectra that there are essentially no pure CHON particles in the 0.1–1 micron size range. Given the uncertainties of the measurement, the presence of pure CHON particles of even smaller sizes can not be unambiguously concluded from the data. The fact that the ^{12}C/^{13}C isotope ratios vary from particle to particle is interpreted that there has been no process leading to chemical homogeneity in the history of cometary dust (Jessberger, Christoforidis & Kissel 1988). Fomenkova & Chang (1997) point out that that some detected carbon-rich particles have ^{12}C/^{13}C isotope ratios similar to AGB stars.

3.2. *Models of Gas and Dust Interactions in the Coma*

Understanding the coma processes is essential for interpretation of observational data, but the complexity of coma phenomena does not allow a simple model to describe all the processes at the same time. Dust is steadily lifted from the surface of the nucleus by the out-flowing gas but also strongly variable dust and gas production occurs. The amount of ejected dust varies. Hale-Bopp produced, for instance, gas jets that were not associated with observable dust signatures (Lederer, Osip, Thomas-Osip *et al.* 2005). Solar radiation pressure force and gas drag force acting on the dust, together with dust structure and composition determine the dynamical evolution of dust. The size dependences of the acting forces induce relative velocities of more than 0.5 km/s within the dust component, as do the complex motions of coma gas (Combi, Kabin, DeZeeuw *et al.* 1997).

Describing the out-gassing of the nucleus and the evolution of its circum-nuclear coma with a steady-state outflow of gas and dust is already numerically complex (Crifo & Rodionov 1999). To describe the extended source of gas due to ice sublimation Crifo (1995) has developed a numerical two-step model, which accounts for surface ejection of fragments of water ice mixed with mineral dust particles, that subsequently sublimate in the coma. The calculations achieve a better agreement with the measured H_2O velocity profiles in comet Halley than steady-state outflow models. Verifying models of dust-gas flows is difficult, since these are not the only parameters to determine dust and gas in the coma. Konno, Hübner & Boice (1993) study, for instance, dust fragmentation in near-nucleus jet-like features at comet Halley and come to the conclusion that dust fragmentation alone does not explain the dust observations.

To achieve a better description of the coma, Hübner & Benkhoff (1999) consider three sources for coma gas: (1) release of water vapor from the surface of the nucleus, (2) release of other, more volatile, species from the porous interior of the nucleus, and (3) a distributed source releasing gases from ices and volatiles contained in the dust. Greenberg & Li (1998) point out that the distribution of CO, C_2, C_3, CN, H_2CO can not be explained as daughter molecules originating from more complex gas phase species. They apply a highly porous silicate core organic refractory mantle model to explain the presence of these coma species and the amount of CO in Comet Halley. By computing the heating of fluffy aggregates of interstellar core-mantle particles they estimate a maximum CO production rate that is still significantly less than the values needed to explain the observed CO. Differences between the model and the observations can arise from various factors such as the assumed dust to gas ratio, the dust fragmentation and sublimation processes, or overestimation of the extended CO abundance.

The different carriers of extended gas sources are currently not quantified. The fragmentation and vaporization of dust is commonly discussed as one of the possible sources of spatially extended coma gases (see cf. Festou 1999). Organic dust components in particular, are suggested to explain observations (e.g. Bockelee-Morvan & Crovisier 2002; Disanti, Mumma & Dello Russo *et al.* 2001). Chemical reactions within the gas phase are also a possible source: Bockelee-Morvan, Crovisier, Mumma *et al.* (2005) explain the increase in CO in comet Hale-Bopp near perihelion (between solar distances of 0.93 AU and 1.5 AU) as a consequence of CO excitation mechanisms. Aside from the other uncertainties outlined in this section, uncertainties of the dust models arise from a lack of knowledge about the chemical appearance of the organic refractory material. It is therefore difficult to estimate the optical properties and the sublimation rates of the cometary dust.

4. Cometary Dust Optical Properties

The optical properties of cometary dust derived from observations are known for many comets and therefore provide a good basis for analysis. Differences in the optical properties of cometary dust can come from particles processing in the coma, particle fragmentation in the coma, as well as differences of material from comet to comet. Nevertheless, visible light observations indicate common characteristics in the albedo, polarization and colour of the particles, which differ from other cosmic dust populations. They agree with models of porous dust consisting of silicates and absorbing, possibly organic refractory materials. Silicate features, appear in several comets and indicate the existence of both, amorpous and crystalline silicates. The high amount of absorbing material and its close mixture to the silicate, might be the special characteristic of the cometary dust.

4.1. *Albedo and Polarization Observations*

Optical and near IR observations allowed to study the albedo and linear polarization of cometary dust (Dobrovolsky, Kiselev & Chernova 1986; Dollfus, Bastien, Le Borgne *et al.* 1988; Kolokolova, Hanner, Levasseur-Regourd *et al.* 2004, Kelley, Woodward, Jones *et al.* 2004). Improved observation techniques and the recent apparitions of the bright comets Hale-Bopp and Hyakutake have further increased the amount of optical data (Hadamcik & Levasseur-Regourd 2003; Kiselev & Velichko 1998).

Regardless of the differences in the properties of comets, the dust particles have common characteristics in their optical properties:

• The albedo is low compared with other atmosphere-less bodies in the solar system. The albedo gradually increases with wavelength from the optical to the near-infrared wavelength range.

• The brightness smoothly changes with phase angle and shows strong and weak enhancements toward large and small phase angles, respectively.

• The linear polarization in relation to phase angle is described as a bell-shaped curve with a broad maximum around a phase angle of $90°$ and a shallow negative branch of the polarization at small phase angles.

• The polarization increases with wavelength, while in the negative branch at small phase angles it is constant or decreases with wavelength.

Several studies have been carried out to fit the observed data for albedo and linear polarization in the visible wavelength range. In many cases light scattering and absorption properties of cometary dust were simulated based on Mie theory that provides rigorous solutions for interaction between electromagnetic radiation and homogeneous spheres (Mukai & Koike 1990; Mukai, Mukai & Kikuchi 1987). Developments of light-scattering theories combined with increased computer capabilities allowed to assume more realistic morphologies (Lumme & Rahola 1994; Petrova, Jockers & Kiselev 2000; Draine 1988; Draine & Flatau 1994).

A model to describe cometary dust as aggregates of submicron monomers achieved, for the first time qualitative simultaneous agreement with the four observed optical characteristics listed above (Kimura, Kolokolova & Mann 2003; Kolokolova, Kimura & Mann 2004; Mann, Kimura & Kolokolova 2004). The size of the constituent monomers in the model is $100\,nm$, which is in accord with the average size of constituent dust particles of IDPs (Brownlee 1978; Jessberger, Stephan, Rost *et al.* 2001). The obtained results are similar for the considered two types of irregular dust structures. The importance of the assumed scattering properties lies in the assumption of the monomer size and of the refractive index. An increase of both n and k, where n and k are the components of the optical constant $m = n + ik$, is most suitable for obtaining the optical properties

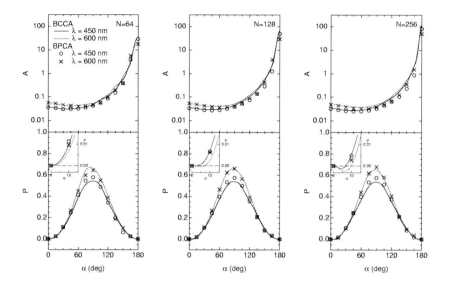

Figure 7. Optical properties of dust: Shown are the geometric albedo A and polarization P calculated for two different models of fluffyaggregate particles. The particles are agglomerates of identical homogeneous spheres. The numbers of monomers are $N = 64, 128, 256$. BPCA particles (symbols) and BCCA (lines) particles denote different structures of the aggregates. The calculations were made at wavelength of 450 and 600 nm, see Mann, Kimura & Kolokolova (2004).

(Mann, Kimura & Kolokolova 2004). Previous studies have shown already, that particles consisting of a silicate core and an organic refractory mantle material can produce a characteristic silicate emission feature (Greenberg & Hage 1990). This model, developed to explain albedo and polarization, may therefore also reproduce thermal properties of cometary dust.

For the calculated cases of particle sizes (see Figure 7), limited by the computer capacities, the obtained maximum polarization is higher than the observed values and the negative polarizaton is small. The authors expect a better agreement for larger particle sizes as well as possibly for aggregates of non-spherical particles. Further studies of this dust model should include a comparison with thermal observations as well as a systematic study of the influence of the dust structure on the results (Mann, Kimura & Kolokolova 2004). Also dedicated laboratory measurements are currently developed (Hadamcik, Renard, Levasseur-Regourd *et al.* 2003).

4.2. *Thermal Emission Data*

Thermal emission in the infrared wavelength range gives an insight into the mineral composition of cometary dust (Becklin & Westphal 1966; Hanner & Bradley 2005). Observing the mineralogy of cometary silicates has interesting implications for the formation of the comets: Cometary silicates are most likely a mixture of (1) remnants of interstellar silicates which are amorphous and (2) crystalline silicates. The crystalline silicates have two possible origins: (a) solar nebula condensates or (b) amorphous (interstellar silicate) annealed in the solar nebula. Crystalline silicates are formed in the solar nebula if cooling is slow enough. On the other hand, if the cooling is fast, amorphous silicate will condense even in the solar nebula.

Characteristic for silicate particles are emission features in the $10\,\mu$m and in the 16–$35\,\mu$m spectral regime. The structure and intensity of the emission features depends on the particle size, the structure and even the temperature of particles and therefore the interpretation of the observed spectra is difficult. Spectroscopic observations show evidence for certain silicate minerals in the infrared spectra (Campins & Ryan 1989; Hanner, Gehrz, Harker *et al.* 1997; Hanner 1999; Hanner, Lynch & Russell 1994). According to observations, silicates in comets appear to comprise both crystalline Mg-rich silicates and Mg-Fe glassy or amorphous silicates (Wooden, Harker, Woodward *et al.* 1999). Strong features have been observed for the majority of Oort cloud comets while weak emission features have been observed for only a few short period comets. Hanner (2003) points out that this absence is either due to a different composition or due to a lower abundance of submicron sized dust particles in the short period comets. Kelley, Woodward, Jones *et al.* (2004) point out that from their polarimetry and near-infrared photometry for six comets the high-polarization comets are characterized by moderate to strong mid-infrared silicate emission, while the low-polarization comets have a weak silicate emission feature or the feature is absent. The mineralogy of cometary dust underlying the different observational results was also discussed by Wooden, Harker & Woodward (2000); Wooden (2002); Harker, Wooden, Woodward *et al.* 2002, 2004) Hanner (2003), Hanner & Bradley (2005), and Wooden, Charnley & Ehrenfreund (2005). Valuable information about the properties of pristine dust ejected from the interior of the nucleus are expected from observation campaigns connected to the Deep Impact mission (see A'Hearn, this issue).

The mineral identifications from the 10 and 20 μm cometary spectra are consistent with the composition of anhydrous chondritic porous aggregate IDPs (Hanner 1999; Bradley 1988; Bradley, Keller, Snow *et al.* 1999; Wooden 2002). This again support the hypothesis that the latter are of cometary origin.

4.3. *Classification of Comets by Optical Properties*

In spite of the similarities of the optical properties when comparing the cometary dust to other dust particles in the solar system, the cometary dust shows a broad range of different properties. Dobrovolsky, Kiselev & Chernova (1986) and subsequently Levasseur-Regourd, Hadamcik & Renard 1996, suggest the existence of two classes of comets for which the optical, linear polarization vs. phase angle relations fall into two distinct groups: best-distinguished at large phase angles, the high polarization comets (including West 1976 VI, P/Halley 1986 III and Levy 1990 XX) have maximum polarizations $P_{max} \approx 30\%$ while the low polarization comets (e.g. Kobayashi-Berger-Milon 1975 IX, Austin 1990 V) have $P_{max} \approx 20\%$. The recent numerous observations of comet Hale-Bopp, however, do not clearly fit within these two groups Kiselev & Velichko 1999; Manset & Bastien 2000; Hadamcik & Levasseur-Regourd (2003). Moreover, the two classes of comets that Levasseur-Regourd, Hadamcik & Renard (1996) suggest show no clear relationship to the dynamical properties of the comets. It is therefore doubtful whether the comets can be clearly distinguished into two different groups of different dust properties.

We expect that variation of particle properties within comets to significantly influences observational results. Observations obtained before the 1990's have not been made with imaging detectors (Kiselev & Velichko 1999) and therefore interpretation might be difficult. Gas pollution of early polarization data may hamper the results, as indicated, for instance by recent observations of Comet Encke (Jewitt 2004; Jockers, Kiselev, Bonev *et al.* 2005). There is the possibility that the variation of albedo and colour occurs within the comet, as was seen with spatially resolved observations.

The recent, spatially resolved observations, however, show radial gradients and localized structures in the polarization data (Kolokolova, Jockers, Gustafson, & Lichtenberg 2001; Jockers, Rosenbush, Bonev *et al.* 1999) as well as higher polarization in the region of coma jets (Jockers, Rosenbush, Bonev *et al.* 1999; Furusho, Suzuki, Yamamoto *et al.*

1999; Hasegawa, Ichikawa, Abe *et al.* 1999). Polarization measurements may therefore highlight the difference in properties of dust released by discrete active areas. Jets may release dust from the nuclear subsurface that has not been subjected to weathering on the parent body. If so, polarization measurements may provide additional about different dust components within a comet.

4.4. *Current Models to Describe Optical Properties*

The recent model to describe albedo and polarization of cometary dust contradicts the long-lasting paradigm that cometary dust is dark because it is fluffy. The new model calculations show that absorbing material – possibly organic refractory – is closely connected to silicate into the smallest scales. This might explain the common characteristics of cometary dust compared to other dust species. The model of core-mantle particles as the sub-structure of the cometary dust is intriguing since it provides a direct connection to the interstellar core-mantle particles as they possibly exist in molecular clouds. The molecular cloud material, however is processed during the solar system formation and subsequent evolution. Therefore this picture of the cometary dust as agglomerate of interstellar particles is too simple. It is quite plausible that aggregates build out of monomers with different material composition provide a more realisitc description of the cometary dust. On the other hand, the simple model may be adequate for describing the average optical properties: Since the carbonaceous monomers are relatively dark, we can expect them to determine the scattering properties of an aggregate even when other types of monomers exist.

It still has to be investigated, how properties of cometary dust in the Zodiacal cloud would appear. We expect, that the particle processing in the interplanetary medium, though possibly present, will cause comparably small changes. Reach, Morris, Boulanger *et al.* (2003) studied the Zodiacal emission with ISOCAM observations on ISO. They report excess emission of 6% of the continuum in the 9–11 μm range, which could be matched by a mixture of Mg-rich amorphous silicate, dirty crystalline olivine and a hydrous silicate (montmorillonite). In the data that range from solar elongations 68° to 113° and from the ecliptic plane to the pole, they note a tendency that the strength of the features increases toward the Sun and toward high latitude above the ecliptic. This possibly indicates that the Zodiacal light is more influenced by cometary dust at high latitudes above the ecliptic and at small distances from the Sun. The difficulties of the line-of-sight inversion, however, do not allow to clearly follow the cometary dust properties in the interplanetary medium.

5. Constituents of the Solar System Dust Cloud

The relative contributions of different sources to the overall solar system dust cloud change within the solar systems as well as they depend on the considered measurement technique. Certainly dust from asteroids provides a large component of the dust near Earth orbit, but it is also preferably seen with most of the detection methods. Dust from comets becomes increasingly important at small distances from the Sun inward from 1 AU where collisions of cometary meteoroids provide the majority of dust production. Destruction of cometary dust and meteoroids including the carbon-bearing species feeds the pick-up ion component of the solar wind.

5.1. *Dust Dynamics*

The major effects that determine the distribution of dust in interplanetary space are solar gravitation, solar radiation pressure, mutual catastrophic collisions, and the influence of the Lorentz force on electrically charged dust particles. After being released dust particles stay initially in orbits similar to (but not identical to) those of their parent bodies. The influence of the planets inward from Jupiters orbit, aside from local effects, causes orbital perturbations as seen in the symmetry plane of the dust cloud. Orbital resonances at the outer planets and subsequent ejection in hyperbolic orbits are common for dust particles originating from the Kuiper belt and limit the amount of Kuiper belt dust that reaches the inner solar system (Liou, Zook, Dermott 1996). The particles that stay in the solar system are decelerated by the Poynting Robertson (P-R) effect and the Plasma Poynting-Robertson effect and therefore approach the Sun. Typical timescales of the P-R lifetime at 1 AU range up to 10^5 years, depending on the mass and scattering properties, which determine the radiation pressure force. Particles are formed and destroyed by mutual catastrophic collisions. Collisions limit the lifetime of large particles and provide a source of smaller particles, with the dividing mass between the smaller and the larger particles at approximately 10^{-10} to 10^{-9} kg.

As a result the particles seen in the Zodiacal light form an approximately rotationally symmetric dust cloud. All the inclination distributions agree with a concentration of the dust cloud toward the ecliptic plane. Namely, dust from asteroids and short-period comets produce the concentration of the dust cloud in the ecliptic plane. The distributions of inclinations derived from different Zodiacal cloud models peak more strongly at small inclinations than does the distribution of the inclinations in the orbits of the sporadic meteors (Kneissel & Mann 1991). Model calculations indicate that mutual collisions of dust inward from 1 AU could shift the size distribution to smaller particles (Ishimoto & Mann 1999; Ishimoto 2000). Since observational data show no evidence of this, the lost particles need to be replenished.

5.2. *Different Sources*

The overall picture of a homogenous dust cloud of particles produced in the outer solar system and drifting toward the Sun as a result of the P-R effect is not fully correct and does only apply to the region near Earth orbit. Composition, structure and size distribution of dust change with latitude and with distance from the Sun within the dust cloud, as does the relative contribution of the different sources. The relative amount of dust from the different sources within the dust cloud is uncertain and varies spatially within the dust cloud and possibly also with time. Also the detection methods are biased and therefore different types of observations indicate the preponderance of different types of particles. The picture of the solar system dust cloud seems to be as follows: The majority of dust outward from 5 AU originates from the Kuiper belt and the local interstellar medium. The dust cloud near 1 AU contains to a large extent dust from comets and asteroids, with most observations, i.e. analysis of collected samples, Zodiacal light analysis, indicating the preponderance of asteroidal dust. While observations of the Zodiacal light indicate the stability of the overall dust cloud near 1 AU over scales of years and decades, this is not the case for the inner solar system dust cloud where observational data are limited. Model calculations of the collision evolution indicate that the dust inward from 1 AU needs to be locally replenished. The collisional fragmentation of cometary meteoroids is the most plausible source. It is also possible that local dust production from cometary meteoroids inside 1 AU leads to changes of the dust cloud composition on time scales of years (Mann, Kimura, Biesecker *et al.* 2004). This would explain differences in the F-corona brightness observed during eclipses in different years (Kimura, Mann & Mukai 1998; Ohgaito *et al.* 2002).

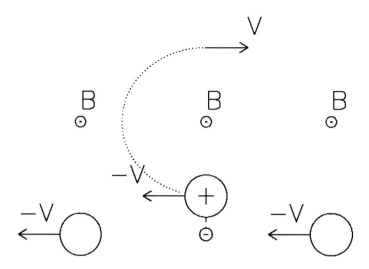

Figure 8. A schematic view of the pick-up process in the solar wind plasma frame. The neutral atoms (large circles) move at the velocity-V relative to plasma. When ionized (the circle with + sign) their motion becomes affected by the magnetic field B carried with the plasma flow. The Figure shows the simplest case where the B field is perpendicular to the velocity V. The ions motion in the plasma frame immediately after pick-up consists then of rotation (with the original speed V) around the magnetic field direction. If the B field is inclined, the ions motion is composed of rotation and sliding along the magnetic field. Soon after the pick-up, scattering off magnetic field irregularities causes the ion velocity distribution to become partly isotropic ("shell" distribution).

5.3. *Evidence for Collisional Evolution near the Sun*

Evidence for the collisional evolution in the solar system dust cloud can be found in the minor species of the interplanetary medium plasma, namely in the pick-up ions that are produced when neutrals are ionized and then carried with the solar wind. Material released from dust particles by various mechanisms including vaporization, sublimation, desorption or direct collisions provide a source of neutral and ionized molecules, atoms and ions in the solar-wind plasma of the interplanetary medium. The neutral gas is quickly ionized by the solar wind and photons and picked up by the plasma of the solar wind. As illustrated in Figure 8, the freshly ionized particles start gyrating around the magnetic field that is carried in the solar wind plasma. As a consequence ions move outward with the solar wind. Since, at larger heliocentric distances, further ionization by the solar photons is unlikely and collisions with other species which could cause ionization are less frequent, the ions in majority keep their single charge state. Pick-up ions are distinguished from the solar-wind ions by their single charge state as well as by their velocity distribution (see Figure 9). These dust-related, or meteoritic production of ions gains importance in the inner solar system, where the number density and relative velocities of dust are the highest. These dust generated ions are proposed as an explanation of the inner source component of the pick-up ions discovered by Ulysses. The main component of pick-up ions in the solar wind is formed by the ionization of interstellar neutral gas that streams into the solar system. This inner source contains especially carbon, which is not a component of the interstellar neutral gas (Geiss, Gloeckler & Steiger 1996). The inner source increases towards the Sun, which suggests it is correlated to the dust distri-

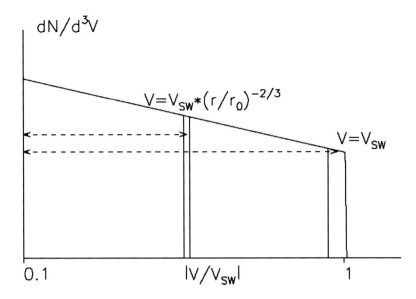

Figure 9. Pick-up ion velocity distribution at the distance r from the Sun (velocity in the plasma frame). The distribution includes the ions picked up at all distances less than r. Freshly picked ions have the speed $|V|$ equal to the solar wind speed V_{SW}. The ions picked up previously (at the distance $r_0 < r$) appear with reduced speed ($V = V_{SW} * (r/r_0)^{-2/3}$ for isotropic pick-up ion velocity distribution and constant solar wind speed V_{SW}) because of adiabatic cooling caused by plasma expansion (solar wind density decreases as $1/r^2$).

bution (Geiss, Gloeckler & Steiger 1996; Gloeckler & Geiss 1998, 2001). Noble gases and light elements in the inner source pick-up ions have abundances similar to that of the slow solar wind. Molecular ions in the mass range up to 40 amu have also been detected (Gloeckler & Geiss 2001). The similarity to solar-wind abundances suggests desorption of the solar-wind constituents is an important mechanism for the origin of these pick-up ions. However, the fluxes of dust required to account for the amounts of observed pick-up ions exceed by orders of magnitude the fluxes deduced from Zodiacal light observations (Mann, Kimura, Biesecker *et al.* 2004) and therefore surface interactions on the dust can not explain the observed ions.

In a recent study Mann & Czechowski (2005) have shown that collisional vaporization of dust and meteoroids can account for the observed fluxes of heavy inner source pick-up ions and that the ion production from this process exceeds the production from other dust-related processes (see Figure 10). (Aside from the collisional destruction, significant amounts of ions are produced by dust sublimation near the Sun and since sublimation takes place in the most inner regions, the produced ions are multiply charged.)

The observed fraction of carbon among the pick-up ions cannot be explained by the fragmentation of materials with meteoritic element abundances, but rather with cometary dust. The carbon content for carbonaceous chondrites as the most primitive meteorite material is clearly below the values for cometary dust at comet Halley. This agrees with studies of the near solar dust cloud that have shown that it is most likely locally resupplied by collisional fragmentation of cometary meteoroids. The model of the pick-up ion production from dust cannot fully explain the present few data, but the detection of inner source ions implies that cometary dust contains carbon bearing species that can survive high temperatures in the vicinity of the Sun (Mann & Czechowski 2005).

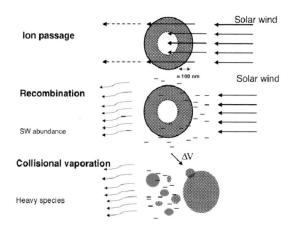

Figure 10. Dust interactions with the solar wind plasma of the interplanetary medium and the calculated amount of ions released by these processes for a typical model of dusta densities in the inner solar system. The collisional destruction provided the largest source of ions (Mann & Czechowski 2005)

Nevertheless, it is not possible at present to directly derive dust compositions from the pick-up-ion measurements. The charge states of the ions depend on the distance from the Sun at which they are released as well as on the actual solar-wind conditions at that time. For ions released inward from 0.1 AU of the Sun, multiple ionization frequently occurs, so that depending on the atomic species and the solar-wind condition only about half of the total ions appear in the singly charged state.

6. Summary

The overall structure of the solar system dust cloud results from the contribution of asteroids, comets, Kuiper Belt objects, and interstellar dust. The two latter components produce the dust seen outward from about 5 AU and are less important in the inner solar system. The dust near Earth orbit results mainly from asteroids and comets. Estimates of their relative contributions vary, which is partly a result of the different detection techniques, many of which are biased to the dust produced from asteroids. Dust from asteroids has typically smaller velocities relative to Earth and hence larger survival probability and therefore is overabundant in collected samples. Dust from asteroids also has a higher albedo and therefore more contributes to the Zodiacal light. This bias may not occur for meteors and indeed the orbital distribution of the sporadic meteors is different from orbital distributions derived from Zodiacal light models. Consideration of the collisional evolution in the dust cloud show that local sources need to replenish the cloud inward from 1 AU, and that cometary meteoroids are a likely source. Collision processes and dust sublimation feed ions into the solar wind plasma and it is expected that these species make up the heavy inner source pick-up ions that are seen in solar wind measurements.

Of the different constituents of the solar system dust cloud, the cometary dust undergoes the most complex alteration. Most of the alteration takes place in the coma and is connected to the evolution of the gas component. Fragmentation of dust in the coma is inferred from Stardust measurements as well as from brightness observations of the cometary dust. The fragmentation process is unclear but may include dust collisions, sublimation of volatiles and other effects. The break-up of larger particles due to sublimation of ice inclusions seems particularly plausible and would explain the irregular dust fluxes. All these processes may provide a source of coma gas, but yet carbon-bearing species will remain in the dust component. Particularly the CHON component detected with in-situ experiments of cometary dust is refractory and does not account for extended gas sources in the coma. The mass spectra obtained from in-situ measurements show no indication that the CHON component changes with distance from the comet, and particles were measured at distances from the nucleus where ices were sublimated already.

Both the optical properties of dust particles and the laboratory analyses of IDPs of likely cometary origin point to the fact that cometary dust is porous and irregular in shape, with the size of single monomers of the order of 100 nm. The low albedo of the cometary dust, however, results from darkness of the material forming the single monomers rather than from the fluffy structure. These properties are best explained when optically dark components, possibly organic refractories, are a constituent of the single monomers. This cometary dust model is consistent with the model assumption that the interstellar grains incorporated into the cometary nuclei form large aggregates of core-mantle particles. Nevertheless, this similarity should be viewed with some caution, since it is plausible to assume that only a fraction of the dust in the solar nebula may survive and be built directly into the cometary nuclei. Processing of the dust will occur prior to, during and after incorporation into the cometary nucleus.

The models of optical properties, the detection of carbon in the inner source pick-up ions generated by cometary meteoroids as well as the in-situ measurements of dust at comet Halley indicate that carbon-bearing species are intimately related to the silicate component of the cometary dust. We expect that a considerable fraction of the organic materials that are assumed to be present in the cometary dust formed refractory species. Traces of these remnants of the organic refractories are possibly seen in the ion composition of the solar wind.

Acknowledgements

We thank Dr. David Jewitt, Dr. George Flynn, and Dr. Masateru Ishiguro for helpful discussions and Dr. Diane Wooden for the careful review of the manuscript. Part of this review results from collaboration during I.M.'s and T.M.'s stay at the Institute of Low Temperature Science, Hokkaido University, the financial support for this stay is acknowledged. This research has been supported by the German Aerospace Center, DLR (project 'Rosetta: MIDAS, MIRO, MUPUS' RD-RX-50 QP 0403) and by the Japanese Ministry of Education, Culture, Sports, Science and Technology, MEXT, (Monbu Kagaku Sho) under Grant-in-Aid for Scientific Research on Priority Areas "Development of Extra-Solar Planetary Science" (16077203).

References

Baggaley, W.J. & Neslusan, L. 2002, *A&A* 382, 1118
Baum, W.A., Kreidl, T.J., & Schleicher, D.G. 1992, *AJ* 104, 1216
Becklin, E.E. & Westphal, J.A. 1966, *ApJ* 145, 445
Bertaux, J.L. & Blamont, J.E. 1976, *Nature* 262, 263
Bockelée-Morvan, D. & Crovisier, J. 2002, *Earth Moon & Planets* 89, 53
Bockelée-Morvan, D., Crovisier, J., Mumma, M.J., & Weaver, H.A. 2005, in: M. Festou, H.U. Keller, & H.A. Weaver (eds.) *Comets II* (Tucson: University of Arizona Press), p. 391

Bradley, J.P., Brownlee, D.E., & Fraundorf, P. 1984, *Science* 226, 1432

Bradley, J.P. 1988, *Geochimica et Cosmochimica Acta* 52, 889

Bradley, J.P. 1994, *Science* 265, 925

Bradley, J.P., Keller, L.P., Snow, T.P., Hanner, M.S., Flynn, G.J., Gezo, J.C., Clemett, S.J., Brownlee, D.E., & Bowey, J.E. 1999, *Science* 285, 1716

Brownlee, D.E. 1978, in: J.A.M. McDonnell (ed.), *Cosmic Dust* (Wiley-Interscience) p. 295

Brownlee, D.E. 1985, in: *Properties and interactions of interplanetary dust; Proceedings of the Eighty-fifth Colloquium* (D. Reidel Publishing Co.) p. 143

Campins, H. & Ryan, E.V. 1989, *ApJ* 341, 1059

Ceplecha, Z., Borovicka, J., Elford, W.G., Revelle, D.O., Hawkes, R.L., Porubčan, V., & Šimek, M. 1998, *Space Sci. Rev.* 84, 327

Clark, B.C., Green, S.F., Economou, T.E., Sandford, S.A., Zolensky, M.E., McBride, N., & Brownlee, D.E. 2004, *J. Geophys. Res.* 109, Issue E12, CiteID E12S03

Combi, M.R. 1994, *AJ* 108, 304

Combi, M.R., Kabin, K., DeZeeuw, D.L., Gombosi, T.I., & Powell, K.G. 1997, *Earth, Moon, & Planets* 79, 275

Crifo, J.F. 1995, *ApJ* 445, 470–488

Crifo, J.F. & Rodionov, A.V. 1999, *Planet. Space Sci.* 47, 797

Disanti, M.A., Mumma, M.J., dello Russo, N., Magee-Sauer, K., Novak, R., & Rettig, T.W. 1999, *Nature* 399, 662

Disanti, M.A., Mumma, M.J., & dello Russo, N. 2001, *Icarus* 153, 361

Dobrovolsky, O.V., Kiselev, N.N., & Chernova, G.P. 1986, *Earth, Moon & Planets* 34, 189

Dollfus, A., Bastien, P., Le Borgne, J.-F., Levasseur-Regourd, A.C., & Mukai, T. 1988, *A&A* 206, 348

Draine, B.T. 1988, *ApJ* 333, 848

Draine, B.T. & Flatau, P.J. 1994, *Journal of the Optical Society of America A* A11(4), 1491

Eberhardt, P. 1999, *Space Sci. Rev.* 90, 45

Fechtig, H. 1982, in: H. Wilkening (ed.) *Comets* (Tucson: University of Arizona Press) p. 383

Festou, M.C. 1999, *Space Sci. Rev.* 90, 53

Flynn, G.J. 1989, *Icarus* 77, 287

Flynn, G.J. 1996, in: B.A.S. Gustafson & M.S. Hanner (eds.) *Astronomical Society of the Pacific Conference Series; Proceedings of the 150th colloquium of the International Astronomical Union* ASP 104, p. 171

Flynn, G.J., Keller, L.P., Feser, M., Wirick, S., & Jacobsen, C. 2003, *Geochimica et Cosmochimica Acta* 67, 4791

Fomenkova, M. & Chang, S. 1997, in: J.M. Greenberg (ed.) *Proceedings of the NATO Advanced Study Institute* (The Cosmic Dust Connection) 487, 459

Frisch, P.C., Dorschner, J.M., Geiss, J., Greenberg, J.M., Grünn, E., Landgraf, M., Hoppe, P., Jones, A.P., Krtschmer, W., Linde, T.J., Morfill, G.E., Reach, W., Slavin, J.D., Svestka, J., Witt, A.N., & Zank, G.P. 1999, *ApJ* 525, 492

Furusho, R., Suzuki, B., Yamamoto, N., Kawakita, H., Sasaki, T., Shimizu, Y., & Kurakami, T. 1999, *PASJ* 51, 367

Geiss, J., Gloeckler, G., & von Steiger, R. 1996, *Space Sci. Rev.* 78, 43

Gloeckler, G., Fisk, L.A., Geiss, J., Schwadron, N.A., & Zurbuchen, T.H. 2000, *Journal of Geophysical Research* 105, A4, 7459

Gloeckler, G. & Geiss, J. 2001, *Space Sci. Rev.* 97, 169

Green, S.F., McDonnell, J.A.M., McBride, N., Colwell, M.T.S.H., Tuzzolino, A.J., Economou, T.E., Tsou, P., Clark, B.C., & Brownlee, D.E. 2004, *J. Geophys. Res.* 109, Issue E12, CiteID7E12S04

Greenberg, J.M. & Hage, J.I. 1990, *ApJ* 361, 260

Greenberg, J.M. & Li, A. 1998, *A&A* 332, 374

Grün, E., Pailer, N., Fechtig, H., & Kissel, J. 1980, *Planet. Space Sci.* 28, 333

Grün, E., Zook, H.A., Fechtig, H., & Giese, R.H. 1998, *Icarus* 62, 244

Grün, E., Gustafson, B., Mann, I., Baguhl, M., Morfill, G.E., Staubach, P., Taylor, A., & Zook, H.A. 1994, *A&A* 286, 915

Hadamcik, E. & Levasseur-Regourd, A.C. 2003, *A&A* 403, 757

Hadamcik, E., Renard, J., Levasseur-Regourd, A.C., & Worms, J.C. 2003, *JQSRT* 79, 679

Hajduk, A. 2001, in: B. Warmbein (ed.) *Meteoroids 2001* (ESA Publications Division, Noordwijk) ESA SP-495, p. 557

Hanner, M.S., Lynch, D.K., & Russell, R.W. 1994, *ApJ* 425, 247

Hanner, M.S., Gehrz, R.D., Harker, D.E., Hayward, T.L., Lynch, D.K., Mason, C.C., Russell, R.W., Williams, D.M., Wooden, D.H., & Woodward, C.E. 1997, *Earth, Moon & Planets* 79, 247

Hanner, M.S. 1999, *Space Sci. Rev.* 90, 998

Hanner, M.S. 2003, in: T.K. Henning (ed.) *Astromineralogy* (Springer) p. 609

Hanner, M.S. & Bradley, J.P. 2005, in: M. Festou, H.U. Keller, & H.A. Weaver (eds.) *Comets II* (Tucson: University of Arizona Press), p.

Harker, D.E., Wooden, D.H., Woodward, C.E., & Lisse, C.M. 2002, *ApJ* 580, 579

Harker, D.E., Wooden, D.H., Woodward, C.E., & Lisse, C.M. 2004, *ApJ* 615, 1081

Hasegawa, H., Ichikawa, T., Abe, S., Hamamura, S., Ohnishi, K., & Watanabe, J. 1999, *Earth, Moon, & Planets* 78, 353

Hübner, W.F. & Benkhoff,, J.
1999, *Space Sci. Rev.* 90, 117

Hübner, W.F. 2002, *Earth, Moon, & Planets* 89, 179

Hudson, B., Flynn, G.J., Fraundorf, P., Hohenberg, C.M., & Shirck, J. 1981, *Science* 211, 383

Ishiguro, M., Watanabe, J., Usui, F., Tanigawa, T., Kinoshita, D., Suzuki, J., Nakamura, R., Ueno, M., & Mukai, T. 2002, *ApJ* 572, L117

Ishimoto, H. & Mann, I. 1999, *Planet. Space Sci.* 47, 225

Ishimoto, H. 2000, *A & A* 362, 1158

Janches, D., Pellinen-Wannberg, A., Wannberg, G., Westman, A., Häggström, I., & Meisel, D.D. 2002, *J. Geophys. Res.* 107, A11, 14–1

Jessberger, E.K., Christoforidis, A., & Kissel, J. 1988, *Nature* 332, 691

Jessberger, E.K., Stephan, T., Rost, D., Arndt, P., Maetz, M., Stadermann, F.J., Brownlee, D.E., Bradley, J.P., & Kurat, G. 2001, in: E. Grün, B.A.S. Gustafson, S.F. Dermott, & H. Fechtig (eds.) *Interplanetary Dust* (Springer-Verlag) p. 253

Jewitt, D. 2004, *AJ* 128, 3061

Jockers, K., Rosenbush, V., Bonev, T., & Credner, T. 1999, *Earth, Moon, & Planets* 78, 373

Jockers, K., Kiselev, N., Bonev, T., Rosenbush, V., Shakhovskov, N., Kolesnikov, S., Efimov, Y., Shakhovskoy, D., & Antonyuk, K. 2005, *A&A* 441, 773

Joswiak, D.J., Brownlee, D.E., Bradley, J.P., Schlutter, D.J., & Pepin, R.O. 1996, *Lunar and Planetary Institute Conference Abstracts*, p. 625

Keller, L.P., Messenger, S., Flynn, G.J., Wirick, S., & Jacobsen, C. 2002, *Meteoritics & Planetary Sci.* 37, 76

Kelley, M.S., Woodward, C.E., Jones, T.J., Reach, W.T., & Johnson, J. 2004, *AJ* 127, 2398

Kimura, H., Mann, I., & Mukai, T. 1998, *Planet. Space Sci.* 46, 911

Kimura, H., Kolokolova, L., & Mann, I. 2003, *A&A* 407, L5

Kimura, H., Mann, I., & Jessberger, E.K. 2003, *Icarus ApJ*, 314

Kiselev, N.N. & Velichko, F.P. 1998, *Icarus* 133, 286

Kiselev, N.N. & Velichko, F.P. 1999, *Earth, Moon, & Planets* 78, 347

Kissel, J., Brownlee, D.E., Büchler, K., Clark, B.C., Fechtig, H., Grün, E., Hornung, K., Igenbergs, E.B., Jessberger, E.K., Krueger, F.R., Kuczera, H., McDonnell, J.A.M., Morfill, G.M., Rahe, J., Schwehm, G.H., Sekanina, Z., Utterback, N.G., Völk, H.J., & Zook, H.A. 1986, *Nature* 321, 336

Kissel, J., Sagdeev, R.Z., Bertaux, J.L., Angarov, V.N., Audouze, J., Blamont, J.E., Büchler, K., Evlanov, E.N., Fechtig, H., Fomenkova, M.N., von Hoerner, H., Inogamov, N.A., Khromov, V.N., Knabe, W., Krueger, F.R., Langevin, Y., Leonas, V.B., Levasseur-Regourd, A.C., Managadze, G.G., Podkolzin, S.N., Shapiro, V.D., Tabaldyev, S.R., & Zubkov, B.V. 1986, *Nature* 321, 280

Kneissel, B. & Mann, I. 1991, in: A.C. Levasseur-Regourd & H. Hasegawa (eds.) *Origin and Evolution of Interplanetary Dust* (Dordrecht: Kluwer) p. 131

Kolokolova, L., Kimura, H., & Mann, I. 2004, in: G. Videen, Y. Yatskiv, & M. Mishchenko (eds.) *Photopolarimetry in Remote Sensing* (Kluwer Academic Publisher), p. 431

Kolokolova, L., Jockers, K., Gustafson, B., & Lichtenberg, G. 2001, *J. Geophys. Res.* 106, 10113

Kolokolova, L., Hanner, M.S., Levasseur-Regourd, A.-C., & Gustafson, B.A.S. 2005, in: M. Festou, H.U. Keller, & H.A. Weaver (eds.) *Comets II* (University of Arizona Press), p.

Konno, I., Hübner, W.F., & Boice, D.C. 1993, *Icarus* 101, 84

Kurat, G., Koeberl, C., Presper, T., Brandsttter, F., & Maurette, M. 1994, *Geochimica et Cosmochimica Acta* 58, Issue 18, 3879

Landgraf, M. 2000, *J. Geophys. Res.* 105, 10303

Lara, L.M., Schulz, R., Stüwe, J.A., & Tozzi, G.P. 2001, *Icarus* 150, 124

Lawler, M.E. & Brownlee, D.E. 1992, *Nature* 359, 810

Lederer, S.M., Osip, D.J., Thomas-Osip, J.E., DeBuizer, J.M., Mondragon, L.A., Schweiger, D.L., & Viehweg, J., SB Collaboration 2005, *AAS/Division for Planetary Sciences Meeting Abstracts* 37, 4308

Levasseur-Regourd, A.C., Hadamcik, E., & Renard, J. 1996, *A&A* 313, 327

Levasseur-Regourd, A.C., Mann, I., Dumont, R., & Hanner, M.S 2001, in: E. Grün, B.A.S. Gustafson, S.F. Dermott, & H. Fechtig (eds.) *Interplanetary Dust* (Springer-Verlag) p. 57

Liou, J.-C., Zook, H.A., & Dermott, S.F. 1996, *Icarus* 124, 429

Lumme, K. & Rahola, J. 1994, *ApJ* 425, 653

Mann, I., Okamoto, H., Mukai, T., Kimura, H., & Kitada, Y. 1994, *A&A* 291, 1011

Mann, I. 1996, *Space Sci. Rev.* 78, 259

Mann, I. 1998, *Earth, Planets, & Space* 50(6,7), 465

Mann, I. & Kimura, H. 2000, *J. Geophys. Res.* 105, 10317

Mann, I. & Kimura, H. 2001, *Space Sci. Rev.* 97, Issue 1/4, 389

Mann, I. & Jessberger, E.K. 2003, in: T. Henning (ed.) *Astromineralogy* (Berlin: Springer) p. 98

Mann, I., Kimura, H., Biesecker, D.A., Tsurutani, B.T., Grün, E., McKibben, B., Liou, J.C., MacQueen, R.M., Mukai, T., Guhartakuta, L., & Lamy, P. 2004, *Space Sci. Rev.* 110, 269

Mann, I., Kimura, H., & Kolokolova, L. 2004, *JQSRT* 89, 291

Mann, I. & Czechowski, A. 2005, *ApJ* 621, L73

Manset, N. & Bastien, P. 2000, *Icarus* 145, 203

Maurette, M., Hammer, C., Reeh, N., Brownlee, D.E., & Thomsen, H.H. 1986, *Science* 233, 869

Maurette, M., Olinger, C., Michel-Levy, M.C., Kurat, G., Pourchet, M., Brandstatter, F., & Bourot-Denise, M. 1991, *Nature* 351, 44

Meisel, D.D., Janches, D., & Mathews, J.D. 2002, *ApJ* 567, 323

Messenger, S. 2000, *Nature* 404, 968

Mukai, T., Mukai, S., & Kikuchi, S. 1987, *A&A* 187, 650

Mukai, T. & Koike, C. 1990, *Icarus* 87, 180

Mukai, T., Ishiguro, M., & Usui, F. 2002, *Adv. Space Res.* in press.

Ohgaito, R., Mann, I., Kuhn, J.R., MacQueen, R.M., & Kimura, H. 2002, *ApJ* 578, 610

Pellinen-Wannberg, A. & Wannberg, G. 1994, *J. Geophys. Res.* 99, 11397

Petrova, E.V., Jockers, K., & Kiselev, N.N. 2000, *Icarus* 148, 526

Reach, W.T., Morris, P., Boulanger, F., & Okumura, K. 2003, *Icarus* 164, 384

Rietmeijer, F.J.M. 1993, *J. Geophys. Res.* 98, 7409

Rietmeijer, F.J.M. 1999, in: J.J. Papike (ed.) *Planetary Materials* (Mineralogical Society of America), vol. 36. p. 2-1-2-95

Rietmeijer, F.J.M. 2000, *Meteoritics & Planetary Sci.* 35, 1025

Rietmeijer, F.J.M. 2002, in: E. Murad & I.P. Williams (eds.) *Meteors in the Earth's atmosphere* (Cambridge: Cambridge University Press), p. 215

Schulze, H., Kissel, J., & Jessberger, E.K. 2002, *ASP Conf. Ser. 122: From Stardust to Planetesimals*, p. 397

Schleicher, D.G. & Osip, D.J. 2002, *Icarus* 159, 210

Sekanina, Z., Brownlee, D.E., Economou, T.E., Tuzzolino, A.J., & Green, S.F. 2004, *Science* 304, 1769

Sykes, M.V., Lebofsky, L.A., Hunten, D.M., & Low, F. 1986, *Science* 1986, 1115

Sykes, M.V. & Walker, R.G. 1992, *Icarus* 1992, 180

Taylor, A.D., Baggaley, W.J., & Steel, D.I. 1996, *Nature* 380, 325

Tsou, P., Brownlee, D.E., Anderson, J.D., Bhaskaran, S., Cheuvront, A.R., Clark, B.C., Duxbury, T., Economou, T., Green, S.F., Hanner, M.S., Hörz, F., Kissel, J., McDonnell, J.A.M., Newburn, R.L., Ryan, R.E., Sandford, S.A., Sekanina, Z., Tuzzolino, A.J., Vellinga, J.M., & Zolensky, M.E. 2004, *J. Geophys. Res.* 109,

Tsurutani, B.T.*et al.* 2004, *Icarus* 167, 89

Turner, N.J. & Smith, G.H. 1999, *ApJ* 118, 3039

Tuzzolino, A.J., Economou, T.E., Clark, B.C., Tsou, P., Brownlee, D.E., Green, S.F., McDonnell, J.A.M., McBride, N., & Colwell, M.T.S.H. 2004, *Science* 304, 17760

Wooden, D.H., Harker, D.E., Woodward, C.E., Butner, H.M., Koike, C., Witteborn, F.C., & McMurtry, C.W. 2004, *ApJ* 517, 1034

Wooden, D.H., Harker, D.E., & Woodward, C.E. 2000, *ASP Conf. Ser. 196: Thermal Emission Spectroscopy and Analysis of Dust, Disks, and Regoliths* 196, 99

Wooden, D.H. 2002, *Earth Moon & Planets* 89, 247

Wooden, D.H., Charnley, S.B., & Ehrenfreund, P. 2005, in: M. Festou, H.U. Keller & H.A. Weaver (eds.) *Comets II* (Tucson: University of Arizona Press) p.

Woodward, C.E., Shure, M.A., Forrest, W.J., Jones, T.J., Gehrz, R.D., Nagata, T., & Tokunaga, A.T. 1996, *Icarus* 124, 651

Asteroids, Comets, Meteors
Proceedings IAU Symposium No. 229, 2005
D. Lazzaro, S. Ferraz-Mello & J.A. Fernández, eds.

© 2006 International Astronomical Union
doi:10.1017/S174392130500668X

Connections between asteroids and cometary nuclei

Imre Toth

Konkoly Observatory, Budapest, P.O. Box 67, H-1525, Hungary
email: tothi@konkoly.hu

Abstract. We review the recent progress in the exploration of the interrelations between primitive small bodies of the solar system which are preserved the pristine material in their interior: cometary nuclei, Transneptunian Objects, Centaurs, and primitive asteroids, and they are considered as primordial objects. In addition, we discuss the properties of the asteroid-comet transition objects which have really enigmatic behavior. The comets have most primitive, accessible material in the solar system but we do not know what is hidden below the evolved surface layers. Comets must become dormant but we do not know whether the ice is exhausted or sublimation is inhibited (blocked by quenching mechanisms). There must be many dormant comets masquerading as asteroids but we do not know to identify these bodies unless via serendipitous discovery observations. Indeed, there are some asteroids which temporarily show comet-like activity. These are among the Damocloids (C/2001 OG108 (LONEOS)), main belt asteroids (7968 Elst-Pizarro = 133P/E-P) and Near-Earth objects (4015 Wilson-Harrington = 107P/W-H). The important questions are: where is the pristine material in the cometary nuclei and in the asteroid-comet transition objects, do comets lose their ice or seal it in? Both the large survey projects and in-situ space missions will help to answer these questions in the near future.

Keywords. comets: general, Kuiper belt: general, minor planets, asteroids: general, Oort cloud: general, solar system: general

1. Introduction

Investigation of the interrelationships among the primitive small bodies can allow us to analyze the evolution of the almost unprocessed material coming from the frontiers of the solar system: from the Oort cloud and from the Transneptunian regions through Centaurs to ecliptic comets (ECs) and primitive asteroids. In addition, the Transneptunian objects, Centaurs, ecliptic comets and related asteroids are evolutionary linked.

We review new results on Damocloids (Section 2), the objects which are connected to the Oort cloud. The cometary activity of TNOs and Centaurs reveals their physico-chemical characteristics and expected connections to the ecliptic comets (Section 3). The new results on Jupiter Trojan asteroids enlight their origin and possible connections to comets (Sections 4 and 5). We present the new results on the origin of ecliptic comets according to physical ground (sizes, colors) (Section 5). Furthermore we discuss the current status in the connection between main belt asteroids and cometary nuclei (Section 6). Comets and asteroids were previously thought to be two completely distinct groups of solar-system objects, with marked contrast in both physical and dynamical characteristics. A comet is operationally defined by the presence of a coma, while an asteroid has no coma. However, recent observations have shown that comets can sometimes take on asteroidal appearances and even asteroidal photometric behavior. Thus the observational distinction between comets and asteroids is not as clear cut as it once seemed. The possible presence of comets hidden among known asteroids forces us to reconsider the criterion by which we distinguish comets from asteroids and possibly our inventory of both comet and asteroid populations. The existence and behavior of the recurrent comet-like activity of 7968 Elst-Pizarro (133P/E-P) which orbits among the main belt asteroids pose serious problems for our understanding of comets and asteroids and how they relate to one another. We discuss the current situation of the asteroids and comets which reside in the Near-Earth and inner Earth regions (Section 7). In the end we summarize the current status and outstanding questions in the physical properties of cometary nuclei and related asteroids (Section 8).

2. Oort cloud asteroids and asteroid-comet transition objects from the Oort cloud: the Damocloids

Heretofore the best and effectively applicable new taxonomy of comets and other small bodies of the solar system is based on the Tisserand parameter with respect to Jupiter, T_J, in the model of Sun – Jupiter – small body circular restricted three body problem (Levison 1996). The Oort cloud associated comets with small perihelion distances, including the Halley-type comets (HTCs), are the nearly-isotropic comets (NICs) have $T_J < 2$ (Levison et al. 2002). The Damocloids are solar system objects orbiting the Sun thought, on dynamical grounds, to be either inactive NICs or asteroids associated to the Oort cloud (Jewitt 2005). Damocloids have orbit similar to the Halley-family or NICs in general and named after 5335 Damocles. They have high orbital eccentricity and inclination, as well as most of them are without visible signs of outgassing activity, which suggest that the Damocloids are the dead or dormant nuclei of NICs (long-period comets, Asher et al. 1994). Damocloids have T_J less than 2, which falls to the same range as of NICs. Moreover, in correspondence with the description given by Asher et al. (1994) Jewitt (2005) has a specific definition, which restricts the Damocloids to asteroidal appearance: "a Damocloid is any point-source object having T_J with respect to Jupiter less than or equal to 2". On physical ground Jewitt describes the Damocloids as inactive HTCs. In addition, to support this view, the cumulative distribution of the orbital inclination of Damocloids and Halley-type comets (HTCs) show the great similarity between the two populations Jewitt (2005).

Minor planet 1996 PW is unusual in having the orbital characteristics of a long period comet but showing no sign of cometary activity. The discovery of 1996 PW prompted Weissman & Levison (1997) to examine and evaluate its possible origins, including the intriguing possibility that it is an asteroid from the Oort cloud, which is a new conception. Current models for the formation of the Oort cloud argue that most of the material there should be from the Uranus-Neptune region and thus cometary, not asteroidal, in composition. Weissman & Levison (1997) better quantified these models and show that ~1% of the Oort cloud population should be asteroids. They found that 1996 PW has almost certainly been a resident of the Oort cloud. However, they also found it equally likely that 1996 PW is an extinct comet or an asteroid. Although not conclusive, their results represent a significant change in our understanding of the Oort cloud, because they suggest that the ejection process sampled (i) material from as close to the Sun as the asteroid belt in the primordial solar nebula and hence (ii) much warmer formation temperatures than previously thought. They concluded that this diverse sample is preserved in the Oort cloud. Exploring of the physical properties of 1996 PW Davies et al. (1998), and Hicks et al. (2000) derived the physical parameters of this Damocloid and they found that 1996 PW has moderately red, featureless spectra typical of the D-type asteroids, cometary nuclei, and other extinct cometary candidates. With these findings,

1996 PW join the ranks of 3552 Don Quixote and 944 Hidalgo as established candidates for extinct comet nuclei (see also their results on the asteroid 1997 SE5).

The highly eccentric orbit of C/2001 OG108 (LONEOS) takes it closer to the Sun than the Earth, and beyond the planet Uranus. Additionally, its orbit is inclined 80 degrees with respect to the ecliptic. The object is in a Halley-family orbit, but was apparently inactive until January 2002, when it was only about 1.5 AU from the Sun and just 2 to 3 months before perihelion. Abell *et al.* (2003) and Fernández *et al.* (2003a) presented the results from multiwavelength observations of the nucleus of the unusual asteroid-comet. They observed the object in visible, near-IR, and mid-IR wavelengths near its opposition in October and November 2001, while it was still asteroidal. Thus their observations, originally intended to characterize the surface and physical properties of an unusual asteroid, were, in fact, fortuitously of a bare cometary nucleus; perhaps this nucleus is undergoing its last epoch of activity before dormancy. Very few nuclei have been studied in such detail, and even fewer nuclei that belong to comets originating in the Oort cloud. They have constrained the nucleus's size, shape, color, reflectance spectrum, albedo, and rotation period. The (near-IR) spectrum, (visible) colors, and geometric albedo most closely resemble those of a D-type asteroid. There are no absorption bands in the 0.75 to 2.4 micron range at a few percent level, though the spectrum does show a kink near 0.75 micron. The *V*-band geometric albedo of the nucleus is 0.030 ± 0.005, and this is well within the currently-known distribution of albedos for other active comets and extinct-comet candidates. C/2001 OG108's nucleus is both one of the largest known and one of the most slowly rotating. In addition, French (2002) reported the observation of C/2001 OG108 (LONEOS) in October of 2001: *(i)* image profiles consistent with those of background stars, implying no detectable cometary activity in October, and *(ii)* *BVRI* colors consistent with those of inert comet nuclei and the D-class of asteroids common in the Trojan clouds.

Up to now the biggest survey of physical properties of Damocloids was performed by Jewitt (2005). In another work Fernández *et al.* (2005) investigated the albedos of asteroids in comet-like orbits (including Damocloids and Near-Earth Asteroids). The measured Damocloids have effective radii in the $\sim$2 to $\sim$70 km range with a median effective radius of 8.4 km. They are comparable in size to the best-observed nuclei of Jupiter-family comets (JFCs). Where measured, the geometric albedos of the Damocloids are small (0.02 to 0.04), like those of the JFC nuclei, and suggesting a dark, carbon-enriched surface composition. Jewitt (2005) presented optical measurements of 12 such objects, finding that their mean Kron-Cousins colors are $(B - V) = 0.79 \pm 0.01$, $(V - R) = 0.48 \pm 0.01$, and $(R - I) = 0.48 \pm 0.01$. The normalized reflectivity spectra are generally linear, with a mean gradient $S' = 11.9\% \pm 1.0\%$ per 1000 Å. The latter is consistent with the mean $S' = 11.6\% \pm 2.3\%$ per 1000 Å measured for the nuclei of Jupiter-family comets of ECs, a surprising result given the expected very different formation locations and dynamical histories of these two types of body. The Damocloids are devoid of the ultrared matter (with $S' \geqslant 25\%$ per 1000 Å) that is present on many Kuiper belt objects and Centaurs, and the mean colors of the Damocloids are inconsistent with those of the Kuiper belt objects ($S' = 21.1\% \pm 1.4\%$ per 1000 Å). The data suggest that the ultrared matter, widely thought to consist of a complex organic compound processed by prolonged exposure to cosmic rays, cannot survive long in the inner solar system. Timescales for ejection or burial of ultrared matter on the nuclei of both Jupiter-family comets and Damocloids are short. Such material may also be chemically unstable to the higher temperatures experienced in the inner planetary region. Unfortunately, there are not so many Damocloids observed in detail in order to explore their physical properties (size, shape, albedo, color, rotational parameters, surface and thermal properties, bulk interior characteristics, cometary activity) and chemical composition hence the observation of these objects is encouraged.

3. Cometary activity in TNOs and Centaurs: traces of volatile components

There are a few important recent reviews on the orbital dynamics and physical characterization of Centaurs and TNOs, including their surface properties, colors, albedos, spectra, chemistry (modeling the spectra), and comet-like activity: Barucci *et al.* (2002), Luu & Jewitt (2002), Schulz (2002), *ESO Workshop on TNOs* (*Earth, Moon, and Planets* Vol. 92, 2003), Barucci *et al.* (2004). The Centaurs and TNOs being the source of ecliptic comets, it is expected that the nature of Centaurs and TNOs be the same or very similar as that of pristine ecliptic comets, however the surface ecliptic comets altered during their evolution (Campins & Fernández 2001, Luu & Jewitt 2002, Jewitt 2004). The suspected conditions prevailing at the time and place of their formation support the idea that the water ice trapped large quantities of very volatile material while condensing. These hypotheses, together with the growing observations of activity of very distant comets (at heliocentric distances greater than 23 AU) and Centaurs, as well as with the behavior of the Pluto's atmosphere, indicate that the cometary activity in TNOs should indeed be possible. The cometary activity of Chiron inspired Brown & Luu (1998) to investigate model comae (gas and dust) around Centaurs and KBOs. They found that the Chiron's long-lived coma cannot be a result of outburst. To observe a TNO coma requires a massive $\sim 5 \times 10^9$ kg dust coma, and its lifetime varies from < 2 months to ~ 1 year depending on object size and heliocentric distance.

Most Centaurs are inactive but a few display cometary activity: 95P/Chiron (2060 Chiron), 29P/Schwassmann–Wachmann 1, 39P/Oterma, 165P/2000 B4 (LINEAR), 166P/2001 T4 (NEAT), C/2001 M10 (NEAT), and 167P/2004 PY42 (CINEOS). We note that 29P with $q = 5.7$ AU and $a = 6.0$ AU satisfies the definition of Jewitt & Kalas (1998) for Centaurs but that it has $T_J = 2.983$ (while for Centaurs $T_J > 3$). In addition, there is a nomenclature problem in the designation of active Centaurs with cometary identification: the cometary provisional or periodic comet identification is preferred, rather than their names (Green 2005). Recently Bauer *et al.* (2003) presented the results of optical observations of 166P/2001 T4 (NEAT) in 2001 and 2002. Coma was present for each observations but the activity level was variable. Dust production rate was between $\sim 10^{-2}$ and 20 kg s^{-1}, comparable to active Jupiter-family comets (cf. A'Hearn *et al.* 1995).

Aside the well known cometary activity in Centaurs a few years ago there were observations which are interpreted as direct evidences of TNO cometary activity. Most notably, Fletcher *et al.* (2000) (see Delahodde *et al.* 2000) announced the detection of a coma around 1994 TB, using HST. Hainaut *et al.* (2000), Delahodde *et al.* (2000) obtained a detailed portrait of 1996 TO66 using VLT: a dramatic change of that object's rotational lightcurve between 1997 and 1998 is interpreted as the signature of a cometary outburst.

Meech *et al.* (2003) searched for cometary activity in KBO (24952) 1997 QJ4 using the Subaru 8-m telescope. There is a large color diversity among the KBOs. The neutral blue colors with respect to the Sun can be explained by the possible surface outgassing (cometary) activity (it is the most plausible explanation but there are other causes of the surface blueing during the evolution of the outer solar system objects). They selected this target because of its intrinsically blue color $(V - R) = 0.296$. They placed sensitive upper limits on the dust production rate from the object at $Q < 0.01$ kg s^{-1}.

Surveying of the cometary activity of the faint objects in the outer solar system is usually a very difficult task despite of using large telescopes. For example the Scattered Disk object (29981) 1999 TD10 was observed simultaneously in the R, J and H bands in September 2001, and in B, V, R, and I in October 2002 by Mueller *et al.* (2004). But their observations at the same time, with better S/N and seeing, show no evidence of a coma, contrary to the claim by Choi *et al.* (2003). Moreover, Rousselot *et al.* (2003) also did not observed cometary activity on 1999 TD10.

Regarding the cometary activity of Centaurs and TNOs obviously volatile compounds are needed. Brown & Koresko (1998) reported the detection of the 1.5 and 2.0 micron m absorption bands due to water ice in the near-infrared reflection spectrum of the Centaur 10199 Chariklo (1997 CU26). The water ice bands are weaker than those detected on the surface of any other solar system body; the spectrum is well fit with a model surface consisting predominantly of a neutral dark absorbing substance with only ~3% a real coverage of water ice. Luu *et al.* (2000) reported the detection of water ice in the Centaur 2060 Chiron, based on near-infrared spectra (1.0–2.5 micron). The appearance of this ice is correlated with the recent decline in Chiron's cometary activity: the decrease in the coma cross section allows previously hidden solid-state surface features to be seen. They predicted that water ice is ubiquitous among Centaurs and Kuiper belt objects, but its surface coverage varies from object to object and thus determines its detectability and the occurrence of cometary activity. Independently, Foster *et al.* (1999) reported the presence of the water ice 2.03 micron spectral feature in the reflectance spectrum of Chiron and several other Centaurs. In the frame of the *ESO Large Program on Transneptunian Objects and Centaurs* Boehnhardt *et al.* (2003) obtained 12 spectra in the visible region and nine of them for which they obtained also near infrared spectra up to 2.4 microns. The principal reported results obtained are, including possible detection of water ice: (*i*) a wide range of visible slopes, (*ii*) evidence for surface variations on 2001 PT13, and (*iii*) possible detection of few percent of water ice (1999 TC36), 2000 EB173, 1999 DE9, 2001 PT13, 2000 QC243, 1998 SG35).

More accurate and realistic models are important in understanding the surface properties and comet-like activity in the outer solar system objects. In addition, combining the visible and infrared observations, as well as using an adequate thermal model the effective radius and albedo can be determined. Fernández *et al.* (2003b) measured the mid-infrared thermal continuum from inactive 8405 Asbolus and active 2060 Chiron. Using simple thermal models they found that Chiron has a variable dust coma. The surface heterogenity of 10199 Chariklo (1997 CU26) was taken into account in the models developed by Dotto *et al.* (2003) and compared with the near-infrared observations. They used tholins, amorphous carbon and water ice and they confirmed the presence of water ice on the surface of this Centaur as it was detected earlier by Brown & Koresko (1998). A new surface thermal model of Centaurs was developed by Groussin *et al.* (2004) who analyzed visible, infrared, radio and spectroscopic observations of 2060 Chiron (95P/Chiron) in a synthetic way to determine its physical properties. Infrared observations at 25, 60, 100 and 160 micron (i.e., covering the broad maximum of the spectral energy distribution) obtained with the Infrared Space Observatory Photometer (ISOPHOT) in June 1996 when Chiron was near its perihelion are analyzed with a thermal model which considers an intimate mixture of water ice and refractory materials and includes heat conduction into the interior of the nucleus. They found that the observed spectra of Chiron can be fitted by a mixture of water ice (~30%) and refractory (~70%) grains, and that this surface model has a geometric albedo consistent with the above value. They also analyzed the visible, infrared and radio observations of Chariklo (1997 CU26) and concluded that a mixture of water ice (~20%) and refractory (~80%) grains is compatible with the near-infrared spectrum and the above albedo.

First evidence for the presence of water ice was detected in TNO 1996 TO66 in the NIR spectra taken with the Keck telescope Brown *et al.* (1999). The published Keck NIR spectra of 1996 TO66 obtained almost simultaneously with 1998 data obtained by Hainaut *et al.* (2000), showing the presence of water ice on that object, further reinforcing the conviction of its cometary nature.

Later, Bockelée-Morvan *et al.* (2001) searched for rotational lines of CO in Pluto/Charon (the largest known bodies among the KBOs), other KBOs, and Centaurs at radio wavelengths, i.e., the CO as a supervolatile compound was looked for which can drive the outgassing activity at large heliocentric distances beyond the heliocentric distance limit of the water ice sublimation activity. None of the Centaurs or Kuiper belt objects were detected in CO. The CO production rate upper limit obtained for Chiron $(3-5 \times 10^{27}$ mol s$^{-1})$ over 1998–2000 years is a factor of 10 lower than the CO production rate derived from the marginal CO detection obtained in June 1995 by Womack & Stern (1999), using same modeling of CO emission. Upper limits obtained for other Centaurs are typically $\sim 10^{28}$ mol s^{-1}, and between 1 and 5×10^{28} mol s^{-1} for the best observed KBOs. Bockelée-Morvan *et al.* (2001) concluded that the comparison between these upper limits and the CO outgassing rates of comet C/1995 O1 (Hale-Bopp) measured at large distances from the Sun shows that Centaurs and KBOs underwent significant CO-devolatilization since their formation.

Campins & Fernández (2001), Licandro *et al.* (2002), Licandro *et al.* (2003) presented the results of their near-infrared spectroscopic program of TNOs, Centaurs and comet nuclei. TNOs, Centaurs, and Jupiter-family comets are three intimately related populations of minor planet bodies originated in the outer solar system. They probably contain some of the least modified materials remaining from the protosolar nebula. The study of their physical properties and evolution provide invaluable cosmogonical information. Near-infrared spectroscopy is a diagnostic method for remote determination of the surface composition of these objects. In addition to the already published spectra of 38628 Huya and 20000 Varuna, and 28978 Ixion (Licandro *et al.* 2003 amd references therein) and 124P/Mrkos they presented new spectra of the TNOs 38628 Huya, 50000 Quaoar, and 1999 TC36, and the Centaurs 8405 Asbolus, (54598) 2000 QC243, 32532 Thereus, 31824 Elatus, and 2002 PN34. Water ice absorption bands are present in the spectra of several objects. Evidence of surface inhomogeneities is also presented, in particular in the case of 32532 Thereus that was observed during an almost complete rotation. The results are discussed in the framework of the possible resurfacing mechanisms proposed (space weathering, coma activity and collisions).

Fornasier *et al.* (2004a) obtained visible and near infrared spectra of the TNO 90482 Orcus (2004 DW), a few days after its discovery, at the Telescopio Nazionale Galileo (TNG). 90482 Orcus belongs to the plutino dynamical class and has an estimated diameter of about 1600 km, that makes it one of the largest KBO. Data clearly show the 1.5 and 2 micron bands associated to water ice, while the visible spectrum is nearly neutral and featureless. To interpret the available data the best fit model of the surface composition of 90482 Orcus (2004 DW) contains two different mixtures of organics (Titan tholin and kerogen), amorphous carbon and water ice.

During the last years water ice has been reported in a handful of objects in the outer solar system, but most appear spectrally featureless. Most recently Jewitt & Luu (2004) published their infrared spectroscopic observations of the large KBO 50000 Quaoar, which reveal the presence of crystalline water ice and ammonia hydrate. Crystallinity indicates that the ice has been heated at least 110 K. Both ammonia and water ice should be destroyed by energetic particle irradiation on a timescale of about 10^7 yr. They concluded that Quaoar has been recently resurfaced, either by impact exposure of previously buried (shielded) ices or by cryovolcanic outgassing, or by a combination of these processes.

Another possibility to preserve water in small bodies are the aqueously altered minerals. de Bergh *et al.* (2004) obtained visible and near-infrared spectra, as well as photometric data, for two TNOs, (47932) 2000 GN171 and 38628 Huya (2000 EB173), which belong to the dynamical class of plutinos in 2001 and 2002. The features detected in the visible spectra of the two objects are tentatively attributed to the presence of iron oxides or phyllosilicates at the surfaces of the two objects. There are differences between the April 2001 and May 2002 visible spectra, which are attributed to spatial variations at the surfaces of the objects. They proposed the possibilities for aqueous alteration in TNOs, after reviewing what we know about the presence of aqueously altered minerals (silicates) in other small bodies of the solar system. They suggested further studies monitoring the rotation of these two objects are highly desirable.

4. Jupiter Trojan asteroids and comets

The most recent reviews and results on the dynamical characteristics of Trojans are of Beaugé & Roig, Marzari *et al.* (2002), Karlsson (2004), Morbidelli *et al.* (2005), as well as on the physical properties are of Barucci *et al.* (2002), Fernández *et al.* (2003b), Jewitt *et al.* (2000), Licandro *et al.* (2002), Jewitt (2004), Bendjoya *et al.* (2004), Emery & Brown (2003), Emery & Brown (2004), Fornasier *et al.* (2004b).

The observed physical properties (shape, color, albedo) of the Jupiter Trojans are formally indistinguishable from those of the cometary nuclei (Jewitt & Luu 1990, Fernández *et al.* 2003c). As Jewitt (2004) warned: this suggests but does not prove an intriguing compositional similarity between the two classes of body, at least at the surface level where irradiation and solar heating may play a role. No ices have been spectroscopically detected on the Trojans (Jones *et al.* 1990, Luu *et al.* 1994, Dumas *et al.* 1998, Emery & Brown 2003). Bendjoya *et al.* (2004) observed spectra of 34 Jupiter Trojans with the Danish 1.54-m telescope at ESO. They found that large majority of the objects of the sample have been observed to belong to the D taxonomic class, but they found also objects of P- and C-type. In two cases, they found also evidence of blueish spectral trends. These data are important, since they allow us to substantially enlarge the whole data set of available Trojan spectra. Moreover, the observations are also confirmed the lack of absorption features in the visible (Licandro *et al.* 2002, Fornasier *et al.* 2004), and in the near-infrared (Emery & Brown 2000). The collisional evolution is important in the Trojan swarms. Most recently Melita *et al.* (2005) investigated the connections between the physical and dynamical characteristics of the Jupiter Trojans. They found that the moderately-red and stable objects are the close descendants of the primordial Trojan precursors. The color of the unstable Trojans are uniformly distributed over a wide range the spectral slope (S') from 0 to about 15% per 10^3 Å.

Beneath their refractory mantles, however, the Trojans may be ice rich (Jewitt 2004), so it is difficult to show outgassing activity. Unfortunately, they are too cold to measurably sublimate even if ice is present in the near surface regions. Very recently Emery & Brown (2004) modeled the observed spectra of 17 Jupiter Trojans and they estimated that the surfaces of these asteroids contain at most a few wr% of water ice and no more than 10–30 wr% of hydrated silicates. These findings are in agreement with those of Cruikshank *et al.* (2001) who estimated 3 wr% water ice and ∼40 wr% hydrated silicate content of the surface of 624 Hektor. This indicates that hydrous minerals might be present in the surface material and remain undetected with the quality of near-infrared spectroscopic data what now is available. So, there is some hope to find trace of water in the Jupiter Trojans. In addition, the Trojans have very low albedos, consistent with the C-, P-, and D-type asteroids and with many of the small outer satellites of Jupiter (Jewitt *et al.* 2004). The spectral features at 0.7 and 3 micron in the spectra of many low-albedo C, G, and F class main belt asteroids are associated with hydrous minerals. Hydrous minerals might associate with the heating of interior ice in an earlier epoch, or they might indicate the incorporation of materials that were serpentized by other processes. The highly volatile molecules N_2, CO, and CH_4 were not efficiently trapped and incorporated into the planetesimals formed at the heliocentric distance of Jupiter, but NH_3, refractory organics, and compounds of other heavy elements were trapped (see Barucci *et al.* 2002 and references therein).

Regarding the connection between Jupiter Trojan asteroids and comets the first conceptions are based on celestial mechanics. The similarities observed between the physical properties of the nuclei and those of the Trojan asteroids partially explained by any of the capture hypotheses. Temporary captures of comets at Lagrangian points are known to have occurred in the recent past, moreover, the escapes from the L4 and L5 clouds can supply the JFCs (Rabe 1972). Recently, the formation of the trojan swarms are explained in detail by Marzari & Scholl (1998a), Marzari Scholl (1998b), Fleming & Hamilton (2000), Marzari *et al.* (2003), and Morbidelli *et al.* (2005). The existence of objects can be captured to the 1:1 MMR to Jupiter was proven recently by Karlsson (2004) who showed by performing extensive numerical intergrations that there are a few objects which temporary reside in Jupiter Trojan orbit. In addition, the numerical models by Morbidelli *et al.* (2005) pointed out that there are effective capturing processes to lock outer solar system objects like TNOs, Centaurs in Jupiter Trojan orbits, i.e., the chaotic capture mechanism was effective in the early solar system. The connection between the Jupiter Trojans and comets was revisited by Marzari *et al.* (1997), i.e., the Jupiter Trojans are considered as a source region of Jupiter-family comets. They concluded that the Jupiter Trojans may contribute to the comet population through dynamical instabilities and collisional ejection. Once removed from the vicinity of the Lagrangian L_4 and L_5 points, they quickly lose dynamical traces of their origin. Quantitatively, there are too few Jovian Trojans to supply more than ~10% of the flux of Jupiter-family comets (Marzari *et al.* 1997, Jewitt *et al.* 2000).

5. Origin of the ecliptic comets based on comparison of their size distribution and colors with those of other primitive small bodies

Our recent knowledge is that the TNOs are dynamically related to the Centaurs and the Jupiter-family group of ecliptic comets, even the progenitors of the Centaurs are the TNOs, as well as the parent bodies of the ecliptic comets can be found among either the Centaurs or SDOs. The idea that TNOs may be the source of both, Centaurs and ecliptic comets, is widely supported by dynamical considerations even recently by physical observations (see the results and reviews by Stern 1995, Farinella & Davis 1996, Duncan & Levison 1997, Luu & Jewitt 2002, Schulz 2002 and references therein). Simply due to the observational constraints the available data are statistically limited in the studies of the connections and evolutionary links between different classes of primitive small bodies therefore the comparison of the size distributions and the surface colors are hitherto the best methods.

5.1. *Size distribution of cometary nuclei*

First we discuss the size distribution of the cometary nuclei, especially the size distribution of the ecliptic comets, which is based on the data sets selected and analized by Lamy

Table 1. Power exponents of the Cumulative Size Distribution (CSD) of the ecliptic comets and other small body pouplations (Lamy *et al.* 2004)

Population	CSD	Reference
ECs	2.65 ± 0.25	J. Fernández *et al.* (1999)
ECs	1.6 ± 0.1	Lowry *et al.* (2003)
ECs	2.5^*	Meech *et al.* (2004)
ECs	1.59 ± 0.03	Lowry & Weissman (2003)
ECs	1.79 ± 0.05	Weissman (private comm., 2003)
ECs	1.9 ± 0.3	Lamy *et al.* (2004)
KBOs ($r > 20$ km)	3.45	Gladman *et al.* (2001)
	3.20 ± 0.10	Larsen *et al.* (2001)
	3.15 ± 0.10	Trujillo *et al.* (2001)
Centaurs	2.70 ± 0.35	Larsen *et al.* (2001)
	3.0	Sheppard *et al.* (2000)
ECs + "cometary" NEOs	1.6 ± 0.2	Lamy *et al.* (2004)
Near-Earth objects	1.75 ± 0.10	Bottke *et al.* (2002b)
	1.96	Stuart (2001)
Main belt asteroids	1.25–2.80	Jedicke & Metcalfe (1998)
Jupiter Trojans ($2.2 \leqslant r \leqslant 20$ km)	2.0 ± 0.3	Jewitt *et al.* (2000)
Jupiter Trojans ($r \geqslant 42$ km)	4.5 ± 0.9	Jewitt *et al.*(2000)

*From Monte Carlo simulations after truncation at small radii.

et al. (2004). These represent the largest data sets ever assembled. Figures 1 (left and right panels) present the distribution functions of the volume equivalent effective radius of ECs and NICs. The histograms show several structures, which most likely result from the limited statistics in the data set. The apparent roll-off in the number of small cometary nuclei is very likely an observational selection effect (i.e. incompleteness of the data sample, smaller nuclei are simply harder to detect). A similar effect is often encountered with flux-limited surveys, but additional mechanisms cannot be excluded, for example the hypothesized large population of very small EC nuclei (Brandt *et al.* (1996)) and the observed sungrazing fragments of split comets (Biesecker *et al.* 2002).

Cumulative size distribution function (CSD) is used, which is robust and less prone to artifacts: $N_S(>r_n)$, where N_S is the number of nuclei larger than radius r_n is represented by power law:

$$N_S(>r_n) \propto r_n^{-q_S} \tag{5.1}$$

where r_n is the radius of the nucleus, and q_S is the power-exponent of the size distribution function. Figure 1 (right panel) displays the CSD of the 13 nearly-isotropic cometary nuclei whose effective radii have been determined. Also plotted is the CSD of this population augmented by the 12 asteroidal objects thought to be dormant or extinct NICs on the basis of their Tisserand parameters. Owing to the poor statistics, Lamy et al. (2004) did not attempt to fit power laws to the observed CSDs. Results of q_S for various minor body populations in the solar system are summarized in Table 1.

If the ECs are collisional fragments of TNOs (Stern 1995, Farinella & Davis 1996), then the theoretical value $q_S = 2.5$ for a collisionally relaxed population (Dohnanyi 1969) is expected. In reality, the question is probably more complex. The simple model of Dohnanyi (1969) applies to a population of self-similar bodies having the same strength per unit mass. Several groups have attempted to improve this simplified assumption, with

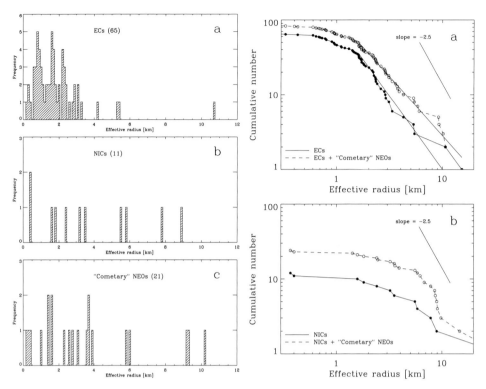

Figure 1. *Left panel:* Size distribution of the cometary nuclei. Distribution of the effective radius for ecliptic comets (ECs) (*a*), for nearly-isotropic comets (NICs) (*b*), for cometary Near-Earth objects (*c*). Note that the largest nuclei are excluded to allow legibility of the historgram at small sizes. Numbers of the objects in the samples are indicated in brackets in each panels. *Right panel:* Cumulative size distribution of the nuclei of ecliptic comets (ECs) (*a*), nearly-isotropic comets (NICs) and cometary Near-Earth objects (*b*) are represented by solid circles while the open circles apply to the population augmented by the "cometary" NEOs. The two solid lines in panel (*a*) correspond to optimum power-law fits according to the Kolmogorov-Smirnov tests, from the cut-off radius 1.6 km up to the largest bodies. Figures from Lamy *et al.* (2004).

O'Brien & Greenberg (2003) presenting the most comprehensive results on steady-state size distributions for collisional populations. In the range of sizes of interest for cometary nuclei, the size distribution of fragments is wavy, and oscillates about the distribution of a population evolved under pure gravity scaling. The differential size distribution of such a population is characterized by a power law with an exponent of -3.04. This translates into $q_S = 2.04$ using our notation for the cumulative distribution. Regarding the size distribution of the KBOs, the progenitors of the Centaurs and ECs, most recently Pan & Sari (2005) revisited the problem of the KBO size distribution. They found that the power exponent $q_S = 2$, which is in agreement with the findings of Bernstein *et al.* (2004), namely the KBOs smaller than $\sim$40 km are effectively strengthless. This does not mean that the ECs are directly related to the KBOs according to the similarity of their size distribution only since there are many size altering processes in the KBO – Centaur – EC evolutionary link (see below).

On the other hand, non–collisional fragmentation (i.e. splitting) is frequent among comets (see Boehnhardt 2004), and nuclei are progressively eroded by their repeated passages through the inner part of the solar system, so that we are certainly not observing a primordial, collisionally relaxed population of TNO fragments (Lowry & Weissman

(2003) Samarasinha (2003)). Mass loss may therefore significantly distort the size distribution of nuclei, particularly at the low end. Moreover there is a difficulty in modeling the evolution of the nuclei due to the chaotic nature of the orbital evolution of ecliptic comets.

In summary, for ECs the q_S could be as small as $\sim$1.6 and as large as $\sim$2.5, with a preferred value of $\sim$2.0. However, we will quote $q_S = 1.9 \pm 0.3$ because that is our result for the CSD that includes all the ECs for which reliable data have been obtained. Table 1 shows that this result is intermediate between those of Lowry *et al.* (2003) ($q_S = 1.6 \pm 0.1$) and Lowry & Weissman (2003) ($q_S = 1.59 \pm 0.03$), recently revised to $q_S = 1.79 \pm 0.05$; Weissman, private communication (2003), on the one hand, and Fernández *et al.* (1999) ($q_S = 2.65 \pm 0.25$), on the other hand.

In addition, the population of asteroidal objects thought to be dormant or extinct comets are considered, on the basis of their Tisserand parameters, or their association with meteor streams. The cometary origin of these NEOs is still highly speculative, and many of them may be asteroids coming from the outer regions of the asteroidal belt, including the Hilda group and Jupiter Trojans (Fernández *et al.* 2002). Selection effects are also different from those of the ECs, and any future unbiasing should reflect these differences. For the purpose of the present exercise, we considered 21 "cometary" NEOs that can be associated with ECs, and whose sizes have been determined (Table 1), thus bringing the data base to 86 objects. The "cometary" NEOs tend to be larger on average than the ECs, thus significantly filling the 2–10 km radius range, but flattening the CSD simply because they are more of them at larger sizes; indeed, we found $q_S = 1.6 \pm 0.2$, P_{KS} reaching 0.85 when including these NEOs.

The power exponent of the CSD of KBOs is quite large, $q_S = 3.15$–3.45, but strictly applies to objects with $r_n > 20$ km. It is not clear whether this value extends down to smaller sizes to allow a meaningful comparison with ECs. In fact, it has been suggested that KBOs follow a broken power law with the larger objects ($r_n \gtrsim 50$ km) retaining their primordial size distribution with the above value of q_S, while the smaller objects represent collisional fragments having a shallower distribution (e.g., Davis & Farinella 1996), which could then be rather similar to that of the ECs. The power exponent of the CSD of Centaurs, $q_S = 2.7$–3.0, is also larger than that of the ECs. However, the statistics are rather poor, and we found that, from the data of 9 Centaurs reported by Barucci *et al.* (2004), it is very difficult to fit a power law to the observed CSD: the exponent can take any value, from 3.1 down to 1.2, depending on the imposed cut–off at small sizes.

The CSD of ecliptic comets is beginning to look remarkably similar to that of Near-Earth objects: note the result of Stuart (2001), $q_S = 1.96$ which is essentially identical to our value. For the main-belt asteroids, size distributions are so well-defined that changes in the power exponent can be recognized in different size regimes (see the details in (Jedicke & Metcalfe 1998), and we have simply indicated the ranges. NEOs and main-belt asteroids are thought to be collisionally dominated populations, yet they have power exponents significantly different from the canonical value of $q_S = 2.5$ obtained by Dohnanyi (1969).

For the Jupiter Trojans, the size distribution exhibits a bimodal structure, i.e., the power exponent can be recognized in two different size regimes (see the details in Jewitt *et al.* 2000), and the separating size is $\sim$40 km (Table 1). The small Trojan asteroids are collisionally produced fragments of large bodies (Marzari *et al.* 1997, Jewitt *et al.* 2000).

A final comparison is that with the CSD of the fragments of comet D/1999 S4 (LINEAR): from water production rates measured following its breakup, Mäkinen *et al.* (2001)

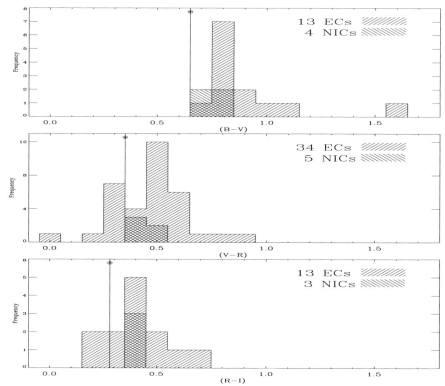

Figure 2. Distribution of the color indices for the cometary nuclei observed with the Hubble Space Telescope: $(B-V)$ (top), $(V-R)$ (middle), and $(R-I)$ (bottom). Numbers of the objects in the samples are indicated in each panels. Figure from Lamy & Toth (2005).

found that the measurements could best be explained by a fragment size distribution having $q_S = 1.74$, which is within the range we estimate for the ECs.

The question of the size distribution of ECs at the lower end, $r_c < 1.6$ km, remains totally open. The possible influence of both observational and evolutionary biases has been mentioned already, but a real depletion cannot be excluded. Indeed, the depletion of small nuclei is supported by the measurements of crater distributions on several airless bodies of the solar system, where cratering from comets is believed to dominate, e.g., Europa (Chapman *et al.* 1997) and Ganymede and Callisto (Zhanle *et al.* 2001).

5.2. *Colors of the cometary nuclei and other primitive small bodies*

Lamy & Toth (2005) extensively explored the statistical properties of the broad-band colors of cometary nuclei and reported color data for 23 active cometary nuclei detected with the Hubble Space Telescope using their well-proven method of nucleus-coma separation. In addition, they supplied the comet nucleus color data with data observed by others, as well as built a color data base of currently available colors of Centaurs, TNOs and other primitive minor bodies.

Figure 2 displays the histograms of the three color indices $(B-V), (V-R)$ and $(R-I)$ constructed with a bin size of 0.1. The gaps must not be overinterpreted as suggesting different groups but most likely result from the (still) limited coverage, particularly for the $(B-V)$ and $(R-I)$ colors. The meager color data we presently have on the nuclei of NICs indicate less diversity of colors and the absence of blue nuclei. It will be interesting to see whether those trends are confirmed when the sample of NICs will expand.

Lamy & Toth (2005) presented implications of colors for searching for the progenitor groups among the primitive minor bodies to determine the "parent-child" relationships. In order to refine the progenitor searching all the Centaurs were considered as single group, moreover, the Centaurs were split into "Centaurs I" (red) and "Centaurs II" (blue) subgroups according to their color bimodality discovered by Peixinho *et al.* (2004). Additionally, the classical KBOs (CKBOs) are split into the low- and high-inclination subgroups (CKBO-LI and CKBO-HI). Use of Kolmogorov-Smirnov (K-S) probability for the cumulative distribution functions of one or two color index samples the parent-child connections are summarized as follows. To take into the blueing in the evolution of primitive small bodies the color shift was applied in the K-S tests to catch the perfect fit of the compared color distributions. For the global population of Centaurs, the K-S test favors the Plutinos as progenitors and rules out the CKBO-LI subgroup. The connection disappears when considering the two populations of Centaurs (Centaur I and II). In order of importance (K-S probability), the connections are Centarur – EC, SDO – EC, and Plutino – EC but the probabilities remain low. The first connection becomes less obvious when Centaurs are split. Significant color shifts are required to boost the probabilities to creditable levels. The probabilities of the EC – dead comets connection are low and again can only be increased by significant shifts.

The present global picture of colors of primitive bodies of the solar system could be explained along the following lines. The coherent and progressive color shifts are consistent with the following non-unique scenarios. CKBO-LI formed in-situ beyond the orbit of Neptune while CKBO-HI, Plutino and SDOs formed inside this orbit, a picture remarkably consistent with that proposed by Gomes (2003). SDOs could have formed at slightly shorter distances than CKBO-HI and Plutinos, possibly explaining their more efficient scattering. Centaurs remains a puzzle. As a single group, the colors suggest they come from Plutinos. As a dual group, Centaurs I could be composed of TNOs (CKBOs, Plutinos, SDOs) and ultra-red bodies from the inner Oort cloud while Centaurs II have already visited the inner solar system and lost their red organics; their origin is unclear on the basis of colors alone. Ecliptic comets lost part of their red organics, most likely thermal alteration, the end-state being reached by the candidate dead comets. This process involves different red compounds having different thermal instabilities and different optical properties. The large range of colors that KBOs, Centaurs and ECs share results from a common large diversity in their composition. It implies limited mixing in the outer part of the protoplanetary disk leading to accretion of unhomogeneous planetesimals.

As a final remark, we stress that the number of objects for which we have $BVRI$ colors remain small, not to say extremely small so that the presently available data sets may simply be not representative of the reality. Recent controversies often resulting from biased data sets give the lesson that we must remain prudent in our interpretations.

6. Connections between comets and main belt asteroids

There are comets, which orbit in the main asteroid belt. The most abundant such a comet population resides in the Hilda asteroid zone in the outer main belt. The most prominent example for this comet population is D/Shoemaker-Levy 9 prior to its capture by Jupiter (Chodas & Yeomans 1995, Tancredi & Sosa 1996). On its possible pre-capture orbit this comet dwelt in the Hilda asteroid zone before, i.e., its probable pre-capture orbits overlap a group of known comets, referred to as quasi-Hildas by Kresák (1979). In addition, Kresák (1979) presented an inventory of other cometary groups which reside in the main asteroid belt. Most recently, Toth (2005a) updated the inventory of the

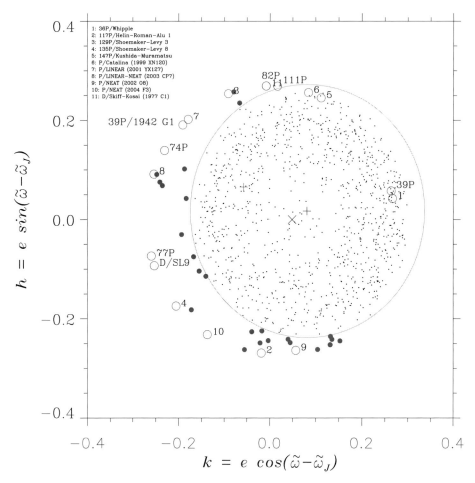

Figure 3. Location of the Hilda asteroid family members and the quasi-Hilda comets in the Lagrangian elements $(e_x, e_y) = (k, h)$ plane is displayed. The Hilda asteroids are shown with black dots and the quasi-Hilda comets are represented by red open circles. The newly identified Hilda asteroid outliers are highlighted with red dots. The mean value of the k and h coordinates of the samples of Hilda asteroids and quasi-Hilda comets are shown with a plus sign $(+)$. Location of Jupiter is shown with a cross $(\times)$. Figure from Toth (2005a).

quasi-Hilda group of possibly comets searching for outliers in the Lagrangian elements space (Fig. 3). New members (11) of the ecliptic comets and Hilda outliers (24) were found. However, there is no observation to explore the physical characteristics of these objects but their orbits belong to the quasi-Hilda comet group. Among these objects can be dormant or extinct cometary nuclei. Physical observations of the Hilda asteroids and quasi-Hilda comets (candidate quasi-Hilda comets) are needed.

There are asteroids which display temporary comet-like activity. These are the asteroid-comet transition objects which can be found among the small body populations in the solar system (Weissman *et al.* 2004, Jewitt 2004, Jewitt 2005). There are two distinguished examples for this behavior: 4015 Wilson-Harrington and 7968 Elst-Pizarro with showing temporary dust tail and trail (see the history of 107P described by Fernández *et al.* 1997, and 133P described by Toth 2000, and Hsieh *et al.* 2004, as well as general reviews by Weissman *et al.* 2002, Jewitt 2004). These two objects are also catalogized as comets: 107P/Wilson-Harrington (Fernández *et al.* 1997, Coradini *et al.* 1997, Jewitt 1996, Jewitt 2004), and 133P/Elst-Pizarro (hereafter E-P in short) but they have typically asteroid orbit with value of Tisserand parameter with respect to Jupiter 3.084 for 107P and 3.184 for E-P.

Elst-Pizarro shows recurrent comet-like activity but it resides in the Themis-zone of the main asteroid belt ($a = 3.158$ AU, $e = 0.165$, $i = 1.38°$) (Boehnhardt *et al.* 1998, Lien 1998, Toth 2000, Weissman *et al.* 2002, Hsieh *et al.* 2004). The hypothesis of recent impact event was among the first explanations of the dust trail (or tail) of E-P, which excavated or activated the surface material and generated the dust mass-loss (Boehnhardt *et al.* 1998, Lien 1998, and Toth 2000). We note that the scenario of the impact event is applied to explain the splitting and outburst events for other comets (Beech 2001, Toth 2001, Gronkowski 2004) but these are only optional and less probable scenarios in the present-day solar system. The intrinsic properties of the asteroid-comet transition objects are the most plausible causes of their unusual behavior. The discovery of the E-P's recurrent activity by Hsieh *et al.* (2004) in 2002 ruled out the one-time impact event hypothesis as a direct cause of the long-lasting recurrent comet-like activity. But in 2003 they have not found any comet-like activity despite that they used the Keck I 10-m telescope. However, very recently Lowry & Fitzsimmons (2005) published their 2002 observations of E-P, which were made on 13 July 2002, and they reported very small dust activity with a small dust trail (or tail). They proposed that the evolving of the dust trail started just prior to their observations in July 2002. *BVRI* observations of E-P obtained by Delahodde *et al.* (2004) in 2000 are consistent with a dormant (inactive) phase of this object, as no coma or tail (trail) were detected.

Hsieh *et al.* (2004) suggested various conceptions to explain the observed recurrent comet-like behavior of E-P. The most preferable idea is connected to the comet hypothesis, i.e., the 1996 and 2002 emission events are consequences of seasonal insolation variation, the area of exposed surface volatiles must be located near one of the rotational poles of the nucleus. This volatile region would then only receive enough solar radiation to become active when the near pole tilts toward the Sun, that is that hemisphere's "summer". This like object can be denominated as "activated asteroid" which remains consistent with the possibility of seasonal modulation of E-P's activity (R.P. Binzel, cf. Hsieh *et al.* 2004). The question is raised whether where E-P exhibited comet-like activity and where it was inactive along its orbit, i.e., how long the periods of activity and dormancy were. For this purpose Toth (2005b) continued the searching for comet-like activity by CCD imaging observations about E-P, which were conducted in January and February of 2005. In addition, mapping of the location of the active and inactive phases of E-P along its orbit (Fig. 4), and taking into account the lightcurve observations which are up to now available in the literature the limitations of the possible rotational pole orientation of this object can be estimated.

In order to find more main belt asteroids which show comet-like activity Hsieh (2004) conducted a deep optical survey of selected main belt asteroids in search of E-P-like objects. Hsieh (2004) presented interim results of this on-going survey. 63 Themis-zone asteroids were selected for this purpose, among others 1999 TU71, but none of them exhibited comet-like activity. Hsieh (2005) (this conference) reviewed the evidence for volatile sublimation as the cause of E-Ps comet-like activity, presented an overview of our ongoing Search for Active Main Belt Asteroids (SAMBA) project that aims to find other E-P-like objects, discussed their progress to date, and considered the implications of this work.

There is another possible explanation, which has not been tested yet for the active asteroids. This mechanism is the removing of the regolithic material from asteroid surface:

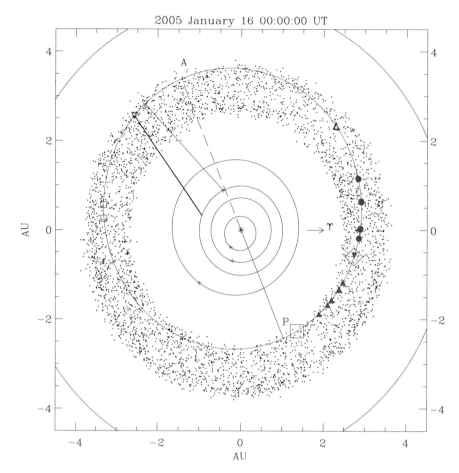

Figure 4. Comparison of the active and dormant phases of the asteroid-comet transition object 7968 Elst-Pizarro (133P/Elst-Pizarro) in its orbit has shown. The orbits and the instantaneous positions of the objects are projected onto the plane of ecliptic for the date of UT 0 h 16 January 2005. The orbit of 7968 Elst-Pizarro is drawn (solid blue line). Direction of its perihelion point (P) is marked by a solid line connecting the Sun and the point in the orbit. Direction of aphelion (A) is marked by a dashed line. Solid line connects Earth and Elst-Pizarro on the date of the observations taken on UT 16 January, 28 February, and 1 March 2005, respectively. Background asteroids in the Themis-zone are also plotted (small black dots). The positions of the major planets from Mercury to Jupiter are marked with crosses. The orbits of major planets from Mercury to Jupiter are drawn with thin solid lines. Vernal Equinox direction is indicated. Observations with reported dust trail (trail) activity are higlighted with solid red symbols: observations taken in 1996 are solid triangles, observations by Hsieh *et al.* (2004) from July to December 2002 are solid dots. Observation taken by Lowry & Fitzsimmons (2005) in July 2002 is shown by a red triangle with downward directed peak, just near to the 1985 observation made by McNaught *et al.* (1996). Observations which did not report any comet-like activity are represented by open black symbols: big open square (1979), open diamond (1985), open triangles with upward peak (1997), small open squares (2000), open circle (2003), and open triangles with downward peak (2005 observations by Toth 2005b).

it is the effect of solar radiation pressure on small regolith particles Scheeres (2005). The radiation pressure can play important role in migration of surface dust particles, whether or not electromagnetic levitation or seismic shaking is present. We are raising now the question, which is still open, that this mechanism can be effective in case of the E-P.

However, there are other mechanisms which may work on the surface of active asteroids. Kargel (1991) proposed a specific activity of asteroids and we quote this idea only for the purpose of completeness of a possible explanation for the activity in the asteroid belt, at least to explain the exposition of icy material on the surface. It is the brine volcanism, in which $MgSO_4$ and Na_2SO_4 are shown by carbonaceous chondrite mineralogies and elemental abundances to be the most important solutes, is presently suggested to account for the resurfacing (it was observed on some icy satellites). The question is that can this process work in small asteroids or only in larger bodies.

Recent progress for modeling of the mass loss from cometary nucleus may adopt and translate to the activity of asteroids analogously. Quite recently, aside the well known and well elaborated models of gas and dust production of comets, there are other new models, which were discussed for the material releasing activity of cometary nuclei: (*i*) one of these the model of subsurface "geysers" for the collimated jets, which was proposed by Yelle *et al.* (2004); (*ii*) another one is the non-active area based interacted hydrodynamic flow model (see the summaries of the advanced new jet 3-D model by Crifo *et al.* (2002), and Rodionov *et al.* (2002)); (*iii*) surface depression and ice releasing model by Laufer *et al.* (2005).

Finally, where is the water in the asteroids, which can drive comet-like activity: outgassing, to develop gas coma and to release dust? In general, the question of the occurrence of water ice in primitive asteroids was considered by Barucci *et al.* (1996). They reviewed the literature dealing with possible signatures of ice in some asteroids (see also various chapters in books *Asteroids, Asteroids II, Asteroids III* of University of Arizona Press). The subject is poorly developed, even if in the last decade several authors started a spectroscopic survey of asteroids in the visible and near-infrared wavelengths in order to find ice features on asteroid spectra. Aqueous alteration of the mineral assemblages has been claimed for a number of B-, F- and for more than 50% of the observed C-type asteroids. The compositional and, consequently, the thermal structure of the asteroid belt supports the hypothesis that ices of volatile elements should be more abundant in the external region of the belt, where pristine materials have not undergone drastic modification processes after the accumulation in the planetesimal swarms between those of Mars and of Jupiter. Still there is no evidence of any water, water ice or aqueous alteration materials on the D-type asteroids, which are considered the least altered objects.

The Dawn mission addresses the long-standing goals of understanding a "wet" relict of the formation of the solar system: it is the 1 Ceres (see Mousis & Alibert 2005, this conference). From the long-exposure IUE spectra A'Hearn & Feldman (1992) already detected OH in the northern limb of Ceres. They concluded that amount of OH is consistent with a polar cap that might be replenished during winter by subsurface percolation, but which dissipates in summer. Under the assumption that most of the volatiles were not vaporized during the accretion phase and thermal evolution of Ceres, utilizing a time-dependent solar-nebula model, Mousis *et al.* (2005) determined tbe volatile abundances (CO_2, CO, CH_4, N_4, NH_3, Ar, Xe, Kr with respect to the water). Recently, high signal to noise spectra of Ceres were obtained by Vernazza *et al.* (2005) with the NASA IRTF. They confirmed the presence of the 3.06 micron absorption feature. Laboratory measurement of ion-irradiated organics and ices suggest that this feature can be reproduced with a linear mixture of crystalline ice and residues of ion-irradiated asphaltite.

Campins & Lauretta (2004) predicted that a significant fraction of cometary solids could be in the form of hydrated silicates. For example, in comet Hale-Bopp the observed spectral features attributed to anhydrous crystalline silicates can be produced by as little as 15% of the particles; the rest of the silicates could be a mixture of hydrated and amorphous grains. This is in part because anhydrous crystalline silicates are easier to identify spectrally than hydrated or amorphous silicates. Campins & Lauretta (2004) quoted the model of chondrule-forming shock waves in icy regions of the early solar nebula, which can produce rapid mineral hydration, thus supporting a nebular origin for the hydrated silicates in the chondrule rims. Previous arguments had suggested hydration had to occur in parent bodies and not in the nebula. Hence, comets may have also accreted hydrated silicates from the solar nebula. Observationally, cometary hydrated silicates could be detected using near-infrared reflectance spectroscopy of cometary surfaces and mid-infrared spectroscopy of cometary dust. There is an ongoing spectroscopic survey in the visible and near-infrared to observe the surface of cometary nuclei and related minor bodies by Licandro *et al.* (2002).

7. Connections between asteroids and comets in the Near-Earth space: recent progress

Asteroids and comets supply the meteoritic material and dust in the interplanetary space (see reviews by Mann *et al.* 2004, Mann 2005, and the relevant chapters of the *Asteroids III* 2002 and *Comets II* 2004 books, University of Arizona Press). In the Near-Earth region the most abundant source of the interplanetary dust and meteoroids is the Taurid-Encke complex, which contains both asteroids, comets and meteor streams (Asher & Clube 1997, see many chapters in the book *Hazards due to comets and asteroids. Univ. of Arizona Press, Tucson, 1995*). Resonances and chaotic motions are important mechanisms to deliver both the asteroids and comets into the inner-Earth regions even sungrazing orbits. So, in the Near-Earth and inner-Earth regions both the comets and asteroids are the abundant sources of the interplanetary dust and meteoritic material but the relative contribution of these sources are not quantified (Biesecker *et al.* 2002, Mann *et al.* 2004). However, based on an ongoing optical/thermal infrared photometric imaging survey of the dust mass loss from comets Lisse (2002) demonstrated that there is an evolution with time of the kind of dust emitted from a cometary nucleus surface. His results indicate that the mass loss rate from the ecliptic comets alone is enough to supply the interplanetary dust complex against losses. Moreover, based on the ISO and MSX (Midcourse Space Experiment) infrared data Lisse *et al.* (2004) concluded that if the dust emission behavior of 2P/Encke is typical of other ECs, then comets are the major suppliers of the interplanetary dust cloud. Regarding the nucleus of 2P, during its 2003 apparition high resolution spectroscopic observations made using the TNG telescope on La Palma by Saba *et al.* (2004). Spectrum of 2P is poor of spectral features: only a few emission lines were found (C_2, NH_2, H_2O^+) as like in ECs but this may indicate low gas production and it is in agreement with previous observations of 2P pointing at a dark and dusty object.

Fernández *et al.* (2002), and Pittich *et al.* (2004) calculated that Jupiter-family comets can reach the Near-Earth region. Pittich *et al.* (2004) modeled the transfer routes from Jupiter-family towards Encke-like cometary orbits. They found that resonances and non-gravitational forces appear to be key factors in the transfer of orbits.

Recent progress in the studies of physical properties of the Near-Earth Objects in connection with comets done by Fernández *et al.* (2005) who presented the results of a mid-infrared survey of 26 asteroids in comet-like orbits, including six Near-Earth asteroids (NEAs). Merging this with their earlier study of low albedo objects (Fernández *et al.* 2001) they analized 32 objects. They defined a "comet-like" orbit as one having a Tisserand invariant T_J under 3 (but only including objects that are NEAs or otherwise unusual). Visible-wavelength data were also obtained, so geometric albedos (in the Cousins R band) and effective radii are presented as derived using the NEA Thermal

Model. Nine of their objects were observed at two or more mid-infrared wavelengths, and in all cases the low-thermal inertia thermal model was found to be applicable, with various values of the beaming parameter. Their work more than quintuples the total number of observationally constrained albedos among $T_J < 3$ asteroids to 32. Defining the "comet-like" albedos as those below 0.075, they found that $64\% \pm 5\%$ of the sample has comet-like albedos. Objects in comet-like orbits with comet-like albedos are candidates for being dormant or extinct comets. They found a very strong correlation between the albedo distribution and T_J, with the percentage of dark $T_J < 3$ asteroids being much greater than that of the $T_J > 3$ NEAs. There are 10 NEAs among the 32 objects, and of those, $53\% \pm 9\%$ have comet-like albedos. With the current crop of NEAs, this implies that about 4% of all known NEAs are extinct comets. A comparison of the histogram of $T_J < 3$ asteroid albedos with that of active cometary nuclei shows that the former has a larger spread.

Hsieh & Jewitt (2005) presented deep optical imaging of Geminid meteor stream parent Apollo asteroid 3200 Phaethon taken in search of low-level cometary activity (i.e., coma or dust trail). Although no unambiguous cometary behavior was observed, they found an upper limit on the object's cometary mass-loss rate of $M_{\rm lim} \sim 0.01$ kg s^{-1}. The corresponding active fraction (the fraction of the surface area that could consist of freely sublimating water ice) is $f \leqslant 7 \times 10^{-6}$, at least 2 orders of magnitude smaller than other known comets.

8. Current status and outstanding questions in connections between asteroids and cometary nuclei

However, there are outstanding questions of the connections between asteroids and cometary nuclei, even the physical characteristics of these objects like size determination, size distribution, surface and bulk interior properties, as well as the question of binary objects among the asteroids and cometary nuclei. The current status and outstanding questions are summarized as follows.

8.1. *Sizes and shapes*

Remarkable progress has been made during the past decade in measuring the sizes of cometary nuclei and related primitive small bodies of the solar system (TNOs, Centaurs, and primitive asteroids) but it is also clear that this field is still in its infancy. Weissman *et al.* (2002), Lowry & Weissman (2003), Barucci *et al.* (2004), Jewitt (2004), and Lamy *et al.* (2004) reviewed the recent status of the sizes and shapes of small bodies in the solar system, including comets and related asteroids. The current best estimate, $q_S = 1.9 \pm 0.3$ for ECs, is conspicuously different from that of the KBO and Centaur populations, but is similar to that of the NEOs (Lamy *et al.* 2004). This value also corresponds to that of a collisionally evolved population with pure gravity scaling, but we re–emphasize that O'Brien & Greenberg (2003) showed that this distribution is, in fact, wavy in the size range relevant to ECs. In addition, Pan & Sari (2005) refined the modeling approximation for the size distribution of the KBOs, which has to be tested for larger observational data samples.

A totally open issue is the nature of the size distribution of cometary nuclei at the small end of the spectrum. Does the relatively steep power law derived from the intermediate-sized objects extend indefinitely to smaller sizes? Or is the size distribution truncated at some value that depends on the physical formation mechanism (e.g., gravitational instability within the solar nebula) or destruction mechanism (e.g., total disruption)?

What is the bias in ecliptic comet discoveries and how does that affect the current distribution of sizes? Why do we observe so few large ($r_n \gtrsim 5$ km) cometary nuclei?

What are the correct size distributions of NICs and Damocloids and how does they relate to that of Damocloids, and what can be the result of the comparison of their other physical properties?

How does evolution affect the physical properties of cometary nuclei? Is there really a continuum of surface properties that is dependent on the activity level and physical evolution (e.g., with a youthful Chiron at one end and an aged 2P/Encke and extinct comets with asteroidal appearance on the other)?

Splitting events (Boehnhardt 2004) obviously affect the size and shape of small bodies (Jewitt 2004), and how do we estimate their effect on the distribution functions? Perhaps better data on the splitting rates of comet nuclei, coupled with a better understanding of the physical mechanism(s) for splitting events, will help to resolve these issues, but that remains to be seen.

8.2. *Surface properties*

Comparing the size and albedo determinations, the situation for albedos of comet nuclei and related asteroids is even worse, in the sense that reliable values are available for only about a dozen objects. Nevertheless, we are struck by the relatively small range in the albedo (0.04 ± 0.02), which suggests that the surfaces of cometary nuclei are exceptionally dark, contrary to the early expectations for these "icy" bodies.

The phase function of atmosphereless bodies, both cometary nuclei and primitive asteroids, offers a powerful means for investigating the properties of surface (e.g., roughness and single-particle albedo). Observations of the detailed phase function of the inactive nucleus of comet 28P/Neujmin 1 by Delahodde *et al.* (2001) yielded important results (cf. reviews by Campins *et al.* 2001, Campins *et al.* 2003). They found that the average colours of the nucleus are similar with those of D-type asteroids and other comet nuclei. The phase function obtained for the nucleus has a linear slope of 0.025 ± 0.006 mag deg^{-1}, less steep than that of the mean C-type asteroids. At smaller phase angles, the function steepens and a strong opposition effect appears at $\alpha \lesssim 1.5°$. However, Lamy *et al.* (2004) noted that this effect, comparable to those found on medium geometric albedo ($\sim$0.15) M-type asteroids, and icy satellites, is quite surprising for a cometary nuclei. As surface ice is excluded on such a low activity nucleus, a high surface porosity could perhaps be invoked, but this possible interpretation has not been investigated. The first results of Licandro *et al.* (2002) showed that the spectrum of the bare nucleus of comet 28P/Neujmin 1, which is the less active in the Jupiter-family, exhibited rotational variation: the spectral change appears in the infrared spectral slope and it is consistent with sporadic changes in the V, R, and I bands reported in this comet by Delahodde *et al.* (2001). Moreover, there is another very recent example which supports the connections between dark C-type asteroids and cometary nuclei. Combining the disk integrated magnitudes calculated from the Deep Space 1 images with the HST and ground-based measurements Soderblom *et al.* (2002) and Buratti *et al.* (2004) determined the phase function of the nucleus of 19P/Borrelly over a large range of phase angle, from 3° to 88°. The phase curve is very similar to that of the dark C-type asteroid 253 Mathilde.

It is also difficult to obtain reliable color data on cometary nuclei. While the color of the nucleus itself does not provide unique information on the physical properties, color data are useful for comet-to-comet comparisons, which may suggest differences in surface properties, and, especially, in making comparisons with other minor bodies in the solar system (e.g., Centaurs, TNOs, and asteroids). The colors of cometary nuclei are diverse, with some being highly reddened compared to solar color, some being neutral, and a few having a slightly blue color.

An important, unresolved issue concerns the interpretation of disk-integrated thermal measurements, which, in principle, provide robust determinations of sizes and albedos. The so-called *Standard Thermal Model* for asteroids is often used to interpret cometary thermal data, although its applicability to objects having a mixture of dust and ice is questionable (see examples of the model in Groussin *et al.* 2004, and review of the thermal infrared models and their application and limitations Lamy *et al.* 2004).

Does the P- and D-type asteroids have organic or silicate surface, or in other words: do we face with a new paradigm? A recent extensive spectroscopic survey of asteroids by Emery *et al.* (2005) in order to study the mineralogy of asteroids from observations with the Spitzer Space Telescope led to a very important implication for the D-type asteroids including Jupiter Trojans among which there are many D-class objects. They suggested a silicate nature for the surface of Trojan asteroids based on the absence of organic absorption in 3- to 4 micron region and on the ability to model the vis-NIR spectra with silicates alone. These new results from Spitzer/IRS support a silicate interpretation. This challenges the old paradigm. i.e., the view dominant since the early 1980's that the low albedo and red spectral slopes of P- and D-type asteroids are due to organics on the surfaces. The middle part of the early solar nebula may not have been as rich in organic material as once thought. Emery *et al.* (2005) also found similar spectrum for a few Centaurs and one of them shows a very similar spectrum to 624 hektor. This can be a new paradigm for P- or D-type asteroids, including the Jupiter Trojans and one part of the Centaur population. However, either the observed or composite phase function of an organic material covered surface and the only silicate material covered surface should be different. The observed phase function of the dark primitive asteroids do not support the only silicate surface material (cf. Belskaya *et al.* 2003)). In addition, the analyses of the Tagish Lake meteorite, which is considered that originated from a P- or D-type asteroid, showed that it is similar to the hydrous minerals (Hiroi *et al.* 2003), and the surface regolith of these asteroids may be made of intermediate materials of the CI/CM chondrites (Hiroi *et al.* 2004).

Regarding the activity of the asteroid-comet transition objects, the recurrent comet-like activity of 133P/Elst-Pizarro still waits for the convincing explanation (see the details and model scenarions in Section 8). Perhaps, the asteroid flyby target of Comet Odyssey mission, the Themis-zone object Elst-Pizarro will be selected.

8.3. *Bulk interior properties*

Among the key scientific objectives of the Deep Impact project there are questions which are connected to the comet-asteroid interrelations. Namely, (*i*) comets have the most primitive, accessible material in the solar system, but we do not know what is hidden below the evolved subsurface layers (material composition, interior structure, bulk density and porosity, etc.), (*ii*) there must be many dormant comets masquerading as asteroids but the questions is that we do not know how to identify these bodies, and (*iii*) comets must become dormant but we do not know whether ice is exhausted of sublimation is inhibited.

Although cometary nuclei generally contain much more ice than asteroids do, perhaps the "evolved" comets share many common characteristics with asteroids. In addition, at least some asteroids, and many cometary nuclei, are thought to have porous, "rubble-pile" physical structures (Davis *et al.* 1985, Weissman 1986). Does this point to commonalities in their formation mechanism? There are various model conceptions on the internal structure of comets (Weissman *et al.* 2004) but we have not verified these yet.

Weidenschilling (2004) reviewed the process of the formation of cometesimals as collisionally processed aggregates, i.e., rubble piles (see also Weissman *et al.* 2004). Although his 1-dimensional models suggest preferential formation of bodies having a characteristic size of ~100 m, his more recent 2-dimensional models have a size distribution that is more closely approximated by a power law with only a small depression centered near 1 m in size. The observed fragments of 73P/Schwassmann-Wachmann 3 (Toth *et al.* 2005) and the numerous small, unobserved fragments whose existence is suggested by the results presented on this split comet, may be consistent with Weidenschilling's model, but the observations cannot be used to confirm that model in detail. However, the observed fragments of C/1999 S4 (LINEAR) (Weaver *et al.* 2001), which is a NIC, may confirm this model.

The Stardust images of 81P/Wild 2 raised the question that the cometary nuclei can be diverse in their internal structure and not only in their surface properties due to evolutionary processes (aging, erosion, volatile depletion, etc.), even the nucleus of 81P/Wild 2 is not a rubble pile (Weaver 2004). According to the Stardust images It was suggested that the nucleus of 81P held together by internal strength rather than by gravity alone. But Asphaug *et al.* (2002) concluded that instead of competing for dominance, strength and gravity collaborate over a wide spectrum of sizes and shapes to produce some of the richest structure in nature. Both 19P/Borrelly (Deep Space 1), 81P/Wild 2 (Stardust images), and 9P/Tempel 1 (Deep Impact) are ecliptic comets but there have different surface and possibly interior structures. 1P/Halley (Vega 1 and 2, Giotto) images of this NIC also show differences from the nuclei of these ecliptic comets. Are there different groups of comets in internal and surface properties, i.e., are there different groups among the ECs, and are the EC nuclei different from those of NICs including HTCs or Damoclods?

Study of the stability against rotational disintegration of the minor bodies allows to constrain their bulk interior properties and possibly their evolution and interrelations between them. This old approach is renewed Toth & Lisse (2005) applying the new models developed by Davidsson 1999. Toth & Lisse (2005) estimated the regions of stability, fragmentation, and destruction for cometary bodies versus rotational breakup in the radius – rotational period plane. They have applied the model of Davidsson (1999) for rotational stability of spherical bodies to study the location of observed comets and cometary asteroids in the radius – rotational period plane. In doing so, they considered the range of plausible cometary nucleus material strengths and bulk densities as found in the literature (Fig. 5). We emphasize here that the spherical, solid, homogeneous, rigid, and incompressible body, which is the starting-point of most physical models, can too easily be oversimplified (Davidsson 1999). While it would be interesting to directly model a cometary nucleus with more complex models, there are no direct measurements of the internal material properties of cometary nuclei and related objects.

Toth & Lisse (2005) found that the boundaries of rotational stability move significantly in the (R, P_{rot}) plane as the material properties change. However, the location on the the rotational period – body radius plane, with respect to the rotational breakup limit curves, still allows us to estimate the possible numerical values for the internal strength and bulk density using a statistical approach. E.g., the existence of 95P/Chiron as a coherent body can be accommodated only if its material strength is characteristic of relatively strongly structured material – e.g., stronger than the weak nuclei advocated by Greenberg *et al.* (1995). Moreover, the bulk density constraint for a shear fractured body yields a somewhat larger mean bulk density of Chiron than the bulk density of an aggregate structured cometary nucleus. The large majority of observed comets lie in the allowed region for all possible nuclear constituent properties (Fig. 6). They suggest a

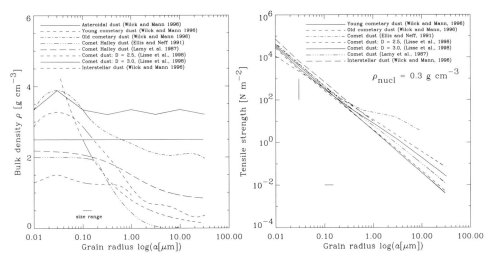

Figure 5. *Left panel:* Size dependence of the bulk density of spherical dust grains as given in the literature (Toth & Lisse 2005 and references therein). The bulk density law is computed for two fractal dimensions: $D = 2.5$ (porous dust grains) and $D = 3$ (solid grains). The size range of ISM dust particles (0.10–0.15 μm) which are presumed to be the basic building blocks of the cometary nucleus is marked by a horizontal bar. The bulk density range of the dust grains is indicated by thick solid parts in the curves. *Right panel:* Tensile strength for aggregate dust grain types as given by Wilck & Mann (1996) and as calculated for the dust grain bulk density functions given in the litarature. The bulk density law is computed for two fractal dimensions: $D = 2.5$ (porous dust grains) and $D = 3$ (solid grains). The size range (0.10–0.15 μm) of dust particles which are presumed to be the basic components of the cometary nucleus is marked by the horizontal bar. The typical tensile strength range is indicated by the vertical bar. The assumed bulk density of the nucleus is $\rho_{nucl} = 0.3$ g cm^{-3}. Figures from Toth & Lisse (2005).

correlation between the proximity to the rotational breakup instability and the surface inhomogenities or activity for the primitive minor bodies.

8.4. *Binary asteroids and satellites of cometary nuclei*

While the occurrence of satellites for both main belt asteroids, Near-Earth asteroids, Transneptunian objects, and Trojan asteroids is steadily growing, there is still no definite, observational evidence that binary cometary nuclei exist. Asteroids do have satellites, and the properties of the known asteroidal binary systems were reviewed by Merline *et al.* (2002). Very recently the problem of the existence of satellites of cometary nuclei was discussed by Lamy *et al.* (2004). The detection of a satellite companion to a cometary nucleus would be of unique value as it would provide access to the mass of primary. If the mass of the nucleus is known, and if the size is independently derived, then the mean bulk density and porosity can be calculated, providing insight into the internal properties of the nucleus. However, a binary system must have long-term orbital stability and the components must be large enough ("bodies"), i.e., we do not consider for example the sub-meteoroid sized debris as a component. But may the binary cometary nuclei exist – or if they do not exist, why? Intuitively, one possible explanation is the rapid divergence of the fragments of split comets due to the strong nongravitational forces (cf. the calculations for the fragments of C/1996 B2 (Hyakutake) by Desvoivres & Levasseur-Regourd 2002 and references therein).

We draw attention that still there is no any binary Centaur observed. The number of known binary TNOs is growing but still there is no known binary among the Centaurs. If

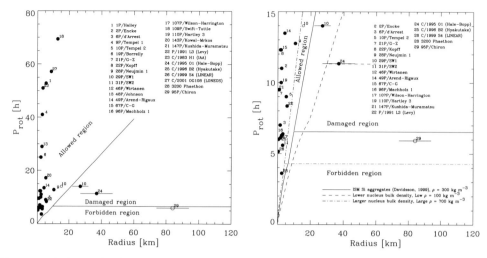

Figure 6. *Left panel:* The critical rotational breakup limits for spherical small bodies assuming a bulk density of $\rho = 300$ kg m^{-3} and material strength according to Greenberg *et al.* (1995). The calculated limits were derived using the model of Davidsson (1999). Measured cometary rotational periods and radii for all the cometary objects in the well-observed dataset are represented by filled circles. The location of the nucleus of disrupted comet C/1999 S4 (LINEAR), before breakup, is also shown (Object No. 26). The two effective radii of 29P/Schwassmann–Wachmann 1 (object No. 4) are reported assuming that the rotational period is the same. The active Centaur 95P/Chiron is shown by a circle (object No. 29). *Right panel:* Same as for the *left panel*, except the vertical axis has been expanded to show the detail of the higher rotation rate objects. Border lines for lower and larger bulk densities of the nucleus are also plotted. Figures from Toth & Lisse (2005).

the Centaurs come from TNOs, among which the duplicity is common, therefore binary Centaurs should exist. The Centaurs are the possible progenitors of ECs therefore the mass and bulk density of binary Centaurs are important and these can be be determined in binary systems. Toth (1999) has already studied the orbital stability and detectability of possible Centaur binaries.

A possible byproduct of the successful Deep Impact experiment could be the discovery of a possible satellite formation from the reaccreted ejected cometary material but the probability of this process is extremely unlikely comparing to the satellite formation from reaccreted material which is an effective process in asteroids as it was described by Michel *et al.* (2001). However, the problem of the satellites to cometary nuclei should be explored via modeling and observations in the future.

9. Concluding remarks

When David Jewitt (Jewitt 1992) discussed the physical characteristics of cometary nuclei and related asteroids in the context of current models of their origin and evolution, and Jane Luu presented the first modern review of the connection between cometary nuclei and related asteroids (Luu 1994) on the basis of physical properties of these objects, more than 10 years ago, their data samples were limited to a few comets, the 2 Centaurs (Chiron and Pholus), a few Jupiter Trojans and other asteroids for which physical measurements were available. Review by Luu (1994) incorporated comparison between the orbital characteristics (groups of minor bodies) and surface properties. At the 1996 Asteroids, Comets, and Meteor (ACM 1996) conference, Karen Meech's review discussed

17 ECs, 7 Halley-type comets (HTC sub-group of the NICs), and 2 Centaurs (Chiron and 29P) in relation with their physical characteristics, even discussing their bulk interior physical properties. At the 2002 Asteroids, Comets, and Meteor (ACM 2002) conference, David Jewitt summarized the results of the origin and evolution of comets including the ecliptic comets and related primitive asteroids according to the data from growing observational samples available as of 2002. For the Asteroids III book (2002), the discussion of the interrelations between comets and related asteroids (TNOs, Centaurs, Jupiter Trojans, and other primitive asteroids) based on large samples of the observations, detailed physico-chemical models, and celestial mechanical studies (see reviews by Barucci *et al.* 2002, Marzari *et al.* 2002, Weissman *et al.* 2002). For the Comets II book (2004), the study of cometary nuclei is divided among 10 chapters, with questions such as the physical properties, rotation, surface properties, internal properties, and the relationship with other minor bodies deserving their own, separate chapters (see reviews by Barucci *et al.* 2004, Jewitt 2004, Lamy *et al.* 2004, Weissman *et al.* 2004) We are far from fully understanding the cometary nuclei and asteroid-comet transition objects (e.g., 133P/Elst-Pizarro), as illustrated above by the list of outstanding issues. On the other hand, the future is bright as new facilities will soon allow us to push the present limits of our observational capabilities even further. Both the large ground-based telescopes (e.g., Keck, Gemini, Very Large Telescope, Pan-STARRS. Large Synoptic Survey Telescope (LSST), ALMA) and infrared space telescopes (Spitzer, Herschel) will provide detection of faint small bodies, measuring the sizes, shapes, albedos, colors, rotational parameters, thermal characteristics. Meanwhile forthcoming cometary space missions (e.g., Stardust sample return in 2006, Comet Odyssey, Rosetta), and some astcroid related missions (Dawn and Gulliver) will provide new results in the asteroid-comet connections. The Rosetta lander will perform a broad range of in-situ observations and analysis of the surface and subsurface regions of the nucleus. These will allow us to study a few cometary nuclei and related asteroids in unprecedented detail.

Acknowledgements

Author acknowledges the beneficial cooperation with Philippe Lamy, as well as Michael A'Hearn, Laurent Jorda, Horst Uwe Keller, Zdenek Sekanina, Nicholas Thomas, Harold Weaver for the collaborations using the Hubble Space Telescope for many years. I. T. thanks Paul Abell, Yanga Fernández, Henry Hsieh, David Jewitt, and Paul Weissman for the discussions on the asteroid-comet transition objects, Michael Belton, Björn Davidsson, and Casey Lisse for the discussions on the bulk interior properties of cometary nuclei, as well as This work was partly supported by the French "Programme National de Planétologie", jointly funded by CNRS and CNES, and of the bilateral French–Hungarian cooperation program. I. T. acknowledges the support of the Université de Provence, of the Hungarian Academy of Sciences through grant No. 9871.

References

Abell, P.A., Fernández, Y.R., Pravec, P., French, L.M., Farnham, T.L., Gaffey, M.J., Hardersen, P.S., Kusnirak, P., Sarounova, L., & Sheppard, S.S. 2003, *LPSC* 28, Abstr. No. 1253
A'Hearn, M.F. & Feldman, P.D. 1992, *Icarus* 98, 54
A'Hearn, M.F., Millis, R.L., Schleicher, D.G., Osip, D.J., & Birch, P.V. 1995, *Icarus* 118, 223
Asher, D.J., Bailey, M.E., Hahn, G., & Steel, D.I. 1994, *Mon. Not. R. Astr. Soc.* 267, 26
Asher, D.J. & Clube, S.V.M. 1997. *Cel. Mech. Dyn. Astr.* 69, 149
Asphaug, E., Ryan, E.V., & Zuber, M.T. 2002, in: W.F. Bottke, A. Cellino, P. Paolicchi & R.P. Binzel (eds.), *Asteroids III*, (Tucson, Arizona: The University of Arizona Press), 463
Barucci, M.A., Fulchignoni, M., & Lazzarin, M. 1996, *Planet. Space Sci.* 44, 1047

Barucci, M.A., Cruikshank, D.P., Mottola, S., & Lazzarin, M. 2002, in: W.F. Bottke, A. Cellino, P. Paolicchi & R.P. Binzel (eds.), *Asteroids III*, (Tucson, Arizona: The University of Arizona Press), 273

Barucci, M.A., Doressoundiram, A., & Cruikshank, D.P. 2004, in: M.C. Festou, H.U. Keller & H.A. Weaver (eds.), *Comets II*, (Tucson, Arizona: The University of Arizona Press)

Bauer, J.M., Fernández, Y.R., & Meech, K.J. 2003, *Publ. Astr. Soc. Pacific* 115, 981

Beaugé, C. & Roig, F. 2001, *Icarus* 153, 391

Beech, M. 2001, *Mon. Not. R. Astr. Soc.* 327, 1201

Belskaya, I.N., Barucci, A., & Shkuratov, Y.G. 2003, *Earth, Moon, and Planets* 92, 201

Bendjoya, P., di Martino, M., & Saba, L. 2004, *Icarus* 168, 374

Bernstein, G.M., Trilling, D.E., Allen, R.L., Brown, M.E., Holman, M., & Malhotra, R. 2004, *Astron. J.* 128, 1364

Biesecker, D.A., Lamy, P., St. Cyr, O.C., Llebaria, A., & Howard, R.A. 2002, *Icarus* 157, 323

Bockelée-Morvan, D., Lellouch, E., Biver, N., Paubert, G., Bauer, J., Colom, P., & Lis, D.C. 2001, *Astron. Astrophys.* 377, 343

Boehnhardt, H., Sekanina, Z., Fiedler, A., Rauer, H., Schulz, R., & Tozzi, G.P. 1998, *Highlights in Astron.* 11A, 233

Boehnhardt, H., Barucci, M.A., Delsanti, A., De Bergh, C., Doressoundiram, A., Romon, J., Dotto, E., Tozzi, G.P., Lazzarin, M., Fornasier, S., Peixinho, N., Hainaut, O., Davies, J., Rousselot, P., Barrera, L., Birkle, K., Meech, K., Ortiz, J.L., Sekiguchi, T., Watanabe, J.-I., Thomas, N., & West, R. 2003, *Earth, Moon, and Planets* 92, 145

Boehnhardt, H. 2004, in: M. Festou, H.U. Keller & H.A. Weaver (eds.), *Comets II*, (Tucson, Arizona: The University of Arizona Press)

Bottke, W.F., Morbidelli, A., Jedicke, R., Petit, J.-M., Levison, H.F., Michel, P., & Metcalfe, T.S. 2002, *Icarus* 156, 399

Brandt, J.C., A'Hearn, M.F., Randall, C.E., Schleicher, D.G., Shoemaker, E.M., & Stewart, A.I.F. 1998, *ASP Conf. Ser.* 107, 289

Brown, W.R. & Luu, J.X. 1998, *Icarus* 135, 415

Brown, M.E. & Koresko, C.C. 1998, *Astrophys. J.* 505, L65

Brown, R.H., Cruikshank, D.P., & Pendleton, Y. 1999, *Astrophys. J.* 519, L101

Buratti, B.J., Hicks, M.D., Soderblom, L.A., Britt, D., Oberst, J., & Hillier, J.K. 2004, *Icarus* 167, 16

Campins, H., Licandro, J., Chamberlain, M., & Brown, R.H. 2001, *Bull. Amer. Astr. Soc.* 33, Abstr. [41.08], 1094

Campins, H. & Fernández, Y.R. 2002, *Earth, Moon, and Planets* 89, 117

Campins, H., Licandro, J., Guerra, J., Chamberlain, M., & Pantin, E. 2003, *Bull. Amer. Astr. Soc.* 35, Abstr. [47.02]

Campins, H. & Lauretta, D.S. 2004, *Bull. Amer. Astr. Soc.* 36, Abstr. [21.08]

Chapman, C.R., Merline, W.J., Bierman, B., Keller, J., & Brooks, S. 1997, *Bull. Amer. Astr. Soc.* 29, Abstr. [12.10]

Chodas, P.W. & Yeomans, D.K. 1995, in: K.S. Noll, H.A. Weaver & P.D. Feldman (eds.), *The Collision of Comet Shoemaker-Levy 9 and Jupiter*, (New York: Cambridge University Press), 1

Choi, Y.J., Brosch, N., & Prialnik, D. 2003, *Icarus* 165, 101

Coradini, A., Capaccioni, F., Capria, M.T., de Sanctis, M.C., Espinasse, S., Orosei, R., Salomone, M., & Federico, C. 1997, *Icarus* 129, 337

Crifo, J.-F., Rodionov, A.V., Szegő, K., & Fulle, M., 2002, *Earth, Moon, and Planets* 90, 227

Cruikshank, D.P., Dalle Ore, C.M., Roush, T.L., Geballe, T.R., Owen, T.C., de Bergh, C., Cash, M.D., & Hartmann, W.K. 2001, *Icarus* 153, 348

Davidsson, B.J.R. 1999, *Icarus* 142, 525

Davidsson, B.J.R. 2001, *Icarus* 148, 375

Davis, D.R., Chapman, C.R., Weidenschilling, S.J., & Greenberg, R. 1985, *Icarus* 62, 30

Davies, J.K., McBride, N., Green, S.F., Mottola, S., Carsenty, U., Basran, D., Hudson, L.A., & Foster, M.J. 1998, *Icarus* 132, 418 de Bergh, C., Boehnhardt, H., Barucci, M.A., Lazzarin, M., Fornasier, S., Romon-Martin, J., Tozzi, G.P., Doressoundiram, A., & Dotto, E. 2004, *Astron. Astrophys.* 416, 791

Delahodde, C.E., Hainaut, O.R., Boehnhardt, H., Dotto, E., Barucci, M.A., West, R.M., & Meech, K.J. 2000, in: A. Fitzsimmons, D. Jewitt & R.K. West (eds.), *Minor Bodies in the Outer Solar System*, (Berlin: Springer-Verlag), 61

Delahodde, C.E., Meech, K.J., Hainaut, O.R., & Dotto, E. 2001, *Astron. Astrophys.* 376, 672

Delahodde, C.E., Hainaut, O.R., Dotto, E., & Campins, H. 2004, *Bull. Amer. Astr. Soc.* 36, Abstr. [34.05]

Desvoivres, E. & Levasseur-Regourd, A.C. 2002, in: B. Warmbein (ed.), *Asteroids, Comets, Meteors – ACM 2002*, (Noordwijk, The Netherlands: ESA Publications Division), *ESA SP-500*, 645

Dohnanyi, J.S. 1969, *J. Geophys. Res.* 74, 2431

Dotto, E., Barucci, M.A., Leyrat, C., Romon, J., de Bergh, C., & Licandro, J. 2003, *Icarus* 164, 122

Dumas, C., Owen, T., & Barucci, M. 1998, *Icarus* 133, 221

Duncan, M. & Levison, H.F. 1997, *Science* 276, 1670

Emery, J.P. & Brown, R.H. 2003, *Icarus* 164, 104

Emery, J.P. & Brown, R.H. 2004, *Icarus* 170, 131

Emery, J.P., Cruikshank, D.P., Van Cleve, J., & Stansberry, J.A. 2005, *LPSC* 36, Abstr. 2072

Farinella, P. & Davis, D.R. 1996, *Science* 273, 938

Fernández, J.A., Tancredi, G., Rickman, H., & Licandro, J. 1999, *Astron. Astrophys.* 352, 327

Fernández, J.A., Gallardo, T., & Brunini, A. 2002, *Icarus* 159, 358

Fernández, Y.R. McFadden, L.A., Lisse, C.M., Helin, E.F., & Chamberlin, A.B. 1997, *Icarus* 128, 114

Fernández, Y.R., Jewitt, D.C., & Sheppard, S.S. 2001, *Astrophys. J.* 553, L197

Fernández, Y.R., Abell, P.A., Pravec, P., French, L.M., Farnham, T.L., Gaffey, M.J., Hardersen, P.S., Kusnirak, P., Sarounova, L., & Sheppard, S.S. 2003a, *Bull. Amer. Astr. Soc.* 35, [47.04]

Fernández, Y., Jewitt, D.C., & Sheppard, S.S. 2003b, *Astron. J.* 123, 1050

Fernández, Y., Sheppard, S. & Jewitt, D. 2003c, *Astron. J.* 126, 1563

Fernández, Y.R., Jewitt, D.C., & Sheppard, S.S. 2005, *Astron. J.* 130, 308

Fleming, H.J. & Hamilton, D.P. 2000, *Icarus* 148, 479

Fletcher, E., Fitzsimmons, A., Williams, I.P., Thomas, N., Ip, W.-H. 2000, in: A. Fitzsimmons, D. Jewitt & R.M. West (eds.), *Minor Bodies in the Outer Solar System*, (Berlin: Springer-Verlag)

Fornasier, S., Barucci, M.A., & Barbieri, C. 2004a, *Astron. Astrophys.* 422, L43

Fornasier, S., Dotto, E., Marzari, F., Barucci, M.A., Boehnhardt, H., Hainaut, O., & de Bergh, C. 2004b, *Icarus* 172, 221

Foster, M.J., Green, S.F., McBride, N., & Davies, J.K. 1999, *Icarus* 141, 408

French, L.M. 2002, *Bull. Amer. Astr. Soc.* 33, [16.01], 368

Gladman, B., Kavelaars, J.J., Petit, J.-M., Morbidelli, A., Holman, M.J., & Loredo, T. 2001, *Astron. J.* 122, 1051

Gomes, R. 2003, *Icarus* 161, 404

Green, D.W.E. 2005, *IAU Circ.* No. 8552

Greenberg, J.M., Mizutani, H., & Yamamoto, T. 1995, *Astron. Astrophys.* 294, L35

Gronkowski, P. 2004, *Astron. Nachr.* 325, 343

Groussin, O., Lamy, P., & Jorda, L. 2004, *Astron. Astrophys.* 413, 1163

Hainaut, O.R., Delahodde, C.E., Boehnhardt, H., Dotto, E., Barucci, M.A., Meech, K.J., Bauer, J.M., West, R.M., & Doressoundiram, A. 2000, *Astron. Astrophys.* 356, 1076

Hicks, M.D., Buratti, B.J., Newburn, R.L., & Rabinowitz, D.L. 2000, *Icarus* 143, 354

Hiroi, T., Kanno, A., Nakamura, R., Abe, M., Ishiguro, M., Hasegawa, S., Miyasaka, S., Sekiguchi, T., Terada, H., & Igarishi, G. 2003, *LPSCI* 34, Abstr. 1425

Hiroi, T., Pieters, C.M., Rutherford, M.J., Zolensky, M.E., Sasaki, S., Ueada, Y., & Miyamoto, 2004, *LPSCI* 35, Abstr. 1616

Hsieh, H.H., Jewitt, D.C., & Fernández, Y.R. 2004, *Astron. J.* 127, 2997

Hsieh, H.H. 2004, *Bull. Amer. Astr. Soc.* 36, Abstr. [28.02]

Hsieh, H.H. & Jewitt, D. 2005, in: D. Lazzaro, S. Ferraz-Mello & J.A. Fernández (eds.), *Asteroids, Comes, Meteors*, (Cambridge: Cambridge University Press)

Hsieh, H.H. & Jewitt, D. 2005, *Astrophys. J.* 624, 1093

Jedicke, R. & Metcalfe, K.J. 1998, *Icarus* 131, 245

Jewitt, D.C. & Luu, J.X. 1990, *Astron. J.* 100, 933

Jewitt, D.C. 1992, in: A. Brahic, J.-C. Gerard & J. Surdej (eds.) *Proc. of the 30th Liége International Astrophysical Colloquium* (Liége: Liége Press), 85

Jewitt, D. 1006, *Earth, Moon, and Planets* 72, 185% From comets to asteroids: wheny hairy stars go bald

Jewitt, D. & Kalas, P. 1998, *Astrophys. J.* 499, L103

Jewitt, D.C., Trujillo, Ch.A., & Luu, J.X. 2000, *Astron. J.* 120, 1140

Jewitt, D.C. 2002, in: B. Warmbein (ed.), *Proceedings of Asteroids, Comets, Meteors – ACM 2002*, (Noordwijk, The Netherlands: ESA Publ. Division), ESA SP-500, 11

Jewitt, D.C. 2004, in: M.C. Festou, H.U. Keller & H.A. Weaver (eds.), *Comets II*, (Tucson, Arizona: The University of Arizona Press), 659

Jewitt, D.C. & Luu, J. 2004, *Nature* 432, 731

Jewitt, D.C., & Sheppard, S.S., & Porco, C. 2004, in: F. Bagenal, T.E. Dowling & W.B. McKinnon (eds.), *Jupiter, The planet, satellites and magnetosphere*, 263

Jewitt, D. 2005, *Astron. J.* 129, 530

Jones, T., Lebofsky, L., Lewis, J., & Marley, M. 1990, *Icarus* 88, 172

Kargel, J.S. 1991, *Icarus* 94. 368

Karlsson, O. 2004, *Astron. Astrophys.* 413, 1153

Kresák, L. 1979, in: T. Gehrels & M.S. Matthews (eds.), *Asteroids*, (Tucson: University of Arizona Press), 289

Lamy, P.L., Toth, I., Fernández, Y.R., & Weaver, H.A. 2004, in: M.C. Festou, H.U. Keller & H.A. Weaver (eds.), *Comets II*, (Tucson, Arizona: The University of Arizona Press)

Lamy, P.L. & Toth, I. 2005, *Icarus*, submitted

Larsen, J.A., Gleason, A.E., Danzl, N.M., Descour, A.S., McMillan, R.S., Gehrels, T., Jedicke, R., Monrani, J.L., & Scotti, J.V. 2001, *Astron. J.* 121, 562

Laufer, D., Pat-El, I., & Bar-Nun, A. 2005, *Icarus*, in press

Levison, H.F. 1996, in: T.W. Rettig & J.M. Hahn (eds.), *Completing the inventory of the solar system*, (Astronomical Society of the Pacific Conference Proceedings), ASP Conf. Ser. 107, 173

Levison, H.F., Morbidelli, A., Dones, L., Jedicke, R., Wiegert, P.A., & Bottke, W.F. 2002, *Science* 296, 2212

Licandro, J., Guerra, J.C., Campins, H., Di Martino, M., Lara, L.M., Gil-Hutton, R., & Tozzi, G.P. 2002, *Earth, Moon, and Planets* 90, 495

Licandro, J., Campins, H., de Leon Cruz, J., Gil-Hutton, R., & Lara-Lopez, L.M. 2003, *Bull. Amer. Astr. Soc.* 35, Abstr. [39.11]

Lien, D.J. 1998, *Bull. Amer. Astr. Soc.* 30, Abstr. [12.07], 1035

Lisse, C.M. 2002, *Earth, Moon, and Planets* 90, 497

Lisse, C.M., Fernández, Y.R., A'Hearn, M.F., Grün, E., Käufl, H.U., Osip, D.J., Lien, D.J., Kostiuk, T., Peschke, S.B., & Walker, R.G. 2004, *Icarus* 171, 444

Lowry, S.C., Fitzsimmons, A., & Collander-Brown, S. 2003, *Astron. Astrophys.* 397, 329

Lowry, S.C., & Fitzsimmons, A. 2005, *Mon. Not. R. Astr. Soc.* 358, 641

Lowry, S.C., & Weissman, P.R. 2003, *Icarus* 164, 492

Lunine, J. *et al.* 2004, in: M.C. Festou, H.U. Keller & H.A. Weaver (eds.), *Comets II*, (Tucson, Arizona: The University of Arizona Press)

Luu, J.X. 1994, *Publ. Astr. Soc. Pacific* 106, 425

Luu, J., Jewitt, D., & Cloutis, E. 1994, *Icarus* 109, 133

Luu, J.X., Jewitt, D.C., & Trujillo, Ch. 2000, *Astrophys. J.* 531, L151

Luu, J.X. & Jewitt, D.C. 2002, *Annu. Rev. Astr. Astrophys.* 40, 63

Mäkinen, J.T.T., Bertaux, J.-L., Pulkkinen, T.I., Schmidt, W., Kyrölä, H., Summanen, T., Quémarais, E., & Lallement, R. 2001, *Astron. Astrophys.* 368, 292

Mann, I., Kimura, H., Biesecker, D.A., Tsurutani, B.T., Grün, E., McKibben, R.B., Liou, J.-C., MacQueen, R.M., Mukai, T., Guhathakurta, M., & Lamy, P. 2004, *Space Sci. Rev.* 110, 269

Marzari, F., Farinella, P., & Vanzani, V. 1995, *Astron. Astrophys.* 299, 267

Marzari, F., Farinella, P., Davis, D.R., Scholl, H., & Campo Bagatin, A. 1997, *Icarus* 126, 39

Marzari, F. & Scholl, H. 1998a. *Icarus* 131, 41

Marzari, F. & Scholl, H. 1998a, *Astron. Astrophys.* 339, 278

Marzari, F., Scholl, H., Murray, C., & Lagerkvist, C. 2002, in: W.F. Bottke, A. Cellino, P. Paolicchi & R.P. Binzel (eds.), *Asteroids III*, (Tucson, Arizona: The University of Arizona Press), 725

Marzari, F., Tricarico, P., & Scholl, H. 2003, *Icarus* 162, 453

Meech, K.J., Hainaut, O.R., Boehnhardt, H., & Delsanti, A. 2003, *Bull. Amer. Astr. Soc.* 35, Abstr. [39.12]

Meech, K.J., Hainaut, O.R., & Marsden, B.G. 2004, *Icarus* 170, 463

Melita, M., Williams, I., & Licandro, J. 2005, *Asteroids, Comets, Meteors – ACM 2005*, Abstract Book

Merline, W.J., Weidenschilling, S.J., Durda, D.D., Margot, J.L., & Pravec, P., Storss, A.D. 2002, in: W.F. Bottke, A. Cellino, P. Paolicchi & R.P. Binzel (eds.), *Asteroids III*, (Tucson, Arizona: The University of Arizona Press), 289

Michel, P., Benz, W., Tanga, P., & Richardson, D.C. 2001, *Science* 294, 1696

Morbidelli, A., Levison, H.F., Tsiganis, K., & Gomes, R. 2005, *Nature* 435, 462

Mousis, O. & Alibert, Y. 2005, *Mon. Not. R. Astr. Soc.* 358, 188

Mueller, B.E.A., Hergenrother, C.W., Samarasinha, N.H., Campins, H., & McCarthy, D.W. 2004, *Icarus* 171, 506

Mumma, M.J., Dello Russo, N., DiSanti, M.A., Magee-Sauer, K., Novak, R.E., Brittain, S., Rettig, T., McLean, I.S., Reuter, D.C., & Xu, Li-H. 2001, *Science* 292, 1334

O'Brien, D.P. & Greenberg, R. 2003, *Icarus* 164, 334

Noll, K.S., Weaver, H.A., & Feldman, P.D 1996, in: K.S. Noll, H.A. Weaver & P.D. Feldman (eds.), *The Collision of Comet Shoemaker-Levy 9 and Jupiter.* IAU Colloquium 156, (Cambridge, UK: Cambridge University Press)

Pan, M. & Sari, R. 2005, *Icarus* 175, 343

Pittich, E.M., D'Abramo, G., & Valsecchi, G.B. 2004, *Astron. Astrophys.* 422, 369

Press, W.H., Flannery, B.P., Teukolsky, S.A., & Vetterling, W.T. 1986, in: *Numerical Recipes – in Fortran The Art of Scientific Computing* (Cambridge: Cambridge University Press)

Rabe, E. 1972, in: G.A. Chebotarev, E.I. Kazimirchak-Polonskaya & B.G. Marsden (eds.), *The Motion, Evolution of Orbits, and Origin of Comets*, IAU Symposium No. 45, (Springer, New York), 55–60

Rodionov, A.V., Crifo, J.-F., Szegő, K., Lagerros, J., & Fulle, M. 2002, *Planet. Space Sci.* 50, 983

Rousselot, P., Petit, J.-M., Poulet, F., Lacerda, P., & Ortiz, J. 2003, *Astron. Astrophys.* 407, 1139

Saba, I., Capria, M.T., & Cremonese, G. 2005, *Mem. S. A. It.* Suppl. Vol. 6, 137

Samarasinha, N.H. 2003, *Adv. Space Res.*, in press.

Scheeres, D.J. 2005, *LPSC* 36, Abstr. No. 1919

Schulz, R. 2002, *Astron. Astrophys. Rev.* 11, 1

Sheppard, S.S., Jewitt, D.C., Trujillo, Ch.A., Brown, M.J.L., & Ashley, M.C.B. 2000, *Astron. J.* 120, 2687

Soderblom, L.A., Becker, T.L., Bennett, G., Boice, D.C.m. Britt, D.T., Brown, R.H., Buratti, B.J., Isbell, C., Giese, B., Hare, T., Hicks, M.D., Howington-Kraus, E., Kirk, R.L., Lee, M., Nelson, R.M., Oberst, J. Owen, T.C., Rayman, M.D., Sandel, B.R., Stern, S.A., Thomas, N., & Yelle, R.V. 2002, *Science* 296, 1087

Stern, S.A. 1995, *Astron. J.* 110, 856

Stuart, J.S. 2001, *Science* 294, 1691

Tancredi, G. & Sosa, A. 1996, *Rev. Mex. Astr. Astrophys.* 4, 118

Toth, I. 1999, *Icarus* 141, 420

Toth, I. 2000, *Astron. Astrophys.* 360, 375

Toth, I. 2001, *Astron. Astrophys.* 368, L25

Toth, I. 2005a, *Astron. Astrophys.*, submitted.

Toth, I. 2005b, *Astron. Astrophys.*, submitted.

Toth, I., Lamy, P.L., & Weaver, H.A. 2005, *Icarus*, in press.

Toth, I. & Lisse, C.M. 2005, *Icarus*, submitted.

Trujillo, Ch.A., Jewitt, D.C., & Luu, J.X. 2001, *Astron. J.* 122, 457

Vernazza, P., Monthé-Diniz, T., Birlan, M., Carvano, J.M., Strazzulla, G., Fulchignoni, M., & Migliorini, A. 2005, *Astron. Astrophys.* 436, 1113

Weidenschilling, S.J. 2004, in: M.C. Festou, H.U. Keller & H.A. Weaver (eds.), *Comets II*, (Tucson, Arizona: The University of Arizona Press)

Weaver, H.A., Sekanina, Z., Toth, I., Delahodde, C.E., Hainaut, O.R., Lamy, P.L., Bauer, J.M., A'Hearn, M.F., Arpigny, C., Combi, M.R., Davies, J.K., Feldman, P.D., Festou, M.C., Hook, R., Jorda, L., Keesey, M.S.W., Lisse, C.M., Marsden, B.G., Meech, K.J., Tozzi, G.P., & West, R. 2001. *Science* 292, 1329

Weaver, H.A. 2004, *Science* 304, 1760

Weissman, P.R. 1986, *Nature* 320, 242

Weissman, P.R. 1989, in: S.K. Atreya, J.B. Pollack & M.S. Matthews (eds.), *Origin and Evolution of Planetary Satellite Atmospheres*, (Tucson, Arizona: The University of Arizona Press), 230

Weissman, P.R., & Levison, H.F. 1997, *Astron. J.* 488, L133

Weissman, P.R., Bottke, W.F. & Levison, H.F. 2002 in: W.F. Bottke, A. Cellino, P. Paolicchi & R.P. Binzel (eds.), *Asteroids III*, (Tucson, Arizona: The University of Arizona Press), 669

Weissman, P.R., Asphaug, E., & Lowry, S.C. 2004, in: M.C. Festou, H.U. Keller & H.A. Weaver (eds.), *Comets II*, (Tucson, Arizona: The University of Arizona Press)

Wilck, M. & Mann, I. 1996, *Planet. Space Sci.* 44, 493

Wilck, & Stern, S.A. 1999, *Astron. Vestnik* 33, 216

Yelle, R.V., Soderblom, L.A., & Jokipii, J.R. 2004, *Icarus* 167, 30

Zahnle, K., Schenk, P., Sobieszczyk, S., Dones, L., & Levison, H.F. 2001, *Icarus* 153, 111

Asteroids, Comets, Meteors
Proceedings IAU Symposium No. 229, 2005
D. Lazzaro, S. Ferraz-Mello & J.A. Fernández, eds.

© 2006 International Astronomical Union
doi:10.1017/S1743921305006691

Dawn mission and operations

C.T. Russel[1], C.A. Raymond[2], T.C. Fraschetti[2], M.D. Rayman[2], C.A. Polanskey[2], K.A. Schimmels[2] and S.P. Joy[1]

[1]Institute of Geophysics and Planetary Physics, University of California, Los Angeles, CA 90095-1567
email: ctrussel@igpp.ucla.edu
[2]Jet Propulsion Laboratory, Pasadena, CA 91109

Abstract. Dawn is the first mission to attempt to orbit two distant planetary bodies. The objects chosen, 4 Vesta followed by 1 Ceres, are the two most massive members of the asteroid belt that appear to have been formed on either side of the dew line in the early solar nebula. This paper describes the present status of the mission development and the plans for operation at Vesta and Ceres.

Keywords. 1 Ceres, 4 Vesta, Dawn Discovery Mission

1. Introduction

The solar system is the epitome of diversity. Mercury, formed close to the Sun, is relatively small and rather dense, indicative of a large iron core covered with a thin silicate crust. Venus, a much larger planet, has a thick atmosphere, is possibly still volcanically active, and is dominated by sulfur chemistry. It too has an iron core and silicate mantle. The Earth provides a much more benign habitat for life, with water on the surface. While the interior may be similar to that of Venus, plate tectonics acts to move the crust and to enhance heat flow from the interior. Mars size lies between Mercury and the Venus-Earth twins. While its weak atmosphere, low temperature and low gravity support neither the sustained presence of surficial liquid water, nor much water vapor in the atmosphere, Mars apparently possesses much subsurface water, at least initially resulting in episodic, wet epochs and significant erosion of the surface. The gas giants, Jupiter and Saturn, present an entirely different style of planet, with relatively small rocky cores, rings and many moons (mostly icy). Uranus and Neptune are significantly smaller than Jupiter and Saturn and are believed to consist of a rocky core surrounded by a water mantle and a hydrogen and helium envelope. They have weak ring systems and multiple moons, again icy. The edge of the solar system brings us Pluto, the Kuiper belt and the Oort cloud, frozen remnants from the earliest days of the solar system.

The planets retain an unsatisfactory record of the early solar system. Their compositions represent an average over a significant radial range in the solar nebula and they have undergone much thermal evolution, continuing to the present time. The comets and Kuiper belt objects also provide an unsatisfactory record of the solar nebula because their present orbits have been greatly altered by the gas giants. Thus it is difficult to relate the measurements of today to the events of yesterday.

The asteroid belt forms a transition from the rocky terrestrial planets to the wet and icy outer solar system. It has been affected by the gravitational stirring of Jupiter that stopped the growth of the planetary embryos in the belt, and that spawned a destructive collisional environment. The early cessation of growth enables us to explore backward

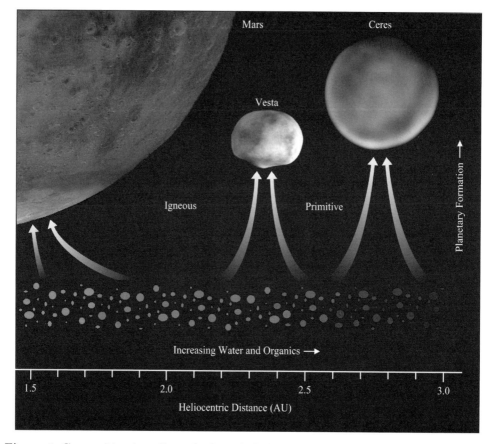

Figure 1. Compositional gradient of solar nebula leads to variation in properties of accreted bodies as a function of heliocentric distance.

in time. Smaller objects evolve less than the larger planets and preserve information on much earlier epochs. The growth of the minor planets stopped early, probably well before the formation of the Earth. The collisions produced by disruptive stirring in the belt are a potential impediment to exploration backward in time. They may have destroyed many of the smaller embryos or scattered them far out of their original orbits. Thus the best candidates for study are the largest asteroids. These objects survived the heavy bombardment and they are the most likely to be near their place of origin. If we wish to determine the original heliocentric gradient, then we should visit the largest asteroids. This strategy underlies the Dawn mission.

The two most massive minor planets are 1 Ceres and 4 Vesta. Figure 1 illustrates their size relative to that of Mars. They clearly merit being labeled planets as they resemble the other planets in many ways. Further Vesta and Ceres differ from each other in a manner not simply explainable by their relative size. They arose at different distances from the Sun at which the composition of the dusty solar nebula was quite different. The differences in the spectrum of reflected light from the surfaces of Vesta and Ceres indicates a significant gradient in the properties of the early solar nebula when the asteroids formed. If we wish to constrain the gradient in the early solar nebula, where should we explore? Figure 2 shows a sketch of the relative locations and relative sizes of the larger minor planets. The distance from the ecliptic plane on this plot is a qualitative indication of both the eccentricity and inclination of their orbits. Five minor planets are most worthy

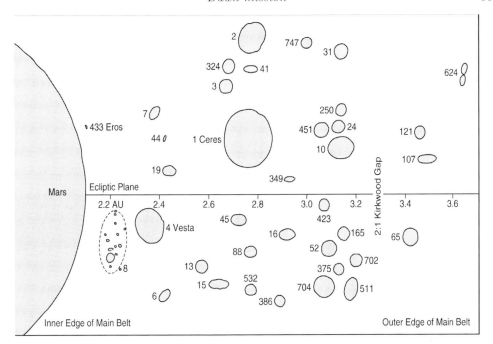

Figure 2. Qualitative sketch of the location and relative sizes of the larger minor planets. The distance from the ecliptic plane is greater for bodies with larger inclination or eccentricity and qualitatively reflects the difficulty in reaching these bodies as compared with others at the same semimajor axis.

of our attention: 1 Ceres, 4 Vesta, 3 Pallas, 10 Hygiea and 16 Psyche. Table 1 lists their semi-major axes, inclinations and masses. These five objects give us a wide range of asteroid types. Ceres is a rocky planet with probably a thick icy crust (McCord & Sotin 2005). Vesta is a rocky body, similar to the Moon, with an iron core and a silicate mantle. Pallas may be very similar to Ceres, but is significantly smaller. Little is known about Hygiea, but, being much further from the Sun than Ceres (and Pallas), it is expected to differ markedly from Ceres. Interest is added by its dynamical family of quite diverse membership, including B, C, D and S type asteroids that might be accessible during a Hygiea mission. Finally, Psyche has all the indications of being an iron core of an originally much larger parent body. It reflects visible light and radar waves like iron does and it has a bulk density close to that of iron (Kuzmanoski & Kovacevic 2002).

It is certainly possible to visit such asteroids with chemical rockets but affordable missions would entail flybys. To make the needed physical, geochemical, geophysical, and geological measurements we need to orbit these bodies. We do not have to visit all the asteroids to determine the nebular gradient, but we do need to visit at least a handful. The problem is to do this affordably.

2. The Mission

2.1. *Ion Propulsion: The Enabling Technology*

An affordable solution to exploring the minor planets is provided by ion propulsion. Since planetary missions are governed by Keplers laws of planetary motion regardless of how they are propelled, and since ion thrusters do not accelerate spacecraft much above the Keplerian velocities, there is a long time over which the weak thrust of ion propulsion can

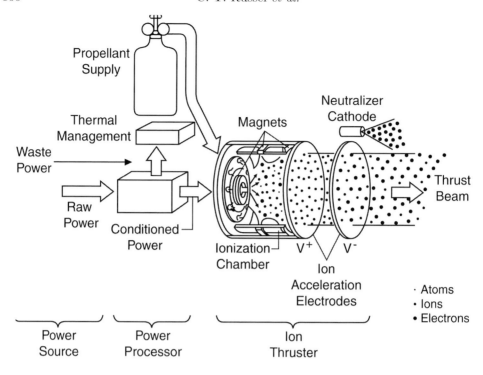

Figure 3. Operating principle of an ion engine.

act. The thrusters accelerate the ions to a very large velocity, currently about $35\,\mathrm{kms}^{-1}$, an order of magnitude greater than the velocity of the fuel expelled by a chemical engine. Thus the fuel carried on board is used with great efficiency. Once in deep space the ion engine (whose thrust is far too weak to lift the spacecraft off the planet) can change the velocity of the spacecraft (after years of thrust) by an amount equal to that provided by the launch vehicle that lifted the spacecraft off the surface of the Earth ($11\,\mathrm{kms}^{-1}$).

The general principles of an ion engine are shown in Figure 3. An ion engine requires much electric power, provided at present by solar arrays. A typical solar-power-driven ion engine requires arrays providing about 10 kW at 1 AU for exploring the asteroid belt. This power is used to ionize the xenon fuel and to accelerate the ions to $35\,\mathrm{kms}^{-1}$, resulting in a thrust of about 90 mN. In order that the ions can escape from the spacecraft, it must not accumulate charge. To maintain charge neutrality electrons are injected into the ion beam. Xenon is used as fuel because it is environmentally benign, can be stored in space as a gas, and is readily ionized. Design of the grids is a special concern as the accelerated ions ablate the grid material limiting the lifetime or throughput of the engine. A single, present-day thruster is rated to process about 150 kg of xenon. Thus if a mission requires 300 kg of xenon it might carry two thrusters. If redundancy were required a third thruster might be included. Other considerations include providing gimbals to adjust the direction of thrust to keep the thrust vector directed through the center of mass of the spacecraft to minimize torques, and the design of the power processing units and power distribution units needed to handle the significant power that such spacecraft require.

Table 1. Linear correlation coefficients for the data sets plotted in figure 1

Name	Mass [kg]	Distance [AU]	Diameter [km]	Comments
1 Ceres	9.4×10^{20}	2.77	974×909	~ 100 km water-ice mantle
4 Vesta	2.7×10^{20}	2.34	570×458	$\sim$ up to 250 km diameter iron core
2 Pallas	2.4×10^{20}	2.77	523	A smaller Ceres?
10 Hygiea	1.0×10^{20}	3.14	407	Dynamical family includes D,B,C,S types
16 Psyche	0.7×10^{20}	2.90	264	Density and reflectivity of iron; could be core of larger parent body

2.2. *Choosing a Target*

Table 1 lists five possible choices: a rocky asteroid about which we know much from the Howardite, Eucrite and Diogenite meteorites (HEDs); the most massive minor planet that appears to be very wet, and remains very much an enigma; a high inclination body that may also be wet and is even more of an enigma; a distant asteroid with an interesting dynamical family (some of which may be adopted); and an asteroid that very much resembles the core of a larger body, stripped of its rocky mantle. Four of these bodies we could probably visit with a present-day, ion-propelled missions, affordable under NASAs Discovery program guidelines. However, it is very difficult to reach the high inclination orbit of Pallas, so it appears to be presently out of contention as a Discovery candidate. It is also not possible to visit all four asteroids with a single spacecraft. The impossibility results from in part the long durations of such a multiple asteroid mission. It is simply too costly to operate for such a long period. This exceeds Discovery cost guidelines, and the spacecraft would require very high reliability to last that long, another cost driver. The spacecraft would have to be quite large with multiple thrusters (six?) and a large xenon load (750 kg?). A relatively inexpensive rocket, like the Delta II, would be unable to launch such a mission to escape velocity. Thus, we must lower our sights.

The idea of multiple rendezvous with two or more asteroids has long intrigued the Dawn team (Russell *et al.* 2004). By the year 2000, the capability of the ion thrusters and the celestial position of Vesta and Ceres had evolved sufficiently to allow a single spacecraft powered by three ion engines, used one at a time, to rendezvous with both Vesta and Ceres. This mission was proposed to NASA and selected for flight in 2001 (Step 1 selection in January and Step 2 in December). The tortuous path to confirmation is described by Russell *et al.* (2005). Figure 4 shows a schematic diagram of the mission trajectory. The spacecraft is launched in summer of 2006; it flies by Mars in 2009; and reaches Vesta in 2011, and Ceres in 2015, with slightly under a year in orbit at each body.

2.2.1. *4 Vesta*

The fourth minor planet, 4 Vesta, was discovered on March 29, 1807 by H. Olbers. Much, much later it was realized that both the reflectance spectrum of Vesta resembled that of a basaltic surface, and that a class of meteorites had that same reflectance spectrum (McCord *et al.*, 1970). Pieces of Vesta (perhaps with the intermediary of Vestoids, collisional fragments from Vesta) had been falling on the Earth. In fact one in twenty falls are of this class, the Howardite-Eucrite-Diogenite (HED) meteorites. Since this discovery many observations have supported this interpretation. Our best estimate of the shape of Vesta is a triaxial ellipsoid with radii 289, 280 and 229 ± 5 km. Its mass is $2.71 \pm 0.10 \times 10^{20}$ kg corresponding to a density of 3750 ± 400 kg/m^3. It has a large southern crater as shown in Figure 5. Evidence for a significant iron core has been obtained from the HED meteorites. HST and adaptive optics observations reveal a non-uniform surface with significant variations in composition (Binzel *et al.* 1997; Drummond *et al.*

Mission Itinerary

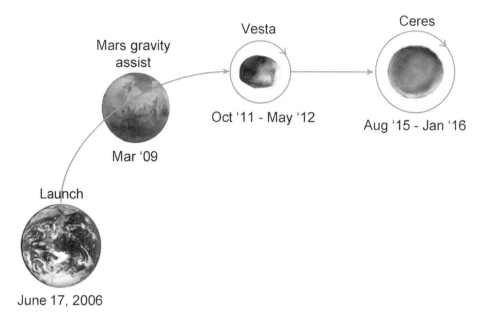

Figure 4. Schematic diagram of mission trajectory.

1998). We presume that these variations are caused by both cratering exposing different materials at depth and lava flows that were extruded at different times and from different reservoirs during Vestas thermal evolution. Since Vesta was formed in the solar nebula near the dew or frost point, the presence of water in or on Vesta remains an interesting question. It is clear from the HED meteorites that Vesta is very dry and that water did not much influence its evolution. However, the discovery of a quartz veinlet in a HED meteorite (Treiman *et al.* 2004) makes a strong case for liquid water at least transiently on Vestas surface. A more thorough review of our knowledge of Vesta has been given by Russell *et al.* (2005).

2.2.2. *1 Ceres*

1 Ceres was discovered on January 1, 1801 by C. Piazzi. In great contrast to Vesta there are no known Ceres-associated meteorites and there are no known Ceroids, collisional fragments of Ceres. This immediately raises the question as to whether the crust of Ceres is rock or some material such as water-ice that would not be stable if removed from Ceres gravitational potential well and the expected dust covering the surface. In fact McCord & Sotin (2005) have constructed a thermal evolution model of Ceres that results in a rocky core covered with an approximately 100 km thick water and ice mantle that may or may not freeze completely in 4.6 billion years.

The mass of Ceres is $9.43 \pm 0.05 \times 10^{20}$ kg with a shape of an oblate spheroid of 487 ± 2 km (equatorial) and 455 ± 2 km (polar) (McCord & Sotin 2005; Thomas *et al.* 2005). Given its 9.075 hr spin period this shape corresponds to a non-uniform, radial distribution of

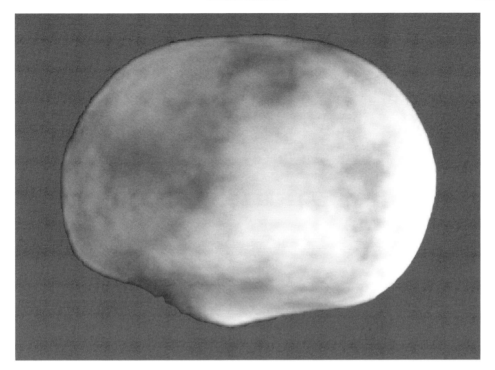

Figure 5. Image made from a computer model of Vesta.

mass, consistent with about 100 km of water ice on top of about 400 km of rock. The same HST observational campaign that produced the shape observations also produced albedo maps at 223 nm, 335 nm and 535 nm (Li *et al.* 2005). A preliminary color map of Ceres surface at 535 nm is shown in Figure 6. Ceres has well defined surface features that can be tracked, enabling the rotation rate and pole position to be determined. However, the average albedo is low and the contrast across the surface small. The surface is smooth at radar wavelength scales and as revealed in the Hapke parameters. However, it is rough at intermediate wavelengths. At planetary scales, as stated above, the body is relaxed to a shape that is in rotational equilibrium, consistent with the presence of a rocky core and a water-ice mantle.

2.3. *Science Objectives*

Dawn' s main objective is to achieve an understanding of the conditions and processes acting at the solar systems earliest epoch. To do this the mission collects images of the surface of Vesta and Ceres and thereby determines their bombardment, thermal, tectonic and possible volcanic history. Dawn determines the topography and internal structure of the two bodies by measuring their mass, shape, volume and spin state with navigation data and imagery. It determines mineral and elemental composition from infrared, gamma ray, and neutron spectroscopy to constrain the thermal history and compositional evolution of Vesta and Ceres. It also provides context for the HED meteorites, samples of the surface already in hand.

These science objectives are allocated to each of the three instruments (Framing Camera, Visible and IR Mapping Spectrometer, and the Gamma Ray and Neutron Spectrometer) and to the gravity science investigation. This allocation in turn levies specific

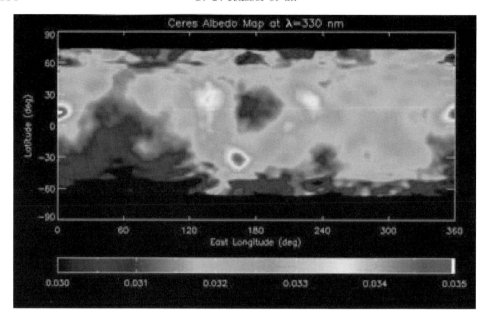

Figure 6. HST albedo map of Ceres (Courtesy of J-Y. Li, 2005).

measurement requirements. In particular the mission is attempting to measure the bulk density of Vesta and Ceres to better than 1% their spin-axis orientations to better than 0.5°, and Vestas gravity field to better than 90 km half-wave length resolution (better than 300 km for Ceres). Over 80% of the surfaces of both bodies will be imaged at better than 100 m per pixel for Vesta and better than 200 m per pixel at Ceres in 3 or more color filters. A topographic map of the surfaces of both bodies will be made with a goal of 10 m vertical accuracy on Vesta and 20 m on Ceres. The mission attempts to map the abundances of the major rock forming elements to $\pm 20\%$ precision with a spatial resolution of ~ 1.5 times the mapping altitude, and to map the abundances of hydrogen and the radioactive elements Th, U and K. The visible and infrared mapping spectrometer will map the mineral composition with spectra between 0.25 and 5 microns with a resolution of 2 to 10 nm. Half of the frames will have a resolution of better than 200 m on Vesta and 400 m on Ceres. Dawn also uses its radiometric tracking system to measure the gravitational field.

2.4. *Scientific Payload*

2.4.1. *Framing Camera*

The two redundant framing cameras (FC) were built by MPS in Lindau, Germany, in cooperation with DLR Berlin and IDA, Braunschweig. The camera shown in Figure 7 uses a $f : 1/8$ radiation-hard refractive optics with a focal length of 150mm. The field of view of 5.5° by 5.5° is imaged by a frame-transfer CCD with 1024 by 1024 pixels for a spatial resolution of 9.3 m/pixel at 100 km. A data processing unit controls the camera and handles the compression and buffering of the data. Compression rates vary from $2 : 1$ (lossless) to $10 : 1$ (lossy). A filter wheel provides 1 clear filter and 7 spectral windows. A more detailed description has been provided by Russell *et al.* (2005).

2.4.2. *Visible and Infrared Mapping Spectrometer*

The mapping spectrometer (VIR) has been provided by the Italian Space Agency and the National Institute for Astrophysics. It combines two data channels in one compact

Figure 7. One of two redundant framing cameras.

instrument. The visible channel covers 0.25 to 1 micron while the infrared channel covers 1 to 5 microns. The use of a single optical chain and overlap in wavelength between two channels facilitates intercalibration. The VIR instantaneous field of view is 250 m/pixel at 100 km while the full field of view is 64 mrad. Figure 8 shows the VIR instrument. The infrared focal place array is cooled to an operating temperature of 70 K by a Stirling cooler. Rice lossless data compression is used to return data with a compression factor of between 2 and 5. Further details on the VIR instrument have been provided by Russell *et al.* (2005).

2.4.3. *Gamma Ray and Neutron Spectrometer*

The Los Alamos National Laboratory provided the gamma ray and neutron spectrometer (GRaND). The gamma-ray sensor is segmented into two parts and the neutron sensor into four parts. On board classification of the multiple signals from each event that can discriminate radiation from the asteroid and from the spacecraft. The gamma-ray sensor is bismuth germanate. Cadmium Zinc Telluride sensors provide a new redundant technology for sensing the gamma rays. Upward (asteroid) and downward (spacecraft) facing boron-loaded plastic segments are optically coupled to a 6Li-loaded glass scintillator to detect neutrons. Figure 9a shows a picture of the instrument. Further details may be found in Russell *et al.* (2005).

2.5. *Spacecraft*

The flight system is built by Orbital Sciences Corporation with oversight and direction by the Jet Propulsion Laboratory. Figure 9b shows the spacecraft. It is large, 19.7 m from tip to tip, three axis stabilized, keeping its solar panel axis perpendicular to the

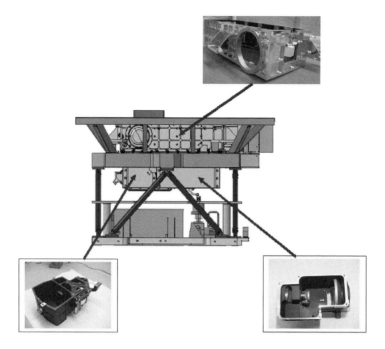

Figure 8. Mapping spectrometer, showing specific subsystems and their location in the instrument.

solar direction and the arrays normal to the star direction. The instruments look in the nadir direction (top of the spacecraft here) that, while in orbit, is kept facing the planet. The main structure of the spacecraft is approximately 1.4 m on a side. The solar panels generate about 10 kW of power at 1 AU. The spacecraft carries 425 kg of xenon. The high gain antenna is mounted on the body of the spacecraft and the spacecraft rotates so that the antenna points at Earth during telemetry sessions.

3. Operations

3.1. *Science Observation Strategies*

Dawn achieves its science objectives by orbiting the main belt asteroids (protoplanets) 1 Ceres and 4 Vesta. Dawn will orbit Vesta for a period of not less than seven months and Ceres for not less than five months, of which at least 1.5 months at Vesta will be below a mean altitude of 200 km, and one month at Ceres will be below 700 km mean altitude.

   The use of ion propulsion results in a highly flexible mission plan, with a long launch period. Dawns launch period opens in the summer of 2006 and is determined by project readiness. The launch period lasts three weeks, which is defined by cost constraints and specification of accurate targeting of the vehicle; however, the acceptable launch period lasts well into the fall of 2006. Dawn will arrive at Vesta in late 2011, leave Vesta mid-2012, and arrive at Ceres in the latter half of 2015, where it will remain. The spacecraft flies by Mars in 2009 enroute to Vesta. A high-level mission timeline is shown in Figure 10. The instruments on the spacecraft are body mounted so the spacecraft rotates to point the instruments at the targets. All data are stored on board for playback to Earth, and in general, only gravity science (Doppler and range measurements) will be collected during tracking sessions. All data are transmitted to Earth within a few days of acquisition.

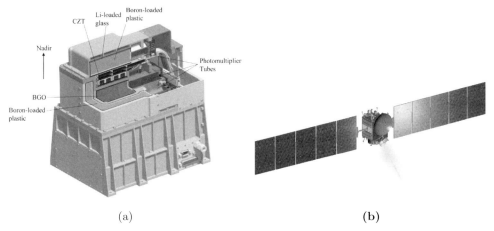

(a) (b)

Figure 9. (a): Gamma ray and neutron spectrometer. (b): Dawn spacecraft.

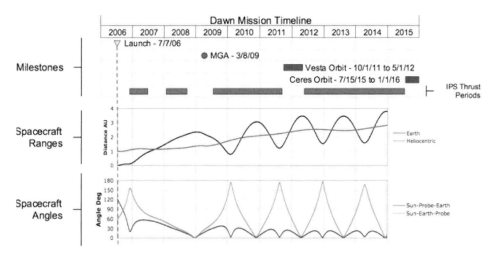

Figure 10. Dawn Mission Timeline.

3.1.1. *Mission Phases*

The mission is divided into phases, including Initial Checkout (Launch+1 to L+60 days), Interplanetary Cruise Phases, and Mars Gravity Assist (closest approach 100 days to CA+30 days), as well as the asteroid encounters (Science Orbits). Each Science Orbit phase begins with an Approach sub-phase starting 85 days before capture. The Science Orbit sub-phases will be: Survey Orbit, High Altitude Mapping Orbit (HAMO), and Low Altitude Mapping Orbit (LAMO). Table 2 describes the orbit sub-phases at Vesta in terms of their durations, beta angles (angle between Sun-body vector and the orbit plane), radii and periods. The minimum mission (performance floor) for the Dawn mission reduces the time spent at Vesta and Ceres from 7 months and 5 months, respectively, to 4 months at each. At Vesta, the lowest altitude orbit is eliminated, and the higher-altitude mapping orbit is lengthened to at least 30 days. At Ceres, the higher-altitude mapping orbit is lowered to 1380 km and lengthened to at least 30 days.

Table 2. Vesta Orbit Sub-phases

Sub-phase	Duration (days)	Beta Angle (deg)	Orbit Raidus (km)	Orbit Period (hrs)
Transfer from Vesta Arrival to Survey Orbit	15			
Vesta Survey Orbit	7 (3 orbits)	30	2700	58.0
Transfer from Survey Orbit to HAMO	21			
Vesta HAMO	26 (52 orbits)	40	950	12.1
Transfer from HAMO to LAMO	30			
Vesta LAMO (includes planned schedule margin)	80	65	~450	4.1
Vesta Departure (including HAMO-2)	68			
TOTAL	**247**			

The current plan allows an eight-month stay at Vesta, exceeding the baseline mission, that is enabled by harvesting of conservative technical margins (mass, power) as the flight system becomes well-defined. This trajectory is called the MGS_Enhanced mission. Flexible mission design enabled by the use of ion propulsion allows continual evolution of the trajectory to reflect actual performance.

3.1.2. *Activities by Mission Phase*

The Initial Checkout (ICO) mission phase is used to turn on and perform initial checkout of the instruments using ground-in-the-loop commanding. Instrument checkout during the 60-day period following launch will not be exhaustive; only a minimal set of checkout activities will be performed during the ICO to minimize interference with critical spacecraft checkouts. Two days have been set aside for Earth-Moon imaging if the contamination environment is deemed sufficiently benign to allow it. Initial instrument calibration tests are done during the early Cruise phase, following ICO.

Seven days of coast (non-thrusting periods) per year have been designated for science calibration activities. These periods will be used to perform functional, performance and calibration tests of the instruments using stellar and planetary targets. The seven days will be split into two 3.5-day periods at six-month intervals. The Framing Camera will perform a functional test each time and an extended calibration test once a year. VIR will do calibrations at each opportunity using stars, star clusters, planets, and planetary nebulae. GRaND may take the opportunity to do adjustments and anneal the instrument.

The purpose of the MGA is to add energy to the spacecraft trajectory to ensure adequate mass and power margins for the designed trajectory. As the flight system matures, and as knowledge of the actual spacecraft mass improves, margin requirements decrease. The MGA may no longer be needed and a direct trajectory to Vesta may be adopted, if this approach is consistent with the required margins.

Science data will be collected during the Approach sub-phase, and in the Science Orbit sub-phases of each encounter, to satisfy the measurement requirements specified above. The Survey Orbit sub-phase will provide an overview of the asteroid and provide

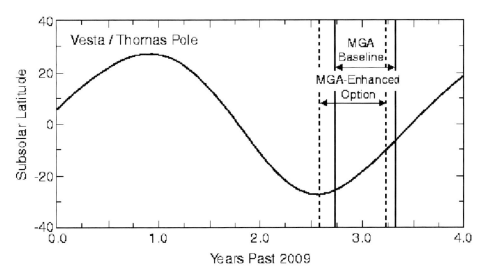

Figure 11. Vesta lighting conditions for baseline and enhanced MGA trajectories

additional observations for the planning of subsequent sub-phases. The Level 1 science required in Survey orbit is 5000 VIR spectral frames. The 5-day Survey Orbit sub-phase begins at the end of the Vesta Approach sub-phase, when the spacecraft has established a circular polar orbit at a radius of 2700 km.

The High Altitude Mapping Orbit (HAMO) will primarily be used for the optical mapping and reflectance spectral sampling of the surface of the asteroid. The required Level 1 science for this sub-phase will be multi-color surface maps (clear and 3 filters) of 80% of the surface, a topographical map of 80% of the surface, and 5000 VIR spectral frames. This sub-phase will last approximately 26 days and will start at the conclusion of the transfer from Survey Orbit to HAMO, when the spacecraft has achieved a circular near-polar orbit at 950 km radius. A second HAMO sub-phase (HAMO-2) will take place during departure, to acquire images of areas that have become illuminated since the first HAMO mapping, and to fill in gaps in coverage.

The Low Altitude Mapping Orbit (LAMO) is the final science sub-phase at Vesta and will primarily be used to collect gamma ray and neutron spectra as well as to determine the gravity field. LAMO will last 80 days at Vesta and will begin at the end of the HAMO to LAMO transfer when the spacecraft is in a 460 km radius circular near-polar orbit. Details of the Vesta Orbit sub-phases are given in Table 2, and the lighting conditions are shown in Figure 11.

3.1.2.1. *Vesta Approach*

During the Approach sub-phase the instruments will go through complete calibration, repeating some of the activities that were done during the post-launch checkout calibration period, including annealing GRaND if necessary. Following the calibration activities, the FC will collect rotation characterization (RC) maps that include nadir and off-nadir viewing, that will be obtained at increasing resolution as the spacecraft approaches Vesta. For the particular trajectory used in this exercise, the phase angle (Sun-Vesta-Probe [Dawn] angle, SVP) varies from $\sim 88°$ for the first rotation map, (32 km/px) to $\sim 108°$ for the second rotation map (6.5 km/px), and captures the minimum phase angle of $\sim 42°$ in the final, highest resolution (~ 0.5 km/px) rotation map.

To create the global maps, one frame of FC data is obtained every 10° of rotation of Vesta. VIR will obtain full-disc spectra coincident with the rotation characterization maps (RCs). Additionally, VIR data are desired at nadir attitude when the polar regions are visible, and the phase angle is below 90°. The three RC maps will be acquired, respectively, when the imaging is at twice the resolution of the Hubble Space Telescope, 4 times HST resolution, and the last is at the point where the body roughly fills the field-of-view (FOV) and the phase angle is lowest. Data obtained in the Approach phase will provide a range of illumination angles to initialize the topographic model, and aid in finalizing the plans for HAMO and LAMO.

During the Vesta approach period several searches for hazards (dust, moons) will be performed in the far- and near-Vesta environment. During the Pre-Approach Cruise sub-phase, at a distance of approximately 1.4×10^6 km, 2×2 image mosaics will be obtained, covering the entire Hill sphere, optimized to detect potential satellites in orbits having periapsis above the Survey orbit. Additional observations will be made during the transit from Vesta Arrival (capture) to the Survey orbit with strategies optimized to detect potential satellites with orbits crossing the Survey, HAMO and LAMO orbits. VIR characterization of discovered satellites will help to define their relationship to specific regions of Vesta, as well as the mechanism and time of their ejection. Dust particles are most easily observed in forward-scattered light. Dawn approaches Vesta at a modestly high phase angle ($\sim 90°$), peaking at $\sim 105°$ during the second rotation characterization map. It is likely that data obtained during the satellite searches can be used to search for dust in the vicinity of Vesta. Further analysis, including a search for stable orbits, is required before the details of hazards characterization observations can be planned at the activity level.

An additional activity in the Approach phase is to exercise the processing streams for the instruments data, mainly the FC and VIR, to verify that quicklook products can be produced on the required timelines, and to check and improve the calibration parameters.

3.1.2.2. *Vesta Survey Orbit*

The science goals for the Survey orbit are to obtain global coverage with VIR, and to create overlapping global images with the FC in all eight filters, centered every 20° of longitude from the equator to the mid-latitudes ($\sim 50°$). In polar latitudes, the images will be taken at 40° intervals. The VIR map constitutes the primary (and perhaps only) global spectral mapping. The FC maps will be 3–4 frames each. The dispersion in viewing angle across the camera field of view at the Survey altitude will provide global stereo coverage between adjacent overlapping sets of images. These VIR and FC global maps will be used for definition of targets to be investigated at lower altitudes, and the FC data will contribute significantly to the topographic model. Cross-calibration of the VIR and FC will be facilitated by concurrent imaging during this sub-phase.

3.1.2.3. *High Altitude Mapping Orbit*

The High Altitude Mapping Orbit (HAMO) is primarily used to create global FC maps of the lit surface in of the body in multiple filters from a nadir attitude, and two maps from two different off-nadir viewing angles, in the clear filter only. The Level-1 requirement is for 80% global coverage in 3 filters; we plan to acquire images of the entire lit surface in the clear plus seven filters. The three FC clear maps from different viewing angles (nadir plus two off-nadir) will be used to create a topographic model, using also limb images, via stereophotogrammetry and photoclinometry. The off-nadir coverage will be at angles of $10 - 20°$ from the surface normal.

VIR will also collect at least 5000 frames as part of the Level 1 requirements, but this will be far short of global coverage. The VIR frames will be collected concurrently with the FC data if possible. The VIR data will be collected to sample the spectral variability at smaller scales than the global Survey map, and to build up high-resolution coverage of areas of interest. The VIR data will be taken at nadir attitude; however, VIR data may also be acquired at off-nadir viewing angles, for which the scan mirror may be used to tailor the viewing angle when the geometry is favorable.

A second HAMO is planned near the end of the encounter, as the spacecraft is spiraling out from Vesta, to capture different lighting conditions, and to fill in gaps in the coverage. This activity is on the order of 3 days (TBC) and will complete global FC mosaics in multiple filters, including off-nadir views of the new illuminated polar region, and VIR spectra of selected targets and the lit pole.

The HAMO plan is the most constrained of any of the orbit sub-phases. It is constrained by the limitations of the data downlink volume, the need for systematic global coverage, and the need for resiliency to interruptions in the plan. The 26-day HAMO is split into two cycles, each 13 days in length, and each cycle is split into two campaigns. Each cycle achieves global coverage on a set of roughly evenly spaced ground tracks. The campaigns are delineated by the operational mode used for polar mapping (i.e., nadir versus fixed off-nadir versus general target pointing).

During the departure phase, the coverage begun in HAMO will be completed with nadir maps and off-nadir views of the newly illuminated polar region obtained concurrently by slewing, as the body-spacecraft relative motion is slow compared to the time needed to slew the spacecraft to obtain the off-nadir views. Gaps at low-latitude will be filled in an opportunistic manner.

3.1.2.4. *Low Altitude Mapping Orbit*

The purpose of LAMO is to obtain spatially resolved neutron and gamma ray spectra of the asteroid and global tracking coverage to determine the gravity field. The strategy is to be nadir pointed, with exceptions for limb imaging and limited target pointing. The bulk of the time will be spent pointed nadir, which is the attitude that GRaND needs to maximize their signal-to-noise. One 5.5 hr downlink pass per day will obtain tracking data for gravity field determination. For both of these objectives, even ground track coverage is needed. The plan is for a repeating pattern of operations: nadir pointing observations, downlink pass, and orbit maintenance once every two days. There will be downlink to spare, so as much nadir imaging (FC and VIR) as can be fit into the data buffers will be obtained. Off-nadir targeted acquisition is anticipated for a small fraction of the time (likely $< 10\%$), including some very high resolution imaging of selected targets.

A dedicated gravity campaign (several days to one week) of continuous tracking is useful for resolving the wobble in the rotation to measure the moment of inertia of the protoplanets. One possibility is to have two 3-day continuous tracks that would empty buffers twice during LAMO (roughly after each third of the sub-phase), allowing more imaging science to be collected in the lowest altitude orbit.

3.2. *Mission Operations*

The Dawn mission operations will be managed by the Jet Propulsion Laboratory with the Dawn Science Center (DSC) at UCLA having the responsibility for the day-to-day co-ordination of science planning and sequencing activities. The Dawn mission is designed to follow heritage JPL Multi-Mission Ground Systems and Services (MGSS) philosophies and processes for mission operations, where feasible. MGSS has evolved over many years of operating JPL and contractor spacecraft, and has developed processes that are

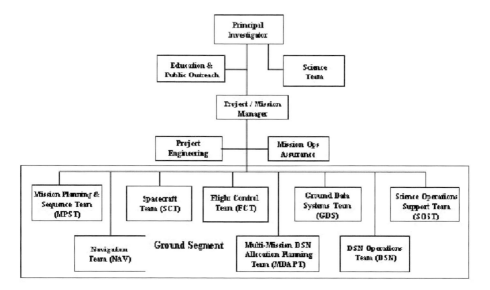

Figure 12. The Dawn Project operations organization showing the teams that are part of the Ground Segment. The Dawn Science Center is included within the SOST.

well formulated, well tested, and flight proven. These processes are adapted for each mission, and lessons learned are incorporated for each new mission. Fitting within the multi-mission operations framework was critical to keeping Dawn operations within the cost constraints of the Discovery program.

The Dawn project operations are the responsibility of the Ground Segment. The Ground Segment is divided into 3 operations organizations: the Navigation Team (NAV), the Mission Operations System (MOS), and the Science team. The NAV team is responsible for the formulation, estimation, and control of the flight systems interplanetary trajectory, as well as orbit determination and control after arrival at each of the asteroids. Prior to launch, the Mission Design and Navigation team is responsible for developing the project mission plan, consistent with the constraints imposed by the flight system design, the trajectory design, the navigation strategy, the science requirements, and the project requirements. The MOS is responsible for the collection of hardware, software, people, processes and procedures needed to operate the Dawn spacecraft from launch through the end of its mission. This same system is also used to support flight system testing during Assembly, Test and Launch Operations (ATLO). The Science team includes the Principal Investigator and all of the members of the project Science Team, as well as the Dawn Science Center, located at UCLA. The Science Team is responsible for planning science data collection, generating science commands and sequences, analyzing the returned science data, disseminating data and results to the broader science community, and the long-term archiving of returned science data. Figure 12 shows the Ground Segment as it is subdivided into the operations teams.

The Dawn project s Ground Segment operations are distributed and global. The Ground Segment infrastructure is encapsulated in the Ground Data System (GDS). The GDS architecture supports the distributed configuration and the interfaces required for flight operations. In effect, the GDS provides the ground-to-space link required to interface the ground system with the flight system. In addition, the GDS supports all mission development and test environments leading up to launch and into operations.

These include the flight system testbeds at Orbital and JPL, ATLO at Orbital, Goddard Space Flight Center, and the Kennedy Space Center, Mission Support Areas at JPL and Orbital, Instrument Sites at DLR, MPS, INAF, ASI and LANL, and science operations at the Dawn Science Center at UCLA.

The Dawn Science Center (DSC) is the primary conduit for the flow of data and information between the Science and Instrument Teams and the MOS. The DSC is collocated with the principal investigator at UCLA so that the PI can provide guidance and assist in prioritizing science observations when conflicts arise due to resource limitations. It has conferencing facilities and office space to Science Team during high activity periods such as flybys and asteroid encounters. In addition to web-accessible computing and data distribution hardware. The DSC also houses a secure flight operations support area where science sequences are generated, validated and submitted to the project and raw downlink telemetry data are received, processed, and rebroadcast to the instrument teams over the public internet.

The operations teams that will be collocated at JPL include the Flight Control Team (FCT), the Mission Planning and Sequence Team (MPST), the Spacecraft Team (SCT), the Ground Data Systems Operations Team (GDST), the Multi-mission Deep Space Network Allocation and Planning Team (MDAPT), the Navigation Team (NAV), and the Science Operations Support Team (SOST). The GDS, NAV, MPST, and MDAPT teams include multi-mission support personnel who are not necessarily collocated with Dawn operations team members in the Mission Support Area. The SOST is the JPL interface the Dawn Science Center.

3.2.1. *Spacecraft Operations*

Although Orbital Sciences Corporation is building the flight system, spacecraft operations will be performed by a JPL Spacecraft Team following the post-launch initial checkout period. During Launch through initial spacecraft checkout, the SCT is an integrated distributed team between both JPL and Orbital. The Orbital spacecraft team will design the initial checkout activities and define the spacecraft flight rules and constraints to be followed once operations have been transitioned to JPL. This is one significant difference between Dawn and other JPL multi-mission projects where the spacecraft operations typically remain the responsibility of the spacecraft contractor.

3.2.2. *Navigation Planning and Analysis*

The Navigation (NAV) Team at JPL performs the orbit determination (OD) analysis and defines the IPS thrust profile to maintain the spacecraft trajectory to meet the mission objectives. During cruise, the OD analysis is performed with traditional radiometric navigation data (two-way Doppler and ranging) received from the Deep Space Network, once to twice weekly during cruise, and more frequently during initial checkout, the Mars Gravity Assist, and asteroid operations. During asteroid operations the radiometric data is supplemented with optical navigation data taken by the Framing Camera. During asteroid operations the NAV Team will make specific requests to the Framing Camera Team to acquire the images needed for optical navigation. The optical navigation data will be given a higher downlink priority over science imaging data to minimize delays in orbit determination.

As the flight system nears the asteroid, IPS thrusting is used to spiral in to the asteroid orbit. On final approach and during the early orbit phase, the navigation team will improve estimates of the asteroid mass, gravity field, and spin state in order to predict the trajectory and compute subsequent orbit correction maneuvers. The navigation team will update the orbit plan once the asteroid physical parameter estimates are adequately determined.

When in orbit about the asteroid, the navigation team relies on a combination of optical navigation and radiometric navigation. Typically, radio tracking will be available 24 – 56 hours/week in science orbits, and 24 hours/week in transfers between those science orbits. Dedicated optical navigation data will be taken in approximately 1-hour sessions, once per day.

Navigation will develop a topographic map sufficient for navigation purposes, using a combined stereophotogrammetry/photoclinometry method that will approach a 10-m accuracy for regions of the surface. The optical landmark tracking using in this method depends on knowledge of Framing Camera pointing and images of surface features (landmarks). Over time, specific landmarks observed in the images provide geometric information that is combined with the optical navigation data and/or radiometric data in the orbit determination filter.

3.2.3. *Mission Planning and Sequencing*

The Mission Planning & Sequencing Team (MPST) is designed to support multiple projects in an efficient manner. By utilizing a combination of project dedicated and multi-mission personnel, the team is able to relatively easily shift resources to support the varying needs of a mission throughout their development and operations life cycles. In addition, the cross training that is required to support this approach, provides for a pool of well trained personnel, experienced in operations on multiple projects. The MPST will also be using a suite of multi-mission planning and sequencing tools that are being adapted for the Dawn Mission. The JPL MOS as well as the DSC at UCLA will use this tool set. The use of common tools simplifies the interface between the MOS and the DSC and minimizes costs.

The Dawn uplink process expands on the heritage uplink process used by other MGSS missions, with some adaptation for Dawn. The uplink process is guided by the Mission Plan and Science Plan both completed prior to launch, and updated 18 and 6 months, respectively prior to arrival at each asteroid. Instrument sequence and mission sequence plans are generated to define in detail sequence load boundaries and resource allocations, in order to be ready to build instrument sequences and science sequences for uplink. The fixed antenna on the Dawn spacecraft drives the need to select telecommunication passes with the Deep Space Network based on the geometry of the orbit and science objectives. Because the DSN is shared with other flight projects, requests for tracking coverage need to be requested far in advance, but whether or not they will be available will not be known with complete certainty until two months before they are used. Therefore, the science sequences need to be flexible enough to shift with changes in the DSN schedule. The science sequence generation is lead by the Dawn Science Center (DSC) and is described in the next section on Science Operations. The DSC will deliver the science sequences as conflict-free products ready to be processed into uplink products by the MPST, as well as other supporting products. Finally, the stored sequence is created by the MPST following a heritage MGSS process used for Odyssey and MGS, with minor changes for Dawn. The details of the spacecraft engineering activities will be added at this point along with the detailed specification of the spacecraft pointing.

All of the instruments are mounted directly to the spacecraft so the instrument pointing involves turning the spacecraft. The instruments also have radiators mounted on the -X side of the spacecraft and so all turns must be modeled to insure that they do not violate the sun constraints. Because the Dawn uplink process is success oriented and there is not

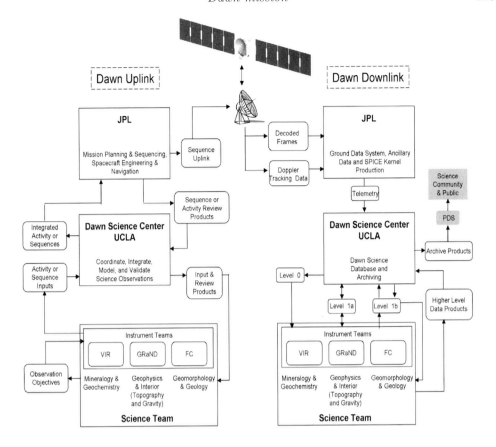

Figure 13. Dawn Uplink and Downlink process schematic. This diagram shows the interactions of the primary organizations associated with the Dawn uplink and downlink processes. Uplink products flow up the left-hand side of the diagram from the Science Team to the spacecraft and downlink products flow down the right-hand side.

time in the schedule for late changes, the science pointing strategies are being developed well in advance of the uplink process.

3.3. *Science Operations*

The DSC plays a central role in science planning and sequencing (uplink) and data distribution within the Science Team and the science community at large (downlink). Cost-capped missions such as Dawn cannot afford large mission operations teams. In order to work within the staffing limits, science observation strategies must be planned well in advance and only minimally revised during the science sequencing process to account for slight shifts in trajectory or timing. The left side of Figure 13 depicts the Dawn uplink process. The Science Team defines the science objectives and creates an observation plan to be implemented by the instrument teams. The instrument teams initially define observations (activities) to meet the objectives that are then passed to the DSC. The DSC integrates the activities and verifies that the integrated plan can be implemented within available spacecraft resources (memory buffers, downlink capability, etc.). Mission planners from the MPST review the plan to verify that sufficient margin remains for spacecraft engineering activities such as orbit maintenance, optical navigation, and downlink.

Several iterations are typically required in order to produce a plan that meets both the observation objectives and the spacecraft resource constraints. Once the observation plan is complete, the process is repeated at the instrument command level. The instrument teams design the commands sequences that are necessary to implement the observation plan for each mission phase. The DSC integrates and validates the sequences and produces review products. Once the science sequence has been accepted by the Science Team, it is integrated with the background spacecraft sequence and reviewed again prior to being sent to the spacecraft.

The right side of Figure 14 shows the downlink data flow process. Raw telemetry flows from the spacecraft, to the Deep Space Network (DSN), to the Dawn ground data system at JPL and then on to the DSC. Instrument housekeeping data that are returned in real-time are immediately rebroadcast to the instrument team home institutions for health and safety monitoring. Raw science data packets are stored in the Dawn Science Database (DSDb) after the completion of each DSN pass. Once the data become available, the instrument teams download the level 0 data (raw telemetry). After the data are decompressed, decoded, and formatted into scientifically useful data structures, the level-1a data products are uploaded back into the DSDb. Additional data processing is performed by the instrument teams to produce calibrated (level-1b) data products (radiometrically corrected images and spectra and fluxes from the GRaND instrument) which are then uploaded to the DSDb, along with data set descriptions, calibration data and instrument models, and calibration procedures. The Science Team uses the calibrated data acquired from the DSDb to produce derived data products (maps, mosaics, etc.) which are also deposited in the DSDb. The Dawn Science Center produces archive volumes (DVDs or other approved media) in compliance with PDS archiving standards and submits them to the PDS for archive and distribution to the science community and general public.

3.3.1. *Dawn Science Database*

The Dawn Science Database (DSDb) is a Dawn adaptation of technology developed by the Planetary Plasma Interactions (PPI) Node of the PDS. The MySQL database contains an inventory of external data files that includes metadata derived from the PDS labels associated with each file. The metadata are used to constrain queries that can locate individual files or collections of similar files. As a result, all products within the DSDb must be fully PDS compatible. All file management system systems require some form of metadata in order to perform their functions. The selection of the PDS metadata format for internal data tracking will reduce the cost and improve the efficiency of the downstream archive process.

The DSDb supports both file upload and download actions. All DSDb users are permitted to download any file within the inventory, however, only selected users are allowed to upload or update files. Once the user has constructed a query by selecting various options, a list of matching products is returned. Users may select none, any, or all of the products in the return set. The download process uses standard shopping cart technology to keep track of selected products. Once the user is finished, the contents of the shopping cart are zipped up and transferred to users workstation. Figure 15 shows the top level page of the DSDb after the user has been authenticated and user privileges determined. The web interface to the DSDb is dynamically generated based on user privileges so that users are only presented with options that they are authorized to perform. The view of the Uplink pages shown in Figure 14 only appears to users with instrument manager

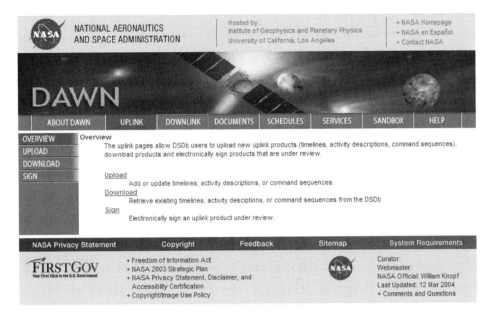

Figure 14. Dawn Uplink and Downlink process schematic. This diagram shows the interactions of the primary organization

privileges. The PDS labels for data products, calibration files, etc. that are required by the DSDb for inventory tracking are uploaded with the products. Many of these products use PDS labels that are physically attached to the data files. PDS labels for other products such as sequences, documents, and presentations are generated automatically after the user enters a few required parameters through the web interface. Raw telemetry data files (level 0) are labeled by a separate automated process at the time they are created.

3.3.2. *Archive Validation, Generation and Schedule*

The primary mechanism by which data products will be validated is through usage by the Dawn Science Team. All products used by the Science Team for analysis or to produce higher level data products are acquired from the DSDb. Errors that are uncovered by users are reported back to the data provider through the DSDb. When corrected files are uploaded to the DSDb the original files are removed from the inventory.

Table 3 gives the basic Dawn data internal distribution and archive schedule. In general, raw telemetry data are provided to the team within 6 hours of each downlink pass and submitted to the PDS to be saved at the end of each mission phase. Raw (level 1a) and calibrated (level 1b) data are then returned by the teams to the DSDb weekly. Raw data are submitted to the PDS for archive at 3 months after the end of each mission phase. Calibrated data are archived within 6 months after the end of each asteroid encounter. This additional delay is needed in order to validate the instrument calibrations. Finally, higher level data products (maps, mosaics, models) must be uploaded into the DSDb within 10 months after departure. This allows the Science Team approximately 2 months to review these products before they are submitted to the PDS for archive. For Ceres, we plan to reduce these delivery times by a factor of two (as indicated in Table 3) in order to archive the data before the end of mission is reached, which occurs 5 months after Ceres departure. This timeline is reasonable because the team will have gained experience with the Vesta deliveries.

Table 3. Dawn data products, processing levels and availability

Processing Level	Product description	Available to Science Team	PDS Archive
Level 0	Raw telemetry data	6 hrs after DSN pass	End of mission phase
Level 1a	Raw data (data numbers) reformatted into images, spectral cubes, time-series	Within 7 days of downlink	3 months after end of phase
Level 1b	Radiometrically calibrated images and spectra, gamma ray and neutron count rate time-series	Within 14 days of downlink	6 (3)[1] months after asteroid departure
Level 2	Geometrically corrected and co-registered image and spectral, gridded gamma ray and neutron count rates, gravity coefficients and uncertainties	Within 6 (3) months of asteroid departure	12 (6) months after asteroid departure
Level 3	Global clear and colored atlases and global mosaic, Olivine and Pyroxene abundance maps, Elemental abundance maps, Free-air gravity map, geoid and uncertainty map	Within 8 (4) months of asteroid departure	12 (6) months after asteroid departure
Level 4	Topographic model, Geologic map, Bouguer gravity map	Within 10 (5) months of asteroid departure	12 (6) months after asteroid departure

1) Numbers in parentheses pertain to Ceres

Dawn data will undergo an additional set of reviews associated with the PDS archive process. Prior to launch, the design of the standard data products (levels 1a, 1b, and 2) will each be formalized in an interface control document (ICD) that will be subject to PDS review and signature. Once the archive design documents are signed, archive volumes will be generated using data from the instrument checkout and the initial in-flight calibration tests. These data will undergo a full PDS peer review and lien resolution process. The peer review panel will verify the data are scientifically useful and adequately documented, and that the data formats are usable to the science community at large. Once this process is complete, the Dawn archive design will be under configuration management and any design changes will require an additional external review. The final validation procedure is to verify that the data sets submitted to PDS conform to the standards set forth in the archive design specification.

4. Concluding Remarks

Dawn makes the first attempt to explore the nebular compositional gradient by exploring two bodies on either side of the dew/frost line in the early solar system. Ion propulsion is the technology that enables this advance in our exploration of the solar system. The spacecraft is nearing completion and plans are well underway to operate the spacecraft and return and archive the data needed to achieve the scientific objectives. Launching a mission of this size is a complex, but very worthwhile undertaking. Discovery missions provide good value for the expenditures required. Nevertheless Dawn is only the first step. There is much left to do in exploring the minor planets.

Acknowledgements

The success of the Dawn project has been
due to the hard work of many individuals at Orbital Sciences Corporation, the Jet Propulsion Laboratory and their subcontractors. In addition the Dawn Science Team has played an essential role in designing the mission and supplying the payload. We are extremely grateful for their help. This work was supported by the Jet Propulsion Laboratory under contract to the National Aeronautics and Space Administration.

References

Binzel, R.P., Gaffey, M.J., Thomas, P.C., Zellner, B.H., Storrs, A.D., Well, E.N. 1997, *Icarus* 128, 95

Drummond, J.D., Fugate, R.Q., Christou, J.C. 1998, *Icarus* 132, 80

Kuzmanoski, M. & Kovacevic, A. 2002, *Astron. & Astro.* 395, L17

Li, J-Y., McFadden, L.A., Parker, J.W., Young, E.F., Stern, S.A., Thomas, P.C., Russell, C.T., Sykes, M.V. 2005, *Icarus* submitted

McCord, T.B. & Sotin, C. 2005, *J. Geophys. Res.* 110, DOI:10.1029/2004JE002377, E05009-1-E05009-14

Russell, C.T., Capaccioni, F., Coradini, A., De Sanctis, M.C., Feldman, W.C., Jaumann, R., Keller, H.U., McCord, T.B., McFadden, L.A., Mottola, S., Pieters, C.M., Prettyman, T.H., Raymond, C.A., Sykes, M.V. Smith, D.E., Zuber, M.T. 2005, in: *Solar System Small Bodies: Synergy between In Situ and Remote Observations*, (Springer)

Thomas, P.C., Parker, J.W., McFadden, L.A., Russell, C.T., Stern, S.A., Sykes, M.V., Young, E.F. 2005 *Nature* 437, 224

Treiman, A.H., Lanzirotti, A. & Xirouchakis, D. 2004 *Earth Planet Sci. Lett.* 219, 189

Asteroids, Comets, Meteors
Proceedings IAU Symposium No. 229, 2005
D. Lazzaro, S. Ferraz-Mello & J.A. Fernánez, eds.

Review of Spitzer Space Telescope observations of small bodies

Y. R. Fernández[1], J. P. Emery[2], D. P. Cruikshank[2] and J. A. Stansberry[3]

[1]Dept. of Physics, Univ. of Central Florida, 4000 Central Florida Blvd., Orlando, FL 32816-2385 U.S.A.; Formerly at Institute for Astronomy, University of Hawai'i at Mānoa, 2680 Woodlawn Drive, Honolulu, HI 96822 U.S.A.
email: yan@physics.ucf.edu

[2]NASA Ames Research Center, MS 245-6, Moffett Field, CA 94035 U.S.A.
email: jemery@mail.arc.nasa.gov, Dale.P.Cruikshank@nasa.gov

[3]Steward Observatory, University of Arizona 933 N. Cherry Avenue, Tucson, AZ 85721 U.S.A.
email: stansber@as.arizona.edu

Abstract. The *Spitzer Space Telescope* (SST), aloft for over two years at time of writing, has so far devoted almost 600 hours of observing time to Solar System science, and small bodies make up a significant fraction of the objects that have been observed. For the first time we now have high accuracy mid-infrared data to study the fundamental mineralogical and physical properties of a large number of objects of different types. In this paper we review some of the exciting recent results derived from SST photometry (in six bands from 3.6 to 70 μm) and spectroscopy (from 5 to 40 μm) of asteroids and comets. The observations reveal their spectral energy distributions (SEDs), and we discuss three important science goals that can be addressed with these data: (1) finding compositional diagnostics of these objects, (2) determining their bulk thermal properties, and (3) understanding the surface evolution of primitive and icy bodies. We focus primarily on comet-asteroid transition objects, low-albedo asteroids, cometary nuclei, Trojans, Centaurs, and trans-Neptunian objects. In particular, we will show: emissivity features in the SEDs and identification of the compositional sources, the constraints on thermal inertia and infrared beaming through the samples of the thermal continuum, and an intercomparison of albedos across dynamically-related bodies.

Keywords. comets: general, infrared: solar system, minor planets

1. Introduction

The *Spitzer Space Telescope* (SST) was launched on 2003 August 25 as the fourth of NASA's "Great Observatories," covering the mid- to far-infrared wavelength regime. A review of the spacecraft design and instrument complement is given by Werner, Roellig, Low, *et al.* (2004). The spacecraft is in a heliocentric orbit that takes it farther and farther from Earth over the course of its expected five to six-year lifespan. The telescope itself, with a 0.85-m diameter primary mirror, is cooled to cryogenic temperatures and for the most part is thermally isolated from the rest of the spacecraft. The combination of innovative design and orbit reduces the heat environment surrounding the instruments, and so gives them spectacular infrared sensitivity in comparison to previous infrared space telescopes.

There are three infrared science instruments: InfraRed Array Camera (IRAC), InfraRed Spectrograph (IRS), and Multiband Imaging Photometry for *Spitzer* (MIPS). All three instruments have imaging capability, and two (IRS and MIPS) have spectroscopic modes. IRAC (Fazio, Hora, Allen, *et al.* 2004) obtains broadband images at four wavelengths $\lambda = 3.6$, 4.5, 5.8, and 8.0 μm. Each wavelength has its own detector with a $(5.2')^2$ field of view. The imaging mode for MIPS (Rieke, Young, Engelbracht, *et al.* 2004) is similar, with three wavelengths – 24, 70, and 160 μm – each having its own detector. The fields of view are $(5.4')^2$, 5.3'-by-2.6', and 5.3'-by-0.5', respectively. Finally, IRS (Houck, Rellig, van Cleve, *et al.* 2004) can perform small field of view (about $1'^2$) imaging at $\lambda = 16$ and 22 μm; this is primarily intended for peak-up imaging to place a target within an IRS slit.

MIPS also has a spectroscopic mode, where a slit spectrum of low resolution ($R \approx 15$ to 25) can be obtained over $\lambda = 60$ to 100 μm. However, the primary spectroscopic instrument is IRS with four modules that cover $\lambda = 5.2$ to 38 μm. Two modules work at low resolution ($R \approx 100$) and two at high ($R \approx 600$). Each module has its own single slit.

2. Observations

2.1. *Telescope Activity*

In practice, observing with SST occurs through scripts that are created by the observer using a software tool ("SPOT") provided by the Spitzer Science Center. A script for a single independently-schedulable observation automatically calculates telescope and instrument overheads, and the duration of the observation includes these overheads.

Counting all observations through 2005 October 5, which includes all of General Observer Cycle 1 and the start of Cycle 2, there have been 10,990 hours' worth of scripts observed. Of those, 3095 hours were for the six Legacy projects, which are virtually complete. Of the remaining 7895 hours, 592 hours (about 7.5%) have been devoted to Solar System science. This fraction of time is completely consistent with the proposal submission fraction from the Solar System community. Including Cycle 2 (which started in June) and including Director's Discretionary Time, there are currently 58 *Spitzer* projects directly related to our Solar System.

Almost three-fourths of the telescope time for Solar System observations has employed MIPS, and most of the rest is IRS. IRAC currently accounts for under 4% of the time. (Note that one cannot observe with multiple instruments simultaneously.) Among non-Solar System programs, IRAC was the most-used instrument (in terms of hours) for the Legacy projects, and has been used for over one-fourth of the other non-Solar System telescope time. The relatively small use of IRAC for Solar System science is partly due to the shape of the spectral energy distributions (SEDs) of many objects; the emission reaches a local minimum where the reflected and thermal components cross, and this often occurs within the IRAC wavelengths. Nonetheless, IRAC is a sensitive instrument and several planetary science programs are making use of it.

2.2. *Imaging and Photometry*

Figure 1 displays what minimum requirements are needed to obtain useful IRAC data in a relatively short amount of observing time. For an object with a geometric albedo of 0.10 and a given heliocentric distance r, we have plotted the requisite effective radius and requisite R-band magnitude necessary to obtain a 4-σ detection of the object in each IRAC band in 1080 seconds of integration. Such an observation would take 50 minutes. The sensitivities of the four IRAC bands (in increasing wavelength) are 0.42, 0.68, 4.5, and 5.3 μJy, respectively, for these integration times. This assumes a "high" background appropriate for the ecliptic. Note that in practice it is possible to achieve

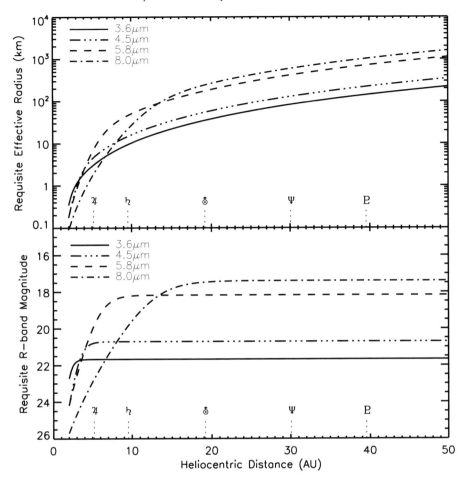

Figure 1. Minimum radius (top) and maximum magnitude (bottom) required to obtain $S/N = 4$ in a given IRAC band for a given heliocentric distance after 1080 seconds of integration. Calculations assume a geometric albedo of 0.10, phase slope parameter of 0.05, solar colors, and an observation occuring at quadrature. Semimajor axes of major planets and Pluto are marked.

sensitivities below the so-called "confusion limit" by obtaining shadow observations, a technique uniquely available to Solar System projects.

One obvious feature in Fig. 1 is the flattening out of the magnitude above a certain distance. This shows at which r the flux at a wavelength is dominated by reflected sunlight and not thermal emission.

We show an analogous graph for MIPS in Fig. 2, where we have assumed 2800 seconds of integration at 24 μm, 600 seconds at 70 μm, and 120 seconds at 160 μm. If all in one observing script, such an observation would take about 115 minutes. The sensitivities of the three MIPS bands (in increasing wavelength) are 0.021, 2.3, and 35.0 mJy, respectively, for these integration times (and a "high" background). As with IRAC, shadowing of MIPS observations can also be used to reach deeper sensitivites. Note that in many Solar System projects, observations at 160 μm are difficult. For objects emitting according to the Rayleigh-Jeans law in the 70 to 160 μm region, the relative sensitivities are such that the S/N at 160 μm will be a factor of about 35 smaller than that at 70 μm for the same exposure time.

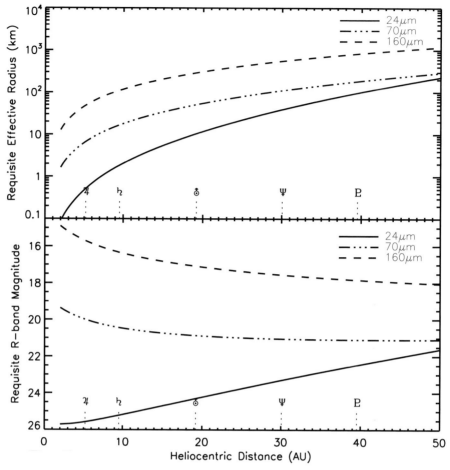

Figure 2. Minimum radius (top) and maximum magnitude (bottom) required to obtain $S/N = 4$ in a given MIPS band for a given heliocentric distance. Exposure times of 2800, 600, and 120 seconds were assumed for the three bands. Calculations assume a geometric albedo of 0.10, phase slope parameter of 0.05, solar colors, and an observation occuring at quadrature. Semimajor axes of major planets and Pluto are marked.

The figure demonstrates that for most Solar System objects, 24 μm is the better choice for obtaining a simple detection. While one could observe longer at 70 μm to bring the crossing point of the 24 μm and 70 μm curves in Fig. 2 to smaller r, it turns out that there is much more overhead for 70 and 160 μm observing.

3. Results and Discussion

3.1. *Topics*

A report on "small bodies" could encompass many types of Solar System objects. SST has already observed many small regular and irregular satellites, comets, near-Earth asteroids, unusual asteroids, small Main-Belt asteroids, zodiacal dust, and Trojans[†].

[†] A list of all projects as well as information on Solar System observing is maintained by one of us (YRF) at URL http://www.physics.ucf.edu/~yfernandez/sss.html.

There are a few projects that we will not discuss here: projects devoted to the outer planets; projects focussing on Pluto and the other, very large TNOs; and projects in Cycle 2 that have yet to receive any observations. This leaves a core group of about 12 projects devoted to dust and 15 projects devoted to the solid bodies themselves. The results of some of these projects will be discussed in the following subsections.

3.2. Dust

There are over twenty projects devoted to Solar System dust – more than one-third of all planetary programs. In particular there are two "medium" projects (i.e. allocation between 50 and 200 hours), run by M. V. Sykes (in Cycle 1) and W. F. Bottke (in Cycle 2), studying the asteroidal dust bands. The combined allocation of these projects is about 270 hours, so they will provide much deeper and more extensive maps of the bands' surface brightness than were possible before. Also of note is a multi-cycle project run by S. Jayaraman to map the spatial extent of the dust cloud in Earth's orbital wake. This is a fascinating and clever application of Spitzer's unique position on an independent heliocentric orbit. During Cycle 2, Spitzer will be from 0.19 to 0.32 AU from Earth, giving us an unprecedented vantage point on the near-Earth dust environment. All of these projects are still in the early stages of analysis.

Reach (2004) and collaborators are doing an extensive survey of the dynamics of cometary dust trails. While trails have been seen in several comets with IRAS (Sykes & Walker 1992) and from ground-based data (Ishiguro, *et al.* 2003), only Spitzer has allowed us to determine the true frequency of their occurrence. Reach (2004) finds that a large fraction of all Jupiter-family comets have trails. Further analysis will reveal if the emission of very large dust grains is correlated with the gas production rate or the orbital anomaly.

Woodward, Kelley, Harker, *et al.* (2005) have several projects devoted to understanding composition of cometary dust in both Jupiter-family and long-period Oort-Cloud comets. Ground-based mid-IR spectroscopy in the 8 to 13 μm region has in the past given us insight into basic silicate composition of dust. However, this technique can only be used on relatively few comets for lack of sufficient signal-to-noise. Spitzer on the other hand has much better sensitivity limitations and a comprehensive survey of comets from all dynamical classes is now possible. Since comets were formed throughout the outer-planet region in the protoplanetary disk, there may be a signature of a compositional gradient within the comets of differing dynamical classes.

Finally, one of the most exciting dust-related projects involves observations of comet Tempel 1. In July 2005, the Discovery-class spacecraft *Deep Impact* delivered a 10-km/s impactor onto the surface of this comet (A'Hearn, Belton, Delamere, *et al.* 2005). Spitzer's IRS observed the exact moment of impact, taking 5.2–8.7 μm spectra with high temporal frequency to watch the rapidly changing post-impact environment (van Cleve, Lisse, Grillmair, *et al.* 2005). Spectra through all of IRS's range were also obtained in the hours following impact. The data are particularly rich in showing emission from a wide variety of compounds in the impact ejecta. Again, analysis is very preliminary, but since there is a good chance that Spitzer was observing relatively unprocessed subsurface material, the data hold great promise for understanding what pristine material comets hold.

3.3. Asteroid Counts

One project performed early in the mission was the Ecliptic Plane Survey (EPS), as discussed by Meadows, Bhattacharya, Reach, *et al.* (2004). The goal was to characterize the sky-plane density of small (kilometer-sized) Main Belt (MB) asteroids, which is helpful for understanding not only the size distribution in the MB but also the contamination

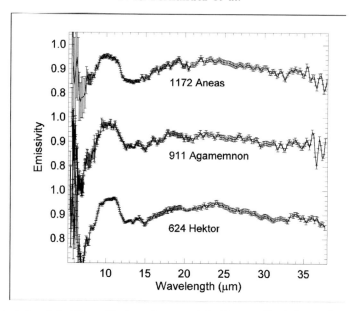

Figure 3. Emissivity plot of three Trojans, from work by Emery, Cruikshank, & van Cleve (2006) using the IRS on SST. Each object's SED has been divided by the continuum as determined using the standard thermal model for slow-rotators (Lebofsky & Spencer 1989).

rate by asteroids in sidereally-tracked fields. Our understanding of these small MB objects is very incomplete. The EPS mapped two fields with IRAC at 8 μm and MIPS at 24 μm, with the fields at two ecliptic latitudes, 0° and +5°. Most of the asteroids that were found were previously unknown even though the exposures were relatively short, demonstrating that Spitzer is a highly efficient asteroid finder. Meadows, Bhattacharya, Reach, *et al.* (2004) report that the ratio of objects found at the two latitudes is nominally higher than expected, although it is not clear yet whether this difference is statistically significant. Poisson statistics of the EPS's asteroid counts coupled with the spread in the predictive models indicate that further observations are necessary at a variety of ecliptic latitudes. Spitzer has given us new clues to solve a problem that was quite infeasible before, and even a modest investment in Spitzer time would yield significant advances in our knowledge of the small end of the asteroid distribution.

3.4. *Composition*

The technique of measuring an object's SED with IRS is a powerful way to find spectral diagnostics that reveal clues to the object's composition. Emery, Cruikshank, & van Cleve (2006) give a detailed analysis of the spectra of three Jovian Trojan asteroids, (624) Hektor, (911) Agamemnon, and (1172) Aneas. As the Trojans are uniquely placed in the Solar System one primary motivation for this study is to understand their constituents in the context of comets and outer Solar System asteroids, and in particular the organic component and the specific proportions of silicates.

The emissivity spectra of the three Trojans are shown in Fig. 3, and spectral structure is clearly evident in all of them. Emery, Cruikshank, & van Cleve (2006) point out the following features: An emission plateau at about 9.1 to 11.5 μm has a spectral contrast of 10 to 15%, and a broader emission high is apparent from about 18 to 28 μm. More subtle features include a possible double peak within the 18 to 28 μm rise and another rise near 34 μm. Differences among the Trojan spectra are also subtle. Hektor and Agamemnon exhibit a small (few % contrast) feature near 14 μm that is absent from the Aneas

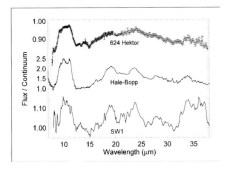

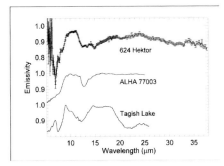

Figure 4. Comparison of the emissivity between a Trojan, Hektor, and two comets, C/1995 O1 Hale-Bopp, and 29P/Schwassmann-Wachmann 1. The figure is taken from the work by Emery, Cruikshank, & van Cleve (2006). Cometary data from Crovisier, Leech, Bockelee-Morvan, *et al.* (1997) and Stansberry, van Cleve, Reach, *et al.* (2004).

Figure 5. Comparison of the emissivity between a Trojan, Hektor, and two meteorites, the CO3-type ALH 77003 and Tagish Lake. The figure is taken from the work by Emery, Cruikshank, & van Cleve (2006).

spectrum. Hektor's 10-μm plateau is slightly tilted, with a peak at about 11.2 μm, whereas those of Agamemnon and Aneas are relatively flat and perhaps even a bit rounded.

Note that care must be taken near wavelengths where spectra from two different orders have been joined. This occurs over the 14.0-to-14.5 μm interval and the 19.5-to-21.3 μm interval. However Emery, Cruikshank, & van Cleve (2006) state that the 14-μm feature in Fig. 3 is likely real, and that the 20.5-μm local minimum is broader than would be expected if it were due to poor order stitching.

It is interesting to compare these spectra with that of cometary dust, meteorites, and laboratory analogues. This is shown in Figs. 4 and 5. The overall similarities with both a long-period comet (Hale-Bopp) and a short-period comet/active Centaur (P/Schwass\-mann-Wachmann 1) are striking. The grains on the surface of Hektor, though making up a solid body, have a similar SED to the micron-sized grains in these comets' comae. Hektor's surface has fine-grained silicates, and when Emery, Cruikshank, & van Cleve (2006) compare the specific band positions with comet spectra, the conclude that Hektor's is dominated by a Mg-rich, amorphous silicate mineralogy (though they also state that the analysis is hampered by limitations of spectral libraries and scattering theories for the appropriate wavelengths).

In Fig. 5, the closest meteorite match that Emery, Cruikshank, & van Cleve (2006) found was with ALH 77003, a CO3 meteorite. Interestingly, it appears that the spectral characteristics of Tagish Lake meteorite are very unlike that of Hektor. The fact that Tagish Lake is thought to be an analog to the D-type asteroids – of which Hektor is a member – apparently does not imply that the mid-IR spectrum is similar (although the thermal emission spectrum for Tagish Lake is still somewhat uncertain).

The presence of fine-grained silicates in Hektor supports the view that the visible/near-IR spectral characteristics (low albedo, red slope, no absorption features) are due to silicates, not organics. The middle part of the Solar System, the transition region between rocky and icy objects, does not appear to contain an abundance of organic material. Emery, Cruikshank, & van Cleve (2006) suggest that the spectral (taxonomic) trends observed in the Main Belt are due to changes in silicate mineralogy with heliocentric distance. Rather than organics causing red spectral slopes in the outer Main Belt and Trojan

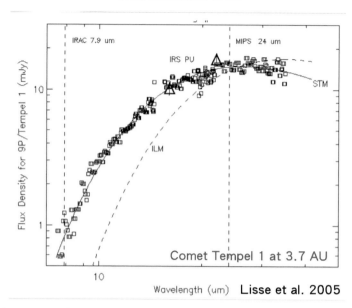

Figure 6. IRS spectrum (squares) of comet 9P/Tempel 1's nucleus when it was at 3.6 AU in March 2004 (Lisse, A'Hearn, Groussin, *et al.* 2005). The slow-rotator thermal model ("STM") fits very well; in contrast, the rapid-rotator model ("ILM") fails. Also plotted (triangles) is photometry from the IRS peak-up camera. Figure courtesy C. M. Lisse.

swarms, it is likely that these surfaces and their spectra are dominated by amorphous silicates. Similarities with comet spectra, and therefore silicate mineralogy, suggest that whereas the silicate fraction in the inner Solar System varies significantly in mineralogy due largely to a temperature gradient during and after accretion, the middle and outer nebular silicate mineralogy may have been more homogenous. Mid-IR spectral observations of a much larger sample of asteroids and comets (as made possible by Spitzer) that span a range of taxonomic types, particularly those without diagnostic absorptions in the visible/near-IR, will test these ideas and provide more insight into silicate mineralogy in the Solar System.

3.5. *Thermal Properties*

One significant mission-oriented result from SST involves comet 9P/Tempel 1. Observations in March 2004 using IRS spectroscopy and imaging allowed once and for all the effective radius (over a rotation period) and the geometric albedo of the nucleus to be pinned down. This was accomplished well before the arrival of the *Deep Impact* spacecraft; understanding the basic physical properties of the nucleus beforehand was crucial for the mission success, in particular for the design and planning of the impactor's autonomous targeting software. The analysis, which showed an effective radius of 3.3 ± 0.2 km and a visible-wavelength geometric albedo of 0.04 ± 0.01, is reported by Lisse, A'Hearn, Groussin, *et al.* (2005). The good signal-to-noise ratio of the spectrum allowed these workers to also constrain the thermal inertia, as shown in Fig. 6. In the past thermal inertia – and hence the applicability of the slow-rotator thermal model (Lebofsky & Spencer 1989) – has been very difficult to quantify in cometary nuclei, since in the large majority of cases ground-based observing yields only broadband photometry at modest S/N. The SST observations confirmed that Tempel 1's thermal inertia was very low, less than that of the Moon, giving confidence that the standard thermal

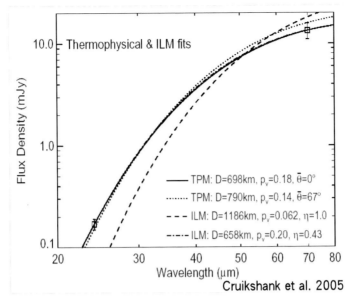

Figure 7. Photometry of TNO 2002 AW$_{197}$ from MIPS at 24 and 70 μm (squares). Various thermal models are plotted as well. The best fitting model is one that is intermediate between the slow- and rapid-rotator models, and that requires some appreciable obliquity of the rotation axis. The visible-wavelength magnitudes are well constrained, and a geometric albedo of about 0.15 was derived.

model for slow rotators is valid. And indeed, the conversion of radiometry to nucleus size was confirmed by the *Deep Impact* flyby, which encountered a nucleus of effective radius 3.0 ± 0.1 km (A'Hearn, Belton, Delamere, *et al.* 2005).

3.6. *Albedo Diversity*

Arguably one of the most significant Solar System results to come out of SST observations is the diversity of albedos among the Centaur and TNO populations. This was tentatively suspected from some earlier low-S/N ground-based thermal measurements (e.g. Fernández, Jewitt, & Sheppard 2002). However the SST results are far superior.

Two projects, whose PIs are G. Rieke and D. P. Cruikshank, have been surveying several of these objects at 24 and 70 μm with MIPS in order to constrain their thermal properties, radii, and albedos. The results have been described at conferences by Stansberry, Cruikshank, Grundy, *et al.* (2005)[†] and Cruikshank, Barucci, Emery, *et al.* (2005). A paper by J. A. Stansberry is in preparation.

A figure from the work by Cruikshank, Stansberry, Emery, *et al.* (2005) on 2002 AW$_{197}$, one of the first TNOs to be observed by SST, is shown in Fig. 7. The analysis comes down to finding what thermal model matches the observed 24-to-70 μm color. In this case, and in virtually all objects at these heliocentric distances, neither the slow-rotator (STM) nor the rapid-rotator (ILM) models is adequate. The STM can be made to fit if the so-called "beaming parameter" is set to 1.2. This is not an unphysical number, but it does indicate that there is some thermal emission from the night side of the object. Therefore to achieve a more robust calculation of the effective radius, an intermediate thermal model is required. Cruikshank, Stansberry, Emery, *et al.* (2005) use a thermophysical model

---

[†] Note that a few of the albedos provided in the published abstract were updated in the oral presentation itself.

(TPM) that accounts for night-side emission and for the obliquity of the rotation axis with respect to the Sun-Earth-object plane. For 2002 AW$_{197}$, an appreciable obliquity was required to achieve the best fit to the MIPS color. In combination with the already-known visible magnitude, Cruikshank, Stansberry, Emery, *et al.* (2005) derive a geometric albedo of 0.15, much higher than the 0.04 canonically assumed in the pre-Spitzer era.

The analysis of 7 TNOs and 7 Centaurs by Stansberry, Cruikshank, Grundy, *et al.* (2005) and Cruikshank, Barucci, Emery, *et al.* (2005) continues the trend. The TNOs and Centaurs have a wide distribution of albedos, with the TNOs ranging from about 0.01 to 0.2, and the Centaurs ranging from about 0.02 to 0.07. Chiron's albedo is already known to be about 0.12 (Campins, Telesco, Osip, *et al.* 1994, Fernández, Jewitt, & Sheppard 2002), extending the Centaurs' range, but so far the Centaurs do not appear to be as reflective as some TNOs. Furthermore there is as yet no correlation observed between the albedo and the radius or the albedo and the V-R color. The observations are continuing however and it is expected that at the end of the project useful radius and albedo results will have been derived for 20 TNOs and 12 Centaurs.

One implication of the wide range of albedos is the connection between surface albedo and cometary activity. Several Centaurs are actively outgassing, and it would be interesting to see if there is a correlation between albedo and activity. Do the active Centaurs all have high albedos? Do Centaurs that have been only recently deactivated have high or low albedos? How long does it take for surface volatiles to disappear? Is albedo correlated with dynamical age? How long would a Centaur need to stay near $\sim$5 to 6 AU before its albedo would drop to the 0.04 to 0.05 that is the average of cometary nuclei and Trojans? Our understanding of the Centaur albedos can give clues to some fundamental problems of cometary evolution.

Lastly, we mention one special object in the sample: 1999 TC$_{36}$, a binary TNO. Measurements of the thermal emission have constrained the radii and thus the volume of the two objects. The masses have been constrained from the orbit, and so Stansberry *et al.* (2006) have calculated a bulk density of 0.3 to 0.9 g/cm^3. This implies that the object would be very porous, with a porosity of at least 50%. While this would require a large amount of void space, the number is not too much higher than the porosity calculated by Jewitt & Sheppard (2002) for the TNO (50000) Varuna.

4. Summary

We have given a brief review of Solar System observations made by the Spitzer Space Telescope since its launch in August 2003. At time of writing, almost five dozen Solar System projects have been allocated time, and they are in various stages of completion. Analysis is still preliminary for many of these projects, and much of the data are still proprietary. The Solar System component has so far accounted for 7.5% of the total non-Legacy SST observing time.

We have also given a brief overview of some of the projects that are devoted to dust, asteroids, and comets. While far from complete, the review of these projects gives a flavor for the type of cutting edge Solar System science from SST.

Acknowledgements

YRF would like to acknowledge the support of a SIRTF Fellowship provided by the Spitzer Science Center and JPL. All the authors appreciate the help of the SSC in supporting the observations reported here, and of the PIs in making their early results available to a wide audience.

References

A'Hearn, M. F.,Belton, M. J. S.,Delamere, A., & Blume, W. H.2005, *Sp. Sci. Rev.* 117, 1

Campins, H.,Telesco, C. M.,Osip, D. J.,Rieke, G. H.,Rieke, M. J.,& Schulz, B.1994, *Astron. J.* 108, 2318

Crovisier, J.,Leech, K.,Bockelee-Morvan, D.,Brooke, T. Y.,Hanner, M. S.,Altieri, B.,Keller, H. U.,& Lellouch, E.1997, *Science* 275, 1904

Cruikshank, D. P.,Barucci, A.,Emery, J. P.,Fernández, Y.,Grundy, W.,Noll, K.,& Stansberry, J. A.2005, in: B.Reipurth, D.Jewitt& K.Keil (eds.), *Protostars and Planets V* (Tucson: U. Arizona), submitted

Cruikshank, D. P.,Stansberry, J. A.,Emery, J. P.,Fernández, Y.R.,Werner, M. W.,Trilling, D. E.,& Rieke, G. H.2005, *Astrophys. J.* 624, L53

Emery, J. P.,Cruikshank, D. P.,& van Cleve, J.2006, *Icarus*, submitted

Fazio, G. G., and 64 colleagues 2004, *Astrophys. J. Supp. Ser.* 154, 10

Fernández, Y. R.,Jewitt, D. C.,& Sheppard, S. S.2002, *Astron. J.* 123, 1050

Houck, J. R., and 34 colleagues 2004, *Astrophys. J. Supp. Ser.* 154, 18

Ishiguro, M.,Kwon, S. M.,Sarugaku, Y.,Hasegawa, S.,Usui, F.,Nishiura, S.,Nakada, Y.,& Yano, H.2003*Astrophys. J.* 589, L101

Jewitt, D. C.,& Sheppard, S. S.2002, *Astron. J.* 123, 2110

Lebofsky, L. A.,& Spencer, J. R.1989, in: R. P.Binzel, T.Gehrels, M. S.Matthews (eds.), *Asteroids II* (Tucson: U. Arizona), p. 128

Lisse, C. M.,A'Hearn, M. F.,Groussin, O.,Fernández, Y. R.,Belton, M. J. S.,van Cleve, J. E.,Charmandaris, V.,Meech, K. J.,& McGleam, C.2005, *Astrophys. J.* 625, L139

Meadows, V. S.,Bhattacharya, B.,Reach, W. T.,Grillmair, C.,Noriega-Crespo, A.,Ryan, E. L.,Tyler, S. R.,Rebull, L. M.,Giorgini, J. D.,& Elliot, J. L.2004*Astrophy. J. Supp. Ser.* 154, 469

Reach, W. T.2004, *Bull. Amer. Astron. Soc.* 36, 43.04 [abstract]

Rieke, G. H., and 42 colleagues 2004, *Astrophys. J. Supp. Ser.* 154, 25

Stansberry, J. A., *et al.* 2006, *Icarus*, submitted

Stansberry, J. A., and 17 colleagues 2004, *Astrophys. J. Supp. Ser.* 154, 463

Stansberry, J. A.,Cruikshank, D. P.,Grundy, W. G.,Margot, J. L.,Emery, J. P.,Fernández, Y. R.,& Rieke, G. H.2005, *Bull. Amer. Astron. Soc.* 37, 52.05 [abstract]

Sykes, M. V.,& Walker, R. G.1992, *Icarus* 95, 108

van Cleve, J. E., and 11 colleagues and 2 teams 2005, *Bull. Amer. Astron. Soc.* 37, 42.05 [abstract]

Werner, W. M., and 25 colleagues 2004, *Astrophys. J. Supp. Ser.* 154, 1

Woodward, C. E.,Kelley, M. S.,Harker, D. E.,Wooden, D. H.,Reach, W. T.,Gehrz, R. D.,& Spitzer GO Comet Team 2005, *Bull. Amer. Astron. Soc.* 37, 16.16 [abstract]

Asteroids, Comets, Meteors 2005
Proceedings IAU Symposium No. 229, 2005
D. Lazzaro, S. Ferraz-Mello & J.A. Fernández, eds.

© 2006 International Astronomical Union
doi:10.1017/S174392130500671X

The molecular composition of comets and its interrelation with other small bodies of the Solar System

Jacques Crovisier

Observatoire de Paris, F-92195 Meudon, France
email: jacques.crovisier@obspm.fr

Abstract. The present status of our knowledge of the composition of cometary nuclei is reviewed and compared with what we know on the composition of other Solar System minor bodies — interplanetary dust, meteorites, asteroids, trans-Neptunian objects. The current methods of investigations — by both in situ analysis and remote sensing — are described. Comets are active objects pouring their internal material to form a dusty atmosphere which can be investigated by remote sensing. This is not the case for minor planets and trans-Neptunian objects for which only the outer surface is accessible. Collected interplanetary dust particles and meteorites can be analysed at leisure in terrestrial laboratories, but we do not know for certain which are their parent bodies.

Considerable progresses have been made from spectroscopic observations of active comets, mainly at infrared and radio wavelengths. We probably know now most of the main components of cometary ices, but we still have a very partial view of the minor ones. The elemental composition of cometary dust particles is known from in situ investigations, but their chemical nature is only known for species like silicates which have observable spectral features. A crucial component, still ill-characterized, is the (semi-)refractory organic material of high molecular mass present in grains. This component is possibly responsible for distributed sources of molecules in the coma. A large diversity of composition from comet to comet is observed, so that no "typical comet" can be defined. No clear correlation between the composition and the region of formation of the comets and their subsequent dynamical history can yet be established.

Keywords. astrochemistry, comets, asteroids, Centaurs, trans-Neptunian objects, molecules, spectroscopy

---

1. Introduction

Small bodies of the Solar System — comets, asteroids, trans-Neptunian objects (TNOs), even some planetary satellites — are interrelated in such a way that in many instances, transition objects have been identified and that a stringent separation between these different categories can no longer be made. Very small bodies — interplanetary dust particles (IDPs), meteors, meteorites —, which are mere decay products of these larger bodies, are also closely related. Even the line between *minor* and *major* planets is difficult to draw.

Indeed, all small bodies of the Solar System were presumably formed from the accretion of planetesimals, whose composition depends upon the epoch and the location of their formation in the primitive Solar Nebula. Therefore, the chemical compositions of all these minor bodies and the chemical processes governing their formation and evolution must be studied globally, in order to assess the interrelation between these objects and to achieve a general view of the Solar System formation and history.

In this context, I can only stress out the frustration of a cometary scientist who attempts to compare the clues to the rich and complex molecular content of comets with the

Table 1. A synopsis of the means of investigation of the composition of comets and other small bodies of the Solar System.

Body[a]	remote sensing (from Earth or Earth orbit)	in situ (space mission)	sample return (lab. analysis)
Comets: nuclei	$\approx 10^b$	*VEGA, Giotto* : 1P/Halley *Deep Space 1* : 19P/Borrelly *Stardust* : 81P/Wild 2 *Deep Impact* : 9P/Tempel 1 *Rosetta* : 67P/Churyumov-G.[c]	—
Comets: comae and tails	many	*ICE* : 21P/Giacobini-Zinner *VEGA, Giotto* : 1P/Halley *Giotto* : 26P/Grigg-Skjellerup *Deep Space 1* : 19P/Borrelly *Stardust* : 81P/Wild 2 *Deep Impact* : 9P/Tempel 1 *Rosetta* : 67P/Churyumov-G.[c]	*Stardust* : 81P/Wild 2[d]
Asteroids	many	*Galileo* : (951) Gaspra *Galileo* : (243) Ida + Dactyl *NEAR* : (253) Mathilde *NEAR* : (433) Eros *Deep Space 1* : (9969) Braille *Stardust* : (5535) Annefranck	*Hayabusa* : (25143) Itokawa[c]
TNOs	many	—	—
IDPs	zodiacal light	—	*Stardust*[d] stratospheric collect
Meteoroids	many	—	—
Meteorites	—	—	many

[a] Planetary satellites are not included.
[b] Comet nuclei with known albedo and colour.
[c] Pending success of mission.
[d] Pending return of sample.

sparse information available on the composition of asteroids and TNOs (Tables 1 & 2). Space missions have explored only six comets and seven asteroids so far, providing chemical information for only a few of them.

The present review relies on other, available reviews (e.g., in the *Comets II* and *Asteroids III* books) and concentrates on recent, new results. This study is not extended here to a comparison with protostellar objects and interstellar matter (this topic is covered in *Comets II* and the *Protostars and Planets* series of books).

Many related reviews are given in this Symposium. The most relevant ones are those by M.A. Barucci (surface properties of TNOs), J. Borovička (composition of meteoroids), C.R. Chapman (physical properties of small bodies), I. Mann (IDPs), R. Schulz (investigations of gas and dust in cometary comae), S.S. Sheppard (relations with irregular planetary satellites), and I. Toth (relations with asteroids).

2. Surfaces and nuclei: albedos, colours, reflectance spectra and densities

For inactive bodies, the only information on their composition available from remote sensing concerns their surface properties: albedo, colour, and reflectance spectrum.

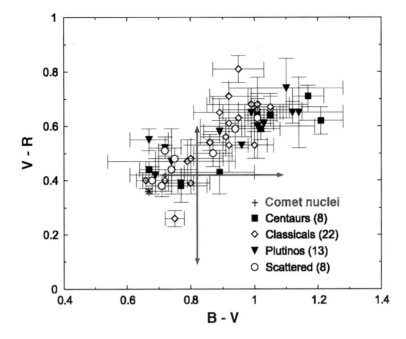

Figure 1. *V–R* vs *B–V* for TNOs and Centaurs (from Barucci et al. 2005) and for cometary nuclei. The cross with arrows indicates the mean value and range of variation from comet to comet (from Lamy *et al.* 2005). The star indicates the solar values.

Ironically, this information can only difficultly be obtained for comet nuclei, because at a distance, the nucleus signal cannot be easily separated from the strong signals of cometary dust and gas. Near-infrared spectroscopy, which is adequate for the identification of minerals and ice species, could only be performed from space by *Deep Space 1* on the nucleus of 19P/Borrelly, by *Deep Impact* on 9P/Tempel 1, and from the ground on a few comets (2P/Encke, 28P/Neujmin 1, 124P/Mrkos, 162P/Siding Spring and C/2001 OG_{108} (LONEOS)). As far as we know, bare comet nuclei do not show any sign of the molecular complexity of these bodies.

For all small bodies, the surface may be quite different from the inner material. This is due to the long-term chemical and physical processing by solar and cosmic radiation. For asteroids and TNOs, this process is know as *space weathering*. For comets, in addition to chemical processing due to irradiation, cometary activity itself leads to sublimation fractionation of the outer ice layers, and to the possible building up of a crust of dust particles too heavy to be dragged away by gas.

2.1. *Albedos and colours*

The albedos of comet nuclei (recently reviewed by Lamy *et al.* 2005) are found in a very narrow range, from 0.02 to 0.06 (except for 29P/Schwassmann-Wachmann 1 for which the albedo is possibly 0.13). TNOs and Centaurs have similar or higher values (when they are known). For instance, the two Centaurs Chiron and Asbolus have $A = 0.17$ and 0.12, respectively.

Some near-Earth asteroids (NEAs) could be disguised comets, as can be suggested by their low albedos (Stuart & Binzel 2004; Fernández *et al.* 2005). 15% of the NEAs could be dead comets.

Table 2. Identified chemical compounds in Solar System small bodies.

Body	volatiles	semi-refractories	refractories
Comets	many (≈25)	indirect	some
TNOs and Centaurs	some (≈5)	indirect	?
Asteroids	—	?	yes
IDPs	—	yes	some
Meteorites (CC)	—	many	many

Figure 1 shows V–R vs B–V for TNOs, Centaurs and 14 comet nuclei. The colours of comet nuclei are not as red as some TNOs, and closer to the solar values. They do not show a diversity as large as that observed for TNOs.

2.2. *Reflectance spectroscopy*

Water ice has been detected at the surface of several Centaurs (2060 Chiron, 5145 Pholus, 10199 Chariklo) and trans-Neptunian objects (1996 TO_{66}, 1999 DE_9, 1999 DC_{36}, 50000 Quaoar, maybe 20000 Varuna). Pluto, Charon and Triton also show ice (de Bergh 2004). Several other ices have been also identified at the surface of these large bodies of the outer Solar System: CO, CO_2, CH_4, N_2, possibly NH_3 and CH_3OH (Cruikshank 2005).

Water ice was *not* detected at the surface of the nucleus of 19P/Borrelly in the 1.3–2.5 μm spectrum observed during its flyby by *Deep Space 1* (Soderblom *et al.* 2004; Fig. 2). The spectrum is featureless except for a puzzling, unidentified feature at 2.39 μm. A possibly related feature was observed by *Cassini/VIMS* at 2.41 μm on Saturn's retrograde satellite Phoebe (Clark *et al.* 2005; Fig. 2) and at 2.44 μm on Iapetus (Buratti *et al.* 2005), tentatively attributed to cyanides.

Other spectra of cometary nuclei are also featureless. So, there is exposed ice on the surface of TNOs and Centaurs, but apparently not on the surface of cometary nuclei. Cometary ice should be below the dust mantle.

2.3. *Densities*

An indirect clue to the composition of small bodies is their bulk density. A low density points to an icy and/or porous material, as expected for comet nuclei. Density evaluations are now available for many asteroids which are double or have satellites (some of them having densities just above 1000 kg m^{-3}), and for a single TNO (1999 TC_{36} which has a density of 550–800 kg m^{-3}; Stansberry *et al.* 2005). For comet nuclei, information is more sparse and present evaluations bear on the modelling of non-gravitational forces or on considerations on the rotation period and shape: they do not lead to stringent estimates (Weissman *et al.* 2005). A density ≈600 kg m^{-3} was derived from the kinematics of the ejecta of 9P/Tempel 1 following the *Deep Impact* experiment (A'Hearn *et al.* 2005b, 2006).

Some planetary satellites could be captured comets, Centaurs or TNOs. This should be the case for the irregular satellites with highly eccentric and/or highly inclined orbits of the giant planets (Jewitt & Sheppard 2005; Sheppard 2006). The case for Phoebe was noted above. Another interesting case is Jupiter's satellite Amalthea. Its density was measured by *Galileo* to be 857 ± 99 kg m^{-3}, suggesting porosity and a high abundance of water ice (Anderson *et al.* 2005).

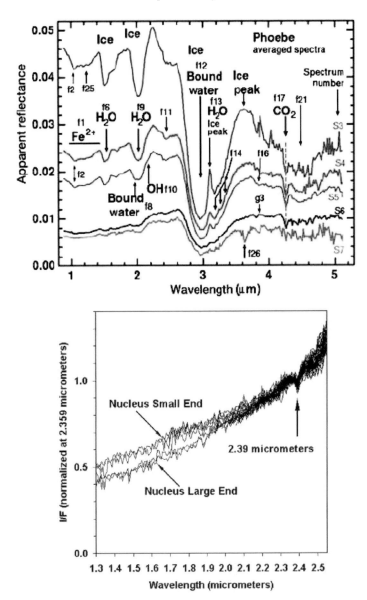

Figure 2. Surface spectroscopy of Saturn's retrograde satellite Phoebe (top) and of the nucleus of 19P/Borrelly (bottom). The satellite spectrum shows the features of several ices, in contrast to the comet spectrum which is almost featureless. Both spectra show a puzzling feature near 2.4 μm. (From Clark *et al.* 2005 and Soderblom *et al.* 2004.)

3. Spectroscopy of released material

For comets, we have the chance, as noted above, to observe material released from the surface or sub-surface of the nucleus. This contrasts with asteroids and TNOs which are inactive objects. Recent reviews on the spectroscopy of cometary comae were published by Bockelée-Morvan *et al.* (2005a), Despois *et al.* (2006) and Feldman *et al.* (2005). See also Crovisier (2004) and Schulz (2006).

Comet nuclei are composed of volatiles (ices, whose sublimation is the motor of cometary activity), refractories (yielding dust tails), and *semi-refractories*. The gross composition

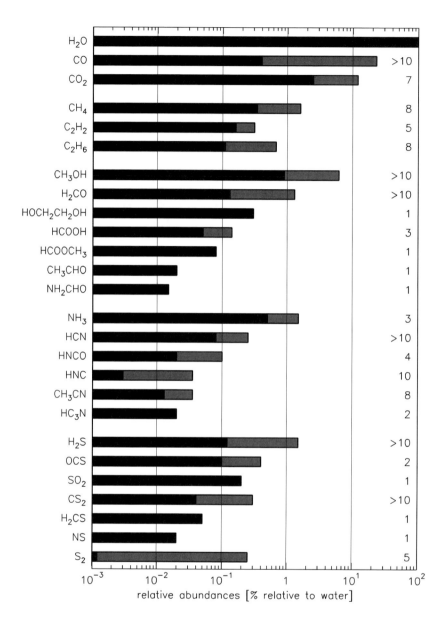

Figure 3. Relative production rates of cometary volatiles and their comet-to-comet variations. These rates are believed to trace the relative abundances in cometary ices. The grey part of each bar indicates the range of variation from comet to comet. On the right, the number of comets in which the species was detected is indicated. CO_2 data include direct infrared measurements as well as indirect measurements from CO prompt emission in the UV. CS_2 data include indirect determinations from UV and radio observations of the CS radical. Some species (e.g., H_2CO) are known to come (in part) from extended sources, not directly from the nucleus ices. The origins of NS and S_2 are ill-understood. (Updated from Bockelée-Morvan et al. 2005a.)

Table 3. Upper limits on the relative abundances of selected species obtained in C/1995 O1 (Hale-Bopp). (Adapted from Crovisier *et al.* 2004.)

Molecule		$X/Q[H_2O]$ [a]
Propyne	CH_3CCH	< 0.045
Ethanol	C_2H_5OH	< 0.10
Ketene	CH_2CO	< 0.032
Acetic acid	CH_3COOH	< 0.06
Dimethyl ether	CH_3OCH_3	< 0.45
Glycolaldehyde	CH_2OHCHO	< 0.04
Glycine I	NH_2CH_2COOH	< 0.15
Cyanodiacetylene	HC_5N	< 0.003
Methyl mercaptan	CH_3SH	< 0.05

[a] For $Q[H_2O] = 100$

of volatiles is now fairly well known from the investigation of the gas coma that formed following their sublimation (see below). Refractories will also be discussed briefly below. *Semi-refractories* are only indirectly known. They are presumably high molecular-mass molecules, responsible for the *extended sources* that release simple molecules in the coma, following pyrolysis (thermal degradation) or UV photolysis. They are probably akin to the so-called *insoluble organic fraction* of meteorites (carbonaceous chondrites — Botta & Bada 2002) and to *tholins*, which are laboratory analogues of ices processed by radiation.

About 25 stable volatile molecules, likely to have sublimated from nucleus ices, are now known. Figure 3 shows a synopsis of the relative production rates of these molecules. Altogether, about 45 molecular species, radicals or molecular ions are identified in cometary atmospheres. This is to be compared with about 130 species (not counting isotopologues) which are known in the interstellar medium. But to be fair we must consider that all these interstellar molecules are not observed in the same classes of objects: some are specific to interstellar hot cores, or dark clouds, or circumstellar envelopes. Indeed, in protoplanetary discs, whose composition is directly relevant to comets, only a handful of molecules are observed in the gas phase (CO, HCN, HNC, CN, CS, H_2CO, HCO^+, $C_2H\ldots$), because these small objects are difficult to investigate with present instrumentation and because most molecules are trapped as ices (Dutrey *et al.* 2005). On the other hand, many more complex organic molecules are identified from laboratory analyses of carbonaceous chondrites (Botta & Bada 2002).

Last news on detected species at radio wavelengths in comet Hale-Bopp, as well as upper limits on some rare species, may be found in Crovisier *et al.* (2004a, b). Limits obtained on a selection of species are listed in Table 3. Biver *et al.* (2005b, 2006) report detections of HC_3N and HCOOH, previously only observed in comet Hale-Bopp, in recent comets. High-resolution infrared cometary spectra are invaluable, especially for the observation of hydrocarbons (e.g., Mumma *et al.* 2003). However, they have not yet been fully exploited. For instance, the lines of ammonia, detected in at least two comets, have not been analysed yet, and upper limits on many species (e.g., C_2H_4) are still to be worked out.

As could be intuitively expected, the abundances of molecules are generally decreasing when the complexity is increasing. This is clear for homologous series of molecules (Table 4). However, some really complex molecules (as ethylene glycol discussed below) have unexpectedly high abundances.

The direct determination of water production rates in comets is now made easier by the observation of water hot bands in the infrared (e.g., Dello Russo *et al.* 2005) and

Table 4. Homologous series of cometary molecules.

Alkanes	CH$_4$	C$_2$H$_6$	C$_3$H$_8$
	methane	ethane	propane
Alcohols	CH$_3$OH	C$_2$H$_5$OH	C$_3$H$_7$OH
	methanol	ethanol	propanol
Aldehydes	H$_2$CO	CH$_3$CHO	C$_2$H$_5$CHO
	formaldehyde	acetaldehyde	propionaldehyde
Carboxylic acids	HCOOH	CH$_3$COOH	C$_2$H$_5$COOH
	formic acid	acetic acid	propionic acid
Cyanopolyynes	HCN	HC$_3$N	HC$_5$N
	hydrogen cyanide	cyanoacetylene	cyanodiacetylene

Molecules framed with a heavy line are those which are detected.
Molecules framed with a light line are those for which an upper limit is available.

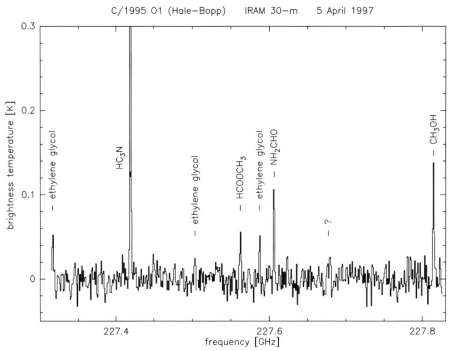

Figure 4. Detection of the radio lines of ethylene glycol and other complex organic species in
C/1995 O1 (Hale-Bopp) with the *IRAM* 30-m telescope (Crovisier *et al.* 2004a).

of the submillimetric 557 GHz line with the orbital observatories *SWAS* (e.g., Bensch
et al. 2004) and *Odin* (Lecacheux *et al.* 2003; Hjalmarson *et al.* 2005; Biver *et al.* 2005a;
Fig. 5). The interpretation of these measurements, for which a radiative transfer treatment of optically thick lines is crucial, now benefits from new-generation modelling of
the rotational lines of water (Bensch & Bergin 2004; Zakharov *et al.* 2005).

3.1. *New species*

Acetaldehyde (CH$_3$CHO) was confirmed by the serendipitous presence of a line in a radio spectrum of comet Hale-Bopp secured with the IRAM interferometer (Crovisier *et al.* 2004b).

Ethylene glycol (HOCH$_2$CH$_2$OH) was identified through the presence of several lines in millimetric spectra of comet Hale-Bopp as soon as molecular data of this molecule were made available (Crovisier *et al.* 2004a; Fig. 4). Its production rate ($\approx 0.25\%$ that of water) makes it one of the most abundant organic molecules in cometary ices, despite its complexity. It is the third "CHO" molecule by order of abundance, after methanol and formaldehyde. Remarkably, this dialcohol is more abundant than ethanol (C$_2$H$_5$OH) or the related molecule glycolaldehyde (CH$_2$OHCHO), with could not be found with respective upper limits $< 0.10\%$ and $< 0.04\%$. In the interstellar medium, glycol aldehyde was recently observed in the Galactic Centre source Sgr B2 (Hollis *et al.* 2002), but its abundance relative to methanol and ethanol is quite smaller.

Molecular hydrogen (H$_2$) was observed by Feldman *et al.* (2002) in comet C/2001 A2 (LINEAR) with *FUSE*. However, this is not a pristine molecule coming from nucleus ices. It is rather a product of the photolysis of water and other molecules.

Carbon disulfide (CS$_2$) has been suspected for a long time to be the progenitor of the CS cometary radical. It was recently tentatively identified in the visible spectrum of comet 122P/de Vico (Jackson *et al.* 2004).

3.2. *Uncomfortable detections*

For a detection to be reliable, several of the following criteria should be met:
- a good signal-to-noise ratio;
- line shape and centre (when a good spectral resolution is available) as expected for the coma kinetics;
- several lines observed simultaneously, with relative intensities that make sense;
- observations at different times and/or in different comets, possibly by different telescopes and/or different teams.

This is not yet the case for some claimed molecular detections in comets, for which confirmation by further observation is needed:
- Thioformaldehyde (H$_2$CS): its detection relies on a single radio line with low signal-to-noise ratio (Woodney *et al.* 1999).
- NS radical: the detection of this unexpected radical relies on the observation of a couple of radio lines (Irvine *et al.* 2000). Neither H$_2$CS nor NS could be confirmed by Crovisier *et al.* (2004b).
- Acetaldehyde (CH$_3$CHO) has been observed in a single radio line with a decent signal-to-noise ratio, plus marginal lines (Crovisier *et al.* 2004b).
- For methyl formate (HCOOCH$_3$), a single radio line, which is a blend of several rotational transitions, has been observed with a low spectral resolution (Bockelée-Morvan *et al.* 2000).
- Diacetylene (C$_4$H$_2$), tentatively detected in the infrared spectrum of 153P/Ikeya-Zhang (Magee-Sauer *et al.* 2002), has not yet been confirmed in other comets.
- The detection of N$_2^+$ in the visible, on which relies the evidence of N$_2$ in cometary ices, appears to be highly controversial, since it could not be confirmed in high-resolution spectra of recent comets (Cochran *et al.* 2000; Cochran 2002). N$_2$ was not detected in *FUSE* spectra either (Feldman *et al.* 2004).
- For carbon disulfide (CS$_2$) tentatively identified in the visible (Jackson *et al.* 2004; see above), no quantitative analysis of the signal have yet been made, pending laboratory measurements of the band strengths.

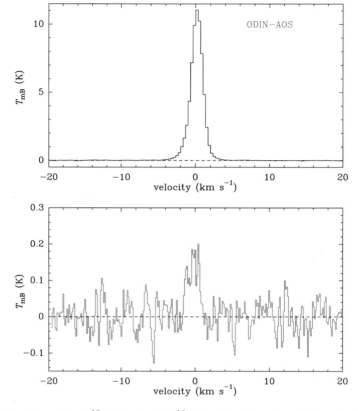

Figure 5. Spectra of the $H_2^{16}O$ (top) and $H_2^{18}O$ (bottom) 1_{10}–1_{01} millimetric lines observed by the *Odin* satellite in comet C/2001 Q4 (NEAT) from 26 April to 1 May 2004 (from Hjalmarson *et al.* 2005).

• Species only detected from mass spectroscopy — e.g., CH_2, C_2H_4, C_3H_2 (Altwegg *et al.* 1999) — are subject to mass ambiguity and to delicate modelling of ion chemistry in the cometary environment.

3.3. *Isotopes*

A detailed review on isotopic analyses of comets was published by Altwegg & Bockelée-Morvan (2003). See also Robert *et al.* (2000) for a discussion of the D/H ratio in the Solar System.

No recent progress has been made on the D/H ratio in cometary water despite sensitive attempts to detect HDO at infrared (Gibb *et al.* 2002) and radio wavelengths (Biver *et al.* 2005b, 2006; significant upper limits D/H < 0.00027 and < 0.00022 were obtained for 153P/Ikeya-Zhang and C/2004 Q2 (Machholz), respectively). Thus the D/H ratio is only precisely known for three comets — 1P/Halley, C/1996 B2 (Hyakutake) and C/1995 O1 (Hale-Bopp) —, all from the Oort cloud, all having D/H $\approx$0.0003. Further progress must await *Herschel*, *ALMA* and *Rosetta* or a really bright new comet. On the other hand, HDO was observed in the envelope of IRAS 16293–2422 and in the protoplanetary disc surrounding DM Tau, which are two solar-type protostars (Parise *et al.* 2005; Ceccarelli *et al.* 2005); D/H ratios 10–100 times higher than in cometary water are found.

For the first time in a comet, atomic deuterium was directly detected through its Lyman α line in C/2001 Q4 (NEAT) line with the *STIS* instrument of the *Hubble Space*

Table 5. Spin temperatures observed in comets. Adapted from the compilation of Kawakita *et al.* (2004), and updated with recent results from Kawakita *et al.* (2005) and Dello Russo *et al.* (2005).

Comet	H$_2$O [K]	NH$_3$ [K]	CH$_4$ [K]	orbital period [yr]
1P/Halley	29 ± 2			76
C/1986 P1 (Wilson)	> 50			dynamically new
C/1995 O1 (Hale-Bopp)	28 ± 2	26^{+10}_{-4}		4 000
103P/Hartley 2	34 ± 3			6.4
C/1999 H1 (Lee)	30^{+15}_{-6}			dynamically new
C/1999 S4 (LINEAR)	$\geqslant 30$	27^{+3}_{-2}		dynamically new
C/2001 A2 (LINEAR)	23^{+4}_{-3}	25^{+1}_{-2}		40 000
153P/Ikeya-Zhang		32^{+5}_{-4}		365
C/2001 Q4 (NEAT)			33^{+3}_{-2}	dynamically new

Telescope (Weaver *et al.* 2004). The interpretation of this measurement, which is still in progress, could provide another determination of the D/H ratio in cometary water, provided the photolysis of deuterated water is well understood, and the contribution (expected to be minor) of other species to deuterium is evaluated.

New, sensitive limits on monodeuterated methane, for which a higher D/H ratio could be expected, were obtained by Kawakita (2005) and Kawakita *et al.* (2005). The limit D/H > 0.01 obtained for comet C/2001 Q4 (NEAT) suggests a methane formation at a temperature higher than 30 K.

Puzzling results were obtained for the ^{14}N/^{15}N ratio. Arpigny *et al.* (2003), Jehin *et al.* (2004), Manfroid *et al.* (2005) and Hutsemékers *et al.* (2005) have consistently observed ^{14}N/^{15}N $\approx$150 from high-resolution visible spectra of the CN radical in several comets, whereas ^{14}N/^{15}N was 300 (close to the terrestrial value) from a radio line of HCN observed in comet Hale-Bopp (Jewitt *et al.* 1998). In contrast, the ^{12}C/^{13}C ratio is found to be 90 ± 4 in CN for the whole sample, close to the terrestrial ratio. This points to an additional source of CN, other than HCN and heavily enriched in ^{15}N, which is still to be identified. High molecular weight organics such as polymerized cyanopolyynes were invoked. But surprisingly, the ^{14}N/^{15}N ratio does not vary from comet to comet, whereas these objects had strongly different dust-to-gas ratios. Other possible evidences of additional sources of CN are reviewed by Fray *et al.* (2005a).

The ^{16}O/^{18}O ratio in water was observed with the *Odin* satellite to be close to the terrestrial ratio ($\approx$500) in four comets (Lecacheux *et al.* 2003; Biver *et al.* 2005c; Fig. 5). However, this ratio, which is evaluated from the comparison of a thin line and a heavily saturated line, is sensitive to modelling issues. This result could put constraints to some chemical models of protosolar nebulae which predict a significant enrichment of ^{18}O in cometary water (up to 20%), due to a self-shielding effect in the photolysis of CO (Qing-zhu Yin 2004).

3.4. *Spin species*

The ortho-to-para ratios (spin temperatures), now measured in three cometary molecules — water, ammonia, and methane —, are puzzling (see Kawakita *et al.* 2004 and Table 5 for a review on recent results; see also see Kawakita *et al.* 2005 for recent results on methane and Dello Russo *et al.* 2005 for further results on water spin temperatures). The observed spin temperatures are remarkably close to 30 K, whatever the molecule, the comet heliocentric distance or its dynamical history. What is the signification of this temperature? Although inter-spin conversions are strongly forbidden during non-destructive

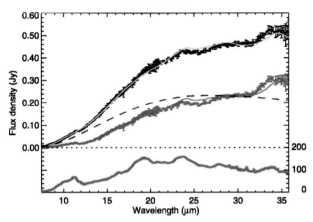

Figure 6. The infrared spectrum of 29P/Schwassmann-Wachmann 1 observed by *Spitzer*, with modelled spectra including forsterite and amorphous olivine. The *ISO* spectrum of comet Hale-Bopp is shown at the bottom for comparison. (From Stansberry *et al.* 2004.)

collisions or radiative transitions, preservation of the spin state over cosmological times is not universally accepted. This is still to be checked by laboratory simulations on analogs of cometary ices. Therefore, the present spin temperatures may not reflect the temperatures at the formation of the molecules. On the other hand, equilibration within the coma or at the comet surface would lead to spin temperatures depending on the heliocentric distance. A spin temperature in equilibrium with the internal temperature of the nucleus would depend upon the comet orbital history and differ between short-period and long-period comets.

3.5. *Unidentified spectral features*

Many unidentified spectral features have been spotted and catalogued in the UV (especially from recent observations by *FUSE*; Feldman *et al.* 2005) and in the visible (see, e.g., comprehensive atlases by Cochran & Cochran 2002, and Capria *et al.* 2005) spectra of comets: they are presumably coming from atoms, radicals and ions rather than from molecules. Identification should benefit from new, reliable databases of molecular lines. The NH_2 radical is still a good candidate for identifications. It has been noted (Wyckoff *et al.* 1999; Kawakita & Watanabe 2002) that some features are correlated with H_2O^+, and thus related to water ion chemistry.

In the infrared, several lines are reported consistently from comet to comet by the Mumma *et al.* group. Some of them could be due to radicals rather than simple small molecules. This is the case of of OH prompt emission, which is still to be fully understood (Bonev *et al.* 2004). Methanol is an important contributor to emission lines in the 3.3 μm region, but a comprehensive modelling of the fluorescence of this molecule is still to be done. No significant progress has been made on the emission of PAHs since the analysis of Bockelée-Morvan *et al.* (1995). (However, PAHs bands at 6–7 μm were observed in 9P/Tempel 1 with *Spitzer* following the *Deep Impact* event; Lisse *et al.* 2005a, 2005b.)

At radio wavelengths, some unidentified lines are still present, but with limited signal-to-noise ratios (Crovisier *et al.* 2004b).

3.6. *Dust*

Mid-infrared observations from space proved to be adequate for the identification of cometary dust silicates. The breakthrough performed by the identification of Mg-rich silicates (forsterite) from observations of the 2.5–45 μm spectrum of comet Hale-Bopp

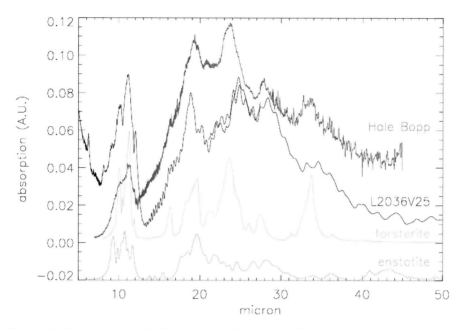

Figure 7. Comparative mid-IR spectra of C/1995 O1 (Hale-Bopp) observed with *ISO* (Crovisier *et al.* 1997) and of the IDP L2036V25 (Molster *et al.* 2003).

(Crovisier *et al.* 1997, 2000; Hanner *et al.* 2005) with the *Infrared Space Observatory* (*ISO*) is now supplemented by observations with the *Spitzer Space Telescope* (e.g., 29P/Schwassmann-Wachmann 1 by Stansberry *et al.* 2004, see Fig. 6; C/2004 Q2 (Machholz) by Wooden et al., *in preparation*; 9P/Tempel 1 by Lisse *et al.* 2005a, b). All these spectra show similar silicate emissions. *Spitzer* observations of Trojans also show similar silicate emissions (e.g., (624) Hektor; Emery *et al.* 2005). *ISO* observations of zodiacal light also revealed the signature of silicates at 10 μm (Reach *et al.* 2003).

Among recent analyses on the composition of cometary dust, one can note a study of the composition and size distribution of the dust of comet Hale-Bopp by Min *et al.* (2005); a comparison between cometary and circumstellar dust from observations of the dust around star HD 69830 by Beichman *et al.* (2005); a comparison of cometary dust and IDPs which emphasizes the similarity of their mid-infrared spectra (Molster *et al.* 2003; Fig. 7).

4. Extended sources of molecules in cometary comae

The case for a possible extended source of cometary carbon monoxide is a highly debated question. Interferometric radio observations of CO in comet Hale-Bopp did not show evidence of an extended source (Henry 2003; Bockelée-Morvan *et al.* 2005b and *in preparation*), in contrast with infrared observations (DiSanti *et al.* 2001). The existence of an extended source of CO in comet 29P/Schwassmann-Wachmann 1 advocated by Gunnarsson *et al.* (2002) is being revisited on the basis of observations obtained with the *HERA* mapping array at the *IRAM* 30-m telescope (Gunnarsson et al., *in preparation*).

The evidence of a distributed source of cometary formaldehyde is more firmly established. Cottin *et al.* (2004) and Fray *et al.* (2004, 2005b) have investigated in the laboratory the release of H_2CO from polyoxymethylene (POM) following UV photolysis or thermal degradation. They showed that this latter process could explain the *Giotto*

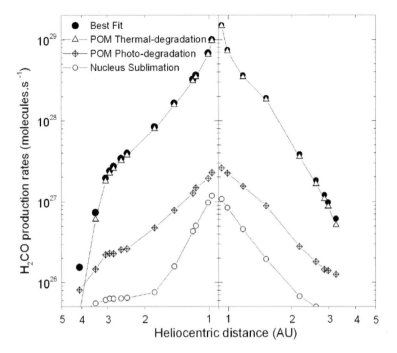

Figure 8. The contributions of various sources of formaldehyde as a function of heliocentric distance for C/1995 O1 (Hale-Bopp), according to the model of Fray *et al.* (2005b).

observations of formaldehyde in 1P/Halley, as well as the production of H_2CO in comet Hale-Bopp and its evolution with heliocentric distance (Fig. 8). Of course, this does not prove that POM is indeed the source of cometary formaldehyde, but it suggests that POM-like polymers might be present on cometary grains.

The case for hydrogen isocyanide HNC is still an open problem. Possible formation mechanisms involving ion–molecule reactions, isomerisation of HCN, or the degradation of cyanopolyyne polymers have been invoked (Rodgers & Charnley 1998; Fray *et al.* 2005a). This problem is discussed by Biver *et al.* 2005b and Bockelée-Morvan *et al.* 2005c), on the basis of observations of the HNC/HCN ratio as a function of heliocentric distance and of interferometric maps of HNC in C/1995 O1 (Hale-Bopp).

5. In situ investigations and sample returns

Direct chemical analyses can be made by in situ exploration or on returned samples.

5.1. *In situ analyses*

In situ analyses of the coma of 1P/Halley were performed by mass spectroscopy with *VEGA* and *Giotto* (e.g., Altwegg *et al.* 1999; Jessberger & Kissel 1991). They will be repeated with improved sensitivity and resolution with the *Rosetta* instrumentation. Much more information is expected on complex cometary molecules from in situ analyses after the landing on the nucleus of 67P/Churyumov-Gerasimenko of *Philae*, which is equipped with mass spectrometers and gas chromatographs.

On the other hand, *NEAR Shoemaker* measured the elemental composition of (433) Eros with its x-ray spectrometer. It was found to be similar to ordinary chondrites (Trombka *et al.* 2000), which was not unexpected, since most NEAs are believed to be related to ordinary chondrites. In the near future, *Hayabusa* will also investigate the composition of (25143) Itokawa.

5.2. *Sample returns*

Meteorites are samples naturally returned to Earth which can be submitted to detailed chemical analyses. Indeed, complex organic molecules, including many aliphatic hydrocarbons, PAHs and amino acids, have been identified in carbonaceous chondrites (Botta & Bada 2002). Could some meteorites (carbonaceous chondrites) be pieces of cometary nuclei? This is still an open question (Campins & Swindle 1998). One should note the relatively high density ($\approx$2000 kg m^{-3}; only a few measurements are available, however; Perron & Zanda 2005) of carbonaceous chondrites compared to the low density of cometary nuclei. This suggests that cometary nuclei may be too fragile to survive their entry in the Earth atmosphere.

IDPs are indeed collected in the stratosphere. Some of them are of cometary origin, but which ones? The sample return mission *Stardust* (Tsou *et al.* 2004), on its way back from 81P/Wild 2, should soon provide ground truth for the link between IDPs and cometary dust, despite the probable loss of all semi-refractory matter in the collect process.

6. Insight from the *Deep Impact* mission

Cometary molecular abundances derived from material released in the coma may greatly differ from real abundances within the nucleus because of sublimation fractionation effects (e.g., Prialnik 2005). How to get under the processed surface of cometary nuclei and verify if the matter released in cometary atmospheres is representative of the inner nucleus? Indeed, inner material is released during partial or total disruption of cometary nuclei, as was the case for D/1993 F2 Shoemaker-Levy 9 or C/1999 S4 (LINEAR), but such unpredictable events can difficultly be thoroughly observed.

Investigating inner nucleus cometary material using a controlled experiment was the goal of the *Deep Impact* mission to comet 9P/Tempel 1 (A'Hearn *et al.* 2005a, b, 2006). A high-energy impact was to excavate matter from $\approx$10–30 m under the nucleus surface. How does the molecular production of 9P/Tempel 1 compare with "standard" comets before and after the impact? The observations are still being analysed and it is premature to draw definite conclusions (Table 6). However, the surge of new gas-phase material following the impact — at least as seen by Earth-based telescopes with large fields of view — was only a small fraction of the quiescent coma (Meech *et al.* 2005, Küppers *et al.* 2005). It will not be easy to extract the contribution due to pristine material.

7. The diversity of comets

The molecular composition of a fairly large number of comets has now been investigated, at both radio and infrared wavelengths (especially from high-resolution IR spectroscopic observations from the group of Mumma et al.).

A first study of this diversity based upon radio spectroscopy was made by Biver *et al.* (2002). It is now updated by Biver *et al.* (2005b, 2006). The sample of comets now amounts to 33 objects. This diversity is shown in Figs 3 & 9.

This diversity can also be studied from the observations of daughter species, which are indirect, but give access to a larger sample of comets (see Schulz 2006).

Jupiter-family comets, which are weaker objects, are still ill-known, despite recent observations of 2P/Encke and 9P/Tempel 1.

Table 6. The top-ten volatile compounds (plus hydrogen cyanide) observed in comets.

	C/1995 O1 (Hale-Bopp)	"standard comet" [a]	9P/Tempel 1 before impact [b,c]		9P/Tempel 1 after impact [c]	
			radio	IR	radio	IR
H_2O	100	100	100	100	100	100
CO	12–23	< 1.7–23				4.3
CO_2	6	—				
H_2CO	1.1	0.13–1.3				
CH_3OH	2.4	< 0.9–6.2	1.7–3.2	1.3	7.2	0.99
H_2S	1.5	0.12–1.5	0.44			
NH_3	0.7	—				
CH_4	1.5	0.2–1.5				0.54
C_2H_2	0.2	0.15–0.3				0.13
C_2H_6	0.6	0.12–0.8			0.19	0.35
HCN	0.25	0.08–0.25	0.08–0.13	0.18	0.17	0.21

[a] A "standard comet" cannot be defined. The range of values for comets observed to date is listed. Jupiter-family comets are only sparsely observed and ill-represented.

[b] A comprehensive characterization of the chemical composition of 9P/Tempel 1 in its quiet state before impact was difficult because this comet was weak, about 1000 times less productive than C/Hale-Bopp. As far as we know, it was a "normal" comet.

[c] Relative abundances from Earth-based observations are listed here. Radio: preliminary values from IRAM observations (Biver *et al.* 2005a); IR: from Keck/NIRSPEC observations (Mumma *et al.* 2005).

The next major step in our understanding of comets — and of the Solar System formation itself — will be to establish to which extent this diversity reflects the primordial chemical composition of comets, and how it relates to the formation sites of these bodies.

8. Conclusion

Some pending problems related to cometary molecules are listed below.

• The origin of molecules/radicals such as NS and S_2 is still unknown (Rodgers & Charnley 2005).

• Many progenitors of the C_2 radical are now known, especially hydrocarbons. Others may still have to be found. On the other hand, progenitors for C_3 are still to be found: HC_3N is not abundant enough, and propyne (CH_3CCH), which has been proposed, is not detected (Table 3). This topic has been recently discussed by Helbert *et al.* (2005).

• Are supervolatiles (N_2, noble gases) present among cometary ices? This would put stringent constraints to the formation temperature of comets.

• Clathrates versus adsorption on amorphous ice is a key issue to the formation of comets, as recently discussed by Gautier & Hersant (2005).

• What is the meaning of the spin temperatures retrieved from the ortho-to-para ratios of several cometary molecules? Why are they so similar?

• Why is the $^{14}N/^{15}N$ isotopic ratio different in HCN and in the CN radical? Have we really missed a significant source of CN? The $^{14}N/^{15}N$ ratio in HCN has only been measured in one comet and should be investigated in other objects.

• What is the D/H ratio in water for Jupiter-family comets? What is their contribution to Earth water?

• How is the comet composition related to their origins? Why do the chemical composition vary so much from comet to comet, whereas other parameters, such as the $^{14}N/^{15}N$ ratio, the D/H ratio and the spin temperatures, seem to be homogeneous?

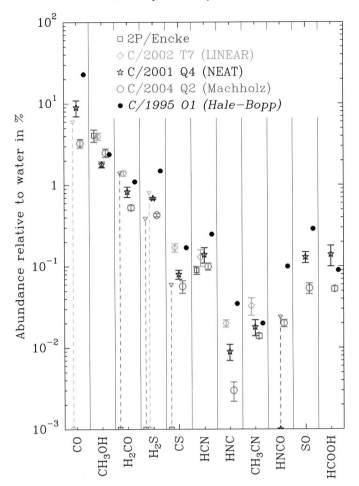

Figure 9. Relative abundances of selected molecules observed at radio wavelengths in a sample of recent comets, with comet Hale-bopp shown for comparison (from Biver *et al.* 2005b, 2006).

• Why are exposed ices obvious on the surface of TNOs and other distant small bodies, but inconspicuous on cometary nuclei?

These topics should be addressed in future observations, taking benefit of improved instrumentation and of space missions.

Acknowledgements

I thank all my colleagues for allowing me to report their results before publication.

References

A'Hearn, M.F., Belton, M.J.S., Delamere, A., & Blume, W.H. 2005a, *Space Scie. Rev.* 117, 1
A'Hearn, M.F., Belton, M.J.S., Delamere, W.A., *et al.* 2005b, *Science* 310, 258
A'Hearn, M.F. & the Deep Impact team 2006, *this volume*
Altwegg, K., Balsiger, H., & Geiss, J. 1999, *Space Sci. Rev.* 90, 3
Altwegg, K. & Bockelée-Morvan, D. 2003, *Space Sci. Rev.* 106, 139
Anderson, J.D., Johnson, T.V., Schubert, G., *et al.* 2005, *Science* 308, 1291
Arpigny, C., Jehin, E., Manfroid, J., *et al.* 2003, *Science* 301, 1522

Barucci, M.A. & Peixinho, N. 2006, *this volume*

Barucci, M.A., Doressoundiram, A., & Cruikshank, D.P. 2005, in: M.C. Festou, H.U. Keller & H.A. Weaver (eds), *Comets II* (Tucson: Univ. Arizona Press), p. 647

Beichman, C.A., Bryden, G., Gautier, T.N., *et al.* 2005, *Astrophys. J.* 626, 1061

Bensch, F. & Bergin, E.A. 2004, *Astrophys. J.* 615, 531

Bensch, F., Bergin, E.A., Bockelée-Morvan, D., Melnick, G.J., & Biver, N. 2004, *Astrophys. J.* 609, 1164

de Bergh, C. 2004, in: P. Ehrenfreund *et al.* (eds), *Astrobiology: Future Perspectives.* Astrophysics and Space Science Library Vol. 305 (Dordrecht: Kluwer Academic Publishers), p. 205

Biver, N., Bockelée-Morvan, D., Crovisier, J., *et al.* 2002, *Earth Moon Planets* 90, 323

Biver, N., Bockelée-Morvan, D., Boissier, J., *et al.* 2005a, *Asteroids, Comets, Meteors 2005, IAU Symp. No 229*, abstract book, 9

Biver, N., Bockelée-Morvan, D., Boissier, J., *et al.* 2005b, *Asteroids, Comets, Meteors 2005, IAU Symp. No 229*, abstract book, 43

Biver, N., Lecacheux, A., Crovisier, J., *et al.* 2005c, *Asteroids, Comets, Meteors 2005, IAU Symp. No 229*, abstract book, 112

Biver, N., Bockelée-Morvan, D., Crovisier, J., *et al.* 2006, *Astron. Astrophys. in press*

Bockelée-Morvan, D., Brooke, T.Y., & Crovisier, J. 1995, *Icarus* 116, 18

Bockelée-Morvan, D., Crovisier, J., Mumma, M.J., & Weaver, H.A. 2005a, in: M.C. Festou, H.U. Keller & H.A. Weaver (eds), *Comets II* (Tucson: Univ. Arizona Press), p. 391

Bockelée-Morvan, D., Boissier, J., Crovisier, J., Henry, F., & Weaver, H.A. 2005b, *BAAS* 37, 633

Bockelée-Morvan, D., Gunnarsson, M., Biver, N., *et al.* 2005c, *Asteroids, Comets, Meteors 2005, IAU Symp. No 229*, abstract book, 111

Bockelée-Morvan, D., Lis, D.C., Wink, J., *et al.* 2000, *Astron. Astrophys.* 353, 1101

Bonev, B.P., Mumma, M.J., Dello Russo, N., *et al.* 2004, *Astrophy. J.* 615, 1048

Borovička, J. 2006, *this volume*

Botta, O. & Bada, J.L. 2002, *Surv. Geophys.* 23, 4511

Buratti, B.J., Cruikshank, D.P., Brown, R.H., *et al.* 2004, *Astrophys. J.* 622, L149

Campins, H. & Swindle, T.D. 1998, *Meteoritics Planet. Scie.* 33, 1201

Capria, M.T., Bhardwaj, A., Cremonese, G., & De Sanctis, M.C. 2005, *Asteroids, Comets, Meteors 2005, IAU Symp. No 229*, abstract book, 114

Ceccarelli, C., Dominik, C., Caux, E., Lefloch, B., & Caselli, P. 2005, *Astrophys. J/.* 631, L81

Chapman, C.R. 2006, *this volume*

Clark, R.N., Brown, R.H., Jaumann, R., *et al.* 2005, *Nature* 435, 66

Cochran, A.L. 2002, *Astrophys. J.* 576, L165

Cochran, A.L., Cochran, W.D., & Barker, E.S. 2000, *Icarus* 146, 583

Cochran, A.L. & Cochran, W.D. 2002, *Icarus* 157, 297

Cottin, H., Bénilian, Y., Gazeau, M.-C., & Raulin, F. 2004, *Icarus* 167, 397

Crovisier, J. 2004, in: P. Ehrenfreund *et al.* (eds), *Astrobiology: Future Perspectives.* Astrophysics and Space Science Library Vol. 305 (Dordrecht: Kluwer Academic Publishers), p. 179

Crovisier, J., Bockelée-Morvan, D., Biver, N., *et al.* 2004a, *Astron. Astrophys.* 418, L35

Crovisier, J., Bockelée-Morvan, D., Colom, P., *et al.* 2004b, *Astron. Astrophys.* 418, 1141

Crovisier, J., Brooke, T.Y, Leech, K., *et al.* 2000, in: M.L. Sitko, A.L. Sprague & D.K. Lynch (eds), *Thermal Emission Spectroscopy and Analysis of Dust, Disks, and Regoliths* (San Francisco: ASP Conf. Series), p. 109

Crovisier, J., Leech, K., Bockelée-Morvan, D., *et al.* 1997, *Science* 275, 1904

Cruikshank, D.P. 2005, *Space Sci. Rev.* 116, 421

Dello Russo, N., Bonev, B., DiSanti, M.A., *et al.* 2005, *Astrophys. J.* 621, 537

Despois, D., Biver, B., Bockelée-Morvan, D., & Crovisier, J. 2006, in: D.C. Lis, G.A. Blake & E. Herbst. (eds), *Astrochemistry throughout the Universe: recent successes and current challenges, IAU Symp. No 231* (Cambridge Univ. Press), *in press*

DiSanti, M.A., Mumma, M.J., Dello Russo, N., & Magee-Sauer, K. 2001, *Icarus* 153, 361

Dutrey, A. Lecavelier des Etangs, A., & Augereau, J.-C. 2005, in: M.C. Festou, H.U. Keller & H.A. Weaver (eds), *Comets II* (Tucson: Univ. Arizona Press), p. 81

Emery, J.P., Cruikshank, D.P., Van Cleve, J., & Stansberry, J.A. 2005, *LPSC* 36, 2072

Feldman, P.D., Weaver, H.A., & Burgh, E.B. 2002, *Astrophys. J.* 576, L91

Feldman, P.D., Weaver, H.A., Christian, D., *et al.* 2004, *BAAS* 36, 1121

Feldman, P.D., Cochran, A.L., & Combi, M.R., 2005, in: M.C. Festou, H.U. Keller & H.A. Weaver (eds), *Comets II* (Tucson: Univ. Arizona Press), p. 391

Fernández, Y.R., Jewitt, D.C., & Sheppard, S.S. 2005, *Astron. J.* 130, 308

Fray, N., Bénilian, Y., Cottin, H., & Gazeau, M.-C. 2004, *J. Geophys. Res.* 109, E07S12

Fray, N., Bénilian, Y., Cottin, H., Gazeau, M.-C., & Crovisier, J. 2005a, *Planet. Space. Scie.* 53, 1243

Fray, N., Bénilian, Y., Biver, N., *et al.* 2005b, *Icarus submitted*

Gautier, D. & Hersant, F. 2005, *Space Sci. Rev.* 116, 25

Gibb, E.L., Mumma, M.J., DiSanti, M.A., Dello Russo, N., & Magee-Sauer, K. 2002, *ACM 2002, ESA SP-500* 705

Gunnarsson, M., Rickman, H., Festou, M.C., Winnberg, A., & Tancredi,G. 2002, *Icarus* 157, 309

Hanner, M.S. & Bradley, J.P. 2005, in: M.C. Festou, H.U. Keller & H.A. Weaver (eds), *Comets II* (Tucson: Univ. Arizona Press), p. 555

Helbert, J., Rauer, H., Boice, D.C., & Huebner, W.F. 2005, *Astron. Astrophys.* 442, 1107

Henry, F. 2003. *La comète Hale-Bopp à l'interféromètre du Plateau de Bure: étude de la distribution du monoxyde de carbone*, PhD thesis, Univ. Paris VI

Hjalmarson, Å, Bergman, P., Biver, N., *et al.* 2005, *Adv. Space Res. in press*

Hollis, J.M., Lovas, F.J., Jewell, P.R., & Coudert, L.H. 2002, *Astrophys. J.* 571, L59

Hutsemékers, D., Manfroid, J., Jehin, E., *et al.* 2005, *Astron. Astrophys.* 440, L21

Irvine, W.M., Senay, M., Lovell, A.J., *et al.* 2000, *Icarus* 143, 412

Jackson, W.M., Scodinu, A., Xu, D., & Cochran, A.L. 2004, *Astrophys. J.* 607, L139

Jehin, E., Manfroid, J., Cochran, A.L., *et al.* 2004, *Astrophys. J.* 613, L61

Jessberger, E.K. & Kissel, J. 1991, in: R.L. Newburn Jr, M. Neugebauer & J. Rahe (eds) *Comets in the Post-Haley Era*(Dordrecht: Kluwer), p. 1075

Jewitt, D.C, Matthews, H.E., Owen, T.C., & Meier, R. 1997, *Science* 278, 90

Jewitt, D. & Sheppard, S. 2005, *Space Sci. Rev.* 116, 441

Kawakita, H. 2005, *Asteroids, Comets, Meteors 2005, IAU Symp. No 229*, abstract book, 109

Kawakita, H. & Watanabe, J. 2002, *Astrophys. J.* 574, L183

Kawakita, H., Watanabe, J., Furusho, R., *et al.* 2004, *Astrophys. J.* 601, 1152

Kawakita, H., Watanabe, J., Furusho, R., Fuse, T., & Boice, D.C. 2005, *Astrophys. J.* 623, L49

Küppers, M., Bertini, I., Fornasier, S., *et al.* 2005, *Nature* 437, 987

Lamy, P.L., Toth, I., Fernandez, Y.R., & Weaver, H.A. 2005, in: M.C. Festou, H.U. Keller & H.A. Weaver (eds), *Comets II*(Tucson: Univ. Arizona Press), p. 223

Lecacheux, A., Biver, N., Crovisier, J., *et al.* 2003, *Astron. Astrophys.* 402, L55

Lisse, C.M., A'Hearn, M.F., & the Deep Impact team 2005a, *Asteroids, Comets, Meteors 2005, IAU Symp. No 229*, abstract book, 8

Lisse, C.M., Van Cleve, J., Fernández, Y.R., & Meech, K.J., the Spitzer Deep Impact team 2005b, *IAUC* No 8571

Magee-Sauer, K., Dello Russo, N., DiSanti, M.A., Gibb, E., & Mumma, M.J. 2002, *ACM 2002, ESA SP-500* 549

Manfroid, J., Jehin, E., Hutsemékers, D., *et al.* 2005, *Astron. Astrophys.* 432, L5

Mann, I. *et al.* 2006, *this volume*

Meech, K.J., Ageorges, N., A'Hearn, M.F., *et al.* 2005, *Science* 310, 265

Min, M., Hovenier, J.W., de Koter, A., Waters, L.B.F.M., & Dominik, C. 2005, *Icarus in press*

Molster, F.J., Demyk, K., d'Hendecourt, L., *et al.* 2003, *LPSC* 34, 1148

Mumma, M.J., DiSanti, M.A., Dello Russo, N., *et al.* 2003, *Adv. Space Res.* 31, 2563

Mumma, M.J., DiSanti, M.A., Magee-Sauer, K., *et al.* 2005, *Science* 310, 270

Parise, B., Caux, E., Castets, A., *et al.* 2005, *Astron. Astrophys.* 431, 547

Perron, C. & Zanda, B. 2005, *C. R. Physique* 6, 345

Prialnik, D. 2006, *this volume*

Qing-zhu Yin 2004, *Science* 305, 1729

Reach, W.T., Morris, P., Boulanger, F., & Okumuraé, K. 2003, *Icarus* 164, 384

Robert, F., Gautier, D., & Dubrulle, B. 2000, *Space Sci. Rev.* 92, 201

Rodgers, S.D. & Charnley, S.B. 1998, *Astrophys. J.* 501, L227

Rodgers, S.D. & Charnley, S.B. 2005, *Adv. Space Res. in press*

Schulz, R. 2006, *this volume*

Sheppard, S.S. 2006, *this volume*

Soderblom, L.A., Britt, D.T., Brown, R.H., *et al.* 2004, *Icarus* 167, 100

Stansberry, J.A., Cruikshank, D.P., Grundy, W.G. *et al.* 2005, *BAAS* 37, 737

Stansberry, J.A., Van Cleve, J., Reach, W.T., *et al.* 2004, *Astrophys. J. Suppl.* 154, 463

Stuart, J.S. & Binzel, R.P. 2004, *Icarus* 170, 294

Toth, I. 2006, *this volume*

Trombka, J., Squyres, S., Brückner, J., *et al.* 2000, *Science* 289, 2101

Tsou, P., Brownlee, D.E., Anderson, J.D., *et al.* 2004, *J. Geophys. Res.* 109, E12S01

Weaver, H.A., A'Hearn, M.F., Arpigny, C., *et al.* 2004, *BAAS* 36, 1120

Weissman, P.R., Asphaug, E., & Lowry, S.C. 2005, in: M.C. Festou, H.U. Keller & H.A. Weaver (eds), *Comets II* (Tucson: Univ. Arizona Press), p. 337

Woodney, L.M., A'Hearn, M.F., McMullin, J., & Samarasinha, N. 1999, *Earth Moon Planets* 78, 69

Wyckoff, S., Heyd, R.S., & Fox, R. 1999, *Astrophys. J.* 512, L73

Zakharov, V., Biver, N., Bockelée-Morvan, D., Crovisier, J., & Lecacheux, A. 2005, *BAAS* 37, 633

Asteroids, Comets, Meteors
Proceedings IAU Symposium No. 229, 2005
D. Lazzaro, S. Ferraz-Mello & J.A. Fernández, eds.

© 2006 International Astronomical Union
doi:10.1017/S1743921305006721

What makes comets active?

Dina Prialnik

Department of Geophysics and Planetary Sciences, Tel Aviv University,
Ramat Aviv 69978, Israel
email: dina@planet.tau.ac.il

Abstract. There are three types of energy sources that affect comet nuclei and may render them active: thermal – solar radiation, nuclear – radioactive decay, and gravitational – through collisions and tidal forces. These sources give rise to processes that, in turn, may release, absorb or transport energy: sublimation or recondensation of volatiles, crystallization of amorphous ice, heat diffusion and advection, gas flow through the porous nucleus. Each of these sources and processes has its own characteristic time scale (or rate) and these may differ by many orders of magnitude. It is the competition between various processes and the interaction between them – as one triggers the other, or else impedes it – that determine the activity pattern and the internal structure of a comet nucleus. Examples of such interactions and their outcome are presented, such as apparently sporadic activity at large heliocentric distances, obtained from numerical simulations of the behavior and evolution of comet nuclei. Confrontation of modeling results with observations provides feedback and constraints for the assumptions and parameters on which models are based. Adjusting the latter to match observations reveals properties of the nucleus that are otherwise inaccessible (except for *in situ* measurements by space missions). However, the interpretation of observations, such as production rates in relation to nucleus abundances, may be misleading. It is shown that monitoring production rates over the active part of a periodic comet's orbit may lead to conclusions regarding the composition and structure of the nucleus.

Keywords. comets: general

---

1. Introduction

Whether comet nuclei contain pristine material, dating back to the time of formation of the solar system, cannot be directly probed; it has to be inferred by studying cometary activity. Comets are observed to be active in many different ways: outgasing of various species, the predominant one being H_2O; ejection of dust grains, ranging in size from submicron to centimeters; outbursts, that is, sudden surges of brightness and ejection of material, on a wide range of intensities and time scales; and occasionally, break-up of fragments or splitting of the nucleus. These are outward manifestations, observed and measured by increasingly sophisticated instruments and methods. Taken at face value, they should reveal the composition and structure of the nucleus. However, it is conceivable that the mechanisms responsible for cometary activity may have altered — at least to some extent — the structure of the nucleus as well. Therefore, the task of deducing the initial structure of the comet nucleus by observing its activity pattern is not a straightforward one. Only if we understand what makes comets active, may we hope to achieve the goal of determining the structure and composition of the nucleus at the time of comet formation.

Any type of cometary activity involves some energy source. These may be divided into primary (independent) and secondary (induced) sources. The dominant primary source is solar radiation, most of which is absorbed at the surface or in a surface layer [Davidsson & Skorov (2002)], since the albedo is relatively low. The important property of this source is its strong dependence on heliocentric distance. Its effect depends on latitude, spin, and spin axis inclination [e.g., Cohen, Prialnik & Podolak (2003)]. Average heating rates may be used for processes that take place below the skin depth (see below). Another source, common in larger solar system bodies, is nuclear in nature and involves radioactive decay. Since comets are assumed to be initially homogeneous, this source is equally distributed over the volume. However, it mainly affects the deep interior of the nucleus, since the outer layers are efficiently cooled by emission of thermal radiation. Given the relatively slow supply of energy — even for short-lived radionuclides, such as ^{26}Al — the effect is a change in structure and composition [Prialnik & Podolak (1995), De Sanctis, Capria & Coradini (2001)], rather than enhanced activity. Energy input of gravitational origin is due to tidal forces exerted by large bodies (e.g., the sun or Jupiter), if a comet comes sufficiently close to them, or to occasional collisions. These are sporadic, of short duration, and rare.

Secondary sources — which may also be negative — are those connected with processes that are triggered by the primary sources, and include latent heat of phase transition, mostly sublimation, heat released in crystallization of amorphous water ice, and perhaps other sources associated with chemical reactions.

The evolution and activity of comets has been intensely studied in the last decade by several research groups through numerical modeling. Active modeler groups are listed below, keeping in mind that there is collaboration among groups or between members listed under different groups:

- M. J. S. Belton, N. H. Samarasinha, W. H. Julian...
- J. Benkhoff, W. F. Huebner, H. Rauer,...
- A. Coradini, M. T. Capria, C. De Sanctis, R. Orosei,...
- F. P. Fanale, J. R. Salvail, N. Bouziani,...
- J. Klinger, A. Enzian, S. Espinasse,...
- E. Kuehrt, H. U. Keller, N. I. Koemle, G. Kargl, Yu. Skorov, G. Steiner,...
- J. Leliwa-Kopystynski, K. J. Kossacki, W. J. Markiewicz,...
- D. Prialnik, M. Podolak, Y. Mekler, E. Beer, Y.-J. Choi, R. Merk, G. Sarid,...
- H. Rickman, B. J. R. Davidsson, M. J. Greenberg, P. J. Gutierrez, G. Tancredi,...
- L. M. Shulman, A. V. Ivanova,...
- T. Yamamoto, S. Sirono, A. Kouchi, S. Yabushita,...

An extensive review of modeling and results in all their aspects has been recently published [Prialnik, Benkhoff & Podolak (2005)]. Radiogenic heating and its effects have been studied by several groups as reviewed by [Merk & Prialnik (2003)] and it will not be addressed in this review. The effect of tidal forces was studied by Davidsson (2001).

In the following sections we shall focus on the link between the nucleus structure and the characteristics of cometary activity, in order to show how the latter can be used in order to deduce the former. Section 2 sets the scene for understanding the complex relations among processes that influence cometary activity, by analysing the characteristic time scales of these processes. Section 3 deals with the relation between production rates of various volatiles and their corresponding abundances in the nucleus. It uses, for illustration, models of comet 67P/Churyumov-Gerasimenko. Section 4 addresses the connection between the stratified structure of the outer layers of the nucleus and the variability of cometary activity, using models of comet 9P/Tempel 1 for illustration. The activity of this comet — continually monitored prior to the impact — was, indeed, observed to be variable [Meech et al. (2005)]. In Section 5 the possibility of outbursts in a distant comet is considered by means of an arbitrary comet model, which is evolved for a relatively long period of time (spanning hundreds of orbital revolutions). Conclusions are briefly summarized in Section 6.

2. Time scales

First, we briefly summarize the equations of evolution of a comet nucleus used — in one form or another — in all model calculations. Given the bulk mass density ρ and porosity Ψ, the equations of mass conservation for H_2O are

$$\frac{\partial \rho_a}{\partial t} = -\lambda \rho_a, \tag{2.1}$$

$$\frac{\partial \rho_c}{\partial t} = (1-f)\lambda \rho_a - q_v, \tag{2.2}$$

$$\frac{\partial \rho_v}{\partial t} + \nabla \cdot \mathbf{J}_v = q_v, \tag{2.3}$$

where the meaning of indices is: a – amorphous water ice, c – crystalline water ice, and v – water vapor; $\lambda(T) = 1.05 \times 10^{13} e^{-5370/T}$ s^{-1} is the rate of crystallization [Schmitt *et al.* (1989)], and f is the total fraction of occluded gas in amorphous ice. Similar equations hold for the other volatiles (α). The energy conservation — or heat diffusion — equation is

$$\sum_\alpha \rho_\alpha \frac{\partial u_\alpha}{\partial t} - \nabla \cdot (K\nabla T) + \left(\sum_\alpha c_\alpha \mathbf{J}_\alpha\right) \cdot \nabla T = \lambda \rho_a \mathcal{H}_{\mathrm{ac}} - \sum_\alpha q_\alpha \mathcal{H}_\alpha, \tag{2.4}$$

where $\mathcal{H}_{\mathrm{ac}}$ is the heat released upon crystallization, and $\mathcal{H}_\alpha$ is the heat of sublimation. The above set of time-dependent equations is subject to constitutive relations: $u(T)$, $\lambda(T)$, $q_\alpha(T, \Psi, r_p)$, $\mathbf{J}_\alpha(T, \Psi, r_p)$, $K(T, \Psi, r_p)$, where r_p denotes pore radius. These relations require additional assumptions for modeling the structure of the nucleus.

The rate of sublimation — mass per unit volume of cometary material per unit time — is given by

$$q_\alpha = S(\Psi, r_p) \left[(P_{\mathrm{vap},\alpha}(T) - P_\alpha)\sqrt{\frac{\mu_\alpha}{2\pi R_g T}} \right], \tag{2.5}$$

where the term in square brackets represents the sublimation rate per unit surface area, and S is the surface to volume ratio, a function of porosity and pore radius. The partial pressure P_α is given by the ideal gas law and the saturated vapor pressure, P_{vap}, by the Clausius – Clapeyron approximation,

$$P_{\mathrm{vap}}(T) = A e^{-B/T}. \tag{2.6}$$

where A and B are constants.

The equations of mass conservation and energy transport require two boundary conditions each: at the center of the nucleus, vanishing (a) heat flux and (b) mass (gas) fluxes; at the surface, (c) energy balance,

$$F(R) = \epsilon\sigma T(R,t)^4 + \mathcal{F}P_{\mathrm{vap}}(T)\sqrt{\frac{\mu}{2\pi R_g T}}\mathcal{H} - (1-A)\frac{L_\odot}{4\pi d_H(t)^2}\cos z, \tag{2.7}$$

where z is the solar zenith angle, and $\mathcal{F} \leqslant 1$ represents the fractional area of exposed ice — since the surface material is a mixture of ice and dust [Crifo & Rodionov (1997)] — and (d) gas pressures exerted by the coma, which — in the lowest approximation — may be assumed to vanish.

Dust is assumed to be, in part, dragged along with the gas flowing through pores and, in part, lifted off the nucleus surface by the sublimating vapor [see Orosei *et al.* (1995)]. For the former, the dust velocity is assumed to be equal to the gas velocity [see Podolak & Prialnik (1996)]. An efficiency factor is calculated, to take account of a dust size distribution that allows only grains up to a critical size to be dragged or lifted off.

156   D. Prialnik

The rest may accumulate to form a dust mantle. The efficiency factor may be adjusted
so as to allow or prevent the formation of a sealing mantle.

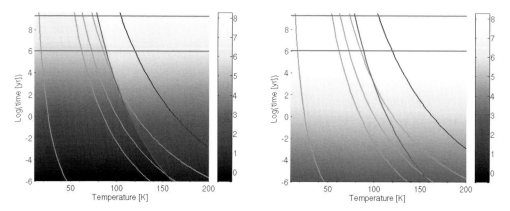

Figure 1. Characteristic times cales of various processes as function of temperature. The depth
scale of diffusion is mapped onto the [temperature, diffusion time] plane. The left panel corresponds to heat diffusion, the right – to vapor diffusion. A sharp shift occurs along the curve
corresponding to the H_2O crystallization time scale. The other curves represent sublimation time
scales for: CO, CO_2, NH_3, HCN and H_2O, from left to right, in order of decreasing volatility.
The horizontal lines are the characteristic decay times of ^{26}Al, and above it, ^{40}K.

The evolution of a comet is determined by several different time scales:

• The thermal time scale, obtained from the energy balance equation, which in its
simplest form, without sources and advection, is a heat diffusion equation. Distinction
must be made between the thermal time scale of amorphous ice τ_a, crystalline ice τ_c, and
dust τ_d. For a layer of thickness Δr and average temperature T we have:

$$\tau_a(\Delta r, T) = (\Delta r)^2 \rho_a c_a / K_a(T),\qquad(2.8)$$

and similar expressions for τ_c and τ_d.

• The time scale of gas diffusion (say, for some representative gas component) τ_{gas},
which is also the time scale of pressure release, obtained from the mass conservation
equation, which (without sources) can be regarded as a diffusion-type equation for the
release of gas pressure.

$$\tau_{gas}(\Delta r, T) = \frac{3}{4}\frac{(\Delta r)^2}{pa}\left(\frac{2\pi\mu}{kT}\right)^{1/2}\qquad(2.9)$$

• The time scale of crystallization τ_{ac}, which is also the time scale of gas-release and
pressure build-up,

$$\tau_{ac}(T) = \lambda(T)^{-1} = 9.54 \times 10^{-14} e^{5370/T}\qquad \text{sec}\qquad(2.10)$$

• The time scales of sublimation of the different volatiles, τ_{sub-H_2O} for water, τ_{sub-CO}
for CO, τ_{sub-CO_2} for CO_2, and so forth.

$$\tau_{sub-H_2O}(T) = \frac{\rho c}{S\mathcal{P}_v\sqrt{\mu_v/2\pi R_g T}},\qquad(2.11)$$

and similar expressions for other volatiles.
To these, the constant characteristic times of decay of the radioactive species may be
added; the only relevant one would be that of ^{26}Al, whose decay time τ_{Al26} is relatively
short.

All these time scales are shown in Figure 1 as function of temperature. The background is a map of the length (depth) scale corresponding to a given diffusion time scale and temperature, since diffusion time scales (for heat and gas) depend on length. We consider two cases: heat diffusion and advective diffusion of a representative gas. In the former, we note a sharp change of depth where ice transforms from amorphous to crystalline, illustrating the fact that crystalline ice is a much better heat conductor than amorphous ice. The change occurs along the curve representing the time scale of crystallization as a function of temperature. The other curves in Fig. 1 represent the sublimation time scales for various volatiles: CO, CO_2, NH_3, HCN and H_2O, in order of increasing time, that is, decreasing volatility.

The relationships between these time scales will determine to a large extent the evolutionary pattern of the comet nucleus. For example, considering the time scales of crystallization, sublimation and heat conduction, we find that at very low temperatures conduction dominates on the relevant length scales, meaning that heat released by a local source will be efficiently removed. Crystalline ice is a much better heat conductor than amorphous ice and hence heat will predominantly flow to the surface through the growing outer crystalline layer. As the crystallization rate is much more sensitive to temperature than the conduction rate (of crystalline ice), it will, eventually surpass the rate of heat conduction. When, due to insolation, the temperature at the crystallization front reaches this critical temperature $T_c \sim 120$ K, the local heat release causes it to rise still further. The higher temperature causes crystallization to proceed even faster and thus a runaway process develops. As the temperature rises, sublimation of the ice from the pore walls becomes important and since it absorbs a large amount of energy per unit mass, the outburst is arrested and proceeds at a controlled steady-state rate.

The time scale of gas diffusion is also characteristic of pressure release. Thus, if gas is released in the interior of the nucleus — due to crystallization of gas-laden amorphous ice or to sublimation — at a depth that surpasses the diffusion length scale (shown in Fig. 1), pressure may build up beyond the material strength and thus breaking and outbursts may result. The critical depth depends on temperature and on the active gas species and hence the outcome of gas release will depend on heliocentric distance and nucleus structure and history.

We note, in particular, that the sublimation curve of HCN intersects the crystallization curve at a temperature slightly above 100 K, where both processes proceed at a characteristic time scale of a few hundred years. At a depth of about 100 m, conduction competes with them, while at a larger depth conduction will be negligible. We shall return to this point when discussing outbursts at large heliocentric distances.

3. From production rates to nucleus composition

A comet nucleus is expected to include many different volatile species, such as are observed in the coma. If cometary water ice is crystalline, these volatiles will be frozen out as separate phases; if cometary ice is amorphous, volatiles may be trapped in the amorphous ice. In the former case, as the heat absorbed at the surface penetrates inward, ices other than water will evaporate and the gas will flow, in part, to the surface and out of the nucleus, and in part, to the colder interior, where it will refreeze. Since evaporation rates are strongly temperature dependent and vary widely among gas species, several distinct evaporation fronts are expected to form, and also several separate layers of refrozen gases (although initially, the ice mixture may have been homogeneous). These layers will evaporate, in turn, when erosion of the nucleus will bring them closer to the surface. Hence, the layered structure may move toward the center, but at the same time remain constant in depth relative to the surface.

Properties of 5 different species are summarized in Table 3. Constants A and B correspond to the coefficients of the Clausius-Clapeyron approximation to the saturated vapor pressure (2.6); they also serve to calculate the typical sublimation temperature,

$$T_s = B/\ln(A/\text{const}). \tag{3.1}$$

The rate of advance of the sublimation front $\dot{z}$ into the nucleus is then estimated for each species by assuming that the conduction flux inward from the surface (typically of order 100W m^{-2}) is absorbed solely in the evaporation of that species,

$$\frac{dz}{dt} = \frac{F_{in}}{\rho X_{ice} \mathcal{H}}. \tag{3.2}$$

Table 1. Volatile properties

Ice	A 10^{10}Nm^{-2}	B K	T_s K	$\mathcal{H}$ 10^6J kg^{-1}	dz/dt cm/d
H_2O	356.	6141.67	133	2.83	0.8
HCN	3.8665	4024.66	97	1.24	37
NH_3	61.412	3603.6	81	1.76	26
CO_2	107.9	3148.	70	0.594	78
CH_4	0.597	1190.2	30	0.617	75
CO	0.1263	764.16	20	0.227	200

If gas is trapped in amorphous ice, it will first escape when the ice crystallizes. It is implied that several gas species can be simultaneously trapped in amorphous ice, and that these gases are released from the ice upon crystallization regardless of their properties (and these assumptions warrant further investigation by laboratory experiments). Thus first, all species will escape together, and secondly, they will escape, generally, at higher temperatures than those typical of evaporation. Once they are released from the ice, these gases will behave similarly to gases that evaporate from the pore walls, flowing in part toward the surface and, in part, toward the interior. In this case, too, a layered structure of refrozen volatiles will emerge, and will eventually evaporate at a later stage. We shall return to this characteristic structure of multi-component nuclei in Section 4.

It has long been recognized that production rates of different volatiles ejected by an active comet should not be taken to reflect the corresponding abundances in the nucleus, as determined at the time of comet formation [e.g., Benkhoff & Huebner (1995)], and thus abundaince ratios in the coma may be vastly different from those of the nucleus, even before molecules are processed by solar radiation. If the source of volatiles is sublimation of the icy phase in the interior of the nucleus, the different rates of outgasing are the direct consequence of the differences in volatility, that is, sublimation temperature and latent heat. In addition, the mobility of the molecules that have to diffuse to the surface is also different. But even if the gases are trapped in amorphous ice and released upon crystallization, in which case they are presumably released in the same proportion, this proportion is not preserved all the way to the surface of the nucleus. The reason, as stated above, is that a fraction of the gas flows inward, refreezes behind the crystallization front and may sublimate later on, independently of crystallization.

This complex behavior is expected to result in production rates that are individual to each species. In order to test this inference and asses the extent of the discrepancy between ejecta and nucleus compositions, several models have been calculated, adopting the characteristic parameters of comet 67P/Churyumov-Gerasimenko, listed in Table 2 below. The compositions adopted for the different models (labelled 1 to 5) are listed in

Table 2. Parameters for comet 67P/Churyumov-Gerasimenko models

Property	Value
Semi-major axis	3.507 AU
Eccentricity	0.6316
Spin period	12.69 hr
Radius	1.98 km
Bulk density	500 kg m^{-3}

Table 3. It includes three types of volatile mixtures: amorphous water ice and trapped gases, crystalline water ice mixed with ices of other volatiles, as well as a combined composition of amorphous water ice and trapped gases mixed with ices of the same species.

Table 3. Initial volatile abundances: first row – frozen; second row – trapped in amorphous ice

Model (T_0)	X_d	X_{ice}	CO	CO$_2$	CH$_4$	HCN	NH$_3$
1.(50 K)	0.50	0.50	—	—	—	—	—
		am	5%	2%	1%	1%	1%
2.(50 K)	0.20	0.80	—	—	—	—	—
		am	5%	2%	1%	1%	1%
3.(50 K)	0.20	0.71	—	0.03	—	0.03	0.03
		am	5%	2%	1%	1%	1%
4.(20 K)	0.25	0.60	0.03	0.03	0.03	0.03	0.03
		cr	—	—	—	—	—
5.(40 K)	0.25	0.60	0.03	0.03	0.03	0.03	0.03
		cr	—	—	—	—	—

Models were run for 6 to 9 consecutive orbits and the resulting production rates and production rate ratios compared to corresponding initial abundance ratios in the nucleus are shown in Figures 2 and 3. Considering ratios of production rates has the advantage that some of the arbitrariness inherent in modeling is thus eliminated. We note that volatile ratios in the ejecta may differ from the corresponding ones in the nucleus by as much as a factor of 100! A noteworthy exception is the ratio CO/CH$_4$, which involves two supervolatile species. These species remain in gaseous form throughout the nucleus, and retain their relative abundances. We also note that the ratios of volatiles do not remain constant, but change along the orbit. If we now integrate production rates over time to obtain the total amount of material ejected during one orbit for each species and then compare ratios of these amounts with the ratios of abundances of the same species in the nucleus, a completely different picture emerges: not only are the differences far smaller, but they converge with repeated orbital revolutions to values which are very closely representative of the nucleus abundance ratios. This is illustrated in Figure 4.

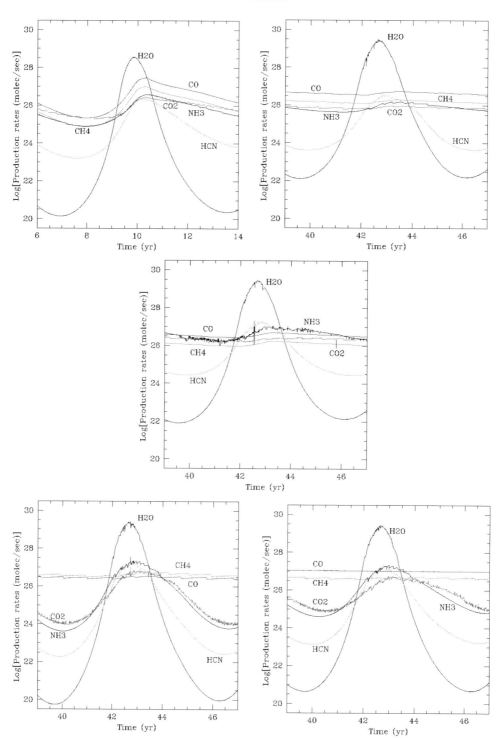

Figure 2. Production rates for one orbital revolution for the models of comet 67P/Churyumov-Gerasimenko as listed in Table 3: 1–top left, 2–top right, 3–middle, 4–bottom left, and 5–bottom right.

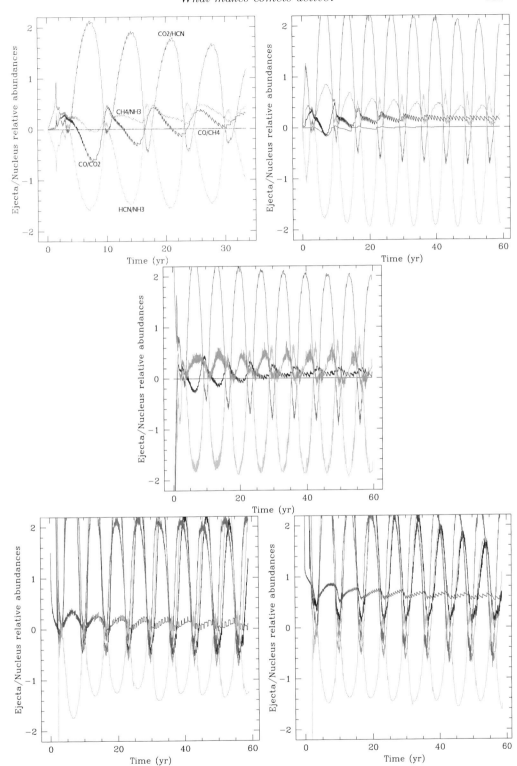

Figure 3. Abundance ratios in the ejecta relative to initial abundance ratio (log scale) as a function of time, along several orbital revolutions for the models listed in Table 3: 1 (top left), 2 (top right), 3 (middle), 4 (bottom left), and 5 (bottom right). Curves are labelled for model 1 (in the color version: blue CO / CO$_2$; green CH$_4$ / NH$_3$; cyan HCN / NH$_3$; magenta CO$_2$ / HCN ; red CO / CH$_4$); results for the other models serve to illustrate the large ratios obtained and the wide variability among species.

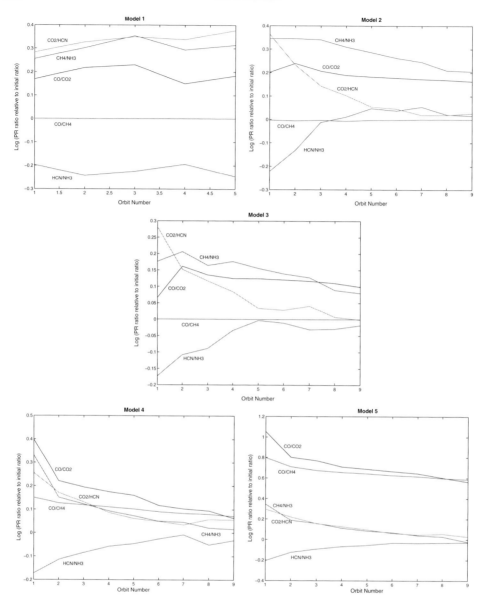

Figure 4. Abundance ratios — obtained by integrating the corresponding production rates over a full orbital revolution — relative to initial abundance ratios in the nucleus (log scale) for several consecutive orbits for the models listed in Table 3: 1 (top left), 2 (top right), 3 (middle), 4 (bottom left, and 5 (bottom right).

This result is of great significance to observations of production rates. It means that if a comet is monitored for a sufficiently large portion of its orbit (where it is highly active) and the obtained production rates are then integrated over that time, the resulting relative abundances should be far closer to those prevailing in the nucleus than they would be at any given point of the orbit. Therefore, continuous observation of a comet (within limits and constraints) are strongly indicated.

4. Variable activity and stratified nucleus structure

Although solar energy is supplied as a smooth function of heliocentric distance, the activity of a comet — as it approaches and then moves away from the sun — is not. The rate of outgasing and dust emission is typically variable, and often outbursts are observed. The most recent and detailed evidence to this variability has been provided by the extensive observations of comet 9P/Tempel 1, the target of the *Deep Impact* mission (Meech *et al.* 2005), but the same behavior was exhibited, for another example, by comet Hale-Bopp [Biver *et al.* (2002)]. In order to understand the variability in outgasing and dust emission, we consider the numerical model calculated by Sarid *et al.* (2005) for this comet. The model assumes a porous, spherical and initially homogeneous nucleus composed of amorphous and crystalline water ice, dust, and 5 other volatile species: CO, CO_2, HCN, NH_3, and C_2H_2. The basic comet nucleus code [Prialnik (1992)] is the same as that used for the calculations described in Section 3 for comet 67P/C-G. In addition, this calculation takes into account rotation of the nucleus and latitudinal variation of insolation. Relevant parameters are listed in Table 4. Structural and compositional parameters were chosen so as to match measured production rates at several points on the orbit [Cochran *et al.* (1992), Osip, Schleicher & Millis (1992)].

Table 4. Parameters for comet 9P/Tempel 1 model

Property	Value
Semi-major axis	3.12 AU
Eccentricity	0.517491
Spin period	41.85 hr
Radius	3.3 km
Bulk density	500 kg m^{-3}
Dust mass fraction	0.5
Ice (a) mass fraction	0.5

The outstanding feature emerging from this calculation is the complicated stratification pattern as a function of depth, where layers enriched in various volatiles alternate. Moreover, several layers enriched in the same volatile may appear at different depths. The effect is illustrated by the mass fraction of amorphous ice and of various volatiles shown in Figure 5.

We recall that the model's composition includes 5 volatile species trapped in amorphous water ice. These volatiles cover a wide range of sublimation temperatures (see Table 3). As the surface of the nucleus is heated and the heat wave propagates inward, the amorphous ice crystallizes and the trapped gas is released. The gas pressure in the pores peaks at the crystallization front. As a result (and as already mentioned earlier), gas flows in part outward and escapes, and in part, inward into colder regions. Eventually, each species reaches a sufficiently cold region for it to recondense. Since recondensation releases heat, it affects the composition of its surroundings and thus a complicated pattern results, of alternating ices mixed with the amorphous water ice. When another heat wave reaches these regions, on a subsequent perihelion passage, the heat is absorbed in sublimation of the recondensed volatiles rather than in crystallization of amorphous ice. This is how alternating layers of crystalline and amorphous ice arise, rather than a single boundary between a crystalline exterior and an amorphous interior.

The stratified layer extends from a depth of about 10 m below the surface and down to a few hundred meters. Since 10 m is roughly the orbital skin depth for 9P/Tempel 1,

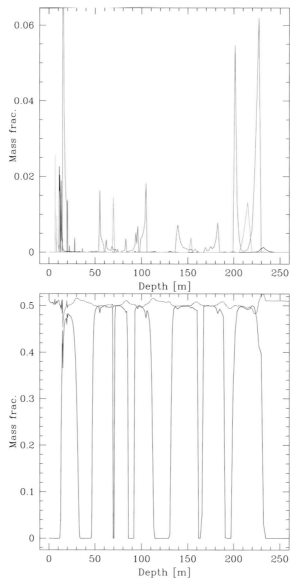

Figure 5. A model of Comet 9P/Tempel 1, near perihelion. *Top* – mass fractions of volatile ices as a function of depth: the outermost peak is HCN, followed by NH_3 and CO_2, then between each two CO_2 peaks there is a lower peak of C_2H_2. *Bottom* – amorphous H_2O ice (oscillating curve) and dust (upper curve) as a function of depth.

this structure should cause activity variations on the orbital time scale. This means that activity may differ from orbit to orbit and occasional spurious outbursts may arise, when the heat wave propagating down from the surface reaches a region enriched in ice of some volatile species. Such outbursts of gas should be accompanied by ejection of dust. Indeed, this variable behavior is apparent in Figure 6, where production rates for all volatiles are shown as function of time, pre- and post-perihelion. Besides the variability encountered for each gas species, we note the marked differences among production rates of differ- ent volatiles (similar to the results obtained for comet 67P/C-G, see Section 3). Thus

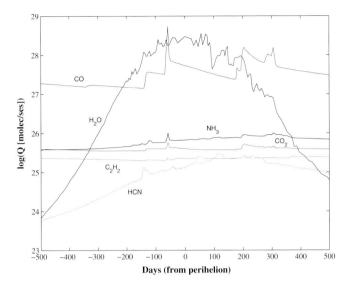

Figure 6. Production rates of volatiles as a function of time measured from perihelion obtained for a model of Comet 9P/Tempel 1.

variability in cometary activity is easily explained. It should be mentioned, however, that the details of the computation results depend on assumptions and adopted parameters. The important theoretical result is the variability *itself*, its general pattern, and the typical amplitude and time scale ranges.

Production rates are shown in more detail in Figure 7, where they are plotted separately, in order of volatility. Changes amounting to factors of up to 5 in outgasing rates can thus be discerned. Since the temperature profile of the nucleus is a decreasing function of depth, the stratified structure follows the inverse order of volatility, with the most volatile species lying deepest, and the least volatile closest to the surface. It is noteworthy that the departure from correlation with heliocentric distance is clearly dependent on volatility. Water production is the most strongly and directly correlated to solar heating, since H_2O sublimates mostly from the surface or a thin subsurface layer. Thus water production is closely correlated with the change in surface temperature, illustrated in Figure 8-left. The next two species in order of volatility, HCN and NH_3, sublimate from correspondingly deeper layers, which have become enriched in these volatiles due to re-condensation of trapped gas. However, both these layers are found within the orbital skin depth, and are hence correlated with solar heating. In addition, we discern a shift in time of the evaporation rate peak. This is due to the thermal lag of the heat front propagating inward from the surface. The more volatile species, C_2H_2 and CO, originate in deeper layers, or are entirely released from amorphous ice, and therefore their production rates are completely independent of the time variation of solar heat. Finally, the right panel of Fig. 8 shows the dust production rate, which reflects the high variability of the total gas production, since dust is dragged by the gas.

5. Distant outbursts and amorphous water ice

The possibility of a comet being active at large heliocentric distances is now illustrated, based on long-term evolutionary calculations, assuming an arbitrary orbit of low

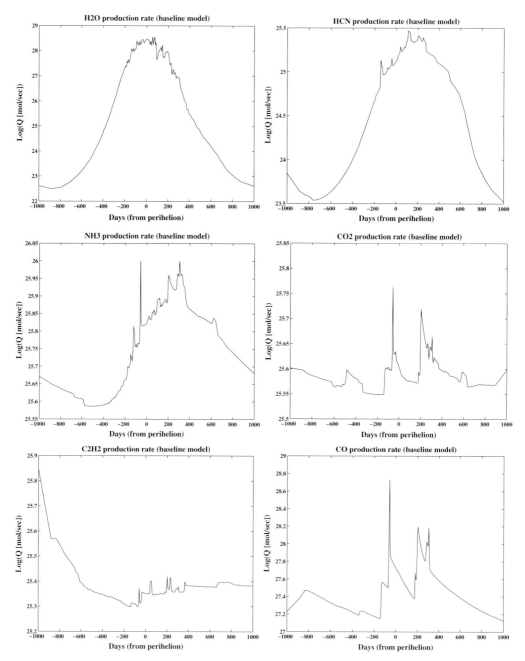

Figure 7. Production rates of various volatiles (see Fig. 6) as a function of time measured from perihelion, as obtained for a model of Comet 9P/Tempel 1: *top left* - H_2O; *top right* - HCN; *middle left* - NH_3; *middle right* - CO_2; *bottom left* - C_2H_2; *bottom right* - CO.

eccentricity, with a semimajor axis of 23 AU, and a composition of amorphous ice and 5 volatile species — as in Section 3 — trapped in it [from Choi (2005)]. Solar radiation is still the primary energy source, but it mainly serves to ignite other, secondary sources. The temperature at the sub-solar point reaches above 100 K, at which point crystallization is triggered and the trapped gases are released from the ice. A new heat wave

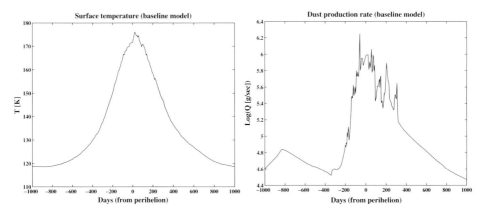

Figure 8. Surface temperature at subsolar point (left) and dust production rate (right) obtained for a model of Comet 9P/Tempel 1.

propagates inward from the surface at the perihelion of each orbit, and thus crystallization is supported by the energy of the heat wave at each perihelion passage, but proceeds extremely slowly. A non-negligible amount of volatile gases is released near perihelion at each orbit, but given the large distance, this activity would remain below detection limit. The moderately volatile gases (CO_2, NH_3 and HCN) recondense — in part — at depths of tens of meters, while the highly volatile ones (CO and CH_4) entirely escape from the nucleus. This is the same effect encountered earlier. Figure 9-top shows the evolution of gas production rates.

After a few tens of orbits a very strong outburst occurs. It arises due to a finely-tuned interaction between recondensation of volatiles and crystallization of amorphous ice on the same time scale (note the intersection of the phase transition times cale of HCN and the crystallization time scale in Fig. 1). As the moderately volatile gases released upon crystallization flow inward into colder regions of the nucleus, recondense and release latent heat, the local temperature rises to the critical value of $\sim$ 100 K. Thus a new crystallization wave is triggered at the densest region of recondensed ice, below the old crystallization front. Crystallization, in turn, releases more heat and occluded gas. For a very brief period of time, in a narrow region, the temperature shoots up to almost 200 K, as shown in the middle panel of Fig. 9. This is close to the maximal temperature reached by the ice on the surface of a comet nucleus near the sun, when sublimation prevents the temperature from rising further. Indeed, the heat of crystallization causes the recondensed ice to sublimate and thus the outburst is due to all the volatile gases: those released by the new crystallization episode as well as those sublimated from recondensed volatile ices by absorption of the crystallization energy. This type of interaction also occurred in the models described in Section 4 and produced variable abundance profiles and short-term, low-amplitude outbursts.

A second outburst, less strong in terms of gas production, but of much longer duration (in fact, a long period of high activity, rather than an outburst), occurs after a large number of quiet orbital revolutions. It originates at a larger depth below the nucleus surface, as shown in the lowest panel of Fig. 9, and involves slow crystallization down to a depth of almost 300 m below the surface. Thereafter the interior cools down and the highest temperature is again obtained at the surface.

It is interesting to consider the possibility of repetitive outbursts. The phase transition time scale of moderately volatile gases is comparable with the diffusion time scale through the orbital skin depth, as shown in Figure 1. Therefore, this kind of outburst may not

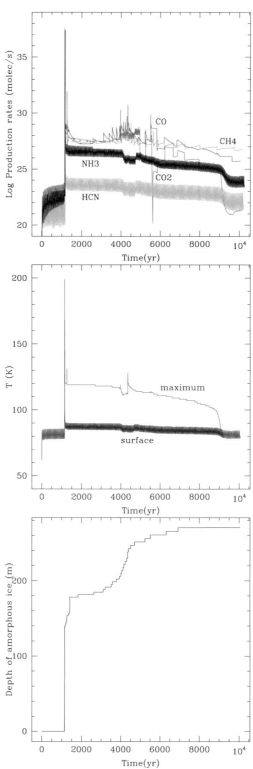

Figure 9. Long-time evolution of a distant comet: *top:* volatile production rates; *middle:* maximal temperature and surface temperature; *bottom:* depth of amorphous ice boundary.

be repetitive. It would be only if the nucleus surface receded due to sublimation — as in short-period comets — but this effect is negligible for the orbit of a distant object. A change of orbit, however, could trigger new outbursts.

6. Conclusions

The activity of a comet nucleus, as predicted by numerical model computations, is variable on different scales, from frequent low-amplitude outbursts, to major outbursts that may be observed at large heliocentric distances. This pattern of behavior results from the inhomogeneity inherent to the outer layers of the nucleus, extending down to possibly a few hundred meters below the surface. The composition of these layers is stratified, with alternating layers enriched in different volatiles. Given that the the characteristic pattern of activity of a comet is determined to a large extent by the structure of the nucleus, it should be possible to use observations of production rates in order to deduce properties of the nucleus interior.

Thus, we have shown that integrated production rate ratios can be used to derive the corresponding abundance ratios of the nucleus composition. Fluctuations in production rates can be used to infer a stratified structure of the outer layers of the nucleus. Furthermore, they may indicate whether the water ice is amorphous or crystalline, or mixed. The systematic differences in production rate dependence on heliocentric distance for various volatiles may, perhaps, be used to get a rough estimate of the relative depths below the nucleus surface where these species originate. We are still far from fully understanding the elusive nature of comet nuclei, but these findings may guide us towards achieving this goal by close collaboration between observations and modeling.

Acknowledgements

I would like to thank Fanny Gur and Gal Sarid for their help with model computations and analysis of the data. Support for this work was provided by Israel Science Foundation grant No.942/04.

References

Benkhoff, J. & Huebner, W.F. 1995, *Icarus* 114, 348
Benkhoff, J. & Boice, D.C. 1996, *Planet. Space Sci.* 47, 665
Biver, N., & 21 collaborators. 2002, *Earth, Moon & Planets* 90, 5
Bouziani, N. & Fanale, F.P. 1998, *Astrophys. J.* 499, 463
Capria, M.T., Capaccioni, A., De Sanctis, M.C., Espinasse, S., Federico, C., Orosei, R., & Salomone, M. 1996, *Planet. Space Sci.* 44, 987
Capria, M.T., Coradini, A., & De Sanctis, M.C. 2002, *Earth Moon & Planets* 90, 217
Capria, M.T., Coradini, A., De Sanctis, M.C., & Orosei, R. 2000, *Astron. J.* 119, 3112
Choi, Y.-J. 2005, PhD thesis, Tel Aviv University, Tel Aviv
Choi, Y.-J., Cohen, M., Merk, R., & Prialnik, D. 2003, *Icarus*
Cochran, A.L., Barker, E.S., Ramseyer, T.F., & Storrs, A.D. 1992, *Icarus* 98, 151
Cohen, M., Prialnik, D., & Podolak, M. 2003, *New Astron.* 8, 179
Coradini, A., Capaccioni, A., Capria, M.T., De Sanctis, M.C., Espinasse, S., Federico, C., Orosei, R., & Salomone, M. 1997, *Icarus* 129, 317
Crifo, J.F. & Rodionov, A.V. 1997, *Icarus* 129, 72
Davidsson, B.J.R. 1999, *Icarus* 142, 525
Davidsson, B.J.R. 2001, *Icarus* 149, 375
Davidsson, B.J.R. & Skorov, Yu.V. 2002, *Icarus* 159, 239
Davidsson, B.J.R. & Skorov, Yu.V. 2004, *Icarus* 168, 163
De Sanctis, M.C., Capria, M.T., & Coradini, A. 2001, *Astron. J.* 121, 2792

Enzian, A., Klinger, J., Schwehm, G., & Weissman, P.R. 1999, *Icarus* 138, 74

Fanale, F. & Salvail, J.R. 1997, *Icarus* 125, 397

Greenberg, J.M., Mizutani, H., & Yamamoto, T. 1995, *Astron.& Astrophys* 295, L35

Gutirrez, P.J., Ortiz, J.L, Rodrigo, R., & Lopez-Moreno, J.J. 2000, *Astron.& Astrophys* 355, 809

Huebner, W.F., Benkhoff, J., Capria, M.T., Coradini, A., De Sanctis, C., Enzian, A., Orosei, R., & Prialnik, D. 1999, *Adv. Space res* 23, 1283

Ivanova, A. & Shulman, L. 2002, *Earth Moon & Planets* 90, 249

Julian, W.H., Samarasinha, N.H., & Belton, M.J.S. 2000, *Icarus* 144, 160

Klinger, J., Levasseur-Regourd, A.C., Bouziani, N., & Enzian, A. 1996, *Planet.& Space Sci.* 44, 637

Kossacki, K.J., Szutowicz, S., & Leliwa-Kopystynski, J. 1999, *Icarus* 142, 202

Kuehrt, E. 2002, *Earth Moon & Planets* 90, 61

Meech, K.J, A'Hearn, M.F., Fernández, Y.R., Lisse, C.M., Weaver, H.A., Biver, N., & Woodney, L.M. 2005, *Space Sci. Rev.* 117, 297

Merk, R. & Prialnik, D. 2003, *Earth, Moon & Planets* 92, 359

Orosei, R., Capaccioni, F., Capria, M.T., Coradini, A., Espinasse, S., Federico, C., Salomone, M., & Schwehm, G.H. 1995, *Astron.& Astrophys* 301, 613

Osip, D.J., Schleicher, D.G., & Millis, R.L. 1992, *Icarus* 98, 115

Podolak, M. & Prialnik, D. 1996, *Planet.& Space Sci.* 44, 655

Prialnik, D. 1992, *ApJ* 388, 196

Prialnik, D. 1999, *Earth Moon & Planets* 77, 223

Prialnik, D., Benkhoff, J., & Podolak, M. 2005, in: M. Festou *et al.* (eds.), *Comets II* (Tucson: Univ. Arizona Press), p 359

Prialnik, D. & Podolak, M. 1995, *Icarus* 117, 420

Sarid, G., Prialnik, D., Meech, K.J., Pittichova, J., & Farnham, T. 2005, *Publ. Astron. Soc. Japan* 117, 796

Schmitt, B., Espinasse, S., Grim, R. J. A., Greenberg, J. M., & Klinger, J. 1989, *ESA SP* 302, 65

Shoshany, Y., Heifetz, E., Prialnik, D., & Podolak, M. 1997, *Icarus* 126, 342

Shoshany, Y., Podolak, M., & Prialnik, D. 1999, *Icarus* 137, 348

Shoshany, Y., Prialnik, D., & Podolak, M. 2002, *Icarus* 157, 219

Sirono, S. & Greenberg, J.M. 2000, *Icarus* 145, 230

Sirono, S. & Yamamoto, T. 1997, *Planet.& Space Sci.* 45, 827

Skorov, Yu.V., Keller, H.U., Jorda, L., & Davidson, B.J.R. 2002, *Earth Moon & Planets* 90, 293

Skorov, Yu.V. & Rickman, H. 1998, *Planet.& Space Sci.* 46, 975

Skorov, Yu.V. & Rickman, H. 1999, *Planet.& Space Sci.* 47, 935

Asteroids, Comets, Meteors
Proceedings IAU Symposium No. 229, 2005
D. Lazzaro, S. Ferraz-Mello & J.A. Fernández, eds.

© 2006 International Astronomical Union
doi:10.1017/S1743921305006733

Trans–Neptunian objects' surface properties

M. A. Barucci[1] and N. Peixinho[1,2]

[1] LESIA-Observatoire de Paris, 92195 Meudon cedex, France
email: Antonella.Barucci@obspm.fr

[2] CAAUL, Observatório Astronómico de Lisboa, PT-1349-018 Lisboa, Portugal
email: peixinho@oal.ul.pt

Abstract. Recent observations in visible photometry have provided B, V, R and I high quality colors for more than 130 objects. Color diversity is now a reality in the TNOs population. Relevant statistical analyses have been performed and all possible correlations between optical colors and orbital parameters have been analyzed. A taxonomy scheme based on multivariate statistical analysis of a subsample of 51 objects described by the 4 color indices (B-V, V-R, V-I and V-J) has been obtained. A tentative interpretation of the obtained groups in terms of surface characteristics is given. Moreover, an extension of this taxonomy to the other 84 objects for which only three colors indices (B-V, V-R, and V-I) are available, is also presented.

The faintness of these objects limits the spectroscopic observations. Despite this, our group provided visible and infrared spectra for 18 objects using the Very Large Telescope (ESO, Paranal, Chile). The wavelength region ranging 0.4–2.3 microns encompasses diagnostic spectral features to investigate organic compounds, minerals and ices present on the surface of the TNOs. The investigation of the surface variation can be an identifier of possible composition diversity and/or different evolution with different physical processes affecting the surface.

The current knowledge of the surface properties and composition of the population will be presented, analyzed and interpreted.

Keywords. Trans-Neptunian Objects, Colors, Photometry, Spectroscopy, Surface composition

---

1. Introduction

Since the first discovery in 1992 (Jewitt & Luu 1993) the ice bodies located beyond the orbit of Neptune, are named Trans-Neptunian objects (TNOs) or Edgeworth Kuiper belt (EKB) objects. They are believed to represent the most pristine and thermally unprocessed objects of the Solar System that are accessible to ground based observations. They are presumed to be remnants of the external planetesimal swarms. More than 1000 objects have been detected to date and their number increases continuously. These objects can be divided dynamically in several classes: Classical, Resonants, Scattered and Extended scattering disk objects. Classical objects have orbits with low eccentricities and low inclinations and semi-major axes between about 42 and 48 AU; Resonants objects which are trapped in resonances with Neptune, the majority are located in or near the 3:2 mean motion resonance; Scattered objects have high-eccentricity, high-inclination orbits and a perihelion distance near q = 35 AU; while the Extended scattering disk objects are located out of interacting gravitational encounters with Neptune. Only two objects have been discovered up to date belonging this dynamical class: 2000 CR$_{105}$ with semimajor axis at 224 AU, perihelion distance at 44 AU and aphelion at 401 AU; and 90377 Sedna with a semimajor axis at 501 AU and perihelion and aphelion distance at 76 and 927 AU, respectively. The Centaurs can also be associated with the TNO population. Centaurs seem to come from TNOs and are injected into their present orbit by gravitational instabilities and collisions. With a less clear dynamical link, giant planets' irregular satellites may also originate from TNOs.

The study of physical and chemical properties of TNOs and Centaurs can provide essential information about the conditions present in the early Solar System environment. Studies of properties of these icy bodies are still limited by the faintness of these objects, even if observed with the world's largest telescopes. Knowledge about them is still very limited (see Barucci *et al.* 2004, for a review), particularly very few information is available on the compositional properties of their surfaces. Spectroscopy is the best method to investigate the surface composition of these remote objects however given their faintness, the visible and near-Infrared spectra are available only for few of them, and in general with very low S/N. Photometry is the only available technique which provides data for a large number of objects particularly in the visible region. Recent large programs have been carried out at Paranal (ESO, Chile) and at Mauna Kea (CFHT, Hawaii) by our group at the Paris Observatory and have provided excellent data on photometry and spectroscopy. Relevant statistical analyses have been performed and all possible correlations between optical colors and orbital parameters have been analyzed.

The analysis of color diversity is important for investigating surface composition diversity and for helping to understand the different evolution with different physical processes effecting the surface.

2. Photometry

Presently, given the faintness of TNOs, multicolor photometry is the most adequate observational technique to search for a statistically relevant characterization of these objects.

Several teams carried out statistical analysis of colors from their own observational surveys and/or compiled published data sets. Tegler & Romanishin (1998); Tegler & Romanishin (2000, 2003); Tegler *et al.* (2003), provided a BVR color data sample of 91 objects, representing one of the most relevant survey. Their works were at the origin of one the most intense debates in TNO surface properties. Another reference work was performed by Hainaut & Delsanti (2002). By compiling the available visible colors for 104 Minor Objects of the Outer Solar System (MBOSS), *i.e.* Centaurs, TNOs and comet nuclei, Hainaut & Delsanti (2002) published the first thorough statistical analysis in the field.

We will focus our discussion on the results and observational strategies from our team surveys, namely: a) The ESO Large Program (LP, PI = H. Boehnhardt): executed from April 2001 until March 2003, at VLT and NTT ESO telescopes, covering both visible/near–infrared photometry and spectroscopy. Visible BVRI colors for 71 objects and JHK colors for ∼20 objects were obtained under this program.

b) The Meudon Multicolor Survey (2MS, PI = A. Doressoundiram): started in 1997, using almost exclusively the CFHT telescope, with the aim of collecting an homogeneous set of color data, this program has obtained visible BVRI colors for another 71 objects.

2.1. Specificities of TNO observations

Typically, TNOs trail at a rate of $3''/hour$ (Centaurs trail typically at $6''/hour$). Therefore, exposure times have to be optimized to minimize trailing and maximize S/N ratio. Moreover, since TNOs rotate and may exhibit significant brightness variations in short time–scales (∼1 hour) colors should not be computed from exposures separated over too long of a time period. Since real simultaneous observations are generally impracticable, to derive quasi-simultaneous colors the filter photometric sequence $RVBIV$ is generally adopted. In some cases the same object is observed several times averaging their values.

Due to the faintness of TNOs, hence low S/N ratios, the use of classical photometry gives frequently large photometric errors and misestimated magnitudes. In order to maximize the S/N ratio the aperture correction (or growth-curve correction) technique is usually applied (Howell 1989; Stetson 1990). Additionally, since TNOs trail, the critical aperture radius below which the moving object aperture correction diverges from the untrailed stars by more than 1% are taken into account according to McBride *et al.* (1999).

2.2. *Absolute magnitude, phase effects and size estimation*

At any constant heliocentric and geocentric distance, an atmosphereless body presents a brightness increase with decreasing phase-angle (α). Phase effects vary smoothly and almost linearly for $10° < \alpha < 90°$. For lower α values an opposition effect occurs. For the case of TNOs phase angles are generally $\alpha < 2°$ and the opposition effect plays a dominant role. Analysis of phase effects on TNOs (Sheppard & Jewitt 2002; Belskaya *et al.* 2003) have shown almost linear and fairly steep phase curves for the $0.2° - 2°$ phase angle range. With a phase curve slope ($\beta[mag/°]$) modal value for TNOs of $\beta_{TNOs} = 0.14\pm0.03$ (Sheppard & Jewitt 2002) and a $\beta_{Centaurs} = 0.11\pm0.01$ for Centaurs, obtained from Bauer *et al.* (2002) data, the phase effect is to be regarded. We compute absolute magnitudes (H) from R–filter magnitudes using the linear approximation phase function $\phi(\alpha) = 10^{-0.4\beta\alpha}$, hence:

$$H_R = R - 5 \log (r\Delta) - \beta\alpha \qquad (2.1)$$

where R is the R–band calibrated magnitude, $r\,[AU]$ is the object's heliocentric distance, $\Delta[AU]$ is the object's geocentric distance, $\alpha[°]$ is the phase angle during the observation and $\beta[mag/°]$ is the phase curve slope. Diameters (D, in kilometers) are estimated using Russell (1916) equation:

$$D = 2 \sqrt{\frac{2.24 \cdot 10^{16} \cdot 10^{0.4\,(R_\odot - H_R)}}{p_R}} \qquad (2.2)$$

where $R_\odot = -27.1$ is the Sun's R–magnitude, H_R the object's absolute magnitude and p_R the geometric albedo in the R–band. Albedos for Centaurs and TNOs are still widely unknown. It has been standard procedure to use the average value for comet nuclei $p_R = 0.04$. However, after a few albedo measurements were available, Brown & Trujillo (2004) suggested $p_R = 0.09$ to be more adequate, so we opted this value in our most recent work. Recent analysis by Grundy *et al.* (2005) conclude a better median value of $p_R = 0.10$ or a mean value of $p_R = 0.14$. The uncertainties on size estimations are evidently high. Nevertheless, size has an albedo dependence of $D \propto 1/\sqrt{p_R}$, consequently if an albedo is underestimated by 50% sizes will be overestimated only by 25%.

3. Search for correlations

The minor bodies of the outer solar system are an immensely large population and can only be studied through samples. From the analysis of such data samples we attempt to infer general properties. One of the primary goals of the color survey was to search for any possible correlation between colors and orbital parameters. We used a statistical approach as the basic tool for our data analysis.

3.1. *Hypothesis testing and its significance*

Statistical hypothesis testing is the well defined process of inferring from a sample whether or not a particular statement about the population appears to be true. Such statement is called *alternative hypothesis*, or H_1. The negation of this alternative hypothesis is called *null hypothesis*, or H_0. It is standard procedure to rate qualitatively the evidence against H_0 when discussing results, however not without controversy. Following the typical rough conventions in statistics, significance levels (SL) throughout this work may be read using the following rating:

$$\begin{cases} SL > 95\% & (\sim\!2.0\sigma) \quad - \text{ reasonably strong evidence against } H_0 \\ SL > 97.5\% & (\sim\!2.2\sigma) \quad - \text{ strong evidence against } H_0 \\ SL > 99\% & (\sim\!2.5\sigma) \quad - \text{ very strong evidence against } H_0 \end{cases}$$

3.2. *Trends and correlations*

Each objects is characterized by several surface colors and orbital parameters. The first and most basic statistical problem we are dealing with when confronted with two or more variables is to find if they are, or appear to be, in some way connected with each other. That is, if there is any trend between variables. The most common test for correlation is Pearson's correlation coefficient (r or r_p) and it has been used by Hainaut & Delsanti (2002). However, it has been common procedure to use its non-parametric equivalent Spearman ρ (or r_s), mainly because it it less sensitive to outliers and allows for very precise estimation of its significance level.

When a correlation value is significant it is often practical to discuss its intensity as ratings. Such is also a rough convention. The following ratings give a general idea of levels of intensity of possible Spearman ρ correlation values:

$$\begin{cases} 0.0 \leqslant \mid \rho \mid < 0.3 & - \text{ absent/negligible correlation} \\ 0.3 \leqslant \mid \rho \mid < 0.6 & - \text{ weak/moderate correlation} \\ 0.6 \leqslant \mid \rho \mid \leqslant 1.0 & - \text{ strong correlation} \end{cases}$$

We have generated composite plots showing color, size, and orbital elements for the Centaurs and TNOs in our sample of 114 objects. Objects are plotted according to orbital eccentricity (e) *versus* semimajor–axis (a) and orbital inclination (i) *versus* semi–major axis (a) — Figs. 1 and 2, respectively. Surface colors are represented using the $B - R$ color index, which measures the ratio of the surface reflectance at B ($\sim\!430\,nm$) and R ($\sim\!660\,nm$) wavelengths. A color palette has been adopted to scale the color spread from neutral (solar like) colors $B - R = 1.0$ (dark blue) to very red colors $B - R = 2.0$ (red). To allow for a more clear color spreading, two objects possessing $B - R > 2.0$ are indicated in black. The size of the symbols are proportional to the corresponding object's diameter (assuming a constant albedo of $p_R = 0.09$) and a legend of size scales is also plotted. Orbital elements were obtained from *IAU Minor Planet Center*†. These two images condense all the trends under discussion here.

3.3. *Classical objects*

The first apparent trend in Fig. 1 is that Classical objects (semi–major axis between the 3:2 and 2:1 resonances) with perihelion distances beyond $40\,AU$ (*i.e.*, below the $q = 40\,AU$ line) are mostly very red (Tegler & Romanishin 2000). Hainaut & Delsanti (2002) reported an explicit correlation between $B - R$ color and orbital eccentricity of $r_p = -0.57$ (Pearson's r). That is, with increasing eccentricity, hence diminishing of perihelion, Classical objects become bluer (Peixinho *et al.* 2004, hereafter PEI04) and

† `http://cfa-www.harvard.edu/iau/mpc.html`

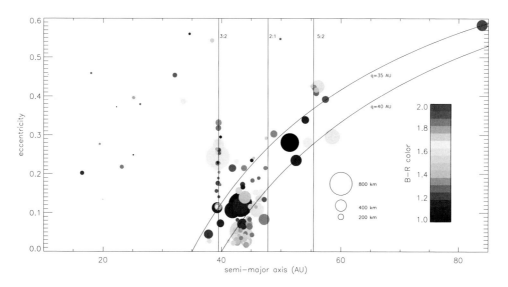

Figure 1. $B - R$ colors of Centaurs and TNOs from our survey for 114 objects in the orbital eccentricity *vs.* semi–major axis plane. Colors are scaled from blue (solar colors) to red (very red colors) indexed to the color palette on the right. Two very red objects are indicated in black. The sizes of the symbols are proportional to the corresponding object's diameter. The 3:2 ($a \sim 39.5\,AU$), 2:1 ($a \sim 48\,AU$) and 5:2 ($a \sim 55.4\,AU$) mean–motion resonances with Neptune are indicated. Perihelion curves for $35\,AU$ and $40\,AU$ are also marked. Figure adapted and updated from Doressoundiram *et al.* (2002).

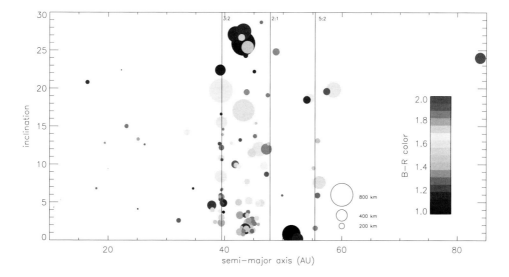

Figure 2. Same as Fig. 1 but for the orbital inclination *vs.* semi–major axis plane.

(Doressoundiram *et al.* 2005b, hereafter DOR05b) obtain $B - R$ color-perihelion corre-lations of $\rho = 0.52^{+0.05}_{-0.05}$ ($SL = 99.96\%$) and $\rho = 0.62$ ($SL = 99.98\%$) (Spearman's ρ), respectively. Such correlation is most probably connected with stronger solar surface pro-cessing (outgassing?) with shorter perihelia. Nevertheless, this color-q correlation is some-what puzzling when confronted with Centaurs. No cometary activity has been detected among Classical objects, nonetheless they show a correlation between surface properties and perihelia. Centaurs, with much closer perihelia and detected cometary activity, do not. The answer may be in the collisional resurfacing model with triggered cometary activity proposed by (Delsanti *et al.* 2004, hereafter DEL04) . Object resurfacing is not only caused by collisional resurfacing but also by the subsequent sublimation of freshly exposed volatiles, which strongly depends on perihelion. Additionally, PEI04 report a H_R dependence (or size if we assume an homogeneous or quasi–homogeneous albedo) of the color-q correlation. Intrinsically brighter (larger) Classical objects exhibit a much stronger color-q correlation. DEL04 model predicts that objects large enough to retain a bound coma (*i.e.*, $D > 100\ km$ assuming $p_R = 0.09$) should be more efficient in rejuvenating their surfaces. Moreover, Lacerda & Luu (2005), based on an analysis of rotational properties of TNOs conclude that objects with

$D < 270\ km$ (assuming $p_R = 0.09$) are better explained as by-products of the collisional evolution. Those with $D > 270\ km$ can be "rubble-piles" which have survived to collisional evolution. Given the correlation results it would be interesting to study if the by-products of the collisional evolution are also less rich in volatiles, hence less efficient in resurfacing.

From the representation with orbital inclination (Fig. 2) we may observe the tendency for Classical objects to be bluer and larger with inclination. The color-inclination trend was first reported by Tegler & Romanishin (2000) and the first explicit correlation values were obtained by Trujillo & Brown (2002); Hainaut & Delsanti (2002); Doressoundiram *et al.* (2002). Levison & Stern (2001) reported a possible size–inclination trend, recently confirmed by Trujillo & Brown (2003); Bernstein *et al.* (2004). Moreover, Levison & Stern (2001); Brown (2001) claimed the existence of two inclination populations among the TNOs: one primordial and dynamically "cold" and another dynamically "hot" probably superimposed.

Based on an analysis of color variance in function of inclination, PEI04 shows that the dynamically "cold" population defines a red cluster of objects for $i < 4.5°$. However, it could extend until $i < 12.0°$. Gomes (2003) proposed a migration model under which the two populations have different origins. Gomes' work distinguishes the disk of primordial objects being scattered by Neptune during its outwards migration in inner disk and outer disk. At the end of the migration phase the inclination distribution of TNOs will result from the superposition of the inner disk objects with highly inclined orbits ("hot" population) into the outer disk objects. The latter maintain their low inclined orbits ("cold" population).

In a nebula whose surface density decreases with heliocentric distance, the size of the largest objects also decrease. Objects formed in the inner disk should be larger than those formed in the outer disk — nevertheless, a population of smaller objects as a result of disruptive collisions should coexist. In the less denser outer regions, collisional evolution should be less significant. Hence, outer disk objects should be more affected by space weathering, exhibiting redder surfaces, while inner disk objects should be more affected by collisional resurfacing, exhibiting bluer surfaces. Inner disk objects may have different compositions, hence different colors. Such reasonings could explain both color–inclination and H_R–inclination trends found for Classical objects (Tegler *et al.* 2003; Morbidelli *et al.* 2003).

Nevertheless, Gomes' model predicts an inclination-eccentricity distribution for Plutinos, in function of their inner or outer disk origin, that when confronted with their color distribution does not seem compatible with the previous reasoning (PEI04). The color-q correlation is still present in the "hot" Classical objects, whereas the color-inclination correlation seems to be a masking effect of the previous one (PEI04). These results, although obtained with small sub-sampling, suggest that the color distribution of Classical objects cannot be exclusively due to eventual intrinsic differences.

3.4. *Plutinos*

Plutinos (objects in the 3:2 resonance) appear to lack any trends between surface colors and orbital parameters. Consequently, processes connected with the color-eccentricity/perihelion and color-inclination trends found for Classical objects seem absent among Plutinos. There is an apparent excess of intrinsically faint (small) blue Plutinos. However, such excess does not result in a significant color-size correlation (Hainaut & Delsanti 2002, PEI04). From the works by Thébault & Doressoundiram (2003) on the collisional resurfacing scenario (Luu & Jewitt 1996a), one of the major drawbacks of such a model was a prediction for a lack of red Plutinos. However, the collisional resurfacing scenario with triggered cometary activity (DEL04) predicts a more efficient surface rejuvenation for larger objects, contrary to what is observed among Plutinos. This subject needs further investigation.

3.5. *Scattered disk objects*

SDOs (eccentric orbits beyond the 2:1 resonance and also those above the $q = 35\,AU$ line) are one of the most undersampled families. Up to date no significant color-orbital correlations have been found for this class. However, Tegler *et al.* (2003) and PEI04 report that SDOs display a lack of very red surfaces. Moroz *et al.* (2003) demonstrate that different components transform neutral (blue) surface colors into red and further to neutral. Given the typically high aphelion values for SDOs, they are exposed to more intense irradiation dosages (Cooper *et al.* 2003) and their bluer colors may be due to ion irradiation saturation. Such results suggest that the present collisional resurfacing models are too simplified
(Strazzula *et al.* 2003).

3.6. *Centaurs*

Centaurs show evidence for a color-eccentricity correlation that does not translate into a color-perihelia correlation (PEI04). There is no obvious interpretation for such trend. However, they subdivide in two color groups. Such puzzling result is discussed in the next section.

3.7. *Bimodality versus Unimodality*

The Centaurs' and TNOs' color distribution has always been very controversial. Tegler & Romanishin (1998); Tegler & Romanishin (2000) reported the identification of two separated color groups for Centaurs and TNOs (mixed together). Other works claimed a continuous color spreading (e.g. Barucci *et al.* 2000, 2001; Jewitt & Luu 2001; Doressoundiram *et al.* 2001; Hainaut & Delsanti 2002). With a new and exhaustive statistical analysis Tegler & Romanishin (2003) — hereafter TR03 — apparently solved the two color controversy.

Peixinho *et al.* (2003), hereafter PEI03, reanalyzing TR03's data sample conclude that mixing both Centaurs and TNOs lead to the erroneous conclusion of a global bimodality, while there is no evidence for two visible color groups in the TNOs population alone. Using Dip Test (Hartigan & Hartigan 1985), from a compiled sample of 20 quasi–simultaneous $B - R$ colors, PEI03 shows that Centaurs divide in two groups with a significance of 99.5%. Such a possibility had already been highlighted by Boehnhardt *et al.* (2001). When removing Centaurs from TR03 sample, significance level for bimodality is on the order ∼55% much below the minimum 95%.

In a latter work Tegler *et al.* (2003) confirm these findings, whereas Mueller *et al.* (2004) — discussing Bauer *et al.* (2003) VRI data for 24 Centaurs — states that this bimodality is not apparent in the VRI color range. Hence, it appears that without the $B - V$ color information the two groups cannot be (easily) distinguished.

Centaurs are presumed to originate among TNOs, although its specific origin is still in debate (Levison & Duncan 1997; Yu & Tremaine 1999). Hence it is physically difficult to understand the existence of these two color groups. It is possible that TNOs possess real intrinsic differences, even if there is no evidence against a continuous spread of visible colors — we will return to this subject in Section 5. Since Centaurs occupy less denser regions and experience fewer collisions, DEL04's model actually produces two color groups among Centaurs, even with a continuos color spreading of TNOs. The model predicts much more red Centaurs than blue ones and considered Centaurs as "native" residents with stable orbits, which is not the case. The mean Centaur's half-live is $\sim$2.76 Myr and one TNO is expected to enter the Centaur region every $\sim$125 yr (Horner *et al.* 2004). More detailed simulation should be carried out in order to understand this behavior.

4. Comparing color distributions

The question of the origins of the several dynamical groups of TNOs and their presumably related populations is still an open debate. Identifying the color distribution compatibilities/incompatibilities, *i.e.* color distribution differences, between each population is a first step to establish their "genetic" links.

Hainaut & Delsanti (2002), from their compiled MBOSS data-sample, had found that: (a) short period comets' (SPCs) colors are incompatible with those of Classical objects, Plutinos and Centaurs, but only marginally incompatible with SDOs; and (b) Plutinos, Classical objects and SDOs do not evidence color incompatibility between them. Jewitt (2002) also reported that the cometary nuclei were bluer than TNOs.

DOR05b, with a different statistical approach analyses the $V - R$ color (in)compatibilities using: (a) LP+2MS data samples for Centaurs and TNOs; (b) Lamy *et al.* (2005) compiled colors of SPC; and (c) colors compiled from the literature of giant planets' irregular satellites. Furthermore, given the lack of strict dynamical definitions for most families of TNOs, two different classifications schemes of Classical objects and SDOs were considered. Peixinho (2005) extends this work including also LP data leading, basically, to coincident conclusions. Apart some subtleties from the two classification schemes (see DOR05b), results may be simplified as:

(a) SPCs: compatible with irregular satellites and marginally incompatible with SDOs.
(b) Irregular satellites: compatible with SDOs and SPCs.
(c) Centaurs: compatible with all families of TNOs, in spite of their bimodal colors.
(d) Plutinos: compatible with SDOs, Centaurs and Classical objects
(e) Classical objects: only compatible with Centaurs and Plutinos.
(f) SDOs: compatible with all classes except with Classical objects and possibly with SPCs.

The differences of trends among each dynamical class of TNOs (SDOs, Plutinos and Classical objects) and Centaurs may be caused by different surface processing. Such eventual processing does not appear to be sufficiently strong to induce color incompatibility between Plutinos and Centaurs with SDOs or with Classical objects. Nevertheless, the incompatibility between SDOs and Classical objects indicates that the red cluster of

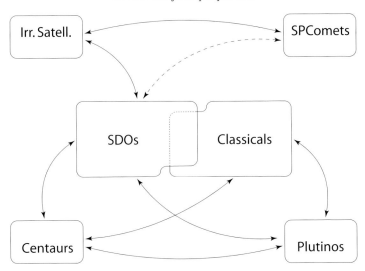

Figure 3. Diagram for Peixinho (2005) and DOR05b's colors compatibility between families. Solid arrows indicate relations between those populations with no evidence for incompatibility, *i.e.*, color compatible populations. Dashed arrows indicate relations with borderline evidence for incompatibility, *i.e.*, possible color (in)compatible families. At the center of the image SDOs and Classicals, which are not compatible, are represented with an overlap to evidence the classification uncertainty of some objects.

"cold" Classical objects may have a different origin. Note, however, that such study was performed only with one color index.

It comes as a surprise that SPCs are not clearly compatible with SDOs and are incompatible with Centaurs, their presumed precursors. Some process modifies SPCs' surfaces when entering into the inner solar system. Jewitt (2002), who had noticed a lack of very red surfaces among SPCs, proposed a surface alteration process in which the formation of a rubble mantle of debris (non-ejected by sublimating stages) or a ballistic mantle (ballistic redeposition of ejected debris) may resurface the object. Considering the irradiation mantles of TNOs as sublimation inhibitors that can be blown off when reaching short perihelion distances, TNOs, with any surface colors, may become Centaurs that will become blue or red, according to the DEL04's model and while migrating to become SPCs loose their irradiation mantles and suffer a complete resurfacing. Under this scenario the color compatibilities of irregular satellites would be the expected ones.

Tegler *et al.* (2003), on the other hand, from a dichotomized $B - R$ color pattern analysis of Centaurs and TNOs conclude a consistency with a primordial origin of their surface colors. Simple dynamical evolution, without post-formation surface evolution of colors, appear also to be a viable approach. The lack of multicolor observations for some parts of the EKB, the low significance of several color correlations and the incapability of the current dynamical and surface evolution models to reproduce some of the observed trends make this subject an open question.

5. Taxonomy

Dealing with a large number of objects it is important to distinguish groups of objects with similar surface properties. Such approach to study physical properties of asteroids was resulted in taxonomy scheme based mostly on surface colors and albedos which became an efficient tool in asteroid investigations. (Barucci *et al.* 2005a, hereafter BA05)

Table 1. The average colors (B-V, V-R, V-I, V-J) and the relative standard deviation for the four taxonomic groups obtained by the G-mode on the sample of 51 objects. The V-H has been computed analyzing the subset of 37 objects.

Class	B-V	V-R	V-I	V-J	V-H
BB	0.70±0.04	0.39±0.03	0.77±0.05	1.16±1.17	1.21±0.52
BR	0.76±0.06	0.49±0.03	0.9±0.07	1.67±0.19	2.04±0.24
IR	0.92±0.03	0.61±0.03	1.20±0.04	1.88±0.09	2.21±0.06
RR	1.08±0.08	0.71±0.04	1.37±0.09	2.27±0.20	2.70±0.24

applied to the 51 TNOs and Centaurs described by the 4 color indices B-V, V-R, V-I and V-J the same statistical analysis used in the '80s to classify the asteroid population: the multivariate statistic analysis (G-Mode) from Barucci *et al.* (1987) and the Principal Component Analysis (Tholen 1984; Tholen & Barucci 1989). BA05 considered all the high quality available colors on TNOs and Centaurs published. They analyzed: i) a set of data for 135 objects observed in B, V, R and I band, ii) a set of 51 objects observed in B, V, R, I, J bands with high quality homogeneous data and iii) a sub-sample of 37 objects also including H band. They selected as a primary sample for the analysis a complete and homogeneous set of 51 objects observed in five filters (B, V, R, I, J), adopting the mean values weighted with the inverse of the error of individual measurement when multiple observations of an object were available. The results of this analysis show that the first principal component (PC1) accounts for most of the variance of the sample (94%) with high weight of V-J (46%). PC2 adds less than 5% to the total variance. Most of the information is concentrated on PC1 which shows four peaks at high density with objects having a neutral color with respect to the Sun (lower PC1 scores) toward the reddest objects of the solar system (higher PC1 scores). As the relationship between the variables is probably not linear, BA05 used a powerful multivariate statistical grouping method (G-mode) to recognize the structure of the found distribution. The G-mode method allows the user to obtain an automatic classification of a statistical sample containing N objects described by M variables (for a total of $M \times N$ degrees of freedom, the d.o.f. number must be >100) in terms of homogeneous taxonomic groups. The method has no a priori grouping criteria, takes into account the instrumental errors in measuring each variable, and also gives indications on the relative importance of the variables in separating the groups. Using the G-mode, a sample of 51 objects and B-V, V- R, V-I and V-J colors were taken as variables ($51 \times 4 = 204$ d.o.f.) and used for the analysis. Using a high confidence level Q (corresponding to 3σ), four homogeneous groups have been obtained. The weight of each variable in separating these groups is 32% for the V-I color, 26% for the V-R, 22% for V-J and 20% for B-V implying that all variables contribute to define the obtained four groups. In Table 1, the color average value and the relative standard deviation for each group are given. The obtained groups are represented in two complementary three dimensional plots (Fig. 4) for the B-V, V-R, V-I and (Fig. 5) for the B-V, V-R, V-J color spaces. The different behavior of the four classes is clearly shown, as well as the role of each color (and particularly V-I and V-J) in assigning the samples to each group. One object, 2000 OK67, does not belong to any group. Moreover, BA05 analyzed a subset of 37 objects for which the V-H color is also available. G-mode analysis applied to this sub-sample described by 5 colors (for a total of $37 \times 5 = 185$ degrees of freedom) provides practically the same well determined four groups. The same objects fall in the same group obtained with 4 variables.

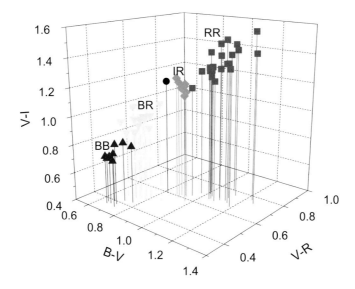

Figure 4. The plot shows the groups in the three dimensional space: B-V, V-R, V-I. Different symbols-colors represent different classes. The black dot represents the objects 2000 OK$_{67}$ which is not belonging to any group.

5.1. *The four obtained classes*

On the basis of this analysis, a taxonomy of TNOs and Centaurs based on their surface broadband colors has been proposed. A two-letter designation of the found groups is introduced to distinguish TNO taxonomy from the asteroid taxonomy. Objects having a neutral colors with respect to the Sun are classified as BB ("blue") group, those having a very high red color are classified as RR ("red"). The BR group consists of objects with an intermediate blue-red color while IR group includes moderately red objects. 2000 OK67, which the G-mode lefts out of this scheme, might be considered as a "single object group". The assignment of each object to one of these groups is reported in the last column of Table 2.

5.2. *Extended taxonomy*

The G-mode has been extended (Fulchignoni *et al.* 2000) to assign to one of the already defined taxonomic groups any object for which the same set of variables become available. Moreover, even if a subset of the variables used in the initial development of the taxonomy is known for an object, the algorithm allows us to give at least a preliminary indication of its appurtenance to a given group. The lack of information on a variable is reflected by the fact that an object could be assigned to two different classes when the lacking variable is the one which operates the discrimination between these classes. The algorithm has been adopted to each of the 84 other TNOs for which the B-V, V-R and V-I colors are available. The obtained preliminary classification for these objects is also reported on the bottom half of Table 2. The appurtenance to a given group has to be considered always with some caution because it is only an indication obtained with an incomplete data set. A double assignment has been obtained for 13 objects, and 15 objects are not classified at all.

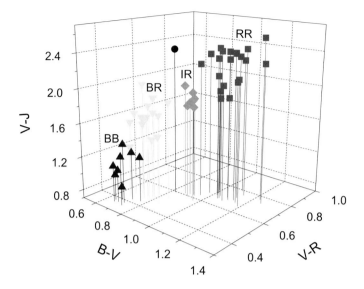

Figure 5. The plot shows the groups in the three dimensional space: B-V, V-R, V-J. Different symbols-colors represent different classes. The black dot represents the objects 2000 OK$_{67}$ which is not belonging to any group.

5.3. *Groups interpretation*

The four groups found by G-mode analysis seem well defined and homogeneous in color properties. The RR group contains the reddest objects of the Solar System. Some well observed objects are members of this group like, 5145 Pholus, 47171 (1999 TC$_{36}$), 55576 (2002 GB$_{10}$), 83982 (2002 GO$_9$) and 90377 Sedna. All these objects seem to contain a few percent of H$_2$O ice on the surface. The reddeness of the group could imply a large amount on the surface composition of Titan tholin and/or ice tholin. Tholins are complex organic solids produced by the irradiation. The BB group contains objects having a neutral reflectance spectra. Typical objects of the group are 2060 Chiron, 90482 Orcus, 19308 (1996 TO$_{66}$) and 15874 (1996 TL$_{66}$). The typical spectra is flat, somewhat bluish in the NIR. The H$_2$O absorption bands seems generally stronger than in the other groups, although the H$_2$O ice presence in the Chiron spectrum seems connected to temporal/orbital variations, and the spectrum of 1996 TL$_{66}$ is completely flat. The presence of large amount of amorphous carbon seems common to the members of this group. The IR group is less red than RR group. Typical members of this class are 20000 Varuna, 38628 Huya, 47932 (2000 GN$_{171}$), 26375 (1999 DE$_9$) and 55565 (2002 AW$_{197}$). Three of these objects seem to contain hydrous silicates on the surface. The BR group is an intermediate group between BB and IR even if its color is closer to the behavior of IR group. The typical members of this class are 8405 Asbolus, 10199 Chariklo, 54598 Bienor and 32532 Thereus. A few percent of H$_2$O is present on the surface of these objects but for Asbolus, Romon-Martin *et al.* (2002) did not find any ice absorption features during its complete rotational period. Passing from neutral (BB group) to very red (RR group) spectra, a higher content of organic material is required to fit appropriately the characteristic spectrum of the single object. Groups BB and BR in general do not require the presence of organic materials or only few percent is needed to achieve their color. H$_2$O ice may be present in all spectra of the groups. The groups BB and BR have color spectra

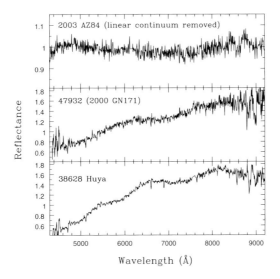

Figure 6. Visible spectra of three TNOs showing aqueous alteration absorption bands. The top spectrum of 2003 AZ84 has been obtained removing the continuum computed with a linear least squares fit to the smoothed spectral data (Fornasier *et al.* 2004). The other two have been obtained by (Lazzarin *et al.* 2003).

very similar to those of C-type and D-type asteroids. Unfortunately we can not associate an albedo range to each taxonomic group because of the lack of albedo data. The few available determinations based from ground observations are very uncertain and do not show any trend with the different classes.

6. Spectroscopy data

Broadband color photometric can provide rough information on the surface of the atmosphereless objects, but the most detailed information on their compositions can be acquired only from spectroscopic observations, especially in the near-infrared spectral region. Most of the known TNOs and Centaurs are too faint for spectroscopic observations, even with the world's largest telescopes, and so far only the brightest bodies have been observed by spectroscopy. The limiting magnitude at the present time, even if observing with 8-10 m telescopes, is about V = 23 mag. The exposure time required is generally long and as the objects rotate around their principal axis the resulting spectra probably arise from signals from both sides of the object. Reflectance spectroscopy (covering the wavelength range between 0.4 and 2.4 μm) provides the most sensitive technique to characterize the major mineral phases and ices present on TNOs. Visible and near-infrared spectroscopy allow us to investigate the presence of the silicate minerals like pyroxene, olivine, and sometimes feldspar, as well as primitive carbonaceous assemblages, and organic tholins. The near infrared region is very important to diagnose the presence of ices and/or hydrocarbons. Weakly active Centaurs or TNOs could show fluorescent gaseous emission bands.

Table 2. On the top, the selected sample of 51 objects and the proposed taxonomical classification based on four color indices (B-V, V-R, V-I, and V-J) are listed. Preliminary classification for the other 84 objects based only on three color indices (B-V, V-R, and V-I) are reported after the separation line.

Object	G-mode class	Object	G-mode class
2060 Chiron	BB	38628 Huya	IR
5145 Pholus	RR	40314 1999 KR_{16}	RR
7066 Nessus	RR	44594 1999 OX_3	RR
8405 Asbolus	BR	47171 1999 TC_{36}	RR
10199 Chariklo	BR	47932 2000 GN_{171}	IR
10370 Hylonome	BR	48639 1995 TL_8	RR
15788 1993 SB	BR	52975 Cyllarus	RR
15789 1993 SC	RR	54598 Bienor	BR
15820 1994 TB	RR	55565 2002 AW_{197}	IR
15874 1996 TL_{66}	BB	55576 2002 GB_{10}	RR
15875 1996 TP_{66}	RR	58534 1997 CQ_{29}	RR
19299 1996 SZ_4	BR	60558 2000 EC_{98}	BR
19308 1996 TO_{66}	BB	63252 2001 BL_{41}	BR
19521 Chaos	IR	79360 1997 CS_{29}	RR
20000 Varuna	IR	83982 2002 GO_9	RR
24835 1995 SM_{55}	BB	90377 Sedna	RR
24952 1997 QJ_4	BB	90482 Orcus	BB
26181 1996 GQ_{21}	RR	91133 1998 HK_{151}	BR
26308 1998 SM_{165}	RR	1996 TQ_{66}	RR
26375 1999 DE_9	IR	1996 TS_{66}	RR
29981 1999 TD_{10}	BR	1998 WU_{24}	BR
32532 Thereus	BR	1999 CD_{158}	BR
32929 1995 QY_9	BR	2000 OJ_{67}	RR
33128 1998 BU_{48}	RR	2000 OK_{67}	—
33340 1998 VG_{44}	IR	2000 PE_{30}	BB
35671 1998 SN_{165}	BB		
15760 1992 QB_1	RR	1996 RR_{20}	RR
15810 1994 JR_1	—	1996 TK_{66}	RR
15836 1995 DA_2	—	1997 QH_4	RR
15883 1997 CR_{29}	BR	1997 RT_5	—
16684 1994 JQ_1	RR	1998 KG_{62}	IR-RR
19255 1994 VK_8	RR	1998 UR_{43}	BR
28978 Ixion	IR-RR	1998 WS_{31}	BR
31824 Elatus	RR-IR	1998 WT_{31}	BB
33001 1997 CU_{29}	RR	1998 WV_{31}	BR
38083 Rhadamanthus	BR	1998 WZ_{31}	BB-BR
38084 1999 HB_{12}	BR	1998 XY_{95}	RR
42301 2001 UR_{163}	—	1999 CB_{119}	RR
49036 Pelion	BR	1999 CF_{119}	BR
52872 Okyrnoe	BR	1999 CX_{131}	RR
55636 2002 TX_{300}	BB	1999 HC_{12}	BR
59358 1999 CL_{158}	BB-BR	1999 HR_{11}	
60454 2000 CH_{105}	RR	1999 HS_{11}	RR
60608 2000 EE_{173}	IR	1999 OE_4	RR
60620 2000 FD_8	RR	1999 OJ_4	RR
60621 2000 FE_8	BR	1999 OM_4	RR
66452 1999 OF_4	RR	1999 RB_{216}	BR
69986 1998 WW_{24}	BR	1999 RE_{215}	RR
69988 1998 WA_{31}	BR	1999 RX_{214}	RR-IR
69990 1998 WU_{31}	—	1999 RY_{214}	BR
79978 1999 CC_{158}	IR-RR	1999 XX_{143}	RR
79983 1999 DF_9	RR	2000 CL_{104}	RR
82075 2000 YW_{134}	BR	2000 FZ_{53}	—
82158 2001 FP_{185}	—	2000 FP_{183}	BB-BR
85633 1998 KR_{65}	RR	2001 CZ_{31}	BR
86047 1999 OY_3	BB	2001 KA_{77}	RR
86177 1999 RY_{215}	BR	2001 KD_{77}	RR
91205 1998 US_{43}	BR	2001 KP_{77}	—
91554 1999 RZ_{215}	—	2001 KY_{76}	—
95626 2002 GZ_{32}	BR	2001 QF_{298}	BB
1993 RO	IR-RR	2001 QY_{297}	BR
1993 FW	RR-IR	2001 UQ_{18}	RR
1994 ES_2	—	2002 DH_5	BR-BB
1994 EV_3	RR	2002 GF_{32}	RR
1994 TA	RR	2002 GH_{32}	—
1995 HM_5	BR	2002 GJ_{32}	—
1995 WY_2	RR-IR	2002 GP_{32}	—
1996 RQ_{20}	IR-RR	2002 GV_{32}	RR

6.1. *Visible spectra*

The visible spectra are generally featureless with a difference in the spectral gradient, ranging from neutral to very red. Pholus and Nessus are the reddest objects known up to now in the Solar System.

The visible wavelength is also an important region and can be used to infer information on the composition, particularly for the especially "red" objects, whose reflectance increases rapidly with wavelength and can be associated to the presence of organic material on their surface. The visible range is also important to detect aqueous altered minerals like for example phyllosilicates.

Few objects (see Fig. 6) have broad absorptions present in their spectra. In the spectrum of 47932 (2000 GN$_{171}$), an absorption centered at around 0.7μm has been detected with a depth of ~8%, while in the spectrum of 38628 (2000 EB$_{173}$) two weak features centered at 0.6μm and at 0.745μm have been detected with depths of ~7% and 8.6%, respectively (Lazzarin *et al.* 2003; de Bergh *et al.* 2004). The spectrum of 2003 AZ$_{84}$ also seems to show a weak absorption centered around 0.7μm and extending from 0.5 to 0.85μm with a depth of about 3% as respect to the continuum (Fornasier *et al.* 2004). These features are very similar to those due to aqueously altered minerals, found in some main belt asteroids (Vilas & Gaffey 1989, and subsequent papers) and are attributed to an Fe^{2+} → Fe^{3+} charge transfer in iron oxides in phillosilicates. How this aqueous alteration process might have effects so far from the Sun is not well understood, but it cannot be excluded that hydrated minerals could have been formed directly in the early solar nebula. Finding aqueous altered materials in TNOs is not too surprising (de Bergh *et al.* 2004) since hydrous materials seem to be present in comets, and hydrous silicates are detected in interplanetary dust particles (IDPs) and in micrometeorites.

6.2. *Near-infrared spectra*

The near-infrared wavelength region carries signatures from water ice (1.5, 1.65, 2.0 μm), other ices (CH$_4$ around 1.7 and 2.3 μm, CH$_3$OH at 2.27 μm, and NH$_3$ at 2 and 2.25 μm), and solid C-N bearing material at 2.2 μm.

The first observed Centaurs were 2060 Chiron and 5145 Pholus (see Barucci *et al.* 2002b, for a reviw on Centaurs) while the first spectrum of a TNO 15789 (1993 SC) was observed by Luu & Jewitt (1996b) in the visible and by Brown *et al.* (1997) in the NIR showing a very noisy reddish spectrum with some features which may be due to hydrocarbon ice (see Barucci *et al.* 2004, for TNO general review).

In the infrared region some spectra are featureless, while others show signatures of water ice, and methanol or other light hydrocarbon ices. Very few of these objects have been well studied in both visible and near infrared and rigorously modeled. In fact these objects are faint and even observations with the largest telescopes (Keck and VLT) do not generally yield good quality spectra. The TNOs are even fainter than Centaurs, and in both cases only few spectra are available, generally with very low S/N. The spectra of TNOs and Centaurs have common behaviour and their surface characteristics seem to show wide diversity. Interpretation is also very difficult because the behavior of models of the spectra depends on the choice of many parameters. In Fig. 7 and 8 we report the visible and NIR spectra for all the objects (9 Centaurs and 9 TNOs) observed at VLT (ESO, Chile) with the best model computed fitting of the data.

6.3. *Surface composition*

Radiative transfer models have been used to interpret the V+NIR spectra of TNOs and Centaurs and several attempts have been performed by modelling the surface composition of these objects with intimate or geographical mixtures of organics, ices and minerals.

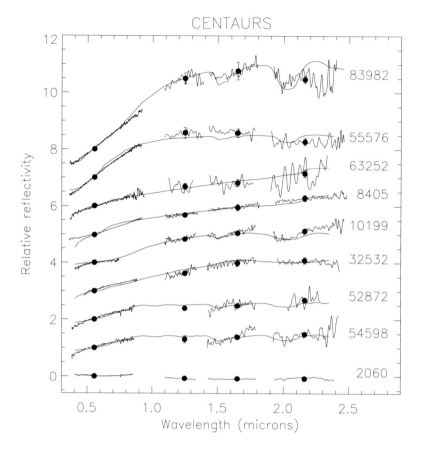

Figure 7. Visible and near-infrared spectra of Centaurs observed at VLT (ESO). The dots represent the B, V, R, I, J, H and K colors, used to adjust the spectral range. The continuous lines superimposed to the spectra are the best-fit models suggested for the surface composition of each objects. All the spectra are normalize at 0.55μm and are shifted by one unit for clarity. The spectra and models are 83982 & 55576 (Doressoundiram *et al.* 2005a); 63252 (Doressoundiram *et al.* 2003); 52872 & 54598 (Dotto *et al.* 2003c); 8405 Romon-Martin *et al.* (2002); 10199 (Dotto *et al.* 2003a); 32532 (Barucci *et al.* 2002a); 2060 (Romon-Martin *et al.* 2003).

The red spectral slopes are in general well reproduced supposing the presence on the surface of organic compounds like tholins or kerogens. Low albedo objects can be well modelled with a high percentage of amorphous carbon. Silicates seem also to be present like the olivine for example on the surface of 5145 Pholus (Cruikshank *et al.* 1998). Water ice is not present in all objects even if ices are the predominant constituent of these distant objects. In Fig. 7 and 8 the computed model is overlapped on the observed spectra. The models present the best current fit to the data, even if they are not unique and depend on many parameters like albedos, different percentages of mixtures and grain size composants. The spectrum of (63252) 2001 BL_{41}, (26181) 1996 GQ_{21} (Doressoundiram *et al.* 2003) and (8405) Asbolus (Romon-Martin *et al.* 2002) are modeled with geographical mixtures of tholins, ice tholin, amorphous carbon. The spectrum of (10199) Chariklo (Dotto *et al.* 2003c) is modeled with a geographical mixture of tholins, amorphous carbon and water ice. Those of (52872) Okyrhoe (1998 SG_{35}) and (54598) 2000 QC_{243} are modeled with geographical mixtures of kerogen, olivine and water ice (Dotto *et al.* 2003b). The spectra of (47171) 1999 TC_{36} (Dotto *et al.* 2003b), (28978) Ixion (Boehnhardt *et al.*

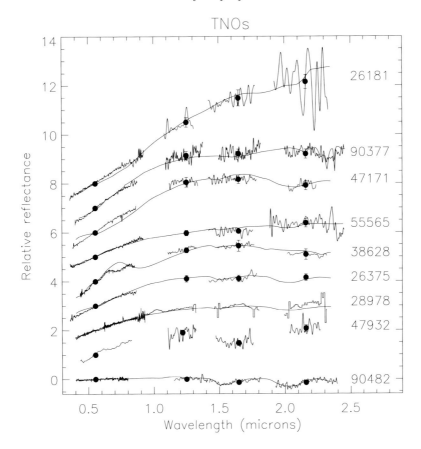

Figure 8. Visible and near-infrared spectra of TNOs observed at VLT (ESO, except the NIR of 28978 that has been observed at TNG (La Palma). The dots represent the B, V, R, I, J, H and K colors, used to adjust the spectral range. The continuous lines superimposed to the spectra are the best-fit models suggested for the surface composition of each objects. All the spectra are normalize at 0.55μm and are shifted by one unit for clarity. The spectra and models are 55565 (Doressoundiram *et al.* 2005a); 47171 (Dotto *et al.* 2003c); 38628 & 47932 (de Bergh *et al.* 2004); 90482 (de Bergh *et al.* 2005); 90377 (Barucci *et al.* 2005b) and 28978 (Boehnhardt *et al.* 2004).

2004), and (32532) 2001 Thereus (Barucci *et al.* 2002a) are modelled with geographical mixtures of tholins, ice tholin, amorphous carbon and water ice. The water ice bands in the spectrum of (28978) Ixion are somewhat uncertain. The spectrum of (38628) Huya (2000 EB$_{173}$) is modelled by a mixture of tholins, amorphous carbon, water ice, and jarosite (a hydrous iron sulfate) (de Bergh *et al.* 2004). The TNO spectra of 55565 2002 AW$_{197}$ and those of the Centaurs 55576 2002 GB$_{10}$ and 83982 2002 GO$_9$ are modelled by intimate mixtures of Triton tholins, amorphous carbon and contaminated water ices (intramixture). A high quality spectra has been obtained for the biggest objects, 90482 Orcus and 90377 Sedna. Orcus are modelled with geographical mixture of kerogen, amorphous carbon and water ice (Fornasier *et al.* 2004; de Bergh *et al.* 2005). Sedna, which is the most distant object has a surface composition completely different from the other TNOs with a total ice abundance >50% (Barucci *et al.* 2005b). Its composition resembles that of Triton, particularly in terms of presence of N$_2$ and CH$_4$. The presence of frozen nitrogen on Sedna implies that there is a thin atmosphere of nitrogen gas surrounding the approximately 200 years of its 10,500-years orbit when it is closest to the Sun. Jewitt & Luu (2004) observed recently Quaoar at Subaru telescope and detected the presence of crystalline water ice on its surface which could imply the existence of cryovolcanism.

A few objects such as 31824 (1999 UG$_5$), 19308 (1996 TO$_{66}$) and 32532 Thereus show surface variations (Merlin *et al.* 2005), but in some cases small spectral difference could be due to low quality data.

7. Conclusions

Even though more than 13 years have passed since the discovery of the first TNO, the knowledge of the TNOs' surface properties is still limited due to the faintness of these distant objects.

Surface color diversity is now confirmed by numerous surveys and the main results from statistical analysis are:

- Strong correlations between optical colors and orbital inclination and perihelion for Classical objects; these two correlations appear connected to each other and the color-perihelion correlation seems size dependent.
- Classical objects subdivide in two inclination groups (dynamically "hot" and "cold") at $i \sim 4.5°$; further studies are needed to clarify if the previous correlations are due to such subdivision.
- Plutinos possess an excess of small blue objects while Scattered Disk Objects a lack of red objects; no clear trends are detected among these dynamical groups.
- Centaurs separate in two visible color groups but there is no evidence of any clear trends with orbital parameters.
- Plutinos' and Centaurs' colors are statistically compatible with those of Scattered Disk Objects and Classical objects, though Scattered Disk Objects and Classical objects are not compatible.
- Colors of Short Period Comets and giant planets' irregular satellites are statistically compatible; irregular satellites are compatible with Scattered Disk Objects but such compatibility is not clear for Short Period Comets.
- Using robust multivariate statistical analysis, four taxonomic groups for TNOs and Centaurs has been obtained, namely: BB, BR, IR, and RR.

The wide diversity of color is confirmed by the different spectral behavior, even though only a few high-quality spectra exist. The spectra show:

- A large range of slope; some are featureless with almost constant gradients over the visible-NIR range, while some show absorption features of ices (most are H$_2$O);
- Three objects show features attributable to the presence of hydrous silicates, but this still needs to be confirmed;
- N$_2$ ice has been detected on the surface of the most distant object 90377 Sedna probably implying a thin atmosphere.

Radiative transfer models have been applied to the obtained spectral reflectance of TNOs and Centaurs, but each is subject to the limitations imposed by the quality of the astronomical spectra, the generally unknown albedo, and to the limited library of materials for which optical constants have been determined.

The models of red objects all use organic materials, such as tholins and kerogen, because common minerals (and ices) cannot provide a sufficiently red color.

H_2O ice is presumed to be the principal component of the bulk composition of outer Solar System objects, but it has been detected (generally with weak absorption) only on few objects. H_2O ice has to be present and could be hidden by low-albedo, opaque surface materials and for these reasons may appear only in few spectra. Additionally, various processes of space weathering (due to solar radiation, cosmic rays and interplanetary dust), can affect the uppermost surface layer and alter their surface chemistry.

The observed surface diversity can be due to different collisional evolutionary states and to different degrees of surface alteration due to space weathering. Collisions can rejuvenate the surface locally by excavating material from the subsurface.

New SPITZER observations will give an important contribution on the albedo knowledge and the NASA's New Horizons mission to the Kuiper Belt and Pluto-Charon, will offer the first close-up views of several solid bodies beyond Neptune.

Note added in proof: The discovery of three new objects – 2003 UB313, 2003EL61, and 2005 FY9 – have been announced at the epoch of ACM meeting. These are very big and 2003 UB313 seems even larger than Pluto. Observations from Gemini Observatory show that the near-infrared spectra of 2003 UB313 and 2005 FY9 are like that of Pluto dominated by the presence of frozen methane, while the spectrum of 2003 EL61 is dominated by crystalline water ice like Charon.

References

Barucci, M. A., Capria, M. T., Coradini, A., & Fulchignoni, M. 1987, Icarus, 72, 304

Barucci, M. A., Romon, J., Doressoundiram, A., & Tholen, D. J. 2000, Astron. J., 120, 496

Barucci, M. A., Fulchignoni, M., Doressoundiram, A., & Birlan, M. 2001, Astron. Astrophys., 471, 1150

Barucci, M. A., Boehnhardt, H., Dotto, E., *et al.* 2002a, Astron. Astrophys., 392, 335

Barucci, M. A., Cruikshank, D. P., Mottola, S., & Lazzarin, M. 2002b, Asteroids III, ed. W. F., Bottke, A., Cellino, P., Paolicchi, & R. P., Binzel, 273

Barucci, M., Doressoundiram, A., & Cruikshank, D. 2004, in Comets II, ed. M. Festou, H. Keller, & H. Weaver, 647

Barucci, M., Belskaya, I., Fulchignoni, M., & Birlan, M. 2005a, Astron. J., 130, 1291

Barucci, M., Cruikshank, D., Dotto, E., *et al.* 2005b, Astron. Astrophys., 439, L1

Bauer, J. M., Meech, K. J., Fernández, Y. R., Farnham, T. L., & Roush, T. L. 2002, Publ. Astron. Soc. Pac., 114, 1309

Bauer, J. M., Fernández, Y. R., & Meech, K. J. 2003, Publ. Astron. Soc. Pac., 115, 981

Belskaya, I. N., Barucci, A. M., & Shkuratov, Y. G. 2003, Earth Moon and Planets, 92, 201

Bernstein, G. M., Trilling, D. E., Allen, R. L., *et al.* 2004, Astron. J., 128, 1364

Boehnhardt, H., Tozzi, G. P., Birkle, K., *et al.* 2001, Astron. Astrophys., 378, 653

Boehnhardt, H., Bagnulo, S., Muinonen, K., *et al.* 2004, Astron. Astrophys., 415, L21

Brown, R. H., Cruikshank, D. P., Pendleton, Y. J., & Veeder, G. J. 1997, Science, 276, 937

Brown, M. E. 2001, Astrophys. J., 121, 2804

Brown, M. & Trujillo, C. 2004, Astrophys. J., 127, 2413

Cooper, J. F., Christian, E. R., Richardson, J. D., & Wang, C. 2003, Earth Moon and Planets, 92, 261

Cruikshank, D. P., Roush, T. L., Bartholomew, M. J., *et al.* 1998, Icarus, 135, 389

de Bergh, C., Boehnhardt, H., Barucci, M. A., *et al.* 2004, Astron. Astrophys., 416, 791

de Bergh, C., Delsanti, A., Tozzi, G., *et al.* 2005, Astron. Astrophys., 437, 1115

Delsanti, A., Hainaut, O., Jourdeuil, E., *et al.* 2004, Astron. Astrophys., 417, 1145

Doressoundiram, A., Barucci, M., Romon, J., & Veillet, C. 2001, Icarus, 154, 277

Doressoundiram, A., Peixinho, N., de Bergh, C., *et al.* 2002, Astron. J., 124, 2279

Doressoundiram, A., Tozzi, G. P., Barucci, M. A., *et al.* 2003, Astrophys. J., 125, 2721

Doressoundiram, A., Barucci, M. A., Tozzi, G. P., *et al.* 2005a, Planet. Space Sci. (in press)

Doressoundiram, A., Peixinho, N., Doucet, C., *et al.* 2005b, Icarus, 174, 90

Dotto, E., Barucci, M. A., Boehnhardt, H., *et al.* 2003a, Icarus, 162, 408

Dotto, E., Barucci, M. A., Boehnhardt, H., *et al.* 2003b, Icarus, 162, 408

Dotto, E., Barucci, M. A., Leyrat, C., *et al.* 2003c, Icarus, 164, 122

Fornasier, S., Doressoundiram, A., Tozzi, G. P., *et al.* 2004, Astron. Astrophys., 421, 353

Fulchignoni, M., Birlan, M., & Antonietta Barucci, M. 2000, Icarus, 146, 204

Gomes, R. S. 2003, Icarus, 161, 404

Grundy, W. M., Noll, K. S., & Stephens, D. C. 2005, Icarus, 176, 184

Hainaut, O. R. & Delsanti, A. C. 2002, Astron. Astrophys., 389, 641

Hartigan, J. A. & Hartigan, P. M. 1985, Ann. Stat., 13, 70

Horner, J., Evans, N. W., & Bailey, M. E. 2004, Mon. Not. R. Astron. Soc., 354, 798

Howell, S. B. 1989, Publ. Astron. Soc. Pac., 101, 616

Jewitt, D. & Luu, J. 1993, Nature, 362, 730

Jewitt, D. C. & Luu, J. X. 2001, Astron. J., 122, 2099

Jewitt, D. 2002, Astron. J., 123, 1039

Jewitt, D. C. & Luu, J. 2004, Nature, 432, 731

Lacerda, P. & Luu, J. 2005, Astrophys. J. (submitted)

Lamy, P., Toth, I., Fernández, Y., & Weaver, H. 2005, in Comets II, ed. M. Festou, H. Keller, & H. Weaver, 223

Lazzarin, M., Barucci, M. A., Boehnhardt, H., *et al.* 2003, Astrophys. J., 125, 1554

Levison, H. F. & Duncan, M. J. 1997, Icarus, 127, 13

Levison, H. F. & Stern, S. A. 2001, Astrophys. J., 121, 1730

Luu, J. & Jewitt, D. 1996a, Astron. J., 112, 2310

Luu, J. & Jewitt, D. C. 1996b, Astron. J., 111, 499

McBride, N., Green, S. F., Hainaut, O., & Delahodde, C. 1999, in Asteroids, Comets, and Meteors 1999 Abstracts Volume

Merlin, F., Barucci, M., Dotto, E., de Bergh, C., & Lo Curto, G. 2005, Astron. Astrophys. (in press)

Morbidelli, A., Brown, M. E., & Levison, H. F. 2003, Earth Moon and Planets, 92, 1

Moroz, L. V., Baratta, G., Distefano, E., *et al.* 2003, Earth Moon and Planets, 92, 279

Mueller, B. E. A., Hergenrother, C. W., Samarasinha, N. H., Campins, H., & McCarthy, D. W. 2004, Icarus, 171, 506

Peixinho, N., Doressoundiram, A., Delsanti, A., *et al.* 2003, Astron. Astrophys., 410, L29

Peixinho, N., Boehnhardt, H., Belskaya, I., *et al.* 2004, Icarus, 170, 153

Peixinho, N. 2005, PhD Thesis, University of Lisbon

Romon-Martin, J., Barucci, M. A., de Bergh, C., *et al.* 2002, Icarus, 160, 59

Romon-Martin, J., Delahodde, C., Barucci, M. A., de Bergh, C., & Peixinho, N. 2003, Astron. Astrophys., 400, 369

Russell, H. N. 1916, Astrophys. J., 43, 173

Sheppard, S. S. & Jewitt, D. C. 2002, Astrophys. J., 124, 1757

Stetson, P. B. 1990, Publ. Astron. Soc. Pac., 102, 932

Strazzula, G., Cooper, J., Christian, E., & Johnson, R. E. 2003, Comptes Rendus de l'Académie des Sciences, Tome 4, Fascicule 7, 791

Tegler, S. C. & Romanishin, W. 1998, Nature, 392, 49

Tegler, S. & Romanishin, W. 2000, Nature, 407, 979

Tegler, S. & Romanishin, W. 2003, Icarus, 161, 181

Tegler, S. C., Romanishin, W., & Consolmagno, S. J. 2003, Astrophys. J., Lett., 599, L49

Thébault, P. & Doressoundiram, A. 2003, Icarus, 162, 27

Tholen, D. 1984, PhD thesis, University of Arizona, Tucson

Tholen, D. J. & Barucci, M. A. 1989, in Asteroids II, 298–315

Trujillo, C. A. & Brown, M. E. 2002, Astrophys. J., Lett., 566, L125

Trujillo, C. A. & Brown, M. E. 2003, Earth Moon and Planets, 92, 99

Vilas, F. & Gaffey, M. J. 1989, Science, 246, 790

Yu, Q. & Tremaine, S. 1999, Astrophys. J., 118, 1873

Asteroids, Comets, Meteors
Proceedings IAU Symposium No. 229, 2005
D. Lazzaro, S. Ferraz-Mello & J.A. Fernández, eds.

© 2006 International Astronomical Union
doi:10.1017/S1743921305006745

Dynamical structure and origin of the Trans-Neptunian population

Rodney S. Gomes[1]

[1]Observatório Nacional, Rua General José Cristino 77 20921-400 Rio de Janeiro, RJ, Brazil
email: rodney@on.br

Abstract. Before the discovery of the first member of the Kuiper belt in 1992, the trans-Neptunian population was supposed to lie on a flat disk and each member would follow a barely eccentric orbit. While less conventional orbits for the trans-Neptunian objects were being discovered, our understanding of its orbital structure and origin was continually changed. A basic classification of the trans-Neptunian population as to their orbits identifies a classical low inclination Kuiper belt population, a resonant population, a high inclination Kuiper belt population, a scattered population and an extended population. Several mechanisms have been proposed to explain the orbital architecture of the Kuiper belt population. Presently, the most plausible scenarios are unequivocally related with the primordial planetary migration induced by a planetesimal disk. Low inclination orbits in the Kuiper belt may have been moderately pushed out from a dynamically cold primordial disk by the resonance sweeping mechanism. The origin of high inclination objects in the classical Kuiper belt is however to be found in a primordial Neptune scattered population, through a perihelion increasing mechanism based on secular resonances. Another push-out mechanism based on the sweeping of the 1:2 resonance with Neptune has also been invoked to explain the low inclination orbits in the classical Kuiper belt. Assuming these last two mechanisms, Kuiper belt objects do not need to have been formed in situ. This kind of formation process would demand a quite large original mass in the Kuiper belt region, which would have brought Neptune beyond its present position at 30 AU. Thus with the exception of the low inclination classical Kuiper belt objects and a few resonant ones, all other trans-Neptunian objects are present or past scattered objects. This notion also includes the case for Sedna, so far the only certain member of the extended population. In its most plausible formation scenario, it was a primordial scattered object by Neptune whose perihelion was increased by the close passage of a star.

Keywords. trans-Neptunian population, Kuiper belt, planetary migration

1. Introduction

The pioneering idea (Kuiper 1951; Edgeworth 1949) that the Solar System would not have an abrupt outer edge at Neptune's orbit was confirmed by the discovery of the first trans-Neptunian object about 13 years ago (Jewitt & Luu 1993). Despite the confirmation that there really existed a Kuiper belt, the very nature of this belt turned out to be quite different from what initially conjectured. In fact, the trans-Neptunian population is much less massive than suggested by the extrapolation from a minimum-mass solar nebula model. It is also much dynamically hotter than naturally supposed for a lightly perturbed disk of planetesimals beyond Neptune that due to its low mass density was not able to accrete into larger bodies. Both the dynamically excited character of the orbits as well as the mass paucity beyond Neptune point to an originally inner disk of planetesimals that was scattered out through a primordial planetary migration. A small fraction of the original disk was then deposited into fairly stable regions in the Kuiper belt and beyond. This review will focus on the mainstream history of the trans-Neptunian

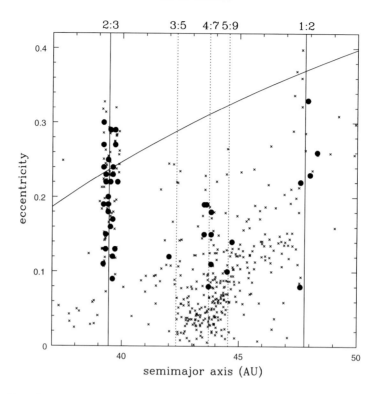

Figure 1. Distribution of semimajor axes and eccentricities for all Kuiper belt objects that have been observed in more than one opposition taken from http://cfa-www.harvard.edu/cfa/ps/lists/trans-Neptunian objects.html as of June/2005. Large circles stand for resonant objects according to Elliot *et al.* (2005).

population dynamics in search of its origin since the discovery of its first member. In section 2, I present a tentative orbital classification of the trans-Neptunian objects and a little about present dynamics of the Kuiper belt follows. The knowledge of this dynamics is essential in order to conclude that the orbital configuration of the trans-Neptunian objects must have an origin beyond present solar system perturbation capabilities. In section 3, I present several theories for the origin of the trans-Neptunian population keeping some chronological order and classifying according to the main theories. In particular, planetary migration theory is given a high priority since I understand that this mechanism is intrinsically related to present trans-Neptunian orbital configuration. Conclusions are drawn in section 4 where I also suggest where we are now and what will come next.

2. Orbital classification of the trans-Neptunian population.

There is not only one possibility of classifying the trans-Neptunian population with respect to their orbits. The simplest classification can distinguish three main groups that we can name as the Kuiper belt, the scattered population and the extended population. Although this last population has so far just one or two representatives, the very specific orbital characteristics of 2003 VB12 (Sedna) undoubtedly distinguishes it from the scattered population. The boundary that separates classical Kuiper belt objects from scattered objects can be the semimajor axis associated with the 1:2 mean motion resonance with Neptune. This number is around 48 AU but 50 AU may work as well. The justification for that limit is the nonexistence of any observed big enough low eccentricity object beyond it (Gladman *et al.* 2001, Allen *et al.* 2002, Bernstein *et al.* 2004). This boundary at least suggests that the 1:2 mean motion resonance with Neptune may have had a decisive influence in its establishment. The nomenclature Kuiper belt is surely suggestive as at first glance it points to a population of objects formed in situ, as some small eccentricity objects might reveal, following the original idea by Kuiper (1951). Nevertheless the Kuiper belt itself is not a uniform population. In fact it can be subdivided into several groups. A useful classification just separates the resonant objects from the non-resonant ones. These latter objects will thus compose the so named classical Kuiper belt. Sometimes the classical belt refers to all objects located between the 2:3 and 1:2 mean motion resonance with Neptune. Anyway the distinction between classical and resonant objects, although theoretically clear, demands a hard task in order to determine a fairly well defined orbit to confirm its resonant status. Recently, Elliot *et al.* (2005) made a comprehensive inventory of Kuiper belt objects placing a good number of the observed objects by their Deep Ecliptic Survey as resonant orbits including high order ones like the 9:5 and 7:4 (see Fig. 1). Anyway many other objects remain to be classified as to its resonant status. This classification can be quite useful since it suggests possible origin scenarios for the Kuiper belt population. A second possible sub-classification for the Kuiper belt orbits distinguishes the high inclination ones from the low inclination ones. Not long after the discovery of the first member of the Kuiper belt, not only eccentric but also very inclined orbits were found. More recently, Brown (2001) determined through statistical inference that there are in fact two different populations in the Kuiper belt, the low inclination and the high inclination ones, sometimes referred to as the cold and hot populations. Correlations of orbital inclinations with color and magnitude (Trujillo & Brown 2002) more accurately established the dual character of the Kuiper belt with respect to the orbital inclination of its members. This new orbital classification more effectively motivated theorists to explain its origin and thus the very origin of the Kuiper belt as a whole.

Figure 1 shows the distribution of semimajor axes with eccentricities for all objects listed in Minor Planet Center electronic pages (available at http://cfa-www.harvard.edu/cfa/ps/lists/trans-Neptunian objects.html and http://cfa-www.harvard.edu/iau/lists/-Centaurs.html) observed in more than one opposition. In this figure we distinguish the resonant semimajor axes and some resonant objects as determined by Elliot *et al.* (2005). Figure 2 depicts the cold and hot population classification, where the objects with inclination smaller than 5° are plotted as circles. One must have in mind that due to observational bias that favors the observation of less inclined orbits we see in Fig. 2 a false ratio between the cold and the hot population. A more realistic ratio between numbers in either population is determined by Brown (2001). In Figure 2, I also plot some of the scattered objects, those closer to the Kuiper belt. It is interesting to note that the cold population seems to invade also the scattered region.

Figure 3 shows the distribution of the semimajor axes and perihelion distances for the scattered and extended populations. It is not difficult to notice that 2000 CR105 and much more specifically Sedna do not belong to the same population as the other scattered objects.

2.1. *Present Kuiper belt dynamics*

A good deal of present orbital configuration of the Kuiper belt can be deduced by its present dynamics (Duncan *et al.* 1995, Morbidelli *et al.* 1995, Malhotra 1996).

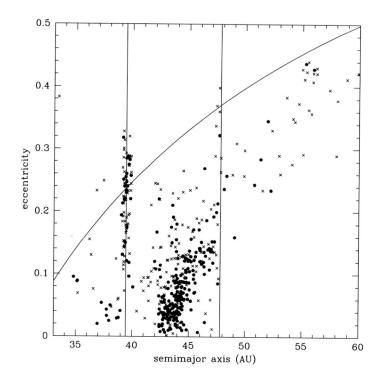

Figure 2. Like Fig.1, now distinguishing the hot (crosses $I > 5°$) and cold (circles $I < 5°$) populations.

A remarkable feature of this orbital distribution is the void of objects just beyond the 2:3 resonance with Neptune. This is caused by the ν_8 secular resonance that can easily empty this region of objects at solar system age. More specific secular and Kozai dynamics can raise eccentricities and inclination in the classical Kuiper belt (Kuchner *et al.* 2002). However this dynamics does not explain the eccentricity and inclination distribution of the observed classical and resonant Kuiper belt. We must find the origin of this high inclination/eccentricity distribution elsewhere.

3. Origin of the trans-Neptunian objects

A consensual point about the origin of trans-Neptunian objects is that they once belonged to an icy planetesimal disk whose members failed to accrete into planet-sized bodies. In fact Kuiper's original conjecture was that these bodies would presently form a disk beyond Neptune. This disk would have been left more or less intact since the end of the formation of the major solar system planets. The continuous discovery of trans-Neptunian objects in excited orbits motivated theorists to seek a way by which these orbits could manage to get excited from a putative dynamically cold disk. It can be instructive to classify these orbit exciting mechanisms by considering two main groups, the static mechanisms group and the migration group. Although it is today almost if not totally consensual that planetary migration of the giant planets once took place in the solar system, it is anyway instructive to review two main non-migration theories for the excitation of the trans-Neptunian orbits. One must bear in mind that these

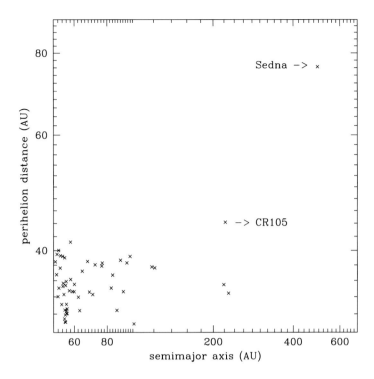

Figure 3. Distribution of semimajor axes with perihelion distances for all trans-Neptunian objects with semimajor axis larger than 50 AU. We notice that 2000 CR105 and Sedna are clearly distinguished as members of a different class of objects.

mechanisms do not in principle negate migration. Nevertheless the consequences of planetary migration on the establishment of the orbital architecture of the trans-Neptunian objects turned out to be so evident that the non-migrating models became of little usefulness.

An important point for the determination of the origin of the trans-Neptunian objects is their present total mass. The classical Kuiper belt mass is now considered to be a few hundredths of an Earth mass. The cold population would have just 0.01 Earth mass and the hot plus scattered population a few times as that (Bernstein *et al.* 2004). The extended population was estimated by Brown *et al.* (2004) to have roughly 5 Earth masses although this estimate is a little crude.

3.1. *The large planetesimal model*

This mechanism was proposed by Morbidelli and Valsecchi (1997) and Petit *et al.* (1999). At that time with a scarce number of known trans-Neptunian objects, this model competed with an already proposed migration model (**?**). The idea of that model is that a large planetesimal (Mars or Earth size) was scattered by the planets and temporarily deposited in the Kuiper region causing the excitation of the Kuiper belt orbits. Eventually the large planetesimal would be scattered out of the solar system by continual perturbations from the planets. The advantage of the large planetesimal model was its generality in producing excited (mainly eccentric) orbits for the Kuiper belt, not only in the resonant regions as in the case for the resonance sweeping model. On the other hand,

it was not so suitable to explain the resonant population including the remarkable plutinos population. Another difficulty with the large planetesimal model (which was not its
own privilege) was its inadequacy to explain the inclinations of the Kuiper belt objects
as nicely as their eccentricities. Later it was also noted that the excitation of such a large
number of objects that would have been formed in situ would unavoidably induce close
encounters of the planetesimals with Neptune thus feeding again its migration towards
the edge of the disk (Gomes *et al.* 2004).

3.2. *The passing star model*

Ida *et al.* (2000) proposed that a passing star with a perihelion near 100 AU could
raise the eccentricities and inclination of the primordial planetesimal disk creating a
distribution of eccentricities and inclinations in the Kuiper region similar to the present
one. The dynamically excited planetesimals would thus start a mass erosion process that
might account for the mass paucity in the belt. The authors also simulated a planetary
migration after the stellar encounter so as to create the resonant Kuiper belt orbits. The
advantage of the passing star model was the creation of some high inclinations for the
orbits in the Kuiper belt region. Nevertheless the distribution of the orbits would hardly
resemble that of the real Kuiper belt. Passing star models have also been invoked to
explain the truncation of the Kuiper belt at 50 AU (Melita *et al.* 2000, Kobayashi & Ida
2001).

3.3. *Migration models*

Fernandez & Ip (1984) first showed that a planetary formation scenario where proto
planets shared its orbital space with smaller planetesimals would induce the migration
of the planets. This model was later improved by the modeling of a planetesimal disk
with a larger number of objects (Hanh & Malhotra, 1999; Gomes *et al.* 2004). These
more accurate models confirmed the main features already suggested in Fernandez & Ip
(1984) which are the outward migration of Neptune, Uranus and Saturn and the inward
much shorter migration of Jupiter. This process is based on the exchange of energy and
angular momentum between the planets and planetesimals (for details see Malhotra *et al.*
2000; Gomes *et al.* 2004). The main migration theories presented to possibly explain the
trans-Neptunian orbital architecture are: the resonance sweeping theory, the Neptune's
aphelion theory and the evader's theory. Another useful classification of the migration
models distinguishes gradual migration models and abrupt migration models. The next
four subsections will deal with these Kuiper belt orbital excitation mechanisms.

3.3.1. *The resonance sweeping model*

In two pioneering papers, Malhotra (1993,1995) showed that Neptune's outward migration into an initially dynamically cold disk of planetesimals would push out along with
the planet many planetesimals trapped in mean motion resonances with the outermost
planet. This mechanism by the adiabatic invariant theory (Malhotra 1995; Gomes 1997)
can also excite the eccentricities of the trapped bodies. By that time many Kuiper belt
objects already found shared the 3:2 mean motion resonance with Neptune like Pluto.
These objects were thus named plutinos after Pluto. By the adiabatic invariant theory, the increment in a plutinos's eccentricity is an analytical well determined function
(Malhotra 1995; Gomes 1997) of the radial displacement of the planetesimal and the
planet. Thus from Pluto's well known eccentricity, it was possible to infer the total radial
shift experienced by Neptune, about 6 AU. With the aid of Fernandez & Ip (1984) results
it was possible to estimate the other planets initial positions, thus many simulations of
planetary migration were undertaken considering a standard set of initial positions for the

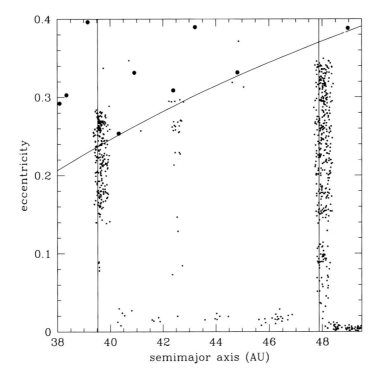

Figure 4. Distribution of semimajor axes with eccentricities of planetesimals from a numerical simulation. The planetesimals were initially distributed in a cold disk outside the orbit of Neptune at 23 AU. The migration was induced by a fictitious force and was made to last for 10 million years after which the planets stopped at their present positions. The migration of Neptune caused planetesimals to be trapped into several mean motion resonances. The orbits with inclinations greater than 5° are represented as the larger circles.

planets which were from Jupiter to Neptune, 5.4 AU, 8.7 AU, 16.3 AU and 23.2 AU. These simulations considered a fictitious force to mimic the planetary migration and the planetesimals assumed massless would experience the planetary gravitational perturbations. Very interesting results came from these simulations showing that plutinos eccentricities could thus be attained. Moreover this model could also account for trapped bodies in other resonances like the 1:2, where objects were really found. Finally the eccentricities of bodies between the 3:2 and 2:1 resonances with Neptune could also be fairly increased due to trapping and release from resonance during migration. Again the difficulty that remained was the explanation of so many orbits with high inclinations in the Kuiper belt that could not be accounted for by the resonance sweeping scenario (see Fig. 4). In particular, plutinos could acquire their inclinations but the process by which this was possible did not create as many high inclination objects as observed (Gomes 2000). Also the adiabatic invariant theory applied to account for Pluto's inclination (through the coupled mean motion plus Kozai resonances) demands a much smaller initial radial distance for Neptune (around 18 AU) if Pluto is to be assumed in an initial dynamically cold orbit (Gomes 1997). This initial position for Neptune (with implied more compact initial orbits for Uranus and Saturn) was considered problematic since many mean motion resonances would be experienced by pairs of planets during migration resulting in a possible destabilization of the system. This was later resolved since the effect of a massive disk on the planets acts to circularize their orbits. The massless disk models considered by then did not show this effect. A more complete model where the disk particles had mass to disturb the planets and to induce their migration was presented in Hanh & Malhotra, 1999. In this work, considering the standard set of initial conditions mentioned above, the authors estimated that the mass for the disk that would bring Neptune to 30 AU should be around 50 Earth masses. The fact that the disk was composed by few particles (1000) implied a too nonuniform migration which impeded the resonance sweeping process, thus very few particles could be trapped in the resonances although the eccentricity raising effect was preserved.

3.3.2. *The Neptune's aphelion model*

The idea that Uranus and Neptune would hardly be formed at their present positions (Levison & Stewart 2001) or even at their shifted location according to the standard initial positions of section 3.3.1 motivated Thommes *et al.* (1999) to propose a model by which Uranus and Neptune formed between the orbits of Jupiter and Saturn, these last ones not very far from their present positions. The orbits of the four planets remained stable while Jupiter and Saturn did not reach the critical mass to start to attract the disk gas. After that, when the gas giants acquired masses similar to present ones, they started to disturb Uranus and Neptune into very elliptical orbits throwing them directly into the planetesimal disk. Because of this great perturbation experienced by the icy giants, the aphelion of the outermost planet temporarily visited the Kuiper belt region, thus exciting the planetesimals orbits in the belt. One of the difficulties with this model had to do with the planets themselves since there is no disk mass big enough to prevent the planets from exiting the solar system due to a strong close perturbation from a gas giant and at the same time small enough to prevent Neptune to go beyond its present position at 30 AU. But as far as the Kuiper belt itself is concerned the temporary passage of Neptune's aphelion at present Kuiper belt region was able to effectively excite again only the eccentricities of the orbits but not their inclinations.

3.3.3. *The evaders model*

Gomes (2003a) envisaged a model where Neptune started its migration between 13.5 and 17.5 AU. A disk of planetesimals would start just beyond Neptune. Several numerical simulations were done for different outer edges of the disk. This disk was simulated with 10000 massive particles that disturbed the planets and induced a planetary migration. The consideration of one order of magnitude higher number of particles allowed a much smoother migration supposedly closer to reality. In fact, in this kind of simulated disk the particles functioned as a fuel for the migration but they also experienced the resonance trapping process proposed by Malhotra (1993). Another consequence of this higher number of simulated particles for the disk was the possibility of allowing the observation in the end of the simulation of some planetesimals deposited in the Kuiper belt as evaders from the scattered population. In fact, mean motion and secular resonances could decrease the eccentricities of a small fraction of all planetesimals that became scattered by the planets during migration. This very migration could also help in erasing the return path for the particles experiencing eccentricity decrease since the planet-particle dynamics would become irreversible. These particles deposited in the Kuiper belt present high inclinations due to its past history of close encounters with the migrating planets, as Fig. 5 shows. These inclinations have a distribution compatible with the inclination distribution of the dynamically hot Kuiper belt population (Gomes 2003a). Although the fraction of planetesimals initially in the disk deposited in the Kuiper belt through this process is

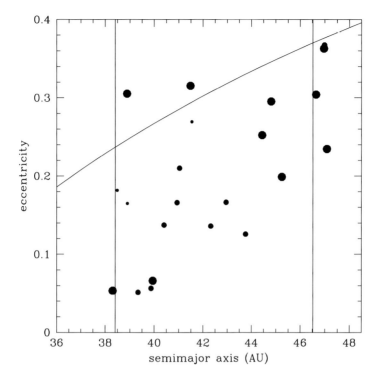

Figure 5. Distribution of semimajor axes with eccentricities of planetesimals from a numerical simulation. The planetesimals were initially distributed in a cold disk outside the orbit of Neptune around 15 AU. The planetesimals perturbed the planetary orbits thus inducing a migration. After one billion years some planetesimals scattered by Neptune had their perihelia increased due to secular and mean motion resonance effects from the planets, thus being deposited in the Kuiper belt (Gomes 2003a). The small circles stand for orbits with inclination lower than 5° Large circles represent planetesimals orbits with inclinations between 5° and 20°, whereas very large circles stand for orbits with inclination greater than 20°

very small (about 0.2 % at solar system age) it is however large enough to account for the present hot Kuiper belt mass. Therefore, another new idea brought up by this scenario is that there is no more need for mass erosion in the Kuiper belt, at least as far as the hot population is concerned, since the present mass of hot Kuiper belt objects turns out to be the same as the original one. However there was now the low inclination population to be explained. Considering them as objects formed in situ and eccentricity-excited by the resonance sweeping process, an estimated mass of 10 Earth masses originally in the Kuiper belt region is required to form the big enough objects in present Kuiper belt (Stern & Colwell 1997; Kenyon and Luu 1998). This requires, just for the cold population, a mass erosion process to present estimated 0.01 Earth mass of around 99.9 %, not well explained by fragmentation theories (Davis & Farinella 1997; Kenyon & Luu 1999). Moreover this high mass in the original outer planetesimal disk would force Neptune to beyond its present position somewhere in the present Kuiper belt (around 45 AU) (Gomes *et al.* 2004). These findings surely suggest a push-out mechanism also for the cold population. Levison & Morbidelli (2003) proposed such a scenario in which objects trapped into the 1:2 mean motion resonance with Neptune during its outward migration are released before the end of migration leaving low inclination objects with moderate

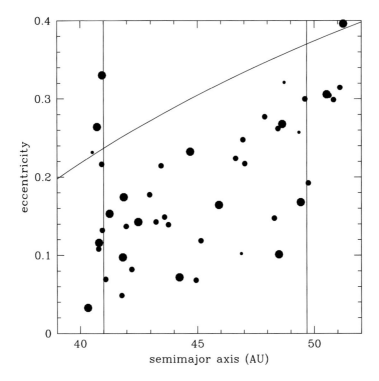

Figure 6. Same as Fig. 5 now for an abrupt migration model in which Uranus and Neptune are thrown into the planetesimal disk starting the late heavy bombardment (Gomes *et al.* 2005a). This orbital distribution stand for 700 million years after the beginning of the bombardment. We still notice the formation of high inclination Kuiper belt objects.

eccentricities in the present Kuiper belt region. This process is based on the effect of secular resonances on the planetesimals trapped in the 1:2 resonance with Neptune. These resonances show up because Neptune's perihelion precession rate is changed due to the perturbation from the great amount of mass trapped in the 1:2 resonance. The decisive consequence of these secular resonances is that planetesimals eccentricities of orbits in the 1:2 resonance do not increase uniformly by the adiabatic theory. Some planetesimals thus remain with low eccentricities and are eventually released from resonance into the classical Kuiper belt.

3.3.4. *Gradual versus abrupt migration models*

Most of the migration models described above is also associated to a gradual migration. The gravitational interaction of a cold disk of small planetesimals with planets will naturally induce a slow migration of the planets with planetary eccentricity damping. This gradual migration is favorable for the adiabatic resonance trapping process (**?**) and the push-out mechanism (Levison & Morbidelli 2003) to create the low inclination objects. It is also compatible with the process that creates the high inclination population (Gomes 2003a). Migration will not proceed smoothly either if the planetesimals are too large†

† using few too large planetesimals to simulate a massive disk that induces migration is an artifice to enable a not too long computation time for the numerical integration; on the other hand real large planetesimals may have existed in the disk (terrestrial planet size, probably in

or if the planets experience close encounters that throw them abruptly into the disk. The Neptune's aphelion model above described can be classified as an abrupt migration model, since the planets experience close encounter due to their initial conditions near the gas giants and the fast gas accretion of Jupiter and Saturn. More recently another abrupt migration model was suggested (Tsiganis *et al.* 2005; Morbidelli *et al.* 2005; Gomes *et al.* 2005a) by which Jupiter and Saturn would have crossed their 1:2 mean motion resonance after an initial slow migration. This resonance passage may have been delayed 700 million years after which Uranus and Neptune would be thrown immediately into the disk triggering the late heavy bombardment in the inner solar system. This process also allows the major planets to stop at their right positions with the right eccentricities and inclinations and also Jupiter Trojans can be captured from the disk population directly into coorbital regions with Jupiter. Abrupt migration theories are not however suitable to explain resonance trapping via adiabatic theory, thus not only Malhotra (1993, 1995) resonance sweeping theory does not apply but also the Levison & Morbidelli (2003) push-out mechanism is weakened, since this theory is based on an initially very smooth migration of Neptune to allow a great number of planetesimals trapped into the 1:2 resonance. On the other hand, the creation of high inclination objects by Gomes (2003a) does not need a long regular migration of Neptune. An abrupt migration model as described above can account for the mechanism that places high inclination objects in the Kuiper belt (see Fig. 6).

3.4. *The origin of the scattered and extended populations*

Besides planetary migration, an unavoidable consequence of the primordial close encounter interaction between the major planets and a disk of planetesimals is the creation of a population of objects scattered by the planets. At solar system age, most objects are scattered out of the solar system but a non-negligible amount remains. These objects have large semimajor axes but their perihelia are not much larger than Neptune's semimajor axis at 30 AU, a clear signature of a past history of close encounters between the planetesimal and Neptune. An estimated mass of a few hundredths of an Earth mass (Bernstein *et al.* 2004) will be left in this population after 4.5 billion years not only by chance but also by mechanisms that turns the orbits stable for the solar system age. A known mechanism is the trapping of scattered objects into coupled mean motion/Kozai resonances that induce an increase of the object's perihelion distance thus placing it far from destabilizing close encounter perturbations from Neptune. If this resonance mechanism happens while Neptune is still migrating, high perihelion fossilized orbits can be created. When migration is ceased, temporary (though long) high perihelion orbits can be also created, as Fig. 7 shows (Gomes 2003b; Gomes *et al.* 2005b). This process will anyway produce a larger number of high perihelion orbits with small semimajor axis as compared with high perihelion orbits with large semimajor axis. Fig. 7 however shows that the two highest perihelion orbits already found for a trans-Neptunian object also have large semimajor axes. This suggests that these objects (Sedna and 2000 CR105) must have had their perihelia increased by some other process than the mean motion + Kozai resonance mechanism from Neptune. Although 2000 CR105 might marginally be created by such a process, Sedna would not (Gomes *et al.* 2005b) so there is more than a statistical inference for a specific origin for these objects. Among several possible mechanisms that can raise the perihelion of a scattered object, the most serious candidate is the passage of one or several stars near a primordial solar system (Fernandez and

its inner part) and a more realistic simulation including both large and small planetesimals is yet to be undertaken

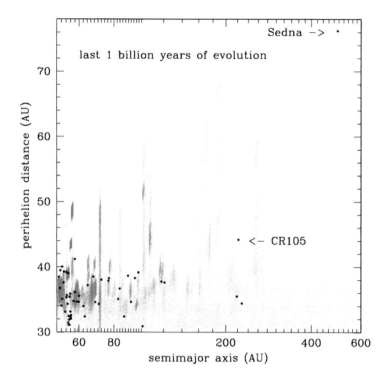

Figure 7. Distribution of semimajor axes and perihelion distances of real scattered and extended population objects (small circles) and from a numerical integration including all four major planets and a disk of planetesimals that induce planetary migration. The integration was carried on to solar system age and orbits were plotted at every million years for the last billion years (small dots). This figure confirms that Sedna cannot be a simple scattered objects and much probably 2000 CR105 is not either.

Brunini 2000; Morbidelli and Levison 2004). This is a reasonable hypothesis as far as we understand that the solar system was formed in a denser star cluster. The creation of populations that include both 2000 CR105 and Sedna can be accomplished by the perihelion raising effect of a passing star (see Fig. 8). A weak point of the passing star theory is the usually low mass of an extended population (about 0.2 Earth mass) created by such a mechanism as compared with the (however crude) estimate of 5 Earth masses for a population of Sedna-like orbits (Brown *et al.* 2004). Another passing star theory worthy commenting (Morbidelli and Levison, 2004) concerns a small (brown dwarf) star that carries with it a disk of planetesimals. Passing near the primordial Sun (200 AU), the brown dwarf system could pass to the Sun some of the brown dwarf's planetesimals, in this case the extended population would be formed by extrasolar bodies. The advantage of this theory is that possibly a substantial fraction of the original brown dwarf disk could have been transferred to the solar system. However we do not know much about brown dwarf disks to give too much a priori credit to this theory. Another interesting theory for the formation of Sedna-like orbits concerns a wide-binary solar system companion (Matese *et al.* 2006). This planet should have a mass larger than Jupiter mass if located in a circular orbit at 5000 AU to create a Sedna-like orbit. To account for 2000 CR105 orbit the planet should have near 10 Jupiter masses at 5000 AU. ..

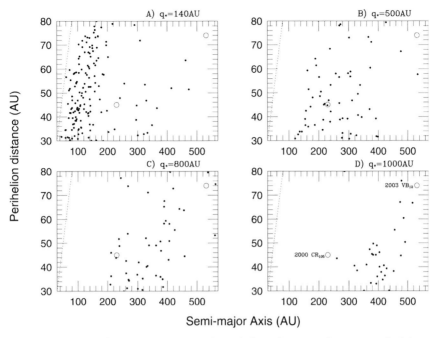

Figure 8. Distribution of semimajor axes and perihelion distances for scattered objects whose orbits were perturbed by a close passing star. The minimum distance of the passing star from the Sun is shown at the right top of each panel (courtesy of Levison and Morbidelli)

4. Conclusions

Since the discovery of the first member of the Kuiper belt, the understanding of the origin of the trans-Neptunian population has been continually challenged. It is now consensual that at least a great part of the trans-Neptunian objects has been shifted outward by the migration of the icy giant planets especially Neptune. This process induced the excited orbits observed in the Kuiper belt and possibly also explains the belt's low mass as an original feature caused by the relatively inefficient processes to bring objects from the scattered population to more stable regions in the Kuiper belt. This is most probably the case for the high inclination objects in the classical and resonant Kuiper belt. The low inclination population in the classical belt can have an explanation in a push-out mechanism from a truncated disk. This mechanism does not however work very well if one must invoke an abrupt migration model as recently suggested (Gomes *et al.* 2005a). Moreover both hot and cold populations would come from about the same region in the primordial planetesimal disk, what is not compatible with the dual character of the Kuiper belt. The formation of the cold belt in situ (or slighted shifted outward by the resonance sweeping mechanism) cannot at this time be ruled out. The most important argument in its favor is the dual character of the classical Kuiper belt population. The correlation of the hot and cold populations with color and magnitude suggests different origins for both populations and the in-situ formation of the cold population would naturally account for that. However this scenario would have also to explain the radical erosion suffered by the mass initially in the Kuiper region that must be reduced to more than two orders of magnitude possibly three orders. Moreover an initial high mass in the Kuiper region would probably have induced an extra migration to Neptune shifting the outermost planet to somewhere in the very Kuiper belt region.

Another challenging characteristic of the trans-Neptunian population concerns the understanding of the origin and orbital distribution of the extended population. Several theories have been proposed that fairly well account for this presently two-member population. As soon as successful observations reveal new members for this population the right theory will be gradually accepted by the scientific community. For now there is plenty of room for theorists.

References

Allen, R.L., Bernstein, G.M., & Malhotra, R. 2002, *Astronomical Journal* 124, 2949

Bernstein, G.M., Trilling, D.E., Allen, R.L., Brown, M.E., Holman, M., & Malhotra, R. 2004, *Astronomical Journal* 128, 1364

Brown, M.E. 2001, *Astronomical Journal* 121, 2804

Brown, M.E., Trujillo, C., & Rabinowitz, D. 2004, *Astrophysical Journal* 617, 645

Davis, D.R. & Farinella, P. 1997, *Icarus* 125, 50

Duncan, M.J., Levison, H.F., & Budd, S.M. 1995, *Astronomical Journal* 110, 3073

Edgeworth, K.E. 1949, *Mon. Not. R. Astron. Soc.* 109, 600

Elliot, J.L., Kern, S.D., Clancy, K.B., Gulbis, A.A.S., Millis, R.L., Buie, M.W., Wasserman, L.H., Chiang, E.I., Jordan, A.B., Trilling, D.E., & Meech, K.J. 2005, *Astronomical Journal* 129, 1117

Fernandez, J.A. & Ip, W.H. 1984, *Icarus* 58, 109

Fernandez, J.A. & Brunini, A. 2000, *Icarus* 145, 580

Gladman, B., Kavelaars, J.J., Petit, J.M., Morbidelli, A., Holman, M.J., & Loredo, T. 2001, *Astronomical Journal* 122, 1051

Gomes, R.S. 1997, *Astronomical Journal* 114, 2166

Gomes, R.S. 2000, *Astronomical Journal* 120, 2695

Gomes, R.S. 2003, *Icarus* 161, 404

Gomes, R. 2003, *Earth, Moon, and Planets* 92, 29

Gomes, R.S., Morbidelli, A., & Levison, H.F. 2004, *Icarus* 170, 492

Gomes, R., Levison, H.F., Tsiganis, K., & Morbidelli, A. 2005, *Nature* 435, 466

Gomes, R.S., Gallardo, T., Fernández, J.A., & Brunini, A. 2005, *Cel. Mech. Dyn. Astron.* 91, 109

Hahn, J.M. & Malhotra, R. 1999, *Astronomical Journal* 117, 3041

Ida, S., Larwood, J., & Burkert, A. 2000, *Astrophysical Journal* 528, 351

Jewitt, D. & Luu, J. 1993, *Nature* 362, 730

Kenyon, S.J. & Luu, J.X. 1998, *Astronomical Journal* 115, 2136

Kenyon, S.J. & Luu, J.X 1999, *Astronomical Journal* 118, 1101

Kobayashi, H. & Ida, S. 2001, *Icarus* 153, 416

Kuchner, M.J., Brown, M.E., & Holman, M. 2002, *Astronomical Journal* 124, 1221

Kuiper, G. 1951, in: Hynek, J.A (ed.), *Proceedings of a topical symposium* (New York: McGraw-Hill) vol. 99, p. 225

Levison, H.F. & Stewart, G.R. 2001. *Icarus* 153, 224

Levison, H.F. & Morbidelli, A. 2003, *Nature* 426, 419

Malhotra, R. 1993, *Nature* 365, 819

Malhotra, R. 1995, *Astronomical Journal* 110, 420

Malhotra, R. 1996, *Astronomical Journal* 111, 504

Malhotra, R., Duncan, M.J., & Levison, H.F. 2000, in: Mannings, V., Boss, A.P., and Russell, S. S. (eds.) *Protostars and Planets IV* (Tucson: University of Arizona Press), p. 1231

Matese, J.J., Whitmire, D.P., & Lissauer, J.J. 2006, *proceedings of the IAU symposium 229*

Melita, M.D.; Larwood, J., Collander-Brown, S., Fitzsimmons, A., Williams, I.P., & Brunini, A. 2000, in: Barbara Warmbein (ed.) *Proceedings of Asteroids, Comets, Meteors - ACM 2002* (ESA Publications Division) p. 305

Morbidelli, A., Thomas, F., & Moons, M. 1995, *Icarus* 118, 322

Morbidelli, A. & Valsecchi, G.B. 1997, *Icarus* 128, 464

Morbidelli, A. & Levison, H.F. 2004, *Astronomical Journal* 128, 2564

Morbidelli, A., Levison, H.F., Tsiganis, K., & Gomes, R. 2005, *Nature* 435, 462

Petit, J.-M., Morbidelli, A., & Valsecchi, G.B. 1999, *Icarus* 141, 367

Stern, S.A. & Colwell, J.E. 1997, *Astronomical Journal* 114, 841

Thommes, E.W., Duncan, M.J., & Levison, H.F. 1999, *Nature* 402, 635

Trujillo, C.A. & Brown, M.E 2002, *Astrophysical Journal* 566, L125

Tsiganis, K., Gomes, R., Morbidelli, A., & Levison, H.F. 2005, *Nature* 435, 459

Asteroids, Comets, Meteors
Proceedings IAU Symposium No. 229, 2005
D. Lazzaro, S. Ferraz-Mello & J.A. Fernández, eds.

© 2006 International Astronomical Union
doi:10.1017/S1743921305006757

Properties of the Near-Earth object population: the ACM 2005 view

Richard P. Binzel[1] and Dimitrij F. Lupishko[2]

[1]Department of Earth, Atmospheric, and Planetary Sciences, Massachusetts Institute
of Technology, Cambridge, MA 02139, USA
email: rpb@mit.edu

[2]Institute of Astronomy of Karazin Kharkiv National University, Ukraine
email: lupishko@astron.kharkov.ua

Abstract. Within the near-Earth object population, one finds asteroids, comets, and meteorites thereby placing the NEO population at the center of the ACM conference. The longstanding gulf between the spectral properties of S-type asteroids and ordinary chondrite meteorites appears to be bridged, where the observational data are consistent with a space weathering type process. As much as 30% of the entire NEO population may reside in orbits having a Jovian Tisserand parameter < 3, and among these roughly half are observed to have comet-like physical properties in terms of their albedos and spectra (taxonomy). Thus $15 \pm 5\%$ of the entire NEO population may be comprised by extinct or dormant comets.

Keywords. steroid, comet, meteorite, spectra, albedo

1. Introduction

Near-Earth objects (NEOs) are defined as asteroids and comets having orbits with perihelion distances of 1.3 AU or less. More and more it is becoming apparent that an understanding of this population requires increasing focus on the interrelationships between the broad classes of objects found in near-Earth space. What are the relationships between asteroids and meteorites? What are the relationships between asteroids and comets? In essence, these are questions at the very heart of the Asteroids Comets Meteors conference themes, placing the study of NEOs at ACMs center stage.

This review, intended to give only a brief overview, will focus primarily on the progress made since the Berlin ACM meeting in 2002. In particular, this review addresses the two fundamental questions: How are meteorites related to asteroids? How are asteroids related to comets? This review will focus on spectroscopic and albedo properties as rotational properties of asteroids, including NEOs, are covered elsewhere.

2. Progress in Measurements

The population of known NEOs is constantly increasing owing to the continuing rapid pace of discovery. Yet observers performing physical studies of NEOs have done an excellent job of keeping pace with discoveries, as shown in Figure 1. A great increase in the number of spectral measurements of NEOs from European sponsored observatories at La Palma and the European Southern Observatory (ESO) has been responsible for much of the growth (e.g. Lazzarin *et al.* 2005; Lazzaro *et al.* 2004; Dandy *et al.* 2003; Vernazza *et al.* 2004). Albedo measurements have similarly kept pace, where measurements from Mauna Kea (e.g. Delbó *et al.* 2003; Wolters *et al.* 2005; Fernandez *et al.* 2005) have been the key to this growth. It will become increasingly challenging for physical

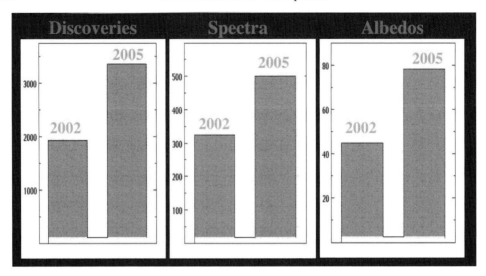

Figure 1. Histograms showing the relative growth in the number of NEO discoveries from 2002 (ACM in Berlin) to 2005 (ACM in Buzios) and the relative growth in measurements of their spectral and albedo properties. As noted in the text, the greatest growth in physical measurements has come from European sponsored observatories.

studies of NEOs to keep pace with discoveries as search surveys push to fainter limiting magnitudes, requiring physical observers to compete for access on larger telescopes. All available data on NEOs confirm an early finding by McFadden *et al.* (1984) that NEOs show the same diversity, if not more, of the main-belt asteroid population (Binzel *et al.* 2002). Such diversity means that unusual taxonomic types present in the main-belt, such as olivine-pyroxene mixture R-types, are also found in the near-Earth population (Marchi *et al.* 2005; Binzel *et al.* 2004a). Additionally, previously rare objects such as V-types, possibly related to asteroid 4 Vesta (which is thought to be differentiated) and eucrite meteorites, represent ~7 percent of all classified NEOs (Marchi *et al.* 2005). Particularly interesting are the physical properties of the smallest objects, observable because of their proximity to Earth. At the small sizes, the likelihood increases that an object samples a distinct geologic unit of a parent body or that the object may be regolith free bare rock. Approaching the Earth gives a unique opportunity to observe them at a much wider range of geometries and illumination than for MBAs, thereby such observations of NEOs also help us to study the MBAs of corresponding types. For example, polarimetric observations of the E-type Aten-object (33342) 1998 WT24 with albedo $p_v = 0.43$ have resulted in a complete phase angle dependence of polarization for high-albedo E-type asteroids (Kiselev *et al.* 2002). That dependence showed an extremely small maximum positive polarization (1.7%), possibly indicative of an absorbing or somewhat peculiar (glassy?) surface. The second peculiarity of polarization-phase dependence of E-type asteroids is a so-called polarization opposition effect, recently revealed for the bright Jovian satellites and E-asteroids (Rosenbush 2005). NEO physical studies, which up till now have largely operated in survey mode, will likely become increasingly focused on understanding the detailed properties of individual objects.

3. Comparisons to Meteorites

By definition, a body entering the atmosphere and delivering meteorites is a near-Earth object prior to its arrival. Thus, the study of NEOs provides an opportunity to examine the population in space that includes precursors to meteorites. By virtue of their proximity, NEOs are the smallest individually observable objects in space. As physical studies have progressed to smaller and smaller sizes, better and better understanding of correlations to meteorite classes have been achieved - even for rare meteorite types. In addition to the eucrites mentioned above, rare types apparently related to aubrite meteorites (enstatite achondrites, consisting of enstatite with very low content of highly reduced Fe) have been increasingly studied. The E-type NEO 3103 Eger is the earliest recognized candidate related to the aubrites (Gaffey *et al.* 1992), with most recent studies focusing on its detailed mineralogy (Burbine *et al.* 2002; Gaffey & Kelley 2004). Additional aubrite candidates include 4660 Nereus (Binzel *et al.* 2004b), a particularly exciting prospect owing to its relatively accessible orbit for spacecraft exploration.

The historically most troublesome asteroid-meteorite connection has been for the most common meteorites, the ordinary chondrites (see the review by Clark *et al.* 2002 and references therein). Most logically, the most common meteorites *should* be related to the most commonly observed asteroids (and NEOs), those falling into the S-class. However, a distinct mismatch in their spectral characteristics, most notably their spectral slopes, has prevented any clear correlation. Pushing to smaller sized objects among NEOs proved fruitful in resolving this problem as the distinct difference between S-type spectra and ordinary chondrite spectra was found to be "filled" by a continuous range of NEO spectral properties when examined over visual wavelengths (Binzel *et al.* 1996). Extending these observations over near-infrared wavelengths continues to confirm this continuum, as shown by Binzel *et al.* (2001; see Figure 6) and further updated in Figure 2, presented here. This continuum is interpreted as being in favor of some type of space weathering process (Clark *et al.* 2002) where ordinary chondrite-like asteroid surfaces are altered over time to look more and more like S-type asteroids. By far the most convincing evidence for a relationship between S-asteroids and ordinary chondrite meteorites is the finding by the NEAR mission of ordinary chondrite-like elemental abundances for the S-type asteroid 433 Eros (Trombka *et al.* 2000). A physical explanation for space weathering, proposed by Hapke *et al.* (1975) and identified in the laboratory by Pieters *et al.* (2000) involves the coating of silicate grains by nanometer scale particles of Fe, created in the vaporization of small Fe particles by micrometeorite impacts. Ongoing laboratory work (Moroz *et al.* 1996; Sasaki *et al.* 2001; Kurahashi *et al.* 2002; Strazzulla *et al.* 2005) shows good support for this process and may explain why ordinary chondrites may be more susceptible to space weathering than carbonaceous chondrites or eucrites. The key ingredient appears to be the availability of metallic Fe in the surface which is present for ordinary chondrites, but not for achondrite meteorites such as eucrites nor for primitive meteorites such as carbonaceous chondrites.

Further support for an age dependence, where young "fresh" surfaces look more like ordinary chondrite meteorites comes from finding ordinary chondrite-like NEOs (denoted as objects having Q-type spectra) preferentially among the smaller-sized NEOs (Angeli & Lazzaro 2002; Dandy *et al.* 2003; Lazzarin *et al.* 2005). Binzel *et al.* (2004a) find an explicit diameter dependence, illustrated in Figure 3, that appears to show NEOs larger than 5 km predominantly being "S-type asteroids" with spectral properties very similar to those observed for S-types among main-belt asteroids. Such a trend is consistent with space weathering models because on average, small objects have shorter collisional lifetimes (before they are destroyed) compared with larger objects. Thus younger "fresher" surfaces are more commonly found among the smallest objects. (In fact, the smallest objects should have the greatest range of variation as this group is most likely to include both fresh surfaces as well as some objects that have been fortunate to survive somewhat longer than average.) While a diameter dependent spectral trend is consistent with a

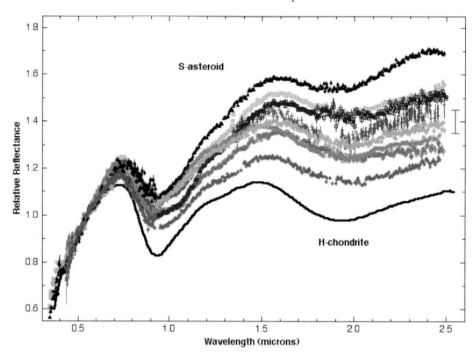

Figure 2. Historically, S-type asteroids have been shown to be distinctly separate from those of ordinary chondrite (OC) meteorites, with the exception of Q-type asteroids matching the OC meteorites. Binzel *et al.* (1996, 2001) found a full continuum of spectral properties between S-type asteroids and ordinary chondrite meteorites. The data displayed here (extending out to 2.5 microns) show a further update on this continuum. These data, obtained with the SpeX instrument on the NASA IRTF, are part of a MIT-Hawaii-IRTF joint program for NEO reconnaissance for which NEO spectral data are made freely available to the community in near-real time after observation. Full information is available via http://smass.mit.edu

surface age space weathering process, outstanding questions remain: How much of this trend is actually the result of increasing gravity improving the retention of a regolith? Under the Hapke *et al.* (1975) - Pieters *et al.* (2000) description of space weathering, the presence of a regolith is a requirement for the process to be effective. The relative roles of the weathering process, gravity, and regolith retention remains an unsolved problem that may be particularly inviting to new researchers.

Overall, it is also possible to suggest that the taxonomic classification and mineralogical interpretation of NEO spectra show evidence of genetic relationship between NEOs and main-belt asteroids. Interestingly, most of the NEOs characterized to date represent differentiated assemblages. Among the NEOs there are bodies with monomineral silicate composition and purely metallic ones. For example, small asteroid 1915 Quetzalcoatl appears to have little or no olivine, and diogenitic meteorites (Mg-pyroxenes) are possible analogs of it. 3199 Nefertiti has the same content of pyroxene and its composition corresponds to that of stony-iron meteorites pallasites (metal + olivine). There are three M-objects, one of which, (6178) 1986 DA has a high radar albedo (0.58) clearly indicating the real metallic composition of this asteroid. 3103 Eger with a very high albedo (0.53) may correspond to assemblages of iron-free silicate minerals, such as enstatite. More than 20 NEOs classified as V-class, have spectra matching the main-belt asteroid 4 Vesta, which is known to be a differentiated body covered by basaltic (pyroxene-rich)

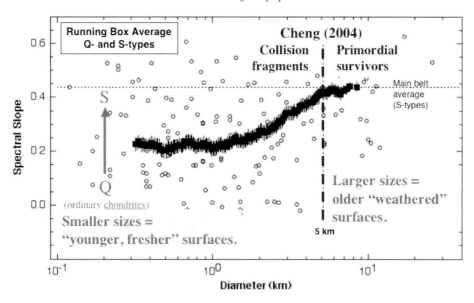

Figure 3. Individual measurements (circles) of spectral slope for Q- and S-type asteroids are plotted versus estimated diameter. A running box mean (box size=50) reveals a diameter dependence where objects trend from low spectral slopes (Q-types, resembling ordinary chondrites) to higher spectral slopes displayed by S-types, typical for asteroids in the main-belt (dashed horizontal line). If the trend is due to a space weathering process, this diagram can be interpreted as revealing increasing weathering with increasing size where the age of the object likely increases (more resilient to collisional disruption) with increasing size. The threshold of 5 km and larger appears to be the size where objects have sufficient age or sufficient gravity for regolith retention so that space weathering processes are complete. Independently, Cheng (2004) finds that this same threshold (5 km) may mark the boundary between primordial survivors and multi-generation fragments among the asteroids.

material. About 30% of NEAs belong to Q-types which are the ordinary chondrite-like objects. Thus the variety of taxonomic classes among NEOs reflects the diversity of their surface mineralogy and an overall analogy with the main belt. Taking into account their small sizes, one might infer that they are the fragments of much larger differentiated bodies which were later injected into the present orbits.

4. Comet Fraction Within the NEO Population

"What is the source of the NEO population?" is one of the most fundamental questions for the field. The question arises because NEOs have relatively short dynamical lifetimes (10^6–10^7 years), meaning that the population we see today must have some source of resupply. (Any primordial objects residing in near-Earth space at the beginning of the solar system have long since been removed). Many observational factors indicate the main-belt as their predominant source: their general matching of taxonomic and mineralogic distributions - especially for differentiated assemblages, their approximate same shapes and rotation, and their overall similar optical properties and surface structure. This conclusion does not contradict the recent results of dynamic considerations, according to which the main asteroid belt can supply a few hundred km-sized NEAs per 1 Myr, well enough to sustain the current population of asteroid-like NEOs (Menichella *et al.* 1996).

While transport of main-belt asteroid fragments via the resonances likely dominates the NEO supply source, some fraction likely enters as nuclei of short-period comets. These objects having a cometary origin, if inactive at the time of discovery, are catalogued as "asteroidal" objects. We seek to answer the question: what fraction of the NEO population, having an "asteroidal appearance" are actually extinct or dormant comets? We apply both dynamical and observational criteria to address this question.

The candidates for cometary origin, as a rule, should satisfy the following conditions: (a) they should be low-albedo objects of D-, P- and C-types (Hartmann *et al.* 1987); (b) their rotational rates should be lower than the mean rates of NEOs (Degewij & Tedesco 1982; Hartmann & Tholen 1990; Weissman *et al.* 2002); (c) they should also be evolved on unstable orbits, and should be associated with meteor streams (Weissman *et al.* 1989). The data available (NEO taxonomy, shapes and rotation parameters, associations with meteors streams) allowed Lupishko & Lupishko (2001) to conclude that no more than 10% of NEAs have cometary origin. A convergence of similar results is coming from many researchers: Fernandez *et al.* (2001) estimate "... at least 9% of NEOs are cometary nuclei"; Whiteley (2001) estimates "... on the order of 5% of cometary origin"; Bottke *et al.* (2004) estimate $\sim 6\%$ of NEOs comes from the Jupiter-family comet region ($2 < T < 3$).

For this review, we use the Jovian Tisserand parameter (T) as our dynamical criterion for identifying potential comet candidates among asteroidal NEOs. We consider dynamical comet candidates to be objects having $T < 3$, because this indicates that they reside in orbits that are strongly influenced by Jupiter, as exemplified by Jupiter family comets. Consistent with this dynamical criterion is the classic discovery by Fernandez *et al.* (2001, 2005) that objects with $T < 3$ have, on average, substantially lower albedos than objects with $T > 3$. The predominance of low albedos among $T < 3$ objects has extremely important implications for statistically evaluating the fraction of comet candidates within the total NEO population: Our discovery statistics are severely biased against discovering these types of objects. The bias arises because a low albedo (and correspondingly fainter apparent magnitude relative to a high albedo object of the same size) decreases the likelihood of $T < 3$ objects being discovered. What's more, the larger orbital eccentricities (a factor in reaching $T < 3$ values) also biases against $T < 3$ NEO discoveries since such objects spend greater fractions of their orbital period at aphelion (where they are more difficult to discover) than at perihelion. Thus a bias corrected estimate for the total population of $T < 3$ objects is required before any overall estimate of the comet fraction can be made. Fortunately, such bias corrected estimates are available through the Ph.D. thesis work of J. Scott Stuart (2003; see also Stuart & Binzel 2004). The debias work of Stuart (2003), based on the extensive search statistics of the LINEAR survey, focuses on determining the size distribution of the total NEO population. One outcome of Stuarts work, key to our analysis, is that the bias correction shows that 30% of all NEOs reside in orbits having $T < 3$.

If 30% of the total NEO population satisfies the $T < 3$ criterion as extinct comet candidates, what fraction of these have physical properties that make them "look" like comets? Based on the fact that measured albedos for comet nuclei also have low values, Fernandez *et al.* (2005) set < 0.075 as a reasonable albedo criterion for identifying objects that look like "comets". Binzel *et al.* (2004a) recognized that relatively few albedo data exist compared with spectral data and used taxonomic class (specifically C-, D-, and P-types) as a proxy for identifying low albedo NEOs within the $T < 3$ population, finding $50 \pm 10\%$ of observed $T < 3$ NEOs have "comet-like" physical properties. Fernandez *et al.* (2005) applied the < 0.075 albedo criterion and found $53 \pm 9\%$ as the observed "comet-like" fraction among $T < 3$ NEOs. These independent physical estimates of comet-like properties for 50% of observed $T < 3$ objects, convolved with the Stuart debiased result that 30% of all NEOs reside in $T < 3$ orbits, yields:

$$0.50 \times 0.30 = 0.15 \qquad (4.1)$$

implying $15 \pm 5\%$ as the total fraction of the NEO population that has both dynamical and physical characteristics consistent with their being dormant or extinct comets.

It is worth emphasizing that this 15% estimate for the extinct or dormant comet fraction is a debiased or diameter-limited estimate. (Many previous, slightly lower estimates are magnitude limited, with no debias criteria applied. Estimates not compensated for the bias against the discovery and measurement of low albedo objects will tend to lead to underestimates.) It is also important to emphasize that our 15% is just a broad characterization of the population as a whole. Certainly there are other criteria (e.g. meteor stream correlations, etc.; see Weissman *et al.* 1989) that allow objects with $T > 3$ to be comet candidates. Objects in the inner solar system, gravitationally interacting with the terrestrial planets, can have their coupling with Jupiter altered so that a $T < 3$ orbit evolves to $T > 3$. Similarly, asteroids entering near-Earth space with $T > 3$ can be perturbed into $T < 3$ orbits. Thus the $15 \pm 5\%$ estimate for comet candidates within all of the NEO population does not tell the complete story of asteroid-comet connections among NEOs, but does likely provide fertile ground for detailed study of possible low level cometary activity or perhaps attractive destinations for mission opportunities.

5. Concluding Remarks

Near-Earth objects have become increasingly central to the study of solar system small bodies by the ACM community. Progress in their measurement and understanding shows they are a diverse population containing both asteroidal bodies from the main-belt as well as a possibly substantial fraction of extinct comet nuclei. Our improved understanding of their meteorite connections, when combined with our understanding of their shapes and rotations (described in this volume), shows their internal structures may be diverse as well. The smallest and fastest rotating bodies may be intact with substantial internal strengths, while the larger and more slowly rotating bodies may be strengthless rubble piles held together only by their mutual gravity. Understanding the internal properties of small bodies, of which NEOs are the most easily accessible to spacecraft missions, is becoming an increasing focus for scientific investigation. Not only does this represent a new area for scientific curiosity, but it is an area of practical responsibility should we some day discover an NEO with a certainty of collision in the coming decades. Thus for both science reasons and practical reasons, the study of NEOs will remain highly important to the ACM community in the future.

Acknowledgements

The research of the first author is supported by grants from NASA and the National Science Foundation. D. Lupishko sincerely thanks the LOC of ACM 2005 (IAU Symposium 229) for the grant allowing him to participate in the meeting.

References

Angeli, C.A. & Lazzaro, D. 2002, *Astron. Astrophys.* 391, 757
Binzel, R.P., Bus, S.J., Burbine, T.H., & Sunshine, J.M. 1996, *Science* 273, 946
Binzel, R.P., Harris A.W., Bus S.J., & Burbine, T.H. 2001, *Icarus* 151, 139
Binzel, R.P., Lupishko D.F., Di Martino M., Whiteley R.J., & Hahn G.J. 2002, in: W. Bottke, A. Cellino, P. Paolicchi, and R. Binzel, (eds.), *Asteroids III* (Tucson: Univ. Arizona Press), p. 255

Binzel, R.P., Rivkin, A.S., Scott, J.S., Harris, A.W., Bus, S.J., & Burbine, T.H. 2004a, *Icarus* 170, 259

Binzel, R.P., Birlan, M., Bus, S.J., Harris, A.W., Rivkin, A.S., & Fornasier, S. 2004b, *Plan. Space Sci.* 52, 291

Bottke, W.F. Jr., Morbidelli, A., & Jedicke R. 2004, in: M.J.S Belton, T.H. Morgan, N.H. Samarasinha, and D.K. Yeomans (eds.), *Mitigation of Hazardous Comets and Asteroids* (Cambridge: Univ. Press), p. 1

Burbine, T.H., McCoy, T.J., Nittler, L.R., Benedix, G.K., Cloutis, E.A., & Dickinson, T.L. 2002, *Met. Plan. Sci.* 37, 1233

Cheng, A.F. 2004, *Icarus* 169, 357

Clark, B.E., Hapke, B., Pieters, C., & Britt, D. 2002, in: W. Bottke, A. Cellino, P. Paolicchi, and R. Binzel (eds.), *Asteroids III* (Tucson: Univ. Arizona Press), p. 585

Dandy, C.L., Fitzsimmons, A., & Collander-Brown, S.J. 2003, *Icarus* 163, 363

Degewij, J. & Tedesco, E.F. 1982, in: L. Wilkening (ed.), *Comets* (Tucson: Univ. Arizona Press), p. 665

Delbó, M., Harris, A.W., Binzel, R.P., Pravec, P., & Davies, J.K. 2003, *Icarus* 166, 116

Fernandez, Y.R., Jewitt, D.C., & Sheppard, S.S. 2001, *Astrophys. J.* 553, L197.

Fernandez, Y.R., Jewitt, D.C., & Sheppard, S.S. 2005, *Astron. J.* 130, 308

Gaffey, M.J., Reed, K.L., & Kelley, M.S. 1992, *Icarus* 100, 95

Gaffey, M.J. & Kelley, M.S. 2004, *Lunar Plan. Sci. Conf.* Abstract 1812

Hartmann, W.K., Tholen, D.J., & Cruikshank, D.P. 1987, *Icarus* 69, 33

Hartmann, W.K. & Tholen, D.J. 1990, *Icarus* 86, 448

Hapke B., Cassidy W., & Wells E. 1975, *Moon* 13, 339

Kiselev, N.N., Rosenbush, V.K., Jockers, K., Velichko, F.P., Shakhovskoj, N.M., Efimov, Yu.S., Lupishko, D.F., & Rumyantsev, V.V. 2002, *Proc. of the Conf. Asteroids, Comets, Meteors 2002* (Berlin:Techn. Univers.), p. 887

Kurahashi, E., Yamanaka, C., Nakamura, K., & Sasaki, S. 2002, *Earth Plan. Space* 54, e5

Lazzarin, M., Marchi, S. Magrin, S., & Licandro, J. 2005, *MNRAS* 359, 1575

Lazzaro, D., Angeli, C.A., Carvano, J.M., Mothe-Diniz, T., Duffard, R., & Florczak, M. 2004, *Icarus* 172, 179

Lupishko, D.F. & Lupishko, T.A. 2001, *Solar System Research* 35, 227

Marchi, S., Lazzarin, M., Paolicchi, P., & Magrin, S. 2005, *Icarus* 175, 170

McFadden, L.A., Gaffey, M.J., & McCord, T.B. 1984, *Icarus* 59, 25

Menichella, M., Paolicchi, P., & Farinella, P. 1996, *Earth, Moon, and Planets* 72, 133

Moroz, L.V., Fisenko, A.V., Semjonova, L.F., Pieters, C.M., & Korotaeva, N. 1996, *Icarus* 122, 366

Pieters, C.A., Taylor, L.A., Noble, S.K., Keller, L.P., Hapke B., Morris, R.V., Allen, C.C., McKay, D.S., & Wentworth, S. 2000, *Meteoritics & Planet. Sci.* 35, 1101

Rosenbush, V.K., Kiselev, N.N., Shevchenko, V.G., Jockers, K., Shakhovskoy, N.M., & Efimov, Y.S. 2005, *Icarus* 178, 222

Sasaki, S., Nakamura, K., Hamabe, Y., Kurahashi, E., & Hiroi, T. 2001, *Nature* 410, 555

Strazzulla, G., Dotto, E., Binzel, R., Brunetto, R., Barucci, M.A., Blanco, A., & Orofino, V. 2005, *Icarus* 174, 31

Stuart, J.S. 2003, PhD thesis, Massachusetts Institute of Technology, Cambridge.

Stuart, J.S. & Binzel, R.P. 2004, *Icarus* 170, 295

Trombka J., Squyres, S., Bruckner, J., Boynton, W., Reedy, R., McCoy, T., Gorenstein, P., Evans, L., Arnold, J., Starr, R., Nittler, L., Murphy, M., Mikheeva, I., McNutt, R., McClanahan, T., McCartney, E., Goldsten, J., Gold, R., Floyd, S., Clark, P., Burbine T., Bhangoo, J., Bailey, S., & Pataev, M. 2000, *Science* 289, 2101

Vernazza, P., Fulchignoni, M., & Birlan, M. 2004, *35th COSPAR Sci. Assembly* 2456.

Whiteley, R.J. 2001, PhD thesis, University of Hawaii.

Weissman, P.R., AHearn, M.F., MacFadden, L.A., & Rickman, H. 1989, in: R.P. Binzel, T. Gehrels, and M.S. Matthews (eds.), *Asteroids II)* (Tucson: Univ. Arizona Press), p. 880

Weissman, P.R., Bottke, W.F., & Levison, H.F. 2002, in: W. Bottke, A. Cellino, P. Paolicchi, and R. Binzel (eds.) *Asteroids III*, (Tucson: Univ. Arizona Press), p. 669

Wolters, S.D., Green, S.F., McBride, N., & Davies, J.K. 2005, *Icarus* 175, 92

Asteroids, Comets, Meteors
Proceedings IAU Symposium No. 229, 2005
D. Lazzaro, S. Ferraz-Mello & J.A. Fernández, eds.

Potential impact detection for Near-Earth asteroids: the case of 99942 Apophis (2004 MN$_4$)

Steven R. Chesley

Jet Propulsion Laboratory, Pasadena, CA 91109, USA
email: Steven.R.Chesley@jpl.nasa.gov

Abstract. Orbit determination for Near-Earth Asteroids presents unique technical challenges due to the imperative of early detection and careful assessment of the risk posed by specific Earth close approaches. This article presents a case study of asteroid 99942 Apophis, a 300-400 meter object that, for a short period in December 2004, held an impact probability of more than 2% in 2029. Now, with an orbit based on radar ranging and more than a year of optical observations, we can confidently say that it will pass safely by the Earth in 2029, although at a distance of only about six Earth radii from the geocenter. However, the extremely close nature of this encounter acts to obscure the trajectory in subsequent years, when resonant returns to the vicinity of the Earth are possible. In particular, an impact possibility in the year 2036 has a roughly 5% probability of persisting through the very favorable 2013 radar and optical observing apparition. In the event that the 2036 potential impact has not been eliminated by 2013, a precise characterization of the Yarkovsky accelerations acting on the asteroid may become an important part of the orbit estimation and impact prediction problem. Even so, the sixteen years available to effect a deflection from 2013 until 2029, after which the problem would become intractable, are sufficient to respond to the threat should a deflection effort become warranted.

Keywords. minor planets, asteroids

1. Introduction

It has long been recognized that an integral part of preventing collisions of a Near-Earth Objects (NEOs) with the Earth is to discover them. This notion is well expressed by the old monster-movie maxim, "It's the ones you don't know about that you should be the most afraid of." Of course, a search and discovery effort alone would be ineffective unless the known objects are monitored continually and carefully for possibilities of future collision, allowing a timely warning of an impending calamity. This early warning capability is presently provided by two independent automatic warning systems, Sentry, operating at NASA's Jet Propulsion Laboratory, and CLOMON2, a part of the NEODyS system operated by the Universities of Pisa (Italy) and Valladolid (Spain). (See Milani *et al.* (2005) for a technical discussion of the operating principles of the two systems.)

During the Christmas holidays of 2004, near-Earth asteroid 2004 MN$_4$ startled the NEO hazard community when both Sentry and CLOMON2 reported a threat of unprecedented magnitude for an impact in 2029. While that particular threat was eventually ruled out by virtue of additional observations, this particular NEO—now named and numbered 99942 Apophis—still poses a moderate risk of impact in the 2030's, and so continues to receive considerable attention from amateur and professional astronomers, and even from the more-interested members of the general public.

This paper summarizes Apophis' history and outlines what the future may hold for this extraordinary asteroid.

2. Background on 99942 Apophis (2004 MN$_4$)

Asteroid 99942 Apophis was discovered by R. Tucker, D. Tholen and F. Bernardi at Kitt Peak, Arizona on June 19, 2004. Originally observed for only two consecutive nights and with observations plagued by astrometric reduction problems, the asteroid, provisionally designated 2004 MN$_4$, was lost until December 2004, when it was serendipitously recovered from Siding Spring, Australia by G. Garradd. At this point Apophis was recognized as a potentially hazardous asteroid, and indeed on December 20, with only a 1.7-day arc of December observations, the first run of Sentry, JPL's automated impact monitoring system, indicated an impact probability (IP) of 2×10^{-4} on April 13, 2029. Adding in the June discovery observations, which allowed a six-month arc of observations, confirmed a significant possibility of impact in 2029, but also revealed substantial astrometric errors in the original June observations. These errors, approaching 2 arcsec, were related to clock errors and a problematical plate solution (D. Tholen, priv. comm.).

Over the next two days D. Tholen obtained accurate remeasurements of the June Kitt Peak observations, and in the meantime a few dozen additional December observations were reported. This observation set, consisting of 55 observations over 187 days, indicated an IP of 0.4% and on December 23 those results were posted to the respective websites of NASA's Sentry and the Italian/Spanish NEODyS. On the subsequent day, ten new observations pushed the IP to 1.6% for an unprecedented rating of +1 on the Palermo Scale and 4 on the Torino Scale. Over the next four days the IP continued to inch upward, reaching a peak of 2.7% on December 27, when pre-discovery observations from March 2004—which eliminated any possibility of an impact in 2029—were reported by the Spacewatch survey.

However, the March 2004 observations, while removing any concern for a 2029 collision, also revealed that Apophis would certainly pass very close to Earth in 2029. An extraordinary flyby, some $10.1 \pm 2.6\ R_\oplus$ (three-sigma limits) from the geocenter, was indicated. ($R_\oplus = 6378$ km is the Earth radius.) Predictably, such a deep encounter allowed a wide range of possible post-2029 semimajor axes, including a number of values that led to resonant return encounters, and even potential impacts, in the years after 2029.

Later, radar ranging and Doppler tracking of Apophis, obtained in late January 2005 from the Arecibo observatory, were found to be inconsistent with then-current predictions (IAUC 8477). Notably, the radar astrometry indicated a much closer 2029 approach, some $5.6 \pm 1.6\ R_\oplus$ (three-sigma limits) from the geocenter (Giorgini, Benner, Nolan, et al. 2005). The discrepancy was traced to systematic errors in the March 2004 pre-discovery observations, and subsequent, independent remeasurements of these five observations brought them into line with the radar-derived orbit.

Optical observations of Apophis continued to be reported until early July 2005, by which time the asteroid was only 46° from the sun and moving into the daytime sky. On Aug. 7, 2005, just as the ACM 2005 Conference was getting underway, L. Benner and colleagues were able to obtain a single Doppler radar measurement at the Arecibo Observatory. Apophis passed within 9° of inferior conjunction on Aug. 11, 2005, during the ACM 2005 conference. The current prediction for the 2029 b-plane intersection, based on all observations received through August 2005, is depicted in Fig. 1. This prediction is for a 2029 encounter distance of $5.89 \pm 0.35\ R_\oplus$ (three-sigma limits).

Schweickart (2005) has suggested that a more or less immediate space mission may be necessary in order to place a radar transponder on the asteroid, which would allow it to be tracked with high-precision, just as interplanetary spacecraft are. The concern he raises is that, if one assumes the unlikely hypothesis that Apophis is indeed on a collision trajectory, it would not be clear that the impact was certain, or even very likely, until

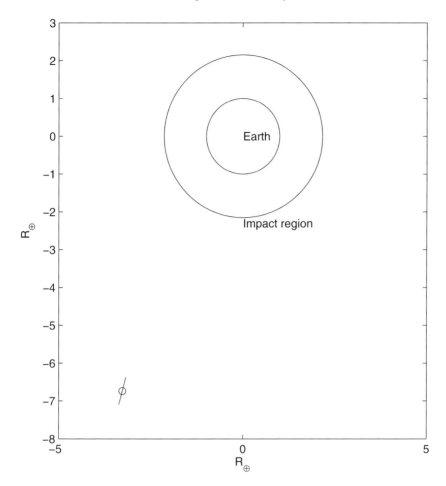

Figure 1. Depiction of the current 2029 b-plane footprint, with three-sigma extent. Note that, according to the definition of the b-plane, the positions shown are unperturbed target plane crossing distances. The fairly low encounter velocity of Apophis (5.86 km/s) causes the actual crossing distances to be somewhat closer. For example, the depicted nominal b-plane distance is 7.2 $R_{\oplus}$, whereas the nominal encounter distance is 5.7 $R_{\oplus}$. Similarly, the impact region is significantly larger than the figure of the Earth, as indicated.

quite late, perhaps too late to mount a deflection mission, which would in any case need to be complete before the 2029 close approach. Thus the call for a transponder mission, which would allow a much earlier recognition of the hypothetical impending impact.

3. Current Risk Analysis

A *resonant return* is an encounter that is, in a sense, spawned by a preceding encounter (Milani, Chesley & Valsecchi 1999). This occurs when the first encounter alters the asteroid's orbit so that its period becomes commensurable with that of the Earth. So, for example, if Apophis were to obtain a 426 day period (7:6 exterior Earth resonance) due to the 2029 encounter it would return near the same point in space 6 revolutions and 7 years later, at which point the Earth would also be there. Thus, a resonant return in 2036 would be obtained. Another key bit of terminology is the *keyhole*, which refers to the

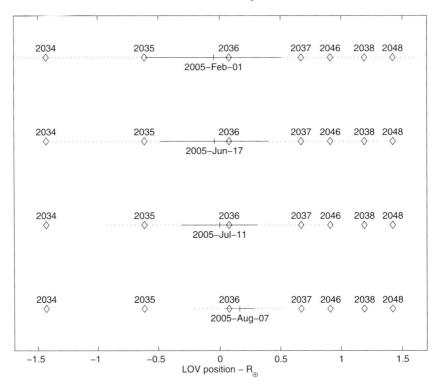

Figure 2. Recent evolution of the 2029 b-plane line of variation (LOV). Each horizontal line shows the one-sigma (solid) and three-sigma (dotted) LOV based on the observational data cut-off printed below the center of the respective LOV. The plot also depicts the LOV positions of the keyholes associated with the primary resonant returns from the 2029 encounter of Apophis. The 2038 keyhole, shown for completeness, allows an encounter distance no closer than 1.42 $R_\oplus$ and thus does not allow an impact. The Earth is located at approximately -7 $R_\oplus$ along the LOV in this depiction. The origin is arbitrarily selected.

small area on the target plane of the preceding encounter, in 2029 for Apophis, through which the asteroid must pass to obtain a resonant return collision (Chodas & Yeomans 1999). In other words, the keyhole is the region of collision orbits in the 2029 target plane and can also be interpreted as the preimage of the Earth in that space (Valsecchi, Milani, Gronchi, *et al.* 2003).

For Apophis there have been several primary resonant returns, tabulated in Table 1, in the years following the 2029 close approach. Additionally, there have been secondary resonant returns, which are spawned by primary resonant returns, as well as occasional non-resonant return impacts in October at the opposite node. Figure 2 depicts the recent evolution of the position and extent of the 2029 b-plane line-of-variation (LOV) with respect to the fixed positions of the keyholes.

Because of Apophis' very close encounter, the 2029 keyholes are far smaller, by orders of magnitude, than the Earth capture cross-section itself, and this is an important point because an impending impact can be averted simply by moving the asteroid trajectory away from the keyhole, and so a deflection becomes far more tractable than the case where there is no leverage from an intervening close approach. Conversely, the small keyholes can make it more challenging to ascertain that the object is actually on an impacting trajectory, if that is indeed the case. If the scale of a keyhole is at the sub-kilometer level, as is the case for Apophis (Table 1), then we need kilometer-level knowledge of the

Table 1. 99942 Apophis Potential Impacts.

Primary Resonant Returns (from 2029)

Year	Resonance	Post-2029 Period (days)	Keyhole Size (km)
2034	5:4	457	0.56
2035	6:5	438	0.56
2036	7:6	426	0.61
2037	8:7	417	0.57
2046	17:15	414	0.66
2048	19:17	408	0.41

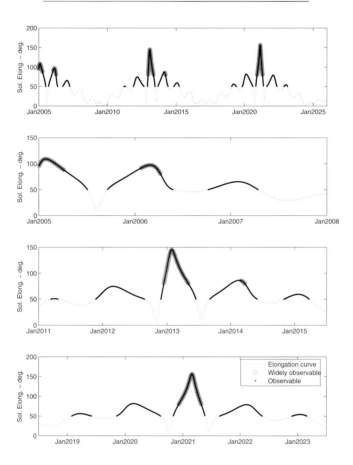

Figure 3. Optical observability of 99942 Apophis. The object is deemed observable if the solar elongation is greater than $50°$ and the visual magnitude is less than 23, and is considered widely observable if the solar elongation is greater than $80°$ and the visual magnitude is less than 20. We simulate one astrometric position with $0''.2$ standard errors on every second night when Apophis is widely observable and one such observation per lunation when otherwise observable.

asteroid state at the keyhole before we can say with some confidence that the object is on or near a collision path. Thus the small sizes of the keyholes associated with the 2029 Earth-Apophis encounter are a two-edged sword: A deflection mission becomes much simpler (if it can be accomplished before 2029), but the knowledge that such a mission would be required does not come until much later. However, since the 2029 keyholes are

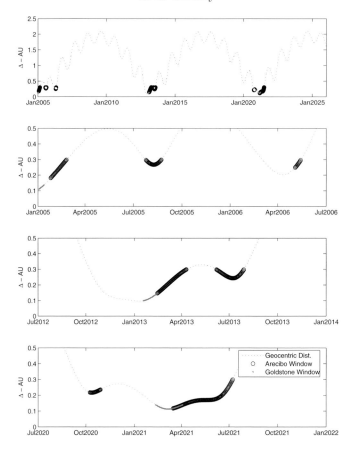

Figure 4. Radar observability of 99942 Apophis. Arecibo observability is assumed when the declination is between $0°$ and $+36°$ and the range is less than 0.3 AU. Similarly, potential Goldstone windows are obtained when the declination is between $-20°$ and $+40°$ and the range is less than 0.14 AU.

widely separated, by 1000 km or more, it may be possible to *rule out* all potential impacts when the 2029 prediction uncertainty is still at the level of several hundred km.

4. Future observations

There are several considerations that need to be addressed in order to fully understand the very complex situation that Apophis presents. Chief among these is the question of how the 2029 prediction uncertainties will evolve under various observational scenarios and to address this issue we will assume a particular future observation schedule.

Figure 3 shows the times when optical observations can be reasonably expected. The optically observable periods are split into two categories according to the level of difficulty, and a less frequent rate of observations is assumed for the more difficult periods, as described in the caption to Fig. 3.

Of course, the possibility of radar ranging is of critical importance, and Fig. 4 depicts the periods when Apophis could potentially be observed by radar from the 305-meter Arecibo Observatory and from the Deep Space Network 70-meter antenna at Goldstone, California. From Fig. 4, five future Arecibo radar observing opportunities are apparent, and the measurements that presumably would be derived from each of those are listed

Table 2. Simulated and Actual Arecibo Radar Apparitions for 99942 Apophis

Dates	Δ (AU)	SNR[†]	Delay sigma (μsec)	No. Range Pts.
2005 Jan 27-30	0.19	15	4.5	3
2005 Aug 07-08	0.27	4	-	0[‡]
2006 May 04-08	0.26	4	4.5	3
2013 Feb 14-20	0.15	40	1.0	7
2013 Jul 06-10	0.24	6	4.5	3
2020 Oct 09-12	0.22	8	4.5	3
2021 Mar 16-20	0.12	90	0.5	10

[†]Future Signal-to-Noise Ratios are inferred from the SNR actually obtained in Jan. 2005 and from the listed value of the topocentric distance Δ.
[‡]Due to transmitter equipment outage, only a single Doppler measurement of uncertainty 0.2 Hz was obtained at the August 2005 opportunity.

in Table 2. The two future Goldstone opportunities were not simulated because their contribution would be nil given the much stronger and contemporaneous echoes expected from Arecibo.

This study also considers the effectiveness of a radio tracking space mission, although, for reasons explained later, this year-long, simulated mission is assumed to begin operations in Jan. 2019, much later than proposed by Schweickart (2005). The mission simulation assumes that 365 daily pseudo-range measurements of 2 m accuracy to the center-of-mass of the asteroid would be obtained.

5. Force modeling

Another important consideration is the effect of force model uncertainties, and of particular concern for Apophis is an acceleration known as the Yarkovsky effect (Bottke, Vokrouhlický, Rubincan, *et al.* 2002). As an asteroid rotates, absorbed solar radiation is re-emitted in the thermal band in a direction offset from the sun direction. This induces a slight along-track acceleration, which in turn leads to an increase or decay in orbital energy and orbital period. Because the Yarkovsky effect causes a steady drift in semimajor axis, it manifests as a drift in the orbital anomaly that grows quadratically with time. The effect has been directly measured on the half-kilometer asteroid 6489 Golevka, which drifted approximately 15 km due to Yarkovsky accelerations in the first twelve years after discovery (Chesley, Ostro, Vokrouhlický, *et al.* 2003). Clearly, for the rather smaller Apophis, the Yarkovsky deflection over the 25 years from the first observations in 2004 to the close approach in 2029 could be substantial, especially in light of the kilometer-level predictions that are required to confirm a post-2029 impact.

The difficulty in predicting, even crudely, the Yarkovsky effect for Apophis is that we have no knowledge of the asteroid's spin pole direction, and yet the obliquity γ of the spin axis is the most crucial element for modeling the Yarkovsky acceleration. Small obliquities (i.e., direct rotation) tend to increase the semimajor axis, large obliquities (i.e., retrograde rotation) tend to decrease the semimajor axis, and intermediate values (near 90°) will typically have a smaller, potentially even negligible, effect. Thus, in the absence of a known obliquity we cannot even say whether the Yarkovsky effect is causing the asteroid to move ahead or fall behind in its orbit. Two additional parameters that are essential to predict the magnitude of the Yarkovsky effect are the bulk density ρ of the object and the thermal conductivity K of the surface material. Figure 5 shows how the mean semimajor axis drift rate depends upon γ and K. The drift rate also varies

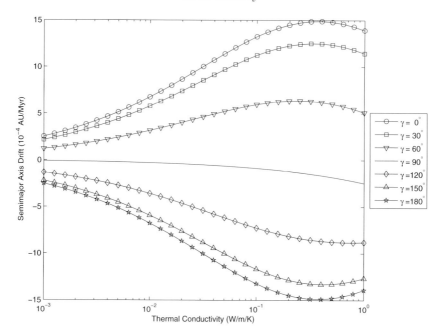

Figure 5. Mean semimajor axis drift rate of 99942 Apophis due to Yarkovsky acceleration. The model assumes a spherically symmetric body with spin period 30.6 hours (Gary & Reddy 2005) and applies a linearized heat diffusion model. See Vokrouhlický, Milani & Chesley (2000) for modeling details. Other assumptions include bulk density $\rho = 2.7$ g/cm^3, surface density $\rho = 2.0$ g/cm^3, surface heat capacity 680 J/kg/K and surface absorptivity (complement of Bond albedo) of 0.9.

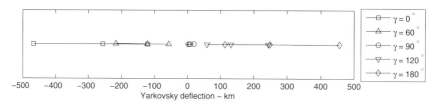

Figure 6. The extent of the Yarkovsky offset for 99942 Apophis in the 2029 b-plane. Five different values of obliquity γ are shown and for each obliquity three cases are plotted: a strong effect ($\rho = 2.0$ g/cm^3 and $K = 3 \times 10^{-2}$ W/m/K), a nominal effect ($\rho = 2.7$ g/cm^3 and $K = 10^{-2}$ W/m/K) and a weak effect ($\rho = 3.3$ g/cm^3 and $K = 3 \times 10^{-3}$ W/m/K).

inversely with ρ. Figure 6 provides an indication of how large the Yarkovsky effect could be when mapped from 2005 to the 2029 b-plane. From that figure it is apparent that deflections of up to 500 km are possible, although deflections less than 200 km appear much more likely. However, if the obliquity were known then the uncertainty related to Yarkovsky accelerations would be much smaller. Fortunately, the 2012–2014 radar and optical observing periods should provide ample opportunity to derive a spin axis orientation, and this paper assumes that that effort will be successful.

6. Prediction Uncertainty

With these pieces in place we are in a position to ask what will be the extent of the b-plane uncertainty region in 2029 as time passes and observations accumulate. Figure 7

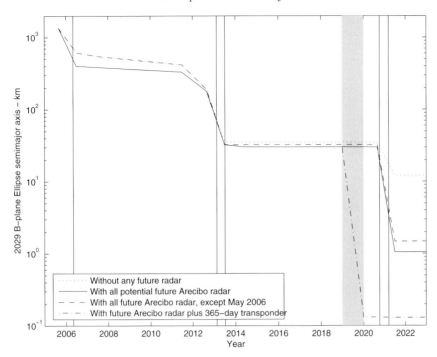

Figure 7. Predicted evolution of the uncertainty extent on the 2029 b-plane for 99942 Apophis. The four curves represent various observation scenarios. The contribution of the uncertainty in Yarkovsky modeling is included as described in the text. The vertical lines indicate the epoch of future Arecibo ranging opportunities. The gray region demarcates the time of a possible radio tracking mission, as described in the text.

shows the evolution of the 2029 uncertainty under the observational assumptions stated above. The uncertainty due to Yarkovsky is also incorporated into the plot by assuming a modest obliquity $\gamma = 60°$ and estimating the bulk density of the body.†

Although the obliquity probably cannot be determined prior to 2013, assuming an obliquity does not materially affect the result because prior to 2013 the uncertainties are dominated by measurement and orbit determination errors. Thus Yarkovsky is not a significant source of uncertainty during the period that the obliquity is unknown. However, after the obliquity is established, around 2013, the estimated uncertainty of the bulk density serves as a proxy for the contributions of the other unknown parameters, notably the thermal conductivity K. With this approach, the uncertainty of the Yarkovsky effect is found to be on the order of 15% in 2014, and this is the predominant source of uncertainty from 2014 to 2021. After the 2021 observations, the uncertainty in the Yarkovsky acceleration is less than 1%. This precision, coupled with the short mapping time from 2022 to 2029, leads to kilometer-level b-plane uncertainties in 2029. The inclusion of Yarkovsky-derived uncertainties does little to change Fig. 7 in the years before 2013 (when the uncertainty is dominated by measurement and mapping errors) and after 2021 (when Yarkovsky is well characterized and the prediction interval is short). In the intervening years from 2013 to 2021, however, Fig. 7 reveals an order of magnitude increase in uncertainty due to Yarkovsky.

While the use of radar imaging may be imperative for determination of the spin axis orientation of Apophis, which will enable improved Yarkovsky modeling and thus better

† The bulk density is estimated with a one-sigma *a priori* constraint 2.5 ± 1.0 g/cm^3.

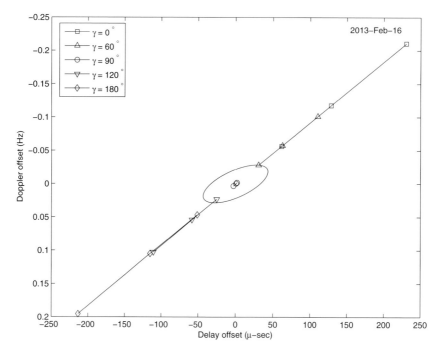

Figure 8. Yarkovsky-induced offsets in the space of radar observables at the Feb. 2013 radar ranging opportunity. The plotted ellipse depicts the 90% confidence region for a prediction with zero Yarkovsky acceleration, assuming expected optical and radar observations through Feb. 2013. The plotted Yarkovsky positions include weak, nominal and strong perturbations, as described in the caption to Fig. 6.

orbital predictions, the orbital improvement provided by future ground-based radar as-trometry is otherwise rather modest. This is typical for asteroids observed over several years (Ostro & Giorigni 2004).

As mentioned above, the observations of 2013 can measure the Yarkovsky accelera-tion with good precision. Figure 8 reveals where the asteroid is predicted to appear in Feb. 2013 in the space of radar observables (delay and Doppler) for various Yarkovsky assumptions. As indicated by the plot, Yarkovsky accelerations may not be outright de-tectable in 2013. But even if the Yarkovsky signal is very weak this can be attributed to an intermediate obliquity, and in any case the magnitude of the effect is constrained.

To a large extent, the focus here has been on the uncertainty induced by the Yarkovsky effect because this will soon become the dominant source of prediction uncertainty. Still, there are other, smaller modeling issues that may need to be considered in the future. As the 2029 prediction uncertainty shrinks, the appropriate force model for making Apophis predictions will evolve. At present, with 1000-km+ uncertainties, the basic model with 13 perturbers (Moon and planets, plus Ceres, Pallas and Vesta) and a simple relativity formulation are sufficient. However, when the uncertainty begins to approach the kilome-ter level, more sophisticated models may become necessary, for example, consideration of additional asteroid perturbers, a more involved relativity formulation, and perhaps direct solar radiation pressure (Giorgini, Benner, Nolan, *et al.* 2005).

Another potential complication, more theoretical than practical, is that the spin state of Apophis is likely to change substantially as a result of tidal interactions during the 2029 close approach (Scheeres, Benner, Ostro, *et al.* 2005). In principle, this different spin state could alter the Yarkovsky accelerations enough to affect the impact prediction. However,

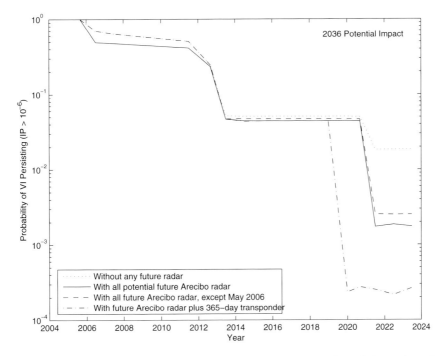

Figure 9. Probability that the 2036 potential impact will not be ruled out, i.e., the probability of persistence at IP $> 10^{-6}$, based on the uncertainties depicted in Fig. 7.

this would only be a concern should the trajectory pass near the edge of a 2029 keyhole, leading to a grazing post-2029 encounter. This because, from the 2029 encounter to, for example, the 2036 potential impact, the Yarkovsky effect can only alter the position by several kilometers or thereabouts. Mapped back to the 2029 b-plane, that effect would show up as a slight (sub-meter) displacement of the actual keyhole.

7. Impact prediction

What is the likelihood that a particular impact possibility will be ruled out or confirmed in future years? This is a crucial question in determining an appropriate course of action that will ensure that we are in a position to deflect Apophis in the unlikely event that it should be necessary. Given the size and current LOV location of a particular keyhole, for example that of the 2036 impact, we can use the 2029 uncertainty information from Fig. 7 to derive a likelihood that the 2036 impact will be ruled out with a given confidence by a given date. The approach is to randomly sample future nominal orbits according to the *current* (relatively large) Gaussian probability distribution, and then apply the smaller *future* Gaussian probability distribution, centered at each new nominal, to obtain a sampling of the future impact probability. The random samples can be collected to obtain a histogram of impact probabilities at a given time in the future. From this approach one can infer the likelihood that a potential impact will persist at a non-negligible IP into the future. This exercise has been done for the 2036 impact, which appears the most interesting at present. Figure 9 indicates a 5% probability that the 2036 impact will persist after 2013 and a 0.2% probability that it will persist beyond the 2021 radar opportunity, assuming a reasonable schedule of future ground-based radar and optical observations.

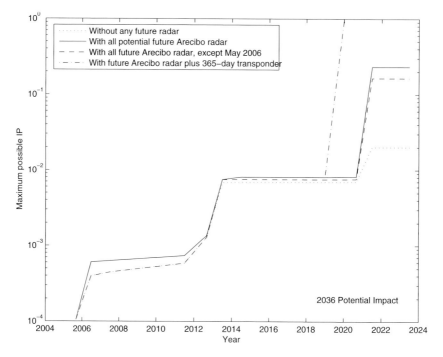

Figure 10. Maximum possible impact probability for the 2036 potential impact, based on the uncertainties depicted in Fig. 7.

Alternatively, if an impact possibility does persist, or if the object is actually on a collision trajectory, how grave might the situation appear at a particular time? In other words, what is the maximum possible IP for a given b-plane uncertainty? Figure 10 addresses this question, revealing that the 2036 IP could approach 1% after 2013, and reach roughly 20% after 2021, based on the assumed ground-based observation schedule.

The possible radio tracking mission reduces the 2029 uncertainty to about 100 m, which likely allow the 2036 impact to be conclusively confirmed with high confidence or ruled out altogether.

8. Deflection Considerations

Based on the foregoing analysis, one can paint a picture of how the situation might evolve, vis-a-vis the known potential impacts of Apophis. Obviously, either all possibilities of impact will eventually be ruled out or one of them will be confirmed. At present we can say with 95% confidence that the 2036 possibility of impact will be substantially eliminated by 2013. If, on the other hand, the post-2013 impact threat is deemed *significant* (a criterion that has yet to be defined) then some basic preparations for a space mission to deflect the asteroid could be initiated at that time. Such efforts could include a reconnaissance mission to the asteroid to characterize its shape, spin state, and material and structural properties. Naturally, such a mission would also be effective in refining the orbital predictions for the object (Fig. 7). Other post-2013 preparations, should they be deemed necessary, would probably include detailed mission and spacecraft design studies so that a deflection mission could be launched more quickly in the event that a decision is made to do so.

The leverage provided by the 2029 Earth encounter substantially eases the technical challenge of actually mounting a deflection mission. In a sense, one need only miss the 2029 keyhole, which means that kilometer-scale deflections would be sufficient if implemented before 2029. The most practical means of deflection for Apophis would likely be a kinetic energy impactor. In rough terms, a kinetic energy deflection mission causes an along-track position shift of

$$\Delta X = 3\,\Delta T\,\Delta V$$

where ΔV is the along-track component of the velocity change due to the impact and ΔT is the time after impact that the effect ΔX is measured (Ahrens & Harris 1994). Thus a modest along-track velocity change of 0.1 mm/s leads in approximately three years to a 25 km change in position, which is more than sufficient to avoid a keyhole. A ΔV of this magnitude can be readily obtained from a 1000 kg impactor with a relative velocity of just a few kilometers per second. Of course there are factors of safety and redundancy issues that need to be considered in detail, but this rough calculation indicates that a kinetic energy impactor would be a feasible option with current technology and available launch vehicles.

The experience of NASA's Deep Impact mission, which was schematically similar and progressed from preliminary planning to launch in only six years and from launch to a successful collision with a comet in six months, is instructive. An Apophis deflection mission could thus reasonably delay launch until the early 2020's, although detailed mission and spacecraft design studies may need to be completed earlier. Important design considerations for a deflection mission would include the possible need for backup missions and the availability of post-deflection radio tracking to verify success. Preliminary work on transfer orbits and deflection considerations has already been reported by Junkins *et al.* (2005) and Gennery (2005).

As an example deflection timeline, should Apophis persist as a significant risk beyond the 2013 apparition, a scientifically-oriented reconnaissance mission to Apophis may be called for in 2014 to arrive around 2019 as indicated in Fig. 7. Additionally, paper studies of deflection options could be conducted, but fabrication of the actual deflection spacecraft could be delayed until around 2020, depending on the results of those studies. After fabrication, launch could reasonably be delayed until after the 2021 apparition, with the actual deflection occurring around 2024, five years before the 2029 b-plane crossing. This example would need to be solidified with comprehensive mission studies, but the key point is that the deflection effort could be called off—at any point in time and after a minimum of investment—if it becomes certain that an impact will not occur. Throughout this scenario, the guiding principle should be to make all decisions and commitments as late as possible while carefully preserving the ability to prevent a potential impact.

Acknowledgements

Helpful discussions with J.D. Giorgini, S.J. Ostro, and D.K. Yeomans are gratefully acknowledged. This research was conducted at the Jet Propulsion Laboratory, California Institute of Technology, under a contract with the National Aeronautics and Space Administration.

References

Ahrens, T.J. & Harris, W.A. 1994, in: T. Gehrels (ed.), *Hazards due to Comets & Asteroids*, (Tucson: Univ.Arizona Press), p. 897.
Bottke, W.F., Vokrouhlický, D., Rubicam, D.P., & Broz, M. 2002, in: W.F. Bottke, A. Cellino, P. Paolicchi, and R.P. Binzel (eds.), *Asteroids III*, (Tucson: Univ. Arizona Press), p. 395.

Chesley, S.R., Ostro, S.J., Vokrouhlický, D., Capek, D., Giorgini, J.D., Nolan, M.C., Margot, J.-L., Hine, A.A., Benner, L.A.M., & Chamberlin, A.B. 2003, *Science* 302, 1739

Chodas, P.W. & Yeomans, D.K. 1999, in: *AAS/AIAA Astrodynamics Specialists Conference*, (Girwood), p. 99

Gary, B.L. & Reddy, V.V. 2005, *http://brucegary.net/MN4*

Gennery, D.B. 2005, *http://www.spaceref.com/news/viewsr.htlm?pid=17666*

Giorgini, J.D., Benner, L.A.M., Nolan, M.C., & Ostro, S.J. 2005, *BAAS* 37, 636

Junkins, J., Singla, P., Mortari, D., Bottke, W., & Durda, D. 2005, in: *International Conference on Computational & Experimental Engineering and Sciences*, (Chennai)

Milani, A., Chesley, S.R., Sansaturio, M.E., Tommei, G., & Valsecchi, G.B. 2005, *Icarus* 173, 362

Milani, A., Chesley, S.R., & Valsecchi, G.B. 1999, *Astron. & Astrophys.* 346, L65

Ostro, S.J. & Giorgini, J.D. 2004, in: M. Belton, T.H. Morgan, N. Samarasinha & D.K. Yeomans (eds.), *Mitigation of Hazardeous Comets and Asteroids*, (Cambridge: Univ. Press) p. 38

Scheeres, D.J., Benner, L.A.M., Ostro, S.J., Rossi, A., Marzari, F., & Washabaugh, P. 2005, *Icarus* in press

Schweickart, R.L. 2005, in: *International Space Development Conference*, (Washington)

Valsecchi, G.B., Milani, A., Gronchi, G.F., & Chesley, S.R. 2003, *Astron. & Astrophys.* 408, 1179

Vokrouhlický, D., Milani, A., & Chesley, S.R. 2000, *Icarus* 148, 118

Asteroids, Comets, Meteors
Proceedings IAU Symposium No. 229, 2005
D. Lazzaro, S. Ferraz-Mello & J.A. Fernández, eds.

© 2006 International Astronomical Union
doi:10.1017/S1743921305006770

Meteoroid streams: mathematical modelling and observations

Galina O. Ryabova[1]

[1]Research Institute of Applied Mathematics and Mechanics of Tomsk State University, Tomsk, 634050, Russia
email: ryabova@niipmm.tsu.ru

Abstract. Mathematical modelling of meteoroid streams formation and evolution is a very fruitfull method for obtaining not only cosmogonical knowledges, but information about the stream up-to-date structure as well. In this review we consider advances of the method, and applications to meteoroid streams of different schemes of formation. We also discuss the part played by observations, and feedback between models and observations. Attention is also drawn to some unresolved problems and promising areas of application.

Keywords. meteors, meteoroids, comets: general, asteroids, methods: numerical, methods: statistical

1. Introduction

Computer simulation of meteoroid streams formation and evolution has gone through several stages, which were dictated mainly by computer power. At the first stage mainly a single orbit was integrated (the mean orbit of a shower † or the orbit of its parent comet, e.g. Weiss (1868)). Or it was not numerically integrated, only the secular perturbation changes were calculated (e.g. Plavec 1950). Or one orbit was integrated, and dispersion of the meteoroid elements was calculated analytically (e.g. Hamid 1951). Or it was a pure analytical method entailed by cumbersome mathematical manipulations (e.g. Murray 1982). Results concerned mainly such problems as the origin of a meteoroid stream, the visibility conditions of a meteor shower, displacement of the shower's node etc.

At the second stage general dynamics of streams was studied by integrating of small amounts of meteoroid orbits (e.g. Kazimirčak-Polonskaya, Belyaev, Astapovič & Terent'eva 1968; Kazimirchak-Polonskaya, Belyaev & Terent'eva 1972; Williams, Murray & Hughes 1979).

The real "ideological" break through has happened after Fox, Williams & Hughes (1983) work. For one thing, for the first time there were obtained model rate profiles of a shower (the Geminids), for another, an idea about approximation of changes in orbital elements by power series of time was new, and, as it turned out later, very fruitful (see Section 5).

With computer power increasing, large scale integration studies have became possible. McIntosh & Jones (1988) modelled the Comet Halley meteoroid streams, Ryabova (1989) modelled the Geminid stream, Brown & Jones (1998) modelled the Perseid stream, Ryabova (2001a) modelled the meteoroid streams of asteroid (1620) Geographos, Vaubaillon, Colas & Jorda (2005b) modelled the Leonid meteoroid stream etc.

The objective to be pursued by the present work is not the historical review of mathematical modelling of meteoroid streams. We shall discuss the general principles

---

† This is not misprint. The mean orbit of the stream is not the same, as the mean orbit of the part of the stream registered at the Earth.

of mathematical modelling (Section 2), and in particular methods of calculations of the stream's age (Section 3) and the ejection velocity (Section 4). At present we have only two more or less full models for meteoroid streams, namely for the Geminid and the Perseid meteoroid streams. These models will be discussed in Sections 5 and 7. In Section 8 we shall consider a model aimed at the Leonids predictions, but having all the potential to grow into full model. The special attention will be called to observations as both base for a model and criterion for its validity. Origin of the Geminid-type streams is still an open question, it is discussed in Section 6. The last Section 9 outlines some results showing that mean motion resonances can cause evident changes in the structure of a meteoroid stream.

2. Mathematical modelling. General

The process of mathematical modelling could be subdivided into four stages.

2.1. *Compiling of physical model for meteoroid stream and cometary decay*

We should know the main structural parameters of a stream and interrelations among them. These parameters are
- period of the visibility of the stream on the Earth;
- outbursts (time, location);
- node regression or progression;
- twin showers, if exist;
- flux density profile of the meteor shower;
- profile of mass distribution index s;
- activity variations along the orbit (or, what is the same, from year to year)
- "windage" of meteoroids, i.e. A/m, where m and A are the mass of the meteoroid and its cross-sectional area, respectively;
- orbital parameters for meteoroids of different masses;
- radiant (location, motion, configuration);
- grouping meteoroids in the stream;
- flickering of meteors light-curve;
- space distribution of meteoroids.

The parameters listed not in importance. As it will be discussed later, even for the most studied streams uncertainty for some of the parameters is too lagre.

It is very important to know the parent body for a meteoroid stream, because the model constructed on the basis of the shower mean orbit (i.e. on the mean orbit of the meteoroids registered at the Earth) cannot give correct spatial distribution of the meteoroid stream matter. If the parent body is known, the very first and key parameter we should determine is age of the meteoroid stream. Let us define "age" as age of meteoroids. If the parent comet have been active for a long time, the stream contains meteoroids of different age. But sometimes a stream can be generated on relatively short time interval, so its meteoroids have approximately the same age. The narrow, practical meaning of this term is "the date from which we begin our modelling". To model decay of the comet we should know the following parameters:
- ephemeris of the comet;
- dust production rate;
- distribution of ejection velocity vectors in magnitude and direction.

2.2. *Choise of mathematical methods*

In the overwhelming majority of cases under the term "model of a meteoroid stream" we imply an ensemble of meteoroids, whose ejections were modelled and evolution followed. So, first of all this is a choise of methods for calculation of the orbital evolution.

In not so remote past the Gauss-Halphen-Goryachev method was used for integration of orbits very intensively, because it is very fast method. With this method only secular perturbations of the first order can be calculated. So, it cannot process correctly close encounters of meteoroids with planets, or resonance cases, therefore, the validity of its application should be carefully verified. For the Geminid meteoroid stream it works perfectly. Moreover, the Geminid stream is so "unperturbed", that we may use polynomial approximation instead of direct integration (see Section 5).

Now, when computing power allows for large scale integration studies, the problem comes to choosing an integrator and algorithm for the most part. There are a number of publications considering algorithms for high-accuracy long-term simulation of the small bodies motion (e.g. Bordovitsyna, Avdyushev & Chernitsov 2001). Numerical methods of integration of the motion equations are applicable to every kind of motion. But they are expensive (even now), and have time limitations because of the accumulation of errors.

To study long-term evolution a symplectic integrator may be applied (Brown & Jones 1998).

It is worth to mention a method developed by Babadzhanov & Obrubov (1987) both in the historical context, and because their results are still referred to. The metod use the following quasi-integrals of motion:

$$(1 - e^2)\cos^2 i \approx const,$$

$$e^2(0.4 - \sin^2 i \sin^2 \omega) \approx const.$$

The first integral was found by Moiseev (1945), and the second one by Lidov (1961). Assuming the longitudes of perihelion of orbits of all the meteoroids to be constant, i.e. $\Omega + \omega = \pi = const$, and assuming that a stream lives long enough that the argument of perihelion makes a full revolution $(0° - 360°)$, we may outline a surface, enclosing all possible meteoroid orbits. Babadzhanov & Obrubov have shown that a meteoroid stream can produce up to 8 meteor showers. My own rough estimate of the time for complete ω circulation is: 16 thousands years for the Quadrantid stream (assuming asteroid 2003EH1 being its parent body), 37 thousand years for the Geminid stream, and > 1 mln years for the Perseid stream. So this method is not applicable to the Geminid meteoroid stream, because its age is about two thousands years (Ryabova 1999).

It is important to define exactly which forces are included in the model. What major planets should be taken into account in the calculations of gravitational perturbations, what radiational perturbations, etc.

A stream model should be statistically relevant, i.e. the results should not fluctuate with increasing number of modelled meteoroids. From the other side, integration of millions of orbits is rather expensive. So, a known method of weighting coefficients could be applied in some cases. The technique consists in ascribing to every model meteoroid a definite weight depending, as a rule, on dust production rate.

2.3. *Comparison with observations and constraining the model*

Parameters, which can be obtained from a stream model to compare with observations are listed in Subsection 2.1. We shall discuss these parameters in details on concrete examples in sections considering specific meteoroid streams.

It is clear, that if there is an agreement between the model and observations within the level of errors, it is a good argument in favour of the model reliability. Sometimes we could try to solve an inverse problem, i.e. to constrain initial model parameters in such a way, that resulting parameters agree with experimental ones.

2.4. Revision of the model

Even the most perfect model of its time has limitations determined by the level of experimental data, and, in a more comprehensive sense, by the level of knowledge. A time inevitably comes, when the model should be "upgraded", or the new model should be constructed.

One practical conclusion: referring to 30-years old results of modeling one should be very careful.

3. Age of meteoroids

The first review of the meteoroid streams age determination was, according to my knowledge, published by Plavec (1955a). Some of the methods described in the work lost their validity. For example, now we know that if a stream is not distributed around all the orbit, it is not necessarily young (Williams 1997). Some others as, for example, the evaluation of the upper limit of the age as lifetime for the largest meteoroids in the stream, still can be used.

A detailed analytical review of the methods for determining the age of a meteoroid stream as applied to the Geminid stream could be found in (Ryabova 1999). The main points from the review will be abstracted below. But the Geminid meteoroid stream seems to be a stream generated during a relatively short period of time. The streams, replenished by new portions of meteoroids for a long time, as, for example, the Leonid or the Perseid streams should be treated differently.

3.1. Retrospective analysis of evolution

Since stream meteoroids and their parent bodies had been once a single whole, an attempt to trace the orbital evolution of the parent body and meteoroids back in time and find moment of their separation suggests itself. But this problem cannot be solved unambiguously and with sufficient accuracy, because of low accuracy of the measured orbital parameters, and uncertainties in physical parameters of meteoroids.

3.2. Mass segregation

The rate of change in orbital elements of meteoroids, in particular, the semimajor axis and the eccentricity depends on the meteoroids windage (A/m). The Poynting-Robertson (PR) effect and its corpuscular analogue (Ryabova 2005) provide the most significant contribution to the phenomenon.

The following equation can be written for the semimajor axis of a meteoroid:

$$a = a_{c0} + \Delta a_0 + (\Delta a)_{gr} + (\Delta a)_{PR} + \varepsilon a, \qquad (3.1)$$

where a_{c0} is the semimajor axis of the parent body orbit at the time of ejection, Δa_0 is the increment due to the ejection velocity and light pressure; $(\Delta a)_{gr}$, $(\Delta a)_{PR}$, and εa are the increments of the semimajor axis due to gravitational perturbations, PR-effect (plus its corpuscular analogue), and all other factors, respectively. The contributions from every term are considered in (Ryabova 1999).

In practice, when estimating age, equation (3.1) simplifies to

$$a - a_c \approx (\Delta a)_{PR}, \qquad (3.2)$$

where a_c is the semimajor axis of the parent body at present time. Ryabova (1999) has shown that such approach leeds to unacceptably large errors. But if the age is estimated from the mean orbits of meteoroids belonging to close mass ranges, the error in age estimation decreases considerably.

3.3. *Mathematical simulation*

If structural characteristics of the observed and simulated streams are in agreement, this is a good argument in favour of the age of the stream assumed by the model. The reverse is true only partially, because the model is defined by many parameters, as important as the age itself but, unfortunately, not known better than the age. As an illustration the models for the Geminid meteoroid stream are compared for ages 2000 and 10000 years in (Ryabova 1999).

Brown & Jones (1998) (their model is described in Section 7) tried to estimate the Perseid stream age comparing model and observed structural characteristics of the shower. From the model the authors estimated the "average" radiant dispersion with time, shift in position of the radiant with time, shift in the position of the main visual maximum with time, width of the ZHR profile, and the full nodal spread of the current shower. The estimates range are in the interval $(7 - 180) \times 10^3$ yr. The dispersion is rather large, and the reason is primarily accuracy and quality of observations. The limitations of the model are in second place.

For streams like the Perseid and Leonid streams, where the parent comet have been active for a long time, it is possible to obtain time of meteoroids ejection from comparing model and observed position and strength of the definite shower's maximum or outburst (e.g. Brown & Jones 1998; Asher, Bailey & Emel'yanenko 1999; Asher 1999).

3.4. *Meteoroids lifetime*

In the interplanetary space meteoroids are subject to destruction from many factors. Estimations of the meteoroid lifetime were improved over the time (Whipple 1967a;Whipple 1967b; Dohnanyi 1978; Tokhtas'ev 1982; Leinert, Röser & Buitrago 1983; Olsson-Steel 1986; Steel & Elford 1986). In all cases it was confirmed that, for particles with mass $\geqslant 10^{-6}-10^{-7}$ g, catastrophic collisions with sporadic meteoroids prevail. The PR-effect plays a secondary role. Comparison between the meteoroid lifetime and the mass distribution of the shower meteoroids allows sometimes make a conclusion on the stream age. For example, if there are no particles of a certain mass range in the stream, it can be suggested that they were removed by one of the destruction factors. The upper limit of the stream's age could be estimated as lifetime of largest meteoroids.

3.5. *Age changes in mass distribution*

An indirect estimate of six (Geminid, D.Arietid, Perseid, Quadrantid, η−Aquarid and ξ−Aquarid) meteoroid stream's age was obtained by Dohnanyi (1970), who studied the effect of impacts with sporadic background on the initial mass distribution in the stream. He revealed that five out of the six streams considered (exclusion is the Quadrantid stream) are sufficiently old that their small particles have reached a steady state distribution under the collisional environment of the sporadics, but at the same time they are sufficiently young that their large particles still have a distribution similar to stream's initial distribution. It is unknown, however, what the actual age that corresponds to the terms "sufficiently young" and "sufficiently old". It is not inconceivable that the described results, based on 35-years old data, need revising.

3.6. *Spin-up time scale*

Beech (2002) suggested a potentially powerful method for finding the time since ejection of meteoroids from their parent body. The approach relies upon being able to estimate the pre-atmospheric spin rates of meteoroids from the effect of light-curve flickering. The basic idea is using the 'windmill effect', which relies on non-isotropic photon scattering by meteoroid surface irregularities to produce a non-zero net torque. For three Geminid meteoroids under consideration the implied age is between 1000 and 4000 yr. Certainly, this method needs further investigation.

4. Velocity of ejection

The first publications devoted to ejection theory of the meteoroid stream formation appear to be the ones by Plavec (1955, 1957). He derived formulae giving the changes of the orbital elements owing to the ejection in the framework of the two body problem. These formulae were used lately in many researches.

A recent review on the determination of the ejection velocity of meteoroids from cometary nuclei could be found in (Williams 2001). Below are some additions.

4.1. *Methods based on ejection theory*

A technique for calculation of the ejection velocity was developed by Babadzhanov, Zausaev & Obrubov (1980). The equations of motion of a parent body and the mean orbit of a meteoroid stream (or an orbit of an individual meteoroid) are integrated backward numerically, and the minimum of the minimal distance between the orbits is computed. The point on the cometary orbit corresponding to this minimum is assumed to be the point of ejection. Taking into account that the cometary and the stream orbits generally are not intersecting, the ejection velocity vector can be determined only approximately in the following way. Let $\Delta a = a_c - a_s$ be the difference between the cometary and stream orbits at time of the ejection, δa the difference between the cometary orbit and an ejected meteoroid orbit at the given ejection velocity vector, calculated from formulae by Plavec (1955), ϵ_a the dispersion in observed meteoroids semi-major axes. The ejection velocity vector c is tabulated in reasonable limits. If the inequality $|\delta a - \Delta a| \leqslant \epsilon_a$ is fulfilled, the analogical inequalities for other orbital elements (e, i, Ω, ω) is checked. In such a way, a vector c obeying the set of inequalities is considered as the ejection velocity vector. Later was found that errors in initial orbital elements are too large to determine the accurate time and point of ejection. And even if they were precise, the estimation of c from non-intersecting orbits is too rough.

For completeness, a mention should be made to determining of the ejection velocity from difference between the parent and meteoroid semi-major axes without consideration on the evolution. This method was used by Williams (1996) for six major streams. But these determination, as noted later by Arter & Williams (2002), are not reliable for the same reason, namely, large errors in the observed meteoroid semi-major axes.

A current difference between the orbits of a parent body and the meteoroids of a stream is due to the ejection process and subsequent changes in orbital parameters due to gravitational and non-gravitational perturbations. One of the parameters, namely the longitude of the ascending node is determined much more accurately than others. So, it suggests itself for the ejection velocity determination.

Such an attempt was made by Andreev (1986, 1987, 1995) who derived equations for change in the orbital elements that arises from an impuls which a meteoroid gets during ejection. A dependence of the ejection velocity from the particle mass was also taken into account. The estimations made by this method are very rough, and can serve only as zero approximation because evolution process was not taken into account.

Another attempt to elaborate a method for the ejection velocities deducing from the observed orbital parameters of meteoroids was made by Ma & Williams (2001). Using the expression for $\Delta\Omega$, which is analogous to the one by Andreev (1986), but was derived by Ma, He & Williams (2001) independently, the authors found a special way for its application. To exclude influence of gravitational and non-gravitational perturbations, the method was applied to meteor outbursts, taking into account that in outbursts we are probably observing very young meteoroids. Ma & Williams also use the fact that if meteoroids are ejected at the nodes, there is no change in Ω. So, the node change $\Delta\Omega$ is interpreted as the difference between the node of the parent comet and the nodal position of the maximum activity in the outburst.

Resume: all the described above methods of the ejection velocity determination give no really reliable results.

4.2. *Comet Halley ejection velocities*

In modelling a meteoroid ejection we mainly use Whipple's formula (1951), which gives the cometocentric velocity of a particle dragged away from the surface of the cometary nucleus by the radial gas flow, or one of its modification (e.g. Hughes 2000). Crifo (1995) and Crifo & Rodionov (1997) formulae are also used.

Could we check these theoretical formula somehow? An opportunity came when a lot of experimental data was obtained during the last approach of Comet Halley to the Sun in 1986. It should be noted that there were not immediate measurements of particles ejection velocities. Moreover, a detailed comparison of different estimates is difficult, because they refer to different particle's masses and different arcs of the cometary orbit. An approximate relation is the following: the ejection velocities obtained by the Finson-Probstein method are 1.5–2 times lower than those obtained from an analysis of the particle's flux measurements by the DUSMA dust-mass analyzers installed on *Vega-1* and *Vega-2* spacecrafts, and these estimations are in turn 1.5–2 times lower than the values obtained from Whipple's formula for comparable conditions. A detailed review can be found in (Ryabova 1997).

Most of the estimations are related to very small particles ($m < 10^{-9}$ g). But Richter, Curdt & Keller (1991) managed to calculate several trajectories of large particles at the last stage of the *Giotto* approach. As a result, the ejection velocities were estimated to be ~ 40 m s^{-1} for particles with $m = 0.001$ g (the ejection time being about 6 – 9 days prior to passage of perihelion, i.e. for true anomaly of the comet $v \approx -20° \ldots - 30°$) and less than 10 m s^{-1} for particles with $m = 0.01$ g (supposly for $v \approx 0° - -25°$). Let us make estimations according to the Whipple's formula: in the first case $c \approx 120 - 180$ m s^{-1}, in the second – $c \approx 80 - 120$ m s^{-1} (estimations were made for spherical particles of density 0.35 g cm^{-3}).

At the time of the *Vega-1* approach to comet Halley, the onboard instrument "Photon" measured the flux of comparatively large dust particles ($m \geqslant 10^{-9}$ g). Ryabova (1997) proposed a probabilistic method that enables one to model the stream of particles ejected from a cometary nucleus into the spacecraft trajectory. The best agreement between the model and the experimental profiles was achieved when the mean ejection velocity was equal to about 8% of the velocity obtained from the Whipple's formula.

Resume: Further development of the theory of particle acceleration by the gas of the cometary atmosphere is needed. New experiments in situ conducted in future space missions, including direct measurements of the velocity of dust grains and the determination of their physical characteristics could give invaluable assistance in this research.

5. The Geminid meteoroid stream

There were a number of the Geminid stream modelling, every of which laid a brick into contemporary understanding of the stream structure and process of formation. Though this work is not, as mention have been made of, an historical review, some of these modelling are direct predessors of the model discussed below, namely works by Fox, Williams & Hughes (1982), Fox, Williams & Hughes (1983), Jones (1985), and Jones & Hawkes (1986). The last version of the Geminid meteoroid stream model was presented on IAU Coll. 197 by Ryabova, but it was not published in full (a breaf could be found in Ryabova (2004)). The previous version of the model could be found in Ryabova (2001b).

5.1. *Orbits*

The orbits of the Geminid meteoroid stream, as well as the one of the asteroid (3200) Phaethon (the Geminid's parent body) are located far inside the Jupiter's orbit (for Phaethon $a = 1.27$ AU, $e = 0.9$, $i = 22°$). Numerical integration of the asteroid's orbit has shown, that its semimajor axis is practically constant (changes are in the third decimal place), and changes in other orbital elements are smooth. The same is true for the most part of the Geminid's meteoroids. So changes in orbital elements may be approximated by a set of nested polynomials of the form

$$b = b_0 + \sum_{j=1}^{n} \sum_{k=0}^{m} b_{jk} a_0^k t^j,$$

where b is one of the Keplerian elements $(a, e, i, \Omega, \omega)$, t is the time from the initial moment, b_0 is the ininial value for an element (at $t = 0$).

Because the radiation pressure and the PR-effect were taken into account, the polynomials, and, respectively, the models were constructed for particles of two masses $(m_3 = 2.14 \times 10^{-3}$ g and $m_4 = 2.14 \times 10^{-4}$ g, if particles are spherical with density of 1 g cm^{-3}). Calculations of evolution for 10 millions of meteoroids take only several minutes.

5.2. *The stream's structure*

Here we consider only the cometary model of the stream formation, i.e. ejection points at the reference orbit are uniformly distributed along the true anomaly, that is approximately consistent with dust production rate proportional to r^{-4}, where r is heliocentric distance. The ejection velocities have been determined using the Whipple (1951) formula, and their directions were distributed uniformly in the solar hemisphere. The stream's age was taken of 2000 years (Ryabova 1999). It was assumed that the stream was generated during a short period of time, maybe even during a single revolution.

In fig. 1 a part of the Geminid's model cross-section in the ecliptic plane for particles of two masses m_3 and m_4 is shown. Each of the two model differential streams† follows its own evolutionary path, because, firstly, influence of PR-effect and radiation pressure depends on A/m, and secondly, because the ejection velocity also depends on meteoroid's mass. That is why the densest parts of the streams, which at ejection moment $(t = 0)$ coincided with the parent (reference) orbit, do not coincide now. At fig. 1 they are designated as "reference orbit m_3" and "reference orbit m_4". The second, as it should be because of the PR-effect, is nearer to the Sun.

The shower activity profiles, i.e. flux density variations along the Earth's orbit is designated by A. Having the profiles for two meteoroids masses we can calculate a profile for

---

† Differential stream is a stream consisted of meteoroids of definite mass, e.g. m_3 or m_4. Cumulative stream consists of meteoroids with masses $m > m_0$.

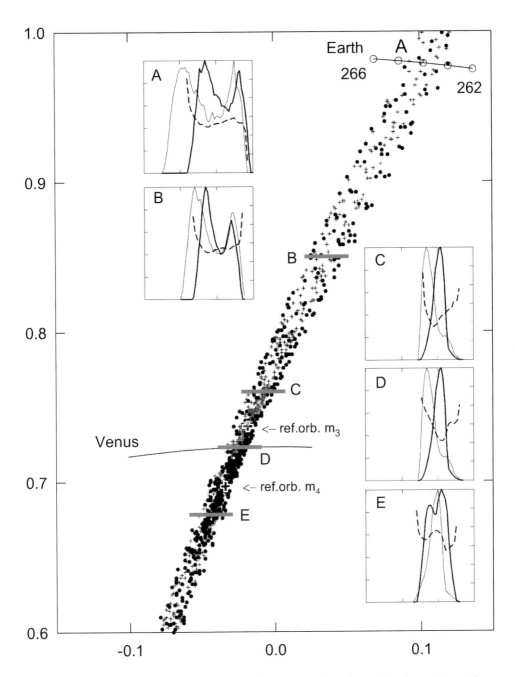

Figure 1. Geminid's model cross-section in the ecliptic plane for orbits of particles with masses m_3 (+) and m_4 (•). The Earth's orbit in the interval $262° - 266°$ in solar longitudes is designated by A. Other sections are designated by $B - E$. In the small panels, designated by $A - E$, activity profiles, i.e. flux density variations along the Earth's orbit, for particles masses m_3 (thick line) and m_4 (thin line), and a profile for mass index s (dashed line) are shown. The profiles are calculated along the corresponding sections.

mass index s. If the Earth's orbit intersects the model stream in other place, for example, along line B, the profile's shape will be different. The problem is that in reality we do not know where exactly the Earth crosses a model stream. To put it more precisely, we do not know the exact location of the stream cross-sections to the ecliptic plane. The main reason is that the moment of the stream generation is unknown. In this model the stream age is taken 2000 years. But if we take for example 2100 years, the stream's structure will be practically the same, but location of the cross-sections in the ecliptic plane will be shifted. Moreover, the accurate orbit of the parent body is not known, because the change of the cometary orbit during its trapping into the current orbit probably was considerable (see Section 6). Using approximations instead of numerical integration also shifts the location of the model stream. But we may use observations to fit or, better to say, to calibrate the model. The experimental profiles obtained by many-years of radar observations looks like something between B and C profiles (Bel'kovich 2005). To calibrate the model we should have flux density profiles with resolution of about $0.1°$ in solar longitude, or better. Now the work on the flux profiles is underway.

Let us assume for the moment that the profiles B are closest to the experimental ones, so the model stream should be moved in such a way that the Earth's orbit passed through the stream along the line designated by B. Then, knowing the particle's space density in the vicinity of the Earth's orbit, and using it as a reference one, we can evaluate quantitatively the particle's space density far from the Earth orbit. Moreover, we can evaluate the mass of the Geminid meteoroid stream.

It is very important that a model adjusting the flux density profile was used, and not a ZHR-profile (Zenith Hourly Rate) or simply a rate profile, because the rate profiles are not clarified from observational selection. The same is true in relation to the profile s. The profile obtained from hourly rates is burden by observational selection.

Besides factors of observational selection inherent to the various methods of observation (radar, visual etc.), there is astronomical selection, i.e. the alteration of the meteoroid flux in the gravitational field of a moving planet (Bel'kovich 1983; Bronshten 1983; Andreev 1984; Zabolotnikov (1984)). It should also be taken into account.

Unfortunately, we have measurements along the single trajectory within the stream (the Earth's orbit). If we had measurements made far from the Earth, their significance hardly could be overestimated.

5.3. Bimodality of cometary streams

Why do some profiles have two maxima, and some profiles have only one? The detailed explanation is given in (Ryabova 2001b) or (Ryabova 2001c). Briefly, the reason is that the orbital parameters of the meteoroids ejected from the parent comet when it approaches perihelion, differs from those of the particles ejected when the comet moves away from perihelion. This difference enhances in time, resulting in the formation of two layers (pre- and after-perihelion) in the stream. The location of the layers will be different for the meteoroids of different masses, mainly because of different ejection velocity, and also due to radiational effects. The Earth in its motion crosses these layers one after the other, we see two activity maxima (see panels A or B in fig. 1). But the layers for each differential stream have intersection along the reference (for this differential stream) orbit. So, if the Earth's way passes near, say, reference orbit m_3, the distance between layers is too small, and we see only one maximum for m_3-activity curve (e.g. panels C or D in fig. 1).

Results of the simulation show that a stream generated by cometary type during one revolution should produce a bimodal shower of specific type, where the distance in solar longitude between the first and the second maxima depends on the mass of meteoroids.

The real Geminid shower possess the feature. So it is reasonable to suggest that it was being produced by Comet Phaethon.

But the flux curve is not the only parameter we could use. It is reasonable to suggest that above mentioned pre- and post-perihelion layers have more or less different orbital characteristics. And it is really so. At any rate it is so for the model (Ryabova 2001b, fig. 6; Ryabova 2004, fig. 3). The problem is that quantity and quality of observed meteoroid orbits are insufficient to use for comparison with the model. We need the observational data on the orbits that meet the following conditions (Ryabova 2001b; Ryabova 2001c).
1. The mass of meteoroids should be less than 10^{-3} g or, better still, 10^{-4} g. Of course, the observations of meteoroids with fixed mass cannot be ensured; however, the mass range should be as narrow as possible, at least within the order of magnitude.
2. The accuracy of determination of orbital elements should be sufficiently high.
3. The number of orbits should be statistically sufficient for regression analysis.

And, again, we meet problem of observational selection. There are two approaches: to exclude observational selection factors from observations or to include them into the model. In the first case we can only outline limits of observability, but not restore missing observations. It could be useful and instructive to plan and carry out a special experiment. For example to do video observations of the Geminids, and to model the observations. Certainly, it is not a simple task.

5.4. *Some possible lines for research*

Besides those mentioned above there are several following interesting problems that could be solved.
1) To use radiant structure for calibration of the model. Certainly, the problem of the observational selection elimination should be the main difficulty.
2) To study influence of the Earth's passage through the stream on fine stream structure.
3) To model grouping in the stream. Data of observations exists (e.g. Karpov 2001; Karpov & Gainullin (2002)).
4) To model the stream's size distribution. There are results of Singer & Stanley (1980) regarding the flux of submicron particles ($m < 10^{-15}$ g) observed in the Geminid stream on satellite *Explorer 46*. Lifiteme for such small particles is about several years, so they are probably a result of collisions in the stream or with sporadical meteoroids.
5) To find a shape of Geminid's meteoroids from light-curve flickering (Beech 2002).

6. Origin of the Arietid and Geminid type meteoroid streams

The question of the Arietid and Geminid type meteoroid streams origin is still an open question. Twenty years ago Lebedinets (1984, 1985) noted that there are about 30 meteoroid streams in the catalogue of radar meteors by Lebedinets, Korpusov & Sosnova (1972) with specific orbits: small size ($a \leqslant 2$ AU) and very small perihelion distance ($0.01 \leqslant q \leqslant 0.1$ AU), including the Daytime Arietids ($q = 0.08$ AU, $a = 2.1$ AU). The orbits of another several tens of streams, including the Geminid stream, have the same small size, but larger perihelion distance ($0.1 \leqslant q \leqslant 0.2$ AU).

The fact that a parent comet (as we discussed earlier, asteroid (3200) Phaethon is probably a dormant comet) was detected with confidence only for the Geminid stream is not surprising, because the lifetime of a comet in such orbits is short. But how can a comet appear on such orbits? It can not be transferred from a long-period orbit under the influence of a strong gravitational perturbation by Jupiter or Saturn, because its aphelion is far inside the Jupiter's orbit ($a_J \approx 5.2$ AU). The gradual shift of the aphelion from the orbit of Jupiter due to repeated weak perturbations is not possible, because the comet should extinct far in advance of generating of an Arietid-type stream. In principle, the transformation of typical cometary orbits in Arietid-type orbits is possible by a single perturbation from the Earth or Venus (Andreev, Terent'eva & Bayuk 1990; Andreev, Terent'eva & Bayuk 1991; Terent'eva & Bayuk 1991). But what is the probability of such event? Is it high enough to form several tens of meteoroid streams?

One of the possible mechanism for the transformation of cometary orbits into Arietid-type orbits could be the reactive drag of the sublimating cometary nucleus. According to estimations of Lebedinets (1984, 1985), who modelled the transformation for a set of cometary orbits with perihelion distances $0.01 \leqslant q \leqslant 0.1$ AU, aphelia near the orbit of Jupiter ($Q = 5.2$ AU), and nuclei radii $0.02 \leqslant R_c \leqslant 2$ km, decresing of aphelion distance from 5.2 AU till 4 AU is quite possible in a time from half to several cometary periods.

The perihelion distance of a comet or a young compact swarm on the Arietid-type orbit is very sensitive to perturbations in aphelion, and can be changed by a weak single perturbation by Jupiter. For example, on changing the velocity in aphelion within ± 1 km s^{-1} the perihelion distance changes in the range $0.01 - 0.14$ AU, and the aphelion distance changes only slightly.

In the case of strong reactive transformation of the cometary orbit, the resulting stream should be more wide, than the one produced under 'normal' conditions. It is so for the Geminid stream.

The described research was only a pilot one and, unfortunalely, was not continued.

Certainly, comparative studies of the Geminid and Arietid meteoroid streams could be helpful in clarifying their origin and formation.

The Arietids is a daytime stream in the Northern hemisphere, that is why it was undiscovered until beginning of radar observations of meteors[†]. In the Southern hemisphere the shower can be observed visually (Jenniskens 1994). It is a wide stream ($\lambda_\odot \approx 68° - 95°$), with maximum at June 9th and activity approximately equal to the Geminids or Quadrantids (Jenniskens 1994; Campbell-Brown 2004).

There were several attempts to identify a parent body for the stream. McIntosh (1990) based on similarities in behaviour of orbital elements suggested that Comet P/Machholtz 1986 VIII, Comet 1491 I, the Quadrantid, δ-Aquarid and D.Arietid meteoroid streams have a common origin. Babadzhanov & Obrubov (1992) using the method of filling volume, described above in Subsection 2.2, simulated the evolution of the Comet P/Machholtz meteoroid stream. The stream may produce eight meteor showers, including six known ones: the D.Arietids, Quadrantids, Ursids, Northern and Southern δ-Aquarids, and α-Cetids. The authors estimated the age of the stream in 7.5 millenia. But according to my own rough estimation, time for full revolution of the node for the Quadrantid stream is about 16 thousand years. At the same conference[‡] results of the Quadrantid meteoroid stream modelling (Jones & Jones 1992) were presented. The stream consisting of 500 test particles was ejected 4000 years ago from the mean orbit of the Quadrantids. Considering quasi-constants of the motion the authors found that the Quadrantids and D.Arietids are unlikely to be members of a single complex. Later the same authors (Jones & Jones 1993) made another simulation based on evolution of the Comet Machholtz and came to a conclusion that the comet probably gave birth to the Quadrantid, D.Arietid, Southern δ-Aquarid meteoroid streams, but strong doubt about D.Arietids exists.

Two years ago results of two researches were published practically simultaneously. Gorbanev & Knyaz'kova (2003) found that a genetic association probably exists for the

---

[†] A historical review can be found in Lovell (1954), Kronk (1988) or Campbell-Brown (2004).
[‡] Asteroids, Comets and Meteors 1991.

meteoroid stream producing the D.Arietids and Southern δ-Aquarids, and the Marsden comet group, discovered in 2002. Ohtsuka, Nakano & Yoshikawa (2003) concluded that comet Machholtz, the D.Arietids, the Marsden group may be generally associated.

The first incident flux density curve for the D.Arietids was obtained recently by Campbell-Brown (2004). It is calculated with constant value for the mass distribution index ($s = 2.1$), so probably will be revised.

To summarize: The D.Arietid meteoroid stream still has no recognized parent body. We have several hypothesis, which need further investigations. Maybe it worth to search among newly-discivered Amor and Apollo asteroids? To construct a preliminary model of the D.Arietid meteoroid stream we need also an estimate of the age.

7. The Perseid meteoroid stream

There are only a few works devoted to the Pereseid modelling. Some of them (Wu & Williams 1993; Williams & Wu 1994; Harris & Hughes 1995; Harris, Yau & Hughes (1995)) are reviewed in the paper by Brown & Jones (1998) discussed below in Subsections 7.1–7.3.

7.1. *Model description*

The formation scheme for the Perseid meteoroid stream and, consequently, the structure of the stream is very different from those of the Geminid stream. The Perseid's parent comet 109P/Swift-Tuttle is still alive and adds meteoroids into the stream at every revolution. Brown & Jones (1998) simulated ejection of particles at each perihelion passage of comet 109P from 59 to 1862 AD and integrated orbits of the meteoroids (18.24 mln meteoroids in sum) till their descending nodes for times closest to 1992, i.e. the last perihelion passage of 109P. Four ejection velocity formulae were used: the first model was based on Crifo (1995) results, the other three are based on Jones (1995) modifications of Whipple's formula. It should be noted that the models of ejection velocities are different, but not dramatically different.

The ejection velocity depends on particle windage, A/m. In the discussed work spherical particles with three densities were used, namely 0.1, 0.8, and 4.0 g cm^{-3}. In total we have 12 distinct models. For each 10000 test meteoroids were ejected at different masses from interval $10^{-5} - 10$ g for each perihelion passage of 109P/Swift-Tuttle.

7.2. *Age of the stream*

To estimate age of the stream long-term integrations was also performed over the interval from 5 to 100 thousands years ago using the SWIFT symplectic integrator. At first to take into account errors in initial conditions, 20 clone orbits were integrated from 1862 backward for 100000 years. Then the two model streams were generated, using the two 'extreme' orbits, and integrated forward. An attempt to estimate age of the stream using the radiant size of the Perseid shower, the photographic radiant locations at maximum activity, the rate of change in the apparent location of the maximum, the width of the ZHR profile gave very different ages in the interval $(7 - 180) \times 10^3$ years. But the long duration of the Perseid shower implies a lower limit for the age of the stream of the order of 10^5 years.

It is interesting that according to the results of modelling some Perseid meteoroids may encounter the Earth at their ascending nodes in mid-March. The Gamma Normids and the Theta Centaurids are the candidates for twin showers of the Perseids. If the existence of streams will be confirmed, that means a stream age of at least 50-75 thousands years.

Babadzhanov & Obrubov (1987) did not consider the Perseid stream. As it was mentioned in Subsection 2.2, my rough estimate has shown that the Perseid stream could produce to 8 showers during 1 mln years. Maybe it worth to check?

7.3. *Models and observations*

From the analysis of the results, the authors have drawn several conclusions regarding role of initial conditions and effects of gravitational and radiational perturbations, and density assumed for meteoroids. I will not retell the conclusions, preferring enlarge on fitting models to observations.

What observational data may be used as criterions for validation of the Perseid models? The main "anchor" is the shower activity, especially locations and strength of outburst peaks. The secondary one is the geocentric radiant distribution. Unfortunately, the present size of the measurement errors makes orbital elements insuitable for this purpose. Data on mass distribution index in the Perseids exist, but they were not used in the research. So, it is a reserve for the future.

The best fit to the observed outbursts from the years 1989-1996 were found from the model using Jones (1995) modification of Whipple's formula with $r^{-0.5}$ dependence and meteoroid densities between 0.1 and 0.8 g cm^{-3}.

The locations and strength of the observed visual peaks associated with the outburst component of the stream and the model one are in good agreement, besides the 1993 and 1994 peak locations, and the 1994 - 1995 peak strength, which were underestimated in the models. As it turned out in 1993 ejections from 1862 and 1610 are predominant in meteoroid population. In 1994 the main contributions are from 1862 and 1479, and in 1995 – from 1479 and 1862. No observations exists for returns of 1479 and 1610, but for 1862 approach there are detailed observations (Sekanina 1981; Fomenkova, Jones, Pina, *et al.* 1995).

It worth to note that activity of the model comet was taken as constant, and ejection velocity vectors were taken isotropically distributed within the sunward hemisphere. There are two possible approaches to fit a model to observations. The direct approach is to change the model, say, to increase the number of ejected particles, until a coincidence with observations of the shower. This approach requires new calculations. Another, the indirect approach, mentioned in Subsection 2.2 as weighting coefficients technique, consist in giving a weight to some model meteoroids contributing in the peak to change it to favourable location or strength, and to analyze coincidence with observations of the comet dust production.

For example, the authors of the model mentioned that it is possible to change the mean nodal longitude of the 1993 and 1994 peaks implying that almost all ejections had a strong component normal to the cometary plane in the north direction. It could be reasonable to look closer at the model meteoroids having "correct" nodal longitudes (certainly, if such meteoroids exist). Maybe their ejection velocity vectors are within registered jets?

In my opinion this model is the best model for the Perseid stream that could be done with existing observational data. Let us imagine, that in the coming years a flux profile of a high resolution together with the profile for mass distribution index s will be obtaided from radar observations of the Perseids. Then, this model could serve once more, if the authors will consider it necessary, and if the files with information will not be lost.

8. The Leonid meteoroid stream

In a sense, the theory of meteoroid streams began from strong Leonid meteor storms of 1799 and 1833. Detailes can be found elsewhere (Lovell 1954; Wu & Williams 1996; Brown 1999). The parent comet of the Leonids is Comet 55P/ Tempel-Tuttle, and storms occur in years of its return to the Sun. In other time, i.e. when the comet is far from perihelion, activity of the Leonids is very low. That is why it had been generally accepted that the Leonid meteoroid stream is young, and the meteoroids had no enough time to spread along the orbit, till Williams (1997) proposed very a interesting explanation of the facts. Gravitational perturbations from planet Uranus, which is on a 5:2 mean motion resonance with the Leonid parent comet, "sweep out" the meteoroids from all the stream orbit except small arc located near the nucleus of the comet.

For a long time efforts of "streams modellers" were concentrated on the Leonids predictions. One of the first advantageous attempt was made by Kazimirčak-Polonskaya, Belyaev, Astapovič & Terent'eva (1968). Besides, it was one of the first meteoroid stream's models. The authors numerically integrated orbits of 17 model meteoroids, and found time of maximum activity for the Leonids of 1966 with accuracy about 1 hr.

The increase in the activity of the Leonids from 1998 to 2002 stimulated many researches to model the Leonid stream with the main aim to predict showers (Brown & Jones 1996; McNaught & Asher 1999; Lyytinen & Van Flandern 2000; Brown & Cooke 2001 etc.). But till now the Leonid stream as a whole was not studied. The last Leonid stream model is already near to that, but still is aimed at predictions. The full description of the method and results are presented in (Vaubaillon, Colas & Jorda 2005a,b), but a brief of the model could be also found in (Vaubaillon & Colas 2005; Vaubaillon, Lyytinen, Nissinen & Asher 2003; Vaubaillon & Colas 2002; Vaubaillon 2002).

The authors apply quite classic approach to meteoroid stream modelling. Using the definite model of the meteoroid ejection from the parent comet, the ejection of 2.5×10^5 particles was modelled for 29 perihelion returns of comet 55P. The dates of perihelion passages are from 1300 to 1998, from 604 to 802, and 1001. The particle's size range $0.1-100$ mm was divided into 5 bins. Evolution of the test particles was calculated till the present time, taking into account gravitational and non-gravitational forces. Then, from all model meteoroids there were selected those with nodes close to the Earth, and their parameters were analyzed. The comparison between predictions and observations for the time of maximum activity shown a good agreement with a difference till a few tens of minutes. The greatest differences are found for years 1999 and 2000. Authors assumed, that another still-unknown stream had been encountered. But certainly other explanations are possible. For example, Brown & Jones (1998) have shown how time of maximum can be shifted by some corrections in initial parameters.

A resume: this model has rather large potential for further study.

9. Resonances in meteoroid streams

Resonances in meteoroid streams is a topic worthy of a separate review. Here I'd like to dwell on some works showing that mean motion resonances can cause evident changes in the structure of a meteoroid stream. Emel'yanenko (2001) also noted that resonant streams are important objects for ground observations. These structures have well determined parameters if the mean motion commensurabilities are known. This can be used to improve physical models of meteor phenomenon.

Investigating the Quadrantid meteoroid stream (more precisely, the part of the stream close to the 2/1 resonance with Jupiter) by Shubart's (1978) averaging method, Froeschlé

& Scholl (1986) found that the stream breaks up into arcs of different sizes with distinctly different dynamical evolution. The other seven meteoroid streams (June Bootids, Annual Andromedids, Librids, June Lyrids, July Phoenicids, Pegasids, and December Phoenicids) known to be located in mean motion resonances with Jupiter do not reveal the same kind of splitting into arcs (Scholl & Froeschlé 1988).

The existence of gaps in the distribution of the semimajor axes of meteoroids in the Perseid meteoroid stream was found by Wu & Williams (1995). The gaps are coincided with the location of low-order mean motion resonances with the major planets. Wu & Williams carried out a numerical integration of 11000 test meteoroids over 1000 yr. The initial orbital elements of the sample were randomly chosen from a range near the value of the same element of P/Swift-Tuttle. These integrations showed that gaps in $1/a$ distribution formed at some of the mean motion resonance locations in a very short interval of 150 yr.

Later the results of Wu & Williams (1995) were discussed by Murray (1996). Murray has shown that the gaps in orbital element distributions of asteroid families and meteoroid streams can be produced by the genuine removal of objects by resonant effects ('real' gaps) and those that could result from a failure to identify some material as belonging to a particular group because of the resonant perturbations ('imaginary' gaps). It was also shown that some gaps in the distribution of the reciprocal semimajor axes of Perseid meteor orbits for visual meteors discussed by Wu & Williams (1995) probably are imaginary. As to results of numerical modelling by Wu & Williams, most of the gaps can be regarded as real ones but the reason for the removal of meteoroids is debatable. It could be both resonant effects and direct perturbation due to a close approach to a planet. So, further studies should be carried out. Murray (1996) also stressed the role of observational selection effect for the orbital data of meteoroids.

A young meteoroid stream generated during an approach of its parent comet to the Sun begins to disperse, as a rule. However librating meteoroids can produce long-lived high-density substreams. This is important from practical point of view: hazard to spacecrafts.

Asher & Clube (1993) described a model of a meteoroid swarm in the Taurid Complex and in $7/2$ resonance with Jupiter. Later Asher & Izumi (1998) made a stronger test of the model using meteor observations of the Nippon Meteor Society over the past six decades. At high statistically significant level a single, dominant concentration of meteoroids in the stream can explain all (to date) observed years of enhancement.

Leonid meteor shower of 1998 surprised astronomers by unexpected high incidence of bright meteors about 16 h before the predicted maximum, i.e. in 1998 November 17.1. Asher, Bailey & Emel'yanenko (1999) have noted that the observed fireball outburst was distinguished by a predominance of very bright meteors originating from large dust grains with sizes ranging up to a few centimeters. The highest level of fireball activity lasted approximately half a day, indicating the presence of a narrow concentrated substream. Investigation have shown that the outburst is explained by the ejection of meteoroids into the $5/14$ mean motion resonance with Jupiter mainly during the perihelion passage of Comet 55P/Tempel-Tuttle in 1333. The meteoroids could not spread around the orbit in a usual way. Instead they produced a concentrated swarm, taking the form of an arc of material close to, but not necessarily identical with, the parent cometary orbit.

It could be extremely interesting to study possible high-concentrated substreams in a full model (like described above the Geminid and the Perseid stream models) of a stream, considering space location of the swarms, and their evolution. As the first candidates for such study I see the Quadrantid and the Leonid meteoroid streams. The Geminid stream also has two very narrow zones located in the mean motion resonances $7/1$ and $8/1$

with Jupiter (Emel'yanenko 2001), so it is desirable to look closer at the behavior of meteoroids in these locations.

Acknowledgements

The author would like to acknowledge the financial support from the Organizers, which make it possible her attendance at the symposium and consequently stimulated the writing of this review.

References

Andreev, G.V. 1984, *Solar System Research* 18, 158
Andreev, G.V. 1986, *Astronomiya i geodeziya* 15, 164 (in Russian)
Andreev, G.V. 1987, *Handbook for MAP* 25, 305
Andreev, G.V., Terent'eva, A.K., & Bayuk O.A. 1990, in: C.-I.Lagerkvist, H.Rickman, B.A.Lindblad & M.Lindgren (eds.), *Asteroids, Comets, Meteors III*, (Uppsala universitet Reprocentralen HSC, Uppsala), p. 493
Andreev, G.V., Terent'eva, A.K., & Bayuk O.A 1991, *Astronomich. Vestnik* 25, 177 (in Russian)
Andreev, G.V. 1995, *Astronomicheskij vestnik* 29, 151 (in Russian)
Arter, T.R., & Williams I.P. 2002, *Mon. Not. R. Astron. Soc.* 329, 175
Asher, D.J. & Clube, S.V.M. 1993, *QJRAS* 34, 481
Asher, D.J. & Izumi, K. 1998, *Mon. Not. R. Astron. Soc.* 297, 23
Asher, D.J., Bailey, M.E., & Emel'yanenko, V.V. 1999, *Mon. Not. R. Astron. Soc.* 304, L53
Asher, D.J. 1999, *Mon. Not. R. Astron. Soc.* 307, 919
Babadzhanov, P.B., Zausaev, A.F., & Obrubov, Yu.V. 1980, *Bull. Inst. Astrophys. of TadzhSSR* 69-70, 45 (in Russian)
Babadzhanov, P.B. & Obrubov, Yu.V. 1987, in: Z.Ceplecha & P. Pecina (eds.), *Interplanetary Matter*, Publ. Astron. Inst. Czechosl. Acad. Sci., Vol. 67(2), (Ond(r)ejov: Czechosl. Acad. Sci.), p. 141
Babadzhanov, P.B. & Obrubov, Yu.V. 1992, in: A.Harris & E.Bowell (eds.), *Proc. Asteroids, Comets and Meteors 1991*, (Lunar and Planetary Inst., Flagstaff, AZ), p. 27
Beech, M. 2002, *Mon. Not. R. Astron. Soc.* 336, 559
Bel'kovich, O.I. 1983, *Sol. Syst. Res.* 17, 83
Bel'kovich, O.I. 2005, *private communication*
Bordovitsyna, T., Avdyushev, V., & Chernitsov, A. 2001, *Cel. Mech. Dyn. Astron.* 80, 227
Bronshten, V.A. 1983, *Sol. Syst. Res.* 17, 175
Brown, P. & Jones, J. 1996, in: B.A.S.Gustafson, M.Hanner (eds.), *Physics, Chemistry, and Dynamics of Interplanetary Dust*, Proc. IAU Coll. No.150 (ASP 104) (San Francisco: Astron. Soc. of the Pasific), p. 113
Brown, P. & Jones, J. 1998, *Icarus* 133, 36
Brown, P. 1999, *Icarus* 138, 287
Brown, P. & Cooke, B. 2001, *Mon. Not. R. Astron. Soc.* 326, L19
Campbell-Brown, M. 2004, *Mon. Not. R. Astron. Soc.* 352, 1421
Crifo, J.F. 1995, *ApJ* 445, 470
Crifo, J.F. & Rodionov, A.V. 1997, *Icarus* 127, 319
Dohnanyi, D. 1970, *J. Geophys. Res.* 75, 3468
Dohnanyi, D. 1978, in: J.A.M. McDonnell (ed.), *Cosmic dust*, (Chichester: Wiley-Interscience), p. 527
Emel'yanenko, V.V. 2001, in: B.Warmbein (ed.), *Proc. Meteoroids 2001 Conf.*, ESA SP-495 (Noordwijk: European Space Agency), p. 43
Fomenkova, M., Jones, B., Pina, R., Puetter, R., Sarmecanic, J., Gehrz, R., & Jones, T. 1995, *AJ* 110, 1866
Fox, K., Williams, I.P., & Hughes, D.W. 1982, *Mon. Not. R. Astron. Soc.* 199, 313
Fox, K., Williams, I.P., & Hughes, D.W. 1983, *Mon. Not. R. Astron. Soc.* 205, 1155
Froeschlé, C. & Scholl, H. 1986, *Astron. Astrophys.* 158, 259

Gockel, C. & Jehn, R. 2001, *Mon. Not. R. Astron. Soc.* 317, L1

Gorbanev, Y.M. & Knyaz'kova, E.F. 2003, *Sol. Syst. Res.* 37, 506

Hamid, S.E. 1951, *AJ* 56, 126

Harris, N.W. & Hughes, D.W. 1995, *Mon. Not. R. Astron. Soc.* 273, 992

Harris, N.W., Yau, K.C.C., & Hughes, D.W. 1995, *Mon. Not. R. Astron. Soc.* 273, 999

Hughes, D.W. 2000, *Planet. Space. Sci.* 48, 1

Jenniskens, P 1994, *Astron. Astrophys.* 287, 990

Jenniskens, P 2001, *WGN, Journal of the International Meteor Organization* 29, 165

Jenniskens, P 2002, in: B.Warmbein (ed.), *Proc. Asteroids, Comets, Meteors – ACM2002 Conf.*, ESA SP-500 (Noordwijk: European Space Agency), p. 117

Jones, J. 1985, *Mon. Not. R. Astron. Soc.* 217, 523

Jones, J. & Hawkes, R.L. 1986, *Mon. Not. R. Astron. Soc.* 223, 479

Jones, J. & Jones, W. 1992, in: A.Harris & E.Bowell (eds.), *Proc. Asteroids, Comets and Meteors 1991*, (Lunar and Planetary Inst., Flagstaff, AZ), p. 269

Jones, J. & Jones, W. 1993, *Mon. Not. R. Astron. Soc.* 261, 605

Jones, J. 1995, *Mon. Not. R. Astron. Soc.* 275, 773

Karpov, A. 2001, in: B.Warmbein (ed.), *Proc. Meteoroids 2001 Conf.*, ESA SP-495 (Noordwijk: European Space Agency), p. 27

Karpov, A. & Gainullin R. 2002, in: B.Warmbein (ed.), *Proc. Asteroids, Comets, Meteors – ACM2002 Conf.*, ESA SP-500 (Noordwijk: European Space Agency), p. 249

Kazimirčak-Polonskaya, E.I., Belyaev, N.A., Astapovič, I.S., & Terent'eva, A.K. 1968, in: L.Kresák & P.M.Millman (eds.), *Physics and Dynamics of Meteors*, Proc. IAU Symp. No.33, (Dordrecht, Reidel), p. 449

Kazimirchak-Polonskaya, E.I., Belyaev, N.A., & Terent'eva, A.K. 1972, in: G.A.Chebotarev, E.I. Kazimirchak-Polonskaya & B.A. Marsden (eds.), *The Motion, Evolution of Orbits, and Origin of Comets*, Proc. IAU Symp. No.45, (Dordrecht, Reidel), p. 462

Kronk, G.W. 1988, *Meteor showers: A descriptive catalog*, (Enslaw Publishers, New Jersey)

Lebedinets, V.N., Korpusov, V.N., & Sosnova A.K. 1972, *Trudy Inst. Eksp. Meteorol.* 1(34), 88 (in Russian)

Lebedinets, V.N. 1984, *Trudy Inst. Eksp. Meteorol.* 14(110), 88 (in Russian)

Lebedinets, V.N. 1985, *Sol. Syst. Res.* 19, 101

Leinert, C., Röser, S., & Buitrago, J. 1983, *Astron. Astrophys.* 118, 345

Lidov, M.L. 1961, *Iskusstvennye sputniki Zemli* 8, 5 (in Russian)

Lovell, A.C.B. 1954, *Meteor astronomy*, (Oxford Clarendon Press)

Lyytinen, E.J. & Van Flandern, T. 2000, *Earth, Moon, & Planets* 82, 149

Ma, Y., He, Y., & Williams I.P. 2001, *Mon. Not. R. Astron. Soc.* 325, 379

Ma, Y., & Williams I.P. 2001, *Mon. Not. R. Astron. Soc.* 325, 457

McIntosh, B.A. & Jones, J. 1988, *Mon. Not. R. Astron. Soc.* 235, 673

McIntosh, B.A. 1990, *Icarus* 86, 299

McNaught, R.H. & Asher, D.J. 1999, *WGN, Journal of the International Meteor Organization* 27, 85

Moiseev, N.D. 1945, *Trudy Gos. Astron. Inst. Moscov. Univ.* 15, 75 (in Russian)

Murray, C.D. 1982, *Icarus* 49, 125

Murray, C.D. 1996, *Mon. Not. R. Astron. Soc.* 279, 978

Ohtsuka, K., Nakano, S., & Yoshikawa, M. 2003, *Publ. Astron. Soc. Japan* 55, 321

Olsson-Steel, D. 1986, *Mon. Not. R. Astron. Soc.* 219, 47

Plavec, M. 1950, *Nature* 165, 362

Plavec, M. 1955, *Bull. Astron. Inst. Czech.* 6, 20

Plavec, M. 1955, in: T.R. Kaiser (ed.), *Meteors*, Proc. Second Intern. Symp. Phys. Meteors., p. 168

Plavec, M. 1957, *On the origin and early stages of the meteoroid streams* (Praha)

Richter, K., Curdt, W., & Keller, H.U. 1991, *Astron. Astrophys* 250, 548

Ryabova, G.O. 1989, *Sol. Syst. Res.* 23, 158

Ryabova, G.O. 1997, *Sol. Syst. Res.* 31, 277

Ryabova, G.O. 1999, *Sol. Syst. Res.* 33, 224

Ryabova, G.O. 2001a, in: B.Warmbein (ed.), *Proc. Meteoroids 2001 Conf.*, ESA SP-495 (Noordwijk: European Space Agency), p. 63

Ryabova, G.O. 2001b, in: B.Warmbein (ed.), *Proc. Meteoroids 2001 Conf.*, ESA SP-495 (Noordwijk: European Space Agency), p. 77

Ryabova, G.O. 2001c, *Sol. Syst. Res.* 35, 151

Ryabova, G.O. 2003, *Mon. Not. R. Astron. Soc.* 341, 739

Ryabova, G.O. 2004, in: M.Triglav-Čekada & C.Trayner (eds.), *Proc. Intern. Meteor Conf. 2003*, (Internat. Meteor. Org.), p. 131

Ryabova, G.O. 2005, in: Z.Knezevic & A.Milani (eds.), *Dynamics of Populations of Planetary Systems*, Proc. IAU Coll. 197, (Cambridge University Press), p. 411

Scholl, H. & Froeschlé, C. 1988, *Astron. Astrophys.* 195, 345

Schubart, J. 1978, in: V.Szebehely (ed.), *Dynamics of Planets and Satellites and Theories of their Motion*, (Reidel, Dordrecht), p. 173

Sekanina, Z. 1981, *AJ* 86, 1741

Singer, S.F. & Stanley, J.E. 1980, in: I.Halliday & B.A.McIntosh (eds.), *Solid Particles in the Solar System*, Proc. IAU Symp. 90, (Reidel, Dordrecht), p. 329

Steel, D.I. & Elford, W.G. 1986, *Mon. Not. R. Astron. Soc.* 218, 185

Terent'eva, A.K. & Bayuk O.A 1991, *Bull. Astron. Inst. Czechosl.* 42, 377

Tokhtas'ev, V.S. 1982, in: O.I.Bel'kovich, P.B.Babadzhanov, V.A.Bronshten & N.I.Suleimanov (eds.), *Meteor Matter in the Interplanetary Space*, (Moscow-Kazan) p. 162

Vaubaillon, J. 2002, *WGN, Journal of the International Meteor Organization* 30, 144

Vaubaillon, J. & Colas, F. 2002, *Proc. Asteroids, Comets, Meteors – ACM2002 Conf.*, ESA SP-500 (Noordwijk: European Space Agency), p. 181

Vaubaillon, J., Lyytinen, E., Nissinen, M., & Asher, D. 2003, *WGN, Journal of the International Meteor Organization* 31, 131

Vaubaillon, J. & Colas, F. 2005, *Astron. Astrophys.* 431, 1139

Vaubaillon, J., Colas, F., & Jorda L. 2005a, *Astron. Astrophys.* 439, 751

Vaubaillon, J., Colas, F., & Jorda L. 2005b, *Astron. Astrophys.* 439, 761

Weiss, E. 1868, *Astronomische Nachrichten* 72, 81

Whipple, F. 1951, *ApJ* 113, 464

Whipple, F. 1967a, *Smithsonian Astrophys. Obs. Spec. Rept.* 239, 1

Whipple, F. 1967b, in: J. Weinberg (ed.), *The Zodiacal Light and the Interplanetary Medium*, NASA SP-150 (Washington: Scientific and Technical Information Division, National Aeronautics and Space Administration), p. 409

Williams, I.P., Murray, C.D., & Hughes, D.W. 1979, *Mon. Not. R. Astron. Soc.* 189, 483

Williams, I.P. & Wu, Z. 1994, *Mon. Not. R. Astron. Soc.* 269, 524

Williams, I.P. 1996, *Earth, Moon, & Planets* 72, 231

Williams, I.P. 1997, *Mon. Not. R. Astron. Soc.* 292, L37

Williams, I.P. 2001, in: B.Warmbein (ed.), *Proc. Meteoroids 2001 Conf.*, ESA SP-495 (Noordwijk: European Space Agency), p. 33

Wu, Z., & Williams, I.P. 1993, *Mon. Not. R. Astron. Soc.* 264, 980

Wu, Z., & Williams, I.P. 1995, *Mon. Not. R. Astron. Soc.* 276, 1017

Wu, Z., & Williams, I.P. 1996, *Mon. Not. R. Astron. Soc.* 280, 1210

Zabolotnikov, V.S. 1984, *Solar System Research* 18, 35

Asteroids, Comets, Meteors
Proceedings IAU Symposium No. 229, 2005
D. Lazzaro, S. Ferraz-Mello & J.A. Fernández, eds.

© 2006 International Astronomical Union
doi:10.1017/S1743921305006782

Physical and chemical properties of meteoroids as deduced from observations

Jiří Borovička

Astronomical Institute of the Academy of Sciences, 251 65 Ondřejov, Czech Republic
email: borovic@asu.cas.cz

Abstract. A review of the current knowledge of physical properties and chemical composition of meteoroids entering the Earth's atmosphere is presented. Meteoroid penetration ability, ablation coefficients, beginning heights, light curves, fragmentation, and spectra are considered. The inferred bulk densities, mechanical strengths, rotation, and atomic elemental abundances are discussed. Cometary meteoroids are effectively grain aggregates with low bulk density (100–1000 kg m^{-3}), high porosity and low cohesivity. A volatile matrix holding the grains together may be present. Presence of large amounts of organic material is not firmly established. Small chunks ($\sim$1 mm) of denser material are sometimes contained in cometary meteoroids. Chemically, cometary grains are similar to CI chondrites but there is a hint of enhancement of Na, Si, and Mg and depletion of Fe, Cr, and Mg. Larger chemical diversity is observed among small meteoroids on cometary orbits not belonging to meteoroid streams. The relatively frequent Na-free meteoroids are probably fragments of cometary irradiation crust. Asteroidal meteoroids exhibit much lower mechanical strengths than stony meteorites, clearly due to the presence of large scale cracks. Iron meteoroids dominate among asteroidal meteoroids smaller than 1 cm.

Keywords. meteors, meteoroids

1. Introduction

All solid bodies in the interplanetary space larger than dust particles and smaller than asteroids are called meteoroids. Meteoroids occupy a significant part of the size spectrum of solar system bodies, from tens of microns to about ten meters. The orbital lifetime of meteoroids in the inner solar system is much shorter than the age of the solar system. Meteoroid population must be therefore continuously replenished. The main sources of meteoroids are asteroids and comets. Only a tiny part comes from the solid surfaces of planets and satellites and from interstellar space. The main motivation of meteoroid study is recognition of different meteoroid populations, establishing of their properties, and contributing this way to the study of asteroids and comets. Another aspect is the direct influence of meteoroids on terrestrial environment.

Meteoroids are too small to be observed remotely in interplanetary space. Only large concentrations of meteoroids can be revealed remotely but these observations provide limited information on meteoroid properties. Very small meteoroids are part of the solar system dust structures such as zodiacal dust cloud and asteroidal dust bands. Larger meteoroids of millimeter to centimeter range form dust trails in the orbits of many short period comets. An encounter of the Earth with a dust trail will manifest itself as a meteor storm.

The terrestrial atmosphere is, in fact, our best detector of meteoroids. The process of meteoroid disintegration in the atmosphere generates electromagnetic radiation including visible light and leads to the formation of an ionized trail. Large meteoroids produce also sonic waves. Since up to 7×10^5 km^2 of atmosphere can be sampled from a single site (for meteor height of 100 km and viewing more than 10 degrees above horizon), small and medium sized meteoroids can be effectively studied from one ground station (although stereoscopic observations from two stations are needed for the determination of meteor trajectory). The study of large meteoroids which enter the atmosphere only rarely must rely on long-term observations using whole networks of stations, or, in the future, on space-borne monitoring of the atmosphere on global scale.

Under favorable circumstances, small part of a meteoroid can survive the atmospheric entry and land as a meteorite. Laboratory studies of meteorites provide, of course, the most detailed information on meteoroid properties. They cover, however, only small part of the whole meteoroid population, since only meteoroids large enough, strong enough, and entering the atmosphere with sufficiently low velocity can drop meteorites.

The atmosphere of the Earth is not the only possible meteoroid detector. Meteoroid impacts have been also detected on the surface of Moon (Cudnik *et al.* 2003) and meteors have been observed in the atmospheres of Mars (Selsis *et al.* 2005) and Jupiter (Cook & Duxbury 1981). A meteorite has been recently found on the surface of Mars (Arvidson & Squyres 2005). Nevertheless, in this paper we will deal only with the meteoroids in the terrestrial atmosphere. Mars, in particular, has a good potential of studying meteoroids closer to the asteroid belt but at the moment the data are very scarce.

Though the terrestrial atmosphere is an efficient detector of meteoroids, inferring physical and chemical properties of incoming meteoroids from remote meteor observation is not a trivial task. In this paper I will review the current state of the problem and outline the open questions. I will concentrate on information gained from the optical observation which provide most complex data. Meteor heights, light curves, deceleration and spectra provide the basis for meteoroid studies. At the same time, pre-encounter heliocentric orbits can be derived. In the case that the meteoroid belongs to a meteor shower, it can be directly linked to the parent body of the shower (if known). This way, various objects can be studied: comet 2P/Encke, Jupiter family comets (e.g. 21P/Giacobini-Zinner), Halley type comets (e.g. 109P/Swift-Tuttle and 55P/Tempel-Tuttle), long period comets (e.g. 1861 I Thatcher), as well as some asteroids (3200 Phaethon and 2003 EH1). For the majority of asteroids, however, no meteoroid streams have been revealed.

2. Classification and ablation of large meteoroids

In the following, I will divide the discussion in two parts, taking separately the "large" and "small" meteoroids. There are at least two reasons for this. First, large meteoroids, larger than about one centimeter in diameter, produce bright meteors (fireballs) which are observed by different techniques than normal meteors. More importantly, large meteoroids spend longer time in the atmosphere before being disintegrated and give rise to more phenomena. In contrast to small meteoroids, they can be significantly decelerated, produce more complex light curves, and can drop meteorites in some cases. From the physical point of view, large meteoroids are loosing most of their mass in the continuum flow regime while small meteoroids are subject to free molecule flow regime of transition flow regime (Popova 2005). There are also evidences that asteroidal and cometary meteoroids are represented in different proportions among small and large meteoroids.

2.1. *Classification*

Meteoroids differ very much in their ability to penetrate the atmosphere. An example is given in Fig. 1, where the light curves two very bright fireballs are plotted. The Šumava and Benešov fireballs are among the brightest well observed fireballs. Their initial velocity was not very different (27 and 21 km s^{-1}) and the initial meteoroid mass was of the order of several metric tons in both cases. The behavior in the atmosphere was very different,

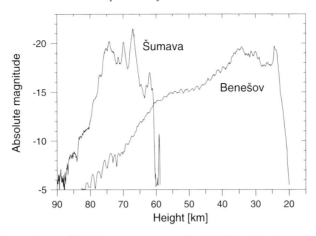

Figure 1. The dependence of brightness on atmospheric height for two very bright fireballs. According to Borovička & Spurný (1996).

nevertheless. Šumava reached the maximum brightness (-21.5 mag) at the height of 67 km, where Benešov was still in process of initial increase of luminosity. The Šumava body was completely destroyed at 58 km, while Benešov radiated below 20 km of height. The difference in terms of air density is by a factor of 400.

The conspicuous differences in fireball end heights form the basis of the classical fireball classification formed by Ceplecha and McCrosky (1976). An empirical PE criterion was created which relates the end height with initial mass, initial velocity and the slope of the trajectory. The fireballs were divided into four groups (or types) which differ greatly in the ablation ability. The most resistant group I was identified with ordinary chondritic or, more generally, stony meteoritic material. This relation is confirmed by the fact that photographed meteorite falls Lost City, Innisfree and Neuschwanstein belong to group I. The first two are ordinary chondrites, the last one is an enstatite chondrite. Group II corresponds to more fragile carbonaceous material. The group IIIA and the most fragile group IIIB are both of cometary origin as evidenced by the fact typical cometary shower fireballs belong to these groups. Perseid fireballs are of type IIIA (Spurný 1995), while most Leonids observed in 1998 were classified as IIIB, although many of them were close to the boundary with IIIA (Spurný *et al.* 2000a). The typical IIIB material is represented by the October Draconids (Ceplecha & McCrosky 1976), which were found to be unusually soft already by Jacchia *et al.* (1950).

The PE criterion is simple and easy to use since only basic geometric and photometric reduction of the data is necessary (note that the original luminous efficiency of Ceplecha & McCrosky 1976 must be used for computing the photometric mass). The classification based on the PE criterion is still valid today. The obvious question, however, is what physical and/or chemical properties of meteoroids cause their different behavior in the atmosphere. To proceed further with this question, we must refer to the so called physical theory of meteors.

2.2. *Ablation and deceleration*

The meteoroid dynamics in the atmosphere is governed by two differential equations – the drag equation and the mass-loss equation (see e.g. Ceplecha *et al.* 1998). The equations contain two independent parameters, the ablation coefficient, σ, defined as $\sigma = \Lambda/(2\xi\Gamma)$, and the shape-density coefficient K defined as $K = \Gamma A \delta^{-2/3}$. Here Λ is the heat transfer coefficient, ξ is the energy necessary for ablation of a unit mass, Γ is the drag coefficient,

Table 1. Average *apparent* ablation coefficients for fireball groups (Ceplecha *et al.* 1993).

Group	Ablation coefficient $[\text{s}^2\,\text{km}^{-2}]$	No. of cases
I	0.02	33
II	0.05	14
IIIA	0.14	3
IIIB	0.6	1

A is the shape coefficient (relating cross-section to volume), and δ is the bulk density of the meteoroid. Assuming that σ and K are constant and the meteoroid does not suffer fragmentation, and knowing the slope of the trajectory, the observed dynamics of the meteoroid can be fitted. The fit provides the initial velocity, v_∞, and the value of σ. The initial mass, m_∞, and K cannot be separated and only the product $Km_\infty^{-1/3}$ is obtained.

The ablation coefficient characterizes the ability of the meteoroid to ablate and can be used to quantify the meteoroid classification. There is strong relation between σ and the meteoroid type according to the PE criterion. Usually, σ is given in $\text{s}^2\,\text{km}^{-2}$, which is equivalent to $\text{kg}\,\text{MJ}^{-1}$. The values of the ablation coefficient computed without considering meteoroid fragmentation are called *apparent* (Ceplecha & ReVelle 2005). In many cases, in particular for well observed type I fireballs, the assumption of no-fragmentation is unsatisfactory since the fireball dynamics cannot be fitted well. This allowed Ceplecha *et al.* (1993) to modify the method for the case of gross-fragmentation, when a sudden mass loss occurs at one point. Taking the gross-fragmentation into account, the resulting ablation coefficients (which describe the mass loss outside fragmentation) decreased. Nevertheless, in Table 1 we present the average *apparent* ablation coefficients for different fireball groups taken from non-fragmentation solutions of Ceplecha *et al.* (1993). These values show the differences between the groups most clearly. Unfortunately, the dynamic method could be used only for few fireballs of groups III, because these bodies suffer only little deceleration.

2.3. *Light curves and the intrinsic ablation coefficient*

To learn more about the meteoroid properties and the process of fragmentation, we can use the information from fireball light curves. According to the physical theory of meteors, the instantaneous meteor luminosity is proportional to the loss of meteoroid kinetic energy (the proportionality factor is called luminous efficiency, τ). In most cases, the energy loss is dominated by mass loss, not deceleration. Fireballs of types IIIA and IIIB often exhibit brief increases of brightness called flares. The flares are caused by the loss of large amount of material in the form of tiny dust particles which evaporate quickly and their energy is radiated out. This was demonstrated directly on an instantaneous photograph showing the correlation of fireballs flare, splitting into two pieces, and formation of long wake made of tiny fragments (Konovalova 2003). Sometimes, the fireball trajectory terminates by a flare caused by complete pulverization of the meteoroid. Fireballs of types I and II do not show bright flares so often. They fragment preferably into macroscopic fragments and smaller amount of dust is released.

Borovička & Spurný (1996) analyzed the light curve of the very bright IIIB fireball Šumava (see Fig. 1). The body of estimated initial mass of 5000 kg and diameter of 4.5 m exhibited five flares, the brightest one having an amplitude of 4 magnitudes, corresponding to 40 fold increase of brightness in 0.1 second. The apparent ablation coefficient of the meteoroid was 0.32 $\text{s}^2\,\text{km}^{-2}$. Nevertheless, the light curve analysis showed that more than 85% of initial mass was lost in five major breakups. During each breakup, only 5%

of lost mass was released in form of macroscopic fragments, the rest was lost in form of dust.

Recently Ceplecha & ReVelle (2005) presented an elaborated fragmentation model (FM) which is able to explain simultaneously fireball dynamics and light curve to amazing details. The parameters σ, K, and τ are taken as variable and two types of fragmentation are considered – into several large fragments and into cluster of small fragments. The controversial aspects of the model formalism is that it allows for "flares" of duration of several seconds and for enormous variations of K, which are difficult to interpret physically. On the other hand, the model for the first time distinguishes the *apparent* and *intrinsic* values of the ablation coefficient and luminous efficiency. The intrinsic values are corrected for the effect of fragmentation and express therefore the meteoroid ablation properties other than fragmentation ability. The surprising result is that the intrinsic ablation coefficient lies the range $0.001 - 0.008\,\mathrm{s^2\,km^{-2}}$ in most fireballs including types IIIA and IIIB and the variations from fireball to fireball do not exceed the variations inside one fireball. Only rarely the intrinsic σ reaches $0.015\,\mathrm{s^2\,km^{-2}}$, which is still less than the typical apparent values. Fragmentation is therefore the dominant process of mass loss for all types of meteoroids. It is more efficient than the evaporation or melting of the meteoroid surface. Moreover, there are no differences in the evaporation properties of meteoroids of various types. The huge differences in the ablation abilities are caused by mechanical properties of the material rather than its composition. This suggests that the cometary meteoroids differ from ordinary chondrites mostly by their bulk density, porosity and mechanical strength.

It is interesting to compare the intrinsic ablation coefficients of stony material obtained empirically by Ceplecha & ReVelle (2005) with theoretical values computed for melting and evaporation of stony meteoroids. Bronshten (1983, p. 123) gives the typical values for large stony meteoroids: $\xi = 8\,\mathrm{MJ\,kg^{-1}}$, $\Lambda = 0.1$, $\Gamma = 0.46$. This gives the ablation coefficient $\sigma = \Lambda/(2\xi\Gamma) = 0.013\,\mathrm{s^2\,km^{-2}}$. However, the heat transfer coefficient, Λ, depends significantly on the meteoroid size, velocity, composition and height of flight. The numerical model of Golub' *et al.* (1996) gives $\Lambda = 0.07$–0.24 for iron meteoroids of radii 0.1–1 m moving at the heights 30–40 km with the velocity 10–20 km s^{-1}. The corresponding values for stony meteoroids should be lower (Bronshten *et al.* 1985). The model of Artemieva & Shuvalov (2001) gives Λ as low as 0.02 and 0.03 for stony body of 1 m radius and 20 km s^{-1} velocity at the heights 50 and 70 km, respectively. If we consider the drag coefficient derived directly from the observation of the Lost City fireball and meteorite fall, $\Gamma = 0.7 \pm 0.1$ (Ceplecha 1996), the possible values of the intrinsic ablation coefficient are between 0.002 and 0.025 s^2 km^{-2}. The observations of fireballs (Ceplecha & ReVelle 2005; Borovička & Kalenda 2003) favor the values closer to the lower edge. Note that small meteoroids in the transition flow regime are expected to have larger σ ($\sigma \sim 0.05\,\mathrm{s^2\,km^{-2}}$) because of larger Λ ($\Lambda \sim 0.8$, $\Gamma \sim 1$). For a recent review of meteoroid ablation models see Popova (2005).

We have not yet included into consideration iron meteoroids, i.e. the nickel-iron metal material which constitutes iron meteorites. This is understandable because iron meteoroids form only few percent of fireballs and they do not appear clearly in a statistical analysis. Irons can be readily identified from their spectra but spectral records are available only for a minority of fireballs. ReVelle & Ceplecha (1994) attempted to identify iron meteoroids from dynamic and photometric data. Because of their large thermal conductivity, irons in the mass range 10^{-3} to 10^5 kg are expected to ablate by melting and to have larger ablation coefficients than chondritic material. The authors found 7 fireballs with smooth light curves, with no dynamically identifiable gross-fragmentation, and with large ablation coefficients (median value 0.05 s^2 km^{-2}), as probable irons. The large ablation coefficients are presumably intrinsic but it would be interesting to analyze those fireballs with the new FM method. According to the end height classification (PE criterion), irons fall into type II fireballs. The lower penetration ability of irons into the atmosphere in comparison with stony meteoroids can be well modeled using material properties of irons (Bronshten *et al.* 1985).

3. Classification and ablation of small meteoroids

3.1. *Classification*

Fainter meteors produced by meteoroids smaller than few centimeters have been classified according to their beginning heights (Ceplecha 1968, 1988). The classification was based on meteors photographed by sensitive Super-Schmidt cameras. These data cover meteoroid masses from about 10^{-5} kg to 10^{-2} kg. The dependence of meteor beginning height on meteoroid mass was found to be not significant and a K_B criterion was created, which relates the beginning height with initial velocity and trajectory slope only. Originally, four groups of small meteoroids have been recognized: A, B, C, D. These groups, however, do not correspond directly to the fireball groups. Group A corresponds to II, C to IIIA, and D to IIIB. The reason for different classifications are different proportions of asteroidal and cometary meteoroids in diferent mass ranges. Asteroidal bodies of group I are virtually absent below 1 gram of mass (Ceplecha 1988). On the other hand, the intermediate group B has no analog among large bodies.

Modern observations of faint meteors with image intensified TV cameras have shown that the meteor beginning height in facts depends on meteoroid mass. Koten *et al.* (2004) studied the beginning heights of meteors of five meteor showers. With the exception of Geminids, they found a definite dependence of the beginning height, h_b, on the meteoroid mass. The dependence could be fitted with a function $h_b = h_0 + k \log m$, with k ranging between 5 and 10 km for Orionids, Taurids, Perseids and Leonids in the mass range 10^{-7} to 10^{-4} kg. Similar dependence with $k = 6$ km was found for 1998 Leonids in the mass range 10^{-8} to 10^{-5} kg by Campbell *et al.* (2000), though the authors considered the dependence as weak. Sarma & Jones (1985) found $k = 5$ and $k = 9$ km, respectively, using two different camera systems on sporadic meteors of the similar mass range. Brown *et al.* (2000) found $k = 9$ km for single station 1999 Leonids.

The remarkable dependence of beginning height on mass does not mean that the meteor groups found by Ceplecha (1968) do not exist. Neither it means that the original data or the analysis were wrong. The Super-Schmidt cameras used blue sensitive X-ray films (Millman 1959), while the image intensifiers have maximum sensitivity in the yellow-green part of the spectrum and their sensitivity extends far into infrared region (e.g. Borovicka *et al.* 1999). By chance, the blue region of the spectrum contains mostly emissions from the evaporated meteoric atoms, while the red and infrared radiation is produced mostly by heated air (Borovicka *et al.* 1999). This may be the basis for different meteor behavior with different detectors. We will discuss meteor begining heights in more detail in Sect. 3.4. In any case, the K_B criterion cannot be used for TV meteors unless the considered mass range is narrow.

3.2. *Ablation coefficients*

Bellot Rubio *et al.* (2002) performed dynamic and photometric analysis of Super-Schmidt meteors. They found that data on 73% of the studied 370 meteors can be fitted with the single body equations, i.e. no fragmentation is needed to explain the data. However, they did not fit directly the length as a function of time but used the interpolated velocities and decelerations from the original catalog. These data were available at only few points along the trajectory for some meteors and photometric data were available at only four points for all meteors. The resulting median ablation coefficients of Bellot Rubio *et al.* (2002) are quite large: $0.07\,\mathrm{s^2\,km^{-2}}$ for C and B groups and $0.11\,\mathrm{s^2\,km^{-2}}$ for A group. Most of C group meteors (85%) could be explained by single body theory, while for A group the percentage was only 44%.

The large ablation coefficients of Super-Schmidt meteors are the result of neglecting fragmentation in my opinion. These are the *apparent* ablation coefficients. The only surprise is the larger median value for the A group than for the B and C groups. The distribution of ablation coefficients was, however, quite wide, from 0.01 to at least $0.20\,\mathrm{s^2\,km^{-2}}$ for all groups.

3.3. *Light curves*

TV observations of meteors provide good light curves, though no deceleration data. The light curves have been a subject of several detailed studies in the recent years (e.g. Murray *et al.* 1999, 2000; Koten *et al.* 2004). These studies concentrated on Leonids and other meteor showers mostly of cometary origin. The light curves of faint meteors are smooth with no flares. Most light curves have only one maximum. On the other hand, the shapes of the light curves vary very much from meteor to meteor, even for meteors of similar brightness belonging to the same shower. The maximum can lie almost anywhere on the light curve. On average, the light curves are nearly symmetrical with the maximum in the middle of the trajectory. These findings contradict the single body theory which produces uniform light curves with gradual increase and rapid decrease of brightness and the maximum at 70% of the trajectory (e.g. Beech & Murray 2003). The observed light curves can be much better interpreted in terms of the dustball model formulated by Hawkes & Jones (1975).

In the dustball model, a cometary meteoroid consists of grains of a high boiling-point material (e.g. stone or iron). The grains are held together by some binding material (glue) of a lower boiling point. The luminosity of the meteor is supposed to be produced only by ablating grains. The grains can start to ablate after they have been released from the dustball, which occurs when the glue surrounding the grain reached the boiling temperature. The grains released high in the atmosphere take some time before they start to ablate and radiate. The grains released deeper in the atmosphere start to radiate very quickly. Meteoroids smaller than certain critical mass release all grains before the grain ablation starts. The main prediction given by Hawkes & Jones (1975) are the meteor heights. The beginning height should be the same for all meteors of given velocity and composition, while the end height and the height of maximum brightness should be constant for meteoroids smaller than the critical mass and should decrease with mass for larger meteoroids.

Beech & Murray (2003) computed synthetic light curves of dustball meteors which release all grains before the onset of ablation. A power law distribution of the grain masses was assumed. More specifically, a $10^{-6}\,\mathrm{kg}$ Leonid meteoroid was assumed to be composed of grains in the mass range 10^{-7} to $10^{-10}\,\mathrm{kg}$. The synthetic light curve was simply the sum of light curves of individual grains. Light curves of various shapes were obtained by varying the mass distribution index, α. For α close to 2, early peaked light curves were obtained (maximum at 35% of the trajectory). For smaller and larger α the light curve approached that of a single body. A few observed double peaked light curves could also be explained by adding a large mass grain to otherwise power law distribution (see also Murray *et al.* 1999).

A similar but more elaborated model was presented by Campbell-Brown & Koschny (2004). Their model allows for gradual release of grains from the dustball and can be

used for direct fitting of observed light curves. The mass distribution of grains is assumed
to be either Gaussian or power law or a combination of both. The temperatures at which
the grains were released in three Leonid meteors were found to lie between 1000–1150 K.
In addition, Koschny et al. (2002) presented a model with Poisson distribution of grain
sizes. The model can generate light curves with quite early maxima.

Another independently developed concept is the analytical model of quasi-continuous
fragmentation (Babadzhanov 2002 and references therein). There are several processes
possible which can cause quasi-continuous separation of small fragments from the meteoroid during the flight – the release of grains from a relatively large dustball is one
of them. Babadzhanov (2002) fitted the light curves of relatively bright photographic
meteors (meteoroid masses $> 10^{-5}$ kg) with this model. 111 out of 197 meteors could be
fitted, the remaining light curves (44%) did not conform with the QCF model. The light
curves of quasi-continuously fragmenting meteoroids are not very different from single
body light curves. Both observations and theory show that large meteoroids have light
curves more similar to a single body. For example, the maxima of all Perseids larger than
10^{-5} kg lie in the second half of the light curve (Koten et al. 2004). This does not mean
that the structure of larger meteoroids is different than smaller ones. Larger meteoroids
are simply not disrupted completely before the grain ablation starts.

Statistically, there are differences among meteoroids of different streams. The average
position of Leonid maximum brightness is in the middle of the light curve. Geminids of
similar mass range reach their maxima at 58% of the trajectory (Koten et al. 2004). This
can mean that Geminids, on average, are less fragile than Leonids. Nevertheless, as noted
above, there is significant spread of light curves within one meteor shower.

3.4. Beginning heights

One prediction of the dustball theory, namely that the meteor beginning height is independent on meteoroid mass, could not be confirmed (see Sect. 3.1). Only for Geminids
is the beginning height nearly constant (Koten et al. 2004). The increasing beginning
heights of Leonids gradually merge into the extreme beginning heights of bright Leonid
fireballs (Koten et al. 2006). The beginnings of Leonid fireballs were found to lie substantially higher when observed with TV systems than photographically observed beginnings
(Fujiwara et al. 1998) and can reach 200 km (Spurný et al. 2000a). There is a continuous
increase of Leonid beginning heights from 110 km for meteoroids of 10^{-7} kg to 200 km
for meteoroids of 1 kg. The high beginnings are not restricted to Leonids but occur also
in other fireballs (Koten et al. 2001, 2006; Spurný et al. 2005), though up to now they
have been observed only in fast fireballs of cometary origin.

The appearance of meteors above the height of 130 km differs from that at lower
heights. Above 130 km the meteors are very diffuse, the luminous volume is several km
wide and shows irregular structures (Spurný et al. 2000b; LeBlanc et al. 2000). Also the
light curve shows irregularities (Spurný et al. 2000b; Koten et al. 2006). The high altitude radiation was originally considered as enigmatic. New studies, however, show that
both the total luminosity and the size of the radiating volume can be explained (Ceplecha
& ReVelle 2005; Popova et al. 2005a; Vinković 2005). The physical mechanism responsible for the effect is sputtering of meteoroid surface by incoming atmospheric atoms
and molecules, as first suggested by Brosch et al. (2001) and described in more detail by
Rogers et al. (2005). This ablation mechanism works for fast meteors before the meteoroid surface reaches boiling temperature. Sputtering is negligible for velocities lower than
30 km s^{-1} (Popova et al. 2005a). At higher velocities, sputtering works for chondritic material but is more efficient for meteoroids containing more volatiles. The luminous volume
is formed by cascade collisions of the sputtered atoms with surrounding atmospheric

species (Vinković 2005). The main contributor to the radiation is oxygen triplet at 777 nm (Popova *et al.* 2005a; Spurný *et al.* 2005). The interaction of the sputtered atoms with the atmosphere can be perhaps more effectively traced by the produced ionization. Brosch *et al.* (2001) detected meteor echoes at heights > 200 km with a powerful phased-array radar during Leonid maxima.

Koten *et al.* (2004) showed that the increase of beginning heights with meteoroid mass can be explained if the beginning height is given by the limiting sensitivity of the instrument and the ablation and radiation in fact starts earlier than the meteor is detected. Indeed, Campbell-Brown & Koschny (2004) predicted a gradual brightening of the meteor after they abandoned the concept of boiling temperature in their model and used the Clausius-Clapeyron equation to calculate the mass loss. The increasing meteor beginning heights are therefore not in contradiction with the dustball model, only with its part which predicts that meteor radiation starts only after the grains reach their boiling temperature. The crucial question is whether the early ablation needs a significant volatile component to be present in meteoroids. In theory, both sputtering and the Clausius-Clapeyron equation allows high altitude ablation of pure stone. On the other hand, a volatile component (e.g. the glue in the dustball model) will enhance the ablation rate and the early meteor luminosity, even if the volatiles contribute to the radiation only indirectly by the collisions with the air. From the observational point of view it is significant that Geminids do not show beginning increase with meteoroid mass (Koten *et al.* 2004). One may argue that this due to their relatively low velocity (35 km s^{-1}). However, even slower Taurids do show the increase, though it is less pronounced than for 71 km s^{-1} Leonids (Koten *et al.* 2004). It seems therefore likely that not only the high velocity sputtering but also gradual evaporation of a volatile component contributes to the luminosity of cometary meteors before the onset of regular grain ablation.

3.5. *Differential ablation*

The situation is complicated by the fact that even the ablation of regular chondritic material may not be uniform. Thermodynamic equilibrium calculation shows that fractionation occurs during melting and vaporization (McNeil *et al.* 2002; Schaefer & Fegley 2005). More volatile atoms, in particular Na, vaporize earlier than Mg, Fe, and Si, while refractory Ca vaporizes later on and not fully. This fact was used to create a model of meteor differential ablation (McNeil *et al.* 1998), which was used to explain much higher abundance of Na than Ca in atmospheric metal layers. The model received some support from probing of fresh meteor trails by atmospheric lidars (von Zahn *et al.* 1999, 2002). The lidars rarely saw two or three elements simultaneously at the same position in the meteor trails. If they did, the observed elemental ratio was different from the expected chondritic value. The differences were larger for smaller meteoroids. The observations were done for K, Fe, and Ca, which have very different volatility.

Differential ablation should be detectable by optical spectroscopy of faint meteors. Sensitive low resolution TV spectrographs detect bright lines of Na and Mg and fainter lines of Fe and Ca emitted by the material vaporized from the meteoroid (Borovicka *et al.* 1999, Abe *et al.* 2000). Potassium is not easily detectable. In Leonids, the Na line was often starts and ends earlier than the Mg line Borovicka *et al.* 1999). This is exactly what the model of differential ablation predicts. However, the effect of early release of Na varies from meteor to meteor and although it is present to some degree in the majority of Leonids, in some cases it is absent (Fig. 2). The situation is more complex if sporadic meteors and other showers are considered. In Taurids, for instance, Na closely follows Mg (Borovička 2001). On the other hand, an excellent example of early Na release was

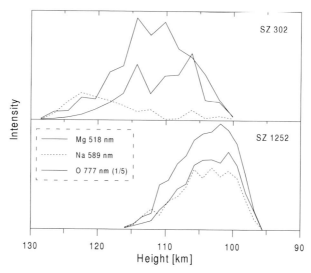

Figure 2. Monochromatic Mg, Na, and O light curves of two Leonid meteors of similar brightness (~ 0 mag). The oxygen line intensity was divided by 5 for better comparison. The heights are approximate.

observed in a photographic Draconid spectrum (Millman 1972). Draconids belong to the slowest meteor showers, so the effect is not a function of velocity.

The effect of early sodium release is not universal and must depend on meteoroid structure. It is reasonable to suppose that the quick evaporation of Na from the whole volume requires initial disruption of the millimeter-sized meteoroid – in accordance with the dustball model. Sodium can be part of the glue. Nevertheless, it can be also part of the constituent grains. For small grains, the complete fractionation and Na evaporation from the whole volume is probably possible. In any case, the early Na release is an indication of meteoroid disruption well before it was heated. The differences of the Na behavior in different meteors are probably due to different disruption height. Detailed correlation of spectral data with other meteor diagnostics – light curve shapes, beginning heights, end heights etc. – is desirable for better understanding of the behavior and structure of small meteoroids.

The differential ablation of calcium will be discussed in Sect. 7.

4. Bulk densities

Meteoroid bulk density is a parameter of large interest. The different penetration ability of asteroidal and cometary meteoroids in the atmosphere may be primarily due to their different densities and it is desirable to know the absolute values of the densities. The derivation of meteoroid density from meteor data is, however, not straightforward and various authors used various approaches.

As noted in Sect. 2.2, the fireball dynamic analysis provides the product $Km^{-1/3}$. If the light curve is taken into consideration and a correct luminous efficiency, τ, is employed, the mass, m can be separated from the shape-density coefficient, K. K itself is a combination of the drag coefficient, Γ, shape coefficient, A, and bulk density, δ. In their most recent model, Ceplecha & ReVelle (2005) were reluctant to compute δ since, strictly speaking, Γ and A are unknown. In their model, K showed large variations and it is difficult to say to which quantity the variations should be ascribed. In fact, the variations of K could also be caused by the inappropriateness of the used τ for the given fireball.

Table 2. Average bulk densities for different meteoroid groups (Ceplecha *et al.* 1993).

Group	Bulk density [g cm^{-3}]
I	3.7
II	2.0
IIIA	0.75
IIIB	0.4

Table 3. Average bulk densities for different meteors showers as determined by two authors.

Shower	Bulk density [g cm^{-3}]	
	Babadzhanov (2002)	Bellot Rubio *et al.* (2002)
Geminids	2.9	1.9
δ Aquarids	2.4	
Quadrantids	1.9	0.8
Taurids	1.5	0.4
Perseids	1.3	0.6
α Capricornids		0.45
Leonids	0.4	

In the earlier gross-fragmentation model, Ceplecha *et al.* (1993) assumed K to be constant and the light curve was not modeled in detail, only the initial mass was computed from the light curve. The resulting average bulk densities for meteoroids of different groups generally confirmed the earlier estimations (Ceplecha 1988) and are given in Table 2. Note that the IIIA and IIIB values are based on one fireball only for each group. They are nevertheless close to the earlier estimations (0.75 and 0.27 g cm^{-3}, respectively). The same gross-fragmentation model was also applied to several Geminid fireballs (Ceplecha & McCrosky 1992) resulting in an average density of 3.3 g cm^{-3} (assuming $\Gamma A = 1.1$). Geminid meteoroids therefore belong to group I, which was also demonstrated on the basis of their apparent ablation coefficients (Spurný 1993). One of the Leonid fireball observed by Spurný *et al.* (2000a) showed measurable deceleration. The bulk density was estimated to 0.7 g cm^{-3} (again for $\Gamma A = 1.1$), close to the IIIA average. The apparent ablation coefficient was 0.16 s^2 km^{-2}. The modeling of the light curve and deceleration of the very bright IIIB Šumava fireball yielded the most probable bulk density as low as 0.1 g cm^{-3} (Borovička & Spurný 1996).

The modeling of Super-Schmidt meteors by Bellot Rubio *et al.* (2002) described in Sect. 3.2 yielded also meteoroid bulk densities. They were computed from the resulting K coefficient assuming $\Gamma = 1$ and $A = 1.21$ (spherical shape). The resulting densities for individual meteoroids show large scatter, from 0.1 g cm^{-3} to 4.5 g cm^{-3}. The mean density was 2.4 g cm^{-3} for A-group meteoroids, 1.4 g cm^{-3} for group B, and 0.4 g cm^{-3} for group C. The authors also computed the mean bulk densities for different meteor showers (Table 3). As noted in Sect. 3.2, the work of Bellot Rubio *et al.* (2002) suffered from relative paucity of data points for individual meteors and from neglecting the fragmentation in the model. We consider their densities underestimated, in particular for group C.

Various attempts were published to determine meteoroid bulk densities using the model of quasi-continuous fragmentation (see Sect. 3.3). The results are sensitive on the tuning of various parameters of the model. As described in Bellot Rubio *et al.* (2002), earlier works used too low specific energy of fragmentation resulting in too high densities. Babadzhanov (2002) used a more realistic energy (2×10^6 J kg^{-1}) and obtained more reasonable densities (Table 3). Still, he had to assume several parameters (shape, heat transfer coefficient, grain density) and he fitted only the light curves, not decelerations of meteors. Similarly, Konovalova (2003) roughly fitted light curves of several Taurid fireballs though deceleration data were available. She obtained average Taurid density of $2.5\,\mathrm{g\,cm^{-3}}$ and fragmentation energy of 5×10^5 J kg^{-1}.

As it can be seen from Table 3, the densities of Bellot Rubio *et al.* (2002) are systematically lower than those of Babadzhanov (2002). Both authors, nevertheless, agree that Geminid stream contains the densest meteoroids of all streams. We consider the absolute Geminid value of Babadzhanov (2002) more realistic since it is close to that of Ceplecha & McCrosky (1992).

In summary, the measured meteoroid bulk densities cover a wide range from about $0.1\,\mathrm{g\,cm^{-3}}$ to $3.7\,\mathrm{g\,cm^{-3}}$. Iron meteoroids with densities of ~ 7 g cm^{-3} certainly also exist but they have not been directly measured. Although the derivation of meteoroid density from meteor data is difficult and involves number of assumptions, I consider the obtained density range as realistic. It correspond to the range of bulk densities covered together by ordinary chondrites, carbonaceous chondrites and interplanetary dust particles (IDP's) (Rietmeijer & Nuth 2000). There are no evidences of significant differences in composition of meteoroids of different bulk densities. As in the case of IDP's, the density differences are most probably due to differences in porosities. The porous aggregate IDP's have a mean density of only $0.1\,\mathrm{g\,cm^{-3}}$ and porosity of 95% (Rietmeijer & Nuth 2000). As shown by the Šumava meteoroid, similar values are possible for bodies up to several meters in diameter. These considerations correspond with the work of ReVelle (2001), who argued that the differences in penetration ability and apparent ablation parameters among meteoroids of different types can be naturally explained by their different porosities. He derived the following porosities and bulk densities: 0% (3.7 g cm^{-3}) for group I, 50% (1.85 g cm^{-3}) for group II, 75% (0.93 g cm^{-3}) for group IIIA, and 91% (0.34 g cm^{-3}) for group IIIB.

5. Mechanical strength

The front surface of a meteoroid passing through the atmosphere with a velocity v is subject to a dynamic pressure $p = \rho v^2$, where ρ is the density of the atmosphere. More precisely, the pressure is $p = \Gamma \rho v^2$ but the drag coefficient Γ is not precisely known and is of the order of unity, so we ignore it. The pressure on the rear surface is nearly zero. If the front pressure, p, exceeds the mechanical strength of the meteoroid, meteoroid break-up occurs. Depending on the internal structure of the body, the break-up can lead to the formation of two macroscopic fragments or to the cloud of tiny dust particles of to something in between. As shown above, there are various methods to reveal different kinds of meteoroid fragmentation: meteor flares, sudden changes in the dynamics, direct imaging of the fragments. As soon as the fragmentation height is determined, it is easy to compute the dynamic pressure at the moment of the fragmentation, since meteor velocity as a function of height is easily measurable.

We have to note that the fragmentation understood in this section is different from the grain release after the disruption of a small dustball meteoroid or the quasi-continuous fragmentation described in Sect. 3. Those processes were driven by the heating of the whole meteoroid or its surface. In this section, the fragmentation is caused by mechanical forces and affects hitherto unaffected meteoroid interior.

5.1. *Cometary meteoroids*

It is not surprising that low dynamic pressures are sufficient to break up porous cometary bodies. The five big disruptions of the Šumava meteoroid occurred under the pressures between 0.025 and 0.14 MPa (Borovička & Spurný 1996). The fragmentation sequence could be described by a simple concept of destruction depth. Under this concept, the dynamic pressure acting on the surface of the meteoroid causes at some moment the destruction of the meteoroid to the depth d, which is proportional to the pressure p. The destroyed part is pulverized while the rest of the body remains more or less intact, suggesting low cohesivity of the material. This process repeats quasi-periodically until d exceeds meteoroid radius and the body is destroyed completely, leaving only few small fragments to continue. The fragments are finally destroyed in the next step. This pattern may be general for IIIB meteoroids with the exception that smaller bodies do not have enough mass for such complex behavior as Šumava and only one flare may be present.

Taurid fireballs were observed to break-up under 0.05–0.18 MPa (Konovalova 2003). The onset of terminal flares of Leonid fireballs typically occurs between the heights 95–85 km (Spurný *et al.* 2000a; Borovička & Jenniskens 2000). This corresponds to $p = 0.007$–0.04 MPa. However, the effective lower limit of the strength of Leonid meteoroids may be close to zero. This statement is based on the observation of meteoroid clusters within the Leonid shower which were shown to be products of meteoroid fragmentation in the interplanetary space few days before the encounter with the Earth (Watanabe *et al.* 2003).

On the other hand, there are evidences that Leonid meteoroids contain also much stronger ingredients. In some cases, small part of the original mass does not participate in the 'terminal' flare and continues the flight deeper in the atmosphere. Examples of Leonid fireballs containing such a strong fragment are the LN98023 fireball of Spurný *et al.* (2000b) and the fireball analyzed by Borovička & Jenniskens (2000). The former fragment penetrated to 73 km of height and experienced the pressure of 0.2 MPa, the latter penetrated down to 56 km and survived the pressure close to 2 MPa, which is quite a lot even in comparison with group I meteoroids. The surviving fragment comprised only a tiny part of the original mass ($\sim 10^{-6}$) and its spectrum did not revealed any anomaly in the chemical composition. The size was of the order of 1 mm. Swindle & Campins (2004) speculated whether this is an evidence for presence of chondrules in cometary matter but considered it unlikely. Rather, this can be regarded as a piece of evidence for the opinion expressed by various authors (see e.g. Lodders & Osborne 1999) that cometary nuclei contain also compact material analogous to (or identical to) CI and CM carbonaceous chondrites.

There are also other evidences for the existence of compact cometary material of different kind. Among the fireballs photographed by the European Fireball Network, the Karlštejn fireball (Spurný & Borovička 1999a,b) is unique. The cometary origin is evident from the retrograde orbit (inclination = 139°), though the semimajor axis was unusually small ($a = 3.5$ AU). The fireball, however, penetrated much deeper than other cometary fireballs of similar velocity and mass and was classified as type I fireball. There is no flare on the light curve and the meteoroid survived dynamic pressure of 0.7 MPa, which is the lower limit of its mechanical strength. The meteoroid density was likely at least 2 g cm^{-3}, the mass was of the order of hundreds of grams and the size reached several centimeters. The fireball spectrum revealed that the meteoroid was completely free of sodium, unlike normal cometary meteoroids. Similar Na-free meteoroids have been later found among small millimeter sized meteoroids on cometary orbits (Borovička *et al.* 2005). These meteoroids are possibly remnants of cometary irradiation crusts that formed during the comet residence in the Oort cloud. Meteor showers do not contain this material because their parent comets have lost their primordial irradiation crust long time ago. Centimeter-sized crust fragments are much rarer than the smaller ones but the Karlštejn meteoroid is a good example of their existence.

5.2. Asteroidal meteoroids

The fragmentation of stronger bodies of groups I and II can be better studied by geometric and dynamic methods because the break-ups are not always accompanied by sizeable flares. Ceplecha *et al.* (1993) applied the dynamic method to 51 fireballs with sufficiently precise photographic records. The method was able to find fragmentation events where more than ~50% of mass was lost. Nearly 60% of fireballs of groups I and II showed at least one fragmentation. (The model of Ceplecha & ReVelle (2005), which takes into account all details on the light curve, often detects a dozen of small fragmetations for one fireball.) The dynamic pressures at fragmentation indicated some grouping around the values of 0.08, 0.25, 0.53, 0.8, and 1.1 MPa, which allowed Ceplecha *et al.* (1993) to define five strength categories, 'a' to 'e'. Three meteoroids survived 1.5 MPa without fragmentation and one meteoroid survived 5 MPa.

The work of Ceplecha *et al.* (1993) concerned ordinary fireballs corresponding to meteoroid sizes up to few decimeters. It is of interest to look at the largest bodies for which data are available, with the sizes of the order of one meter. Large bodies can fragment extensively – dozens of individual fragments can be seen on still photographs and videos of the Peekskill (Brown *et al.* 1994) and Morávka (Borovička & Kalenda 2003) meteorite falls. The fragmentation can occur in several stages. The significant deceleration of the large meteoroids Benešov (Borovička *et al.* 1998) and Morávka (Borovička & Kalenda 2003) at the heights of ~45 km is the evidence for the fact that they have been already disrupted into a dozen of fragments of similar mass (100–300 kg) at that height. The disruption height could not be determined from the data, so only the upper limit for the dynamic pressure can be set: $p < 0.5$ MPa. This limit, nevertheless, is in full accordance with the values for smaller meteoroids (Ceplecha *et al.* 1993). The primary fragments of Benešov and Morávka continued a cascade fragmentation at lower heights. The dynamic pressures at that break-ups were up to 5 MPa for Morávka and up to 9 MPa for Benešov. Not all large bodies, nevertheless, are subject to the initial high altitude disruption. The ~1.3 meter EN 171101 meteoroid (Spurný & Porubčan 2002) survived intact until the pressure of 4 MPa and the daughter fragments later experienced 14 MPa without fragmentation, before reaching the record low terminal height of 13 km.

Popova *et al.* (2005b) recently compiled the fragmentation data of the nine instrumentally observed meteorite falls and compared them with the published strengths of various meteorites. All nine meteoroids experienced fragmentation during the flight. The strongest bodies were Neuschwanstein (10 MPa) and Příbram (14 MPa, uncertain value). The strengths of meteorites measured in the laboratory are significantly larger – the average tensile strength is 30 MPa (range 2–60 MPa) and the compression strength, which is more relevant for atmospheric fragmentation, is 200 MPa on average (20–450 MPa). This confirms that the recovered meteorites represent the strongest structural parts of the incoming meteoroids. The meteoroids, as asteroidal fragments, certainly experienced many collisions in their lifetime and contain cracks and other structural weakness. Most of the meteoroids therefore break easily apart under the dynamic pressures of 0.1–1 MPa. Only the strongest ones can resist up to the pressures of about 5 MPa. The meteoroid strength seems to be not a function of meteoroid mass, rather it depends on individual history of each body.

Of course, the more massive is the body, the more complex can be the atmospheric fragmentation. The most detailed fragmentation study was performed for the Morávka meteoroid (Borovička & Kalenda 2003). The parameters of 15 fragmentation events observed at the heights of 32–24 km were determined from geometric and dynamic data. The complete disruption (mass loss larger than 90%) was not observed. On the other

Table 4. Meteoroid mechanical strengths inferred from fireball fragmentation data

Fireball(s)	Corresponding material/ parent body	Mechanical strength range [MPa]
Šumava	weak cometary material, IIIB	0.025 – 0.14
Taurids	comet 2P/Encke	0.05– 0.18
Leonids	comet 55P/Tempel-Tuttle – bulk material	0 – 0.04
Leonids	– strong constituents	$\geqslant 2$
Karlštejn	primordial cometary crust (?)	$\geqslant 0.7$
type I fireballs	ordinary chondrites	0.08 – 14
Tagish Lake	carbonaceous chondrite, D-type asteroid	0.25 – 0.7

hand, the sum of mass of continuing fragments was always markedly lower than the mass of the original body, indicating that part of mass was always lost in form of dust. One unexplained fact is that successive fragmentations occurred under lower dynamic pressures than the previous events. In addition,

The fragments of the Morávka meteoroid gained lateral velocities up to $300 \, \text{m s}^{-1}$ (Borovička & Kalenda 2003) – an order of magnitude more than the aerodynamic theory (Artemieva & Shuvalov 2001) predicts. The same fact was observed for fragmentation of cometary Taurid meteoroids at much larger heights (65 km) by Konovalova (2003). The Taurid lateral velocity reached 130 m s^{-1}. As suggested by Konovalova (2003), the large lateral velocities could be explained by an explosive nature of the fragmentations, triggered by the exposition of explosive magnesium dust in the atmosphere. The rotational explanation of the fragmentation and fragment dispersion was also proposed , in particular for the Peekskill meteoroid (Beech & Brown 2000), but I consider this less probable because of high rotation rate involved (see Sect. 6).

Among the observed meteorite fall, there is also the unique carbonaceous chondrite Tagish Lake. The bulk density of the meteorite, $1.7 \, \text{g cm}^{-3}$, is the lowest of all known meteorites (Zolensky *et al.* 2002). The material survived the atmospheric entry only owing to the large initial mass of the meteoroid, estimated to 56 tones and corresponding to diameter of 4 meters (Brown *et al.* 2002). The first significant fragmentation occurred under the dynamic pressure 0.25 MPa, the main break-up started at 0.7 MPa. The latter value probably corresponds to the compressive strength of the material (Brown *et al.* 2002). In other words, the fragmentation – unlike ordinary chondrites – cannot be ascribed to the presence of fractures. The meteoroid was in fact disrupted into thousands of pieces in this event. The accompanying flare and dust cloud created in the atmosphere are evidences for enormous dust release. Note that the Tagish Lake meteorite is spectrally similar to D-type asteroids (Hiroi *et al.* 2001).

The data presented in this section are summarized in Table 4.

6. Rotation

The classical equations of meteoroid motion through the atmosphere (e.g. Ceplecha *et al.* 1998) assume that meteoroid cross section decreases monotonically following the decrease of mass. In case of a non-spherical rotating meteoroid, there will be periodic variation of the head cross-section and this will induce periodic variation of both the mass loss rate and the atmospheric drag. In consequence, periodic terms in meteor light curve and decelerations could be observed. In addition, rapid rotation of a meteoroid (even spherical) will delay the onset of intensive evaporation because of slower heating of the surface and can, in theory, lead to meteor beginning heights by 10 km lower in comparison with non-rotating case (Adolfsson & Gustafson 1994). This mechanism does not apply for very small meteoroids which are heated isothermally.

Periodic variation of fireball brightness are really observed from time to time in the form of rapid flickering. The observational data have been summarized by Beech & Brown (2000). The flickering occurs in about 4% of bright fireballs. Typically, the flickering is seen at only a fraction of the trajectory, sometimes near the fireball beginning, sometimes near the end. There is no obvious correlation of height of flickering with meteoroid type or size. The reported flickering frequencies vary from as low as few Hz to as high as 500 Hz. In some fireballs the frequency is constant but often it grows rapidly with time. The amplitudes are fairly constant and less than one magnitude. Beech & Brown (2000) and Beech (2002) interpreted the flickering as a demonstration of meteoroid rotation. On the contrary, Babadzhanov & Konovalova (2004) studied high frequency (> 100 Hz) flickering of three Geminids and discarded the rotation hypothesis on the basis of the observed beginning heights and the fact that flickering started suddenly in the middle of the trajectory. They proposed an autofluctuation mechanism of ablation as the explanation of high frequency flickering. The question of the nature of fireball flickering remains open. I consider a fluctuating ablation mechanism as more likely explanation, at least for fast flickering of large meteoroids. Beech (2001) fitted the light curve of the Innisfree fireball (meteorite fall) with a low frequency (2.5 Hz) periodic term but the original photometric data are sparse and do not indicate periodicity, so this case is not convincing.

Possibly the most reliable value of meteoroid rotation is available for the thoroughly investigated Lost City fireball and meteorite fall (Ceplecha 1996; Ceplecha & ReVelle 2005). The elongated meteoroid of the size 36×17 cm rotated with the period of 3.3 ± 0.3 seconds. This result was obtained by Adolfsson by studying the details of fireball dynamics, namely the residuals in the length as a function of time after applying the fragmentation model. Though the details have not been published, the method seems to be robust.

7. Chemical composition

The derivation of elemental abundances in meteoroids can be in principle done from meteor spectroscopy. Meteor spectra contain emission lines produced by the atoms evaporated from the meteoroid. The derivation of abundances poses some obvious problems. First, estimation of ionization and excitation conditions in meteor plasma is needed for converting line intensities into elemental abundances. Second, the number of observable elements is restricted by the used wavelength region and the quality of the spectrum. Third, because of the effects of differential ablation and incomplete evaporation (Sect. 3.5), the composition of the vapor at a given trajectory point may not reflect the bulk composition of the meteoroid.

The general picture of the excitation conditions in meteor plasma was obtained by Borovička (1993, 1994). Meteor radiation can be fitted by two spectral components, the main component of medium temperature (typically 4000–5000 K), and a high temperature component ($\sim$10,000 K). Both components are assumed to be in thermal equilibrium, at least for larger meteoroids at lower heights (< 100 km). This assumption explains the line intensities reasonably well and was also confirmed theoretically by direct Monte Carlo simulation of a 1 cm Leonid at a height of 95 km (Boyd 2000).

To determine the composition of the radiating vapors, it is desirable to fit the line intensities with thermal equilibrium model and determine the excitation temperature and column densities of observed species. In the next step, it is necessary to estimate the electron density and ionization degree of different elements. The procedure is best applied to high resolution photographic spectra of bright fireballs (Borovička 1993, 2005). Modern CCD detectors can provide comparable resolution but lower wavelength coverage (Jenniskens & Mandell 2004, Jenniskens 2005). Kasuga *et al.* (2005a,b) were courageous enough to apply the procedure to low resolution TV spectra. Note that Kasuga *et al.* (2004) did not apply the ionization correction and they results cannot be considered.

7.1. *Large meteoroids*

The spectra of bright fireballs do not show large diversity in chemical composition of meteoroids. Iron meteoroids have been detected (Halliday 1960; Ceplecha 1966) but they are rare ($\approx 1\%$ of bright meteors). A brief survey of 53 spectra from the Ondřejov archives revealed, besides the Ceplecha's (1966) iron, only one probable diogenite (Borovička 1994b). It is worth to mention that the heliocentric orbit put the diogenite almost exactly in the 3:1 resonance with Jupiter, consistent with the delivery from the Vesta asteroid family (Binzel & Xu 1993). Since then, the Ondřejov observation yielded the unique Karlštejn meteoroid deficient in sodium (Spurný & Borovička 1999b, see also Sect. 5.1), one possible eucrite (unpublished), and a fireball defficient in Mg and Ca (Borovička 2005b). A more extensive survey of fainter photographic meteors (Harvey 1973) yielded fewer than 10% of spectra suggesting anomalous (non-chondritic) chemical composition.

For more detailed consideration, quantitative analysis is needed. Borovička (1993) analyzed an excellent fireball spectrum at 43 points between the heights 57–35 km. The composition of the radiating gas varied significantly along the trajectory. The refractory elements Al, Ti, and Ca showed the most pronounced variations. Their abundances increased at lower heights and in flares but never reached the chondritic values (when compared with Fe or Mg). The effect was even more pronounced in the deeply penetrating Benešov fireball (Borovička & Spurný 1996), where the relative Ca abundance increased by more than two orders of magnitude between the upper (78 km) and lower (24 km) part of the trajectory. The chondritic value was reached in the lower part. It is evident that refractory elements are evaporated incompletely in slow fireballs, unless the fireball reaches dense layers of the atmosphere. For high speed fireballs on Halley-type orbits, various degree of completeness of evaporation of Ca was found (Fig. 3). The incomplete evaporation of refractories is in agreement with thermodynamical calculations (McNeil *et al.* 2002; Schaefer & Fegley 2005) and can explain low abundances (or non-detection) of Ca in meteor trails observed by lidars (von Zahn *et al.* 1999, 2002).

In Fig. 3 the abundances obtained by different authors are compiled. If values at different points along the trajectory were published, the part with most complete evaporation of refractories was taken. All included spectra belong to the 'normal' ones which suggest nearly chondritic composition at the first sight (one spectrum of Trigo-Rodriguez *et al.* (2003) with low Mg abundance was omitted). The quantitative results for cometary meteors on Halley type orbits (mostly Perseids and Leonids) show an interesting pattern. Fe, Cr, and Mn are depleted and Si and Na enhanced relative to Mg when compared with CI abundances. The same trend was measured by mass spectroscopy of the dust of comet Halley (Jessberger *et al.* 1988). The trend, except for the Na enhancement, is less pronounced for cometary meteors on ecliptical orbits. Unfortunately, the scatter of data is large. Trigo-Rodriguez *et al.* (2003) concluded (on basis of Mg-Fe-Si ratio) that cometary meteoroids have CI composition rather than that of Halley dust, while Borovička (2005a) claimed the opposite for Leonids and Perseids. Nevertheless, low Fe/Mg ratio was seen in Leonids also by Abe *et al.* (2005) and Kasuga *et al.* (2005a), and for smaller meteoroids by Borovička *et al.* (2005).

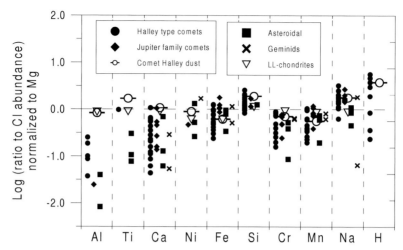

Figure 3. Elemental abundances derived from fireball spectra expressed as deviations from CI abundances, normalized to Mg. The fireballs are divided according to their origin. The volatility of the atoms increases from left to right. The *in situ* measured abundances of the dust of comet 1P/Halley (Jessberger *et al.* 1988) and of LL ordinary chondrites (Wasson & Kallemeyn 1988) are plotted for comparison. CI abundances and volatilities taken from Lodders (2003). Fireball data taken from Borovička (1993), Borovička & Spurný (1996), Borovička & Betlem (1997), Trigo-Rodriguez *et al.* (2003), Kasuga *et al.* (2005a,b), and Borovička (2005a,b).

The difficulty of the interpretation is stressed by the fact that asteroidal fireballs also do not show chondritic composition. The abundances for LL chondrites are also given in Fig. 3 to show the range of chondritic values. The Benešov fireball (Borovička & Spurný 1996) seems to be very Mg-rich. The other two asteroidal bodies show nearly chondritic Fe/Mg and Na/Mg ratios. Cr, however, is underabundant, though to less extent than in Halley-type cometary meteoroids. The fact that Cr shows systematic variations along the fireball trajectory (Borovička 1993) suggests that an unidentified process can cause depletion of atomic Cr in the radiating gas.

The possibility that the composition of Halley dust is representative also for large cometary meteoroids remains open but the decision lies at the boundary of the resolution of present spectroscopic methods.

7.2. *Small meteoroids*

A survey of spectra of meteor produced by small (millimeter sized) meteoroids was published by Borovička *et al.* (2005). Small meteoroids were found much more diverse than large ones. Based on the intensities of the Mg, Na, and Fe lines, only a minority of sporadic meteoroids have chondritic composition. The large diversity of small meteoroids in comparison with larger ones may reflect real inhomogeneity of comets on millimeter scale. Alternatively, environmental effects such as solar wind can alter small meteoroids.

Three populations of Na-free meteoroids were found. The first population consists of iron meteoroids originating in the main asteroid belt, probably related to iron meteorites. This population, surprisingly, forms majority of asteroidal meteoroids in this size range. Millimeter sized ordinary chondrites are rare, which was also found from non-spectral study by Ceplecha (1988). The second population of Na-free meteoroids consist of meteoroids on orbits with small perihelia ($q \leqslant 0.2$ AU). The loss of Na was caused by thermal processes in the vicinity of the Sun and was accompanied by general compaction of the body. The meteoroids of the δ Aquarid stream belong to this population. Geminids, which also have small perihelia, show variable content of Na (see also Fig. 3). This is likely caused by varying ages of the meteoroids. The third population of Na-free meteoroids resides on Halley type orbits. These bodies are probably remnants of cometary irradiation crust and are related to the Karlštejn meteoroid (Spurný & Borovička 1999b; Sect. 5.1).

Fe-poor meteoroids with normal content of Na were also identified. All of them have low material strength and come from comets. Even cometary meteoroids classified as normal, in particular Leonids, showed partial depletion of Fe. Sporadic meteoroids on Halley type orbits are, nevertheless, much more diverse than shower meteoroids. The content of Fe and Na varies widely in cometary meteoroids. The reasons are not quite clear but may include various ages, differences among parent comets, and different processes involved in their release from comets.

7.3. *Organic matter, water*

Cometary meteoroids are a potential source of organic matter (Jenniskens *et al.* 2004). Effort has been made in the recent years to search for organic elements (CHON) in the spectra of meteors, in particular Leonids. Oxygen and nitrogen are readily seen in meteor spectra, their primary source is, however, the atmosphere. Hydrogen can be seen in the high temperature component of the spectra of fast fireballs. The derived H/Mg abundance varies widely, from less than in CI chondrites to somewhat more than in Halley dust (Fig. 3). In addition to data given in Fig. 3, Borovička & Jenniskens (2000) estimated H/Fe = 10–20 and Jenniskens & Mandell (2004) gave H/Si = 4, which both certainly falls into the range given in Fig. 3. The large scatter of H abundances may be partly real. At least some meteoroids contain significant part of hydrogen. The hydrogen may come either from water embedded in partly hydrated minerals Rietmeijer 2005) or from organic material. Water ice is not expected to survive one perihelion passage in 0.1-m sized Leonid or Perseid meteoroids (Beech & Nikolova 2001). Pellinen-Wannberg *et al.* (2004) claimed to detect water in an Leonid but their evidence is very weak and indirect. Carbon has been positively identified in a UV Leonid spectrum (Carbary *et al.* 2003) but the abundance has, unfortunately, not been computed.

In the visual region, carbon is more likely to be detected in molecular form. The search for the main band of CN at 388 nm in Leonid spectra was unsuccessful, yielding a limit CN/Fe ⩽ 0.03 (Jenniskens *et al.* 2004). The authors proposed an explanation that carbon is ablated in form of more complex molecules. Long time ago, Ceplecha (1971) detected CN and possibly also C_2 in the terminal flare of of a bright sporadic cometary fireball. His resolution in the 388 nm region was, however, lower than that of Jenniskens *et al.* (2004). The ultraviolet band of OH at 308 nm was tentatively detected by Jenniskens *et al.* (2002) and Abe *et al.* (2005). Russell *et al.* (2000) detected CO, CO_2, H_2O and probably CH_4 in the mid-infrared spectrum of a meteor train several minutes after the fireball passage. These species are, however, unlikely to come directly from the meteoroid. At the present the content of organic matter in meteoroids is not firmly established.

8. Summary

There are various populations of meteoroids coming to the Earth. Asteroidal fragments capable to drop stony meteorites are well represented among large meteoroids. Their bulk mechanical strength, however, is typically two orders of magnitude smaller than that of recovered meteorites. Stony meteoroids of all sizes are full of cracks and do fragment easily during their atmospheric entry. Atmospheric fragmentation must be taken into account when considering the impacts of small asteroids on the Earth (Bland & Artemieva (2003); Melosh & Collins 2005). Stony meteoroids smaller than about 1 cm are rare in the interplanetary space. On the other hand, iron meteoroids, which form only $\sim 1\%$ of large bodies, are surprisingly well represented among millimeter sized meteoroids.

Cometary meteoroids are present at all sizes. They differ from the asteroidal bodies mainly by their large porosity and low bulk density. Two types of cometary material have been recognized, one with density about 0.7 g cm^{-3} and the other with $0.1 - 0.4$ g cm^{-3}. Both types are formed by loosely bound grains of various sizes and are likely analogical to aggregate IDP's and porous aggregate IDP's, respectively. The grains may be held together by a volatile glue (matrix) but the existence of the glue has not been confirmed with certainty. In any case, the grains can be released under moderate heating. The accumulated pressure acting on very large cometary meteoroids can blow away the material to certain depth but the meteoroid can withstand complete destruction, suggesting that cometary material is not fully cohesive. Fragile cometary meteoroids can, nevertheless, hide small more resistant chunks, possibly similar to carbonaceous chondrites.

The ablation ability and chemical composition of cometary grains does not largely differ from chondritic material. There is a hint that Halley type comets are richer in Na, Si, and Mg and poorer in Fe, Cr, and Mn in comparison with chondrites. The meteoroids coming from active comets are chemically relatively homogenous. Larger diversity is found among sporadic meteoroids of cometary origin and sizes of several millimeters. Part of these are fragments of cometary irradiation crust. They are depleted in volatiles (Na) and are significantly stronger than normal cometary material. The loss of volatiles and general compaction also occurs in the vicinity of Sun. Small meteoroids with perihelia within 0.2 AU are chemically and physically altered. This process can be at least partly responsible for the high density of Geminid meteoroids.

Acknowledgements

This work was supported by grant no. 205/05/0543 from GAČR. My research is conducted under the project ASCR project AV0Z10030501.

References

Abe, S., Yano, H., Ebizuka, N., & Watanabe, J. 2000, *Earth, Moon and Planets* 82-83, 369
Abe, S., Ebizuka, N., Yano, H., Watanabe, J., & Borovička, J. 2005, *Astrophys. J.* 618, L141
Adolfsson, L.G. & Gustafson, B.Å.S. 1994, *Planet. Space Sci.* 42, 593
Artemieva, N.A. & Shuvalov, V.A. 2001, *J. Geophys. Res.* 106 (E2), 3297
Arvidson, R.E. & Squyres, S.W. 2005, *Amer. Geophys. Union, Spring Meeting 2005*, abstract #P31A-02
Babadzhanov, P.B. 2002, *Astron. Astrophys.* 384, 317
Babadzhanov, P.B. & Konovalova, N.A. 2004, *Astron. Astrophys.* 428, 241
Beech, M. 2001, *Mon. Not. R. Astron. Soc.* 326, 937
Beech, M. 2002, *Mon. Not. R. Astron. Soc.* 336, 559
Beech, M. & Brown, P. 2000, *Planet. Space Sci.* 48, 925
Beech, M. & Murray, I.S. 2003, *Mon. Not. R. Astron. Soc.* 345, 696
Beech, M. & Nikolova, S. 2001, *Planet. Space Sci.* 49, 23
Bellot Rubio, L.R., Martínez González, M.J., Ruiz Herrera, L. *et al.* 2002, *Astron. Astrophys.* 389, 680
Binzel, R.P. & Xu, S. 1993, *Science* 260, 186
Bland, P.A. & Artemieva, N.A. 2003, *Nature* 424, 288
Borovička, J. 1993, *Astron. Astrophys.* 279, 627
Borovička, J. 1994a, *Planet. Space Sci.* 42, 145
Borovička, J. 1994b, in: Y. Kozai *et al.* (eds.) *Seventy-Five Years of Hirayama Asteroid Families*, Astron. Soc. Pacific Conf. Ser. 63, p. 186

Borovička, J. 2001, in: B. Warmbein (ed.), *Proc. Meteoroids 2001 Conf.*, ESA-SP 495, p. 203

Borovička, J. 2005a, *Earth, Moon and Planets* (in press)

Borovička, J. 2005b, *IAU Symp. 229* Abstract

Borovička, J. & Betlem, H. 1997, *Planet. Space Sci.* 45, 563

Borovička, J. & Jenniskens,P. 2000, *Earth, Moon and Planets* 82-83, 399

Borovička, J. & Kalenda, P. 2003, *Meteorit. Planet. Sci.* 38, 1023

Borovička, J. & Spurný, P. 1996, *Icarus* 121, 484

Borovicka, J., Popova, O.P., Nemtchinov, I.V., Spurný, P., & Ceplecha, Z. 1998, *Astron. Astrophys.* 334, 713

Borovička, J., Stork, R. & Bocek, J. 1999, *Meteorit. Planet. Sci.* 34, 987

Borovička, J., Koten, P., Spurný, P., Boček, J., & Štork, R. 2005, *Icarus* 174, 15

Boyd, I.D. 2000, *Earth, Moon and Planets* 82, 93

Bronshten, V.A. 1983, *Physics of Meteoric Phenomena* (Dordrecht: Reidel)

Bronshten, V.A., Rabunskij, D.D., & Tertitskij, M.I. 1985, *Astron. Vestnik* 19, 224

Brosch, N., Schijvarg, L.S., Podolak, M., & Rosenkrantz, M.R. 2001, in: B. Warmbein (ed.), *Proc. Meteoroids 2001 Conf.*, ESA-SP 495, p. 165

Brown, P., Ceplecha, Z., Hawkes, R.L., Wetherill, G., Beech, M., & Mossman, K. 1994, *Nature* 367, 624

Brown, P., Campbell, M.D., Ellis, K.J., Hawkes, R.L., Jones, J., Gural, P., Babcock, D., Barnaum, C., Bartlett, R.K., & Bedard, M. and 31 more authors 2000, *Earth, Moon and Planets* 82, 167

Brown, P.G., ReVelle, D.O., Tagliaferri, E., & Hildebrand, A.R. 2002, *Meteorit. Planet. Sci.* 37, 661

Campbell, M.D., Brown, P.G., LeBlanc, A.G. *et al.* 2000, *Meteorit. Planet. Sci.* 35, 1259

Campbell-Brown, M.D. & Koschny, D. 2004, *Astron. Astrophys.* 418, 751

Carbary, J.F., Morrison, D., Romick, G.J., & Yee, J.-H. 2003, *Icarus* 161, 223

Ceplecha, Z. 1966, *Bull. Astron. Inst. Czech.* 17, 195

Ceplecha, Z. 1968, *Smithson. Astrophys. Obs. Spec. Rep.* 279

Ceplecha, Z. 1971, *Bull. Astron. Inst. Czech.* 22, 219

Ceplecha, Z. 1988, *Bull. Astron. Inst. Czech.* 39, 221

Ceplecha, Z. 1996, *Astron. Astrophys.* 311, 329

Ceplecha, Z. & McCrosky, R.E. 1976, *J. Geophys. Res.* 81, 6257

Ceplecha, Z. & McCrosky, R.E. 1992, in: A. W. Harris, E. Bowell (eds.), *Asteroids, Comets, Meteors 1991* (Huston: Lunar Planet. Inst. Houston), p. 109

Ceplecha, Z. & ReVelle, D.O. 2005, *Meteorit. Planet. Sci.* 40, 35

Ceplecha, Z., Spurný, P., Borovička, J., & Keclíková, J. 1993, *Astron. Astrophys.* 279, 615

Ceplecha, Z., Borovička, J., Elford, W.G., ReVelle, D.O., Hawkes, R.L., Porubčan, V., & Šimek, M. 1998, *Space Sci. Rev.* 84, 327

Cook, A.F., & Duxbury, T.C. 1981, *J. Geophys. Res.* 86, 8815

Cudnik, B.M., Dunham, D.W., Palmer, D.M., *et al.* 2003, *Earth, Moon and Planets* 93, 145

Fujiwara, Y., Ueda, M., Shiba, Y. *et al.* 1998, *Geophys. Res. Lett.* 25, 285

Golub', A.P., Kosarev, I.B., Nemchinov, I.V., & Shuvalov, V.V. 1996, *Astron. Vestnik* 30, 213

Halliday, I. 1960, *Astrophys. J.* 132, 482

Harvey, G.A. 1973, in: C.L. Hemenway *et al.* (eds.), *Evolutionary and Physical Properties of Meteoroids*, NASA-SP 319, p. 131

Hawkes, R.L. & Jones, J. 1975, *Mon. Not. R. Astron. Soc.* 173, 339

Hiroi, T., Zolensky, M.E., & Pieters, C.M. 2001, *Science* 293, 2234

Jacchia, L.G., Kopal, Z., & Millman, P.M. 1950, *Astrophys. J.* 111, 104

Jenniskens, P. 2005, *Adv. Space Res.* (submitted)

Jenniskens, P. & Mandell, A.M. 2004, *Astrobiology* 4, 123

Jenniskens, P., Tedesco, E., Murthy, J., Laux, C.O., & Price, S. 2002, *Meteorit. Planet. Sci.* 37, 1071

Jenniskens, P., Schaller, E.L., Laux, C.O., Wilson, M.A., Schmidt, G., & Rairden, R.L. 2004, *Astrobiology* 4, 67

Jessberger, E.K., Christoforidis, A., & Kissel, J. 1988, *Nature* 332, 691

Kasuga, T., Watanabe, J., Ebizuka, N., Sugaya, T., & Sato, Y. 2004, *Astron. Astrophys.* 424, L35

Kasuga, T., Yamamoto, T., Watanabe, J., Ebizuka, N., Kawakita, H., & Yano, H. 2005a, *Astron. Astrophys.* 435, 341

Kasuga, T., Watanabe, J., & Ebizuka, N. 2005b *Astron. Astrophys.* 438, L17

Konovalova, N.A. 2003, *Astron. Astrophys.* 404, 1145

Koschny, D., Reissaus, P., Knöfel, A., Trautner, R., & Zender, J. 2002, in: B. Warmbein (ed.), *Asteroid, Comets, Meteors (ACM2002)*, ESA-SP 500, p. 157

Koten, P., Spurný, P., Borovička, J., & Štork, R. 2001, in: B. Warmbein (ed.), *Proc. Meteoroids 2001 Conf.*, ESA-SP 495, p. 119

Koten, P., Borovička, J., Spurný, P., Betlem, H., & Evans, S. 2004, *Astron. Astrophys.* 428, 683

Koten, P., Spurný, P., Borovička, J. *et al.* 2006, *Meteorit. Planet. Sci.* (submitted)

LeBlanc, A.G., Murray, I.S., Hawkes, R.L., Worden, P., Campbell, M.D., Brown, P., Jenniskens, P., Correll, R.R., Montague, T., & Bavcock, D.D. 2000, *Mon. Not. R. Astron. Soc.* 313, L9

Lodders, K. 2003, *Astrophys. J.* 591, 1220

Lodders, K. & Osborne, R. 1999, *Space Sci. Rev.* 90, 289

McNeil, W.J., Lai, S.T., & Murad, E. 1998, *J. Geophys. Res.* 103 (D9), 10899

McNeil, W.J., Murad, E., & Plane, J.M.C. 2002, in: E. Murad, I.P. Williams (eds.), *Meteors in the Earth's Atmosphere* (Cambridge: Cambridge University Press), p. 265

Melosh, H.J. & Collins, G.S. 2005, *Nature* 434, 157

Millman, P.M. 1959, *J. R. Astron. Soc. Canada* 53, 15

Millman, P.M. 1972, *J. R. Astron. Soc. Canada* 66, 201

Murray, I.S., Hawkes, R.L., & Jenniskens, P. 1999, *Meteorit. Planet. Sci.* 34, 949

Murray, I.S., Beech, M., Taylor, M.J., Jenniskens, P., & Hawkes, R.L. 2000, *Earth, Moon and Planets* 82, 351

Pellinen-Wannberg, A., Murad, E., Gustavsson, B., Brädström, U., Enell, C.-F., Roth, C., Williamns, I.P., & Steen, A. 2004, *Geophys. Res. Lett.* 31, L03812

Popova, O.P. 2005, *Earth, Moon and Planets* (in press)

Popova, O.P., Strelkov, A.S., & Sidneva, S.N. 2005a, *Adv. Space Res.* (submitted)

Popova, O.P. *et al.* 2005b, *IAU Symp.* 229 Abstract

ReVelle, D.O. 2001, in: B. Warmbein (ed.), *Proc. Meteoroids 2001 Conf.*, ESA-SP 495, p. 513

ReVelle, D.O. & Ceplecha, Z. 1994, *Astron. Astrophys.* 292, 330

Rietmeijer, F.J.M 2005, *Earth, Moon and Planets* (in press)

Rietmeijer, F.J.M. & Nuth III, J.A. 2000, *Earth, Moon and Planets* 82, 325

Rogers, L.A., Hill, K.A., & Hawkes, R.L. 2005, *Planet. Space Sci.* 53, 1341

Russell, R.W., Rossano, G.S., Chatelain, M.A., Lynch, D.K., Tessensohn, T.K., Abendroth, E., Kim, D., & Jenniskens, P. 2000, *Earth, Moon and Planets* 82, 439

Sarma, T. & Jones, J. 1985, *Bull. Astron. Inst. Czech.* 36, 9

Schaefer, L. & Fegley Jr., B. 2005, *Earth, Moon and Planets* (in press)

Selsis, F., Lemmon, M.T., Vaubaillon, J., & Bell III, J.F. 2005, *Nature* 435, 581

Spurný, P. 1993, in: J. Štohl & I.P. Williams (eds.), *Meteoroids and Their Parent Bodies* (Bratislava: Astron. Inst. Slovak Acad. Sci.), p. 193

Spurný, P. 1995, *Earth, Moon and Planets* 68, 529

Spurný, P. & Borovička, J. 1999a, in: W.J. Baggaley & V. Porubčan (eds.), *Meteoroids 1998* (Bratislava: Astron. Inst. Slovak Acad. Sci.), p. 143

Spurný, P. & Borovička, J. 1999b, in: J. Svoreň *et al.* (eds.), *Evolution and Source Regions of Asteroids and Comets* (Tatranská Lomnica: Astron. Inst. Slovak Acad. Sci.), p. 163

Spurný, P. & Porubčan, V. 2002, in: B. Warmbein (ed.), *Asteroid, Comets, Meteors (ACM2002)*, ESA-SP 500, p. 269

Spurný, P., Betlem, H., van't Leven, J., & Jenniskenns, P. 2000a, *Meteorit. Planet. Sci.* 35, 243

Spurný, P., Betlem, H., Jobse, K., Koten, P., & van't Leven, J. 2000b, *Meteorit. Planet. Sci.* 35, 1109

Spurný, P., Borovička, J., & Koten, P. 2005, *Earth, Moon and Planets* (in press)

Swindle, T.D. & Campins, H. 2004, *Meteorit. Planet. Sci.* 39, 1733

Trigo-Rodriguez, J.M., Llorca, J., Borovička, J., & Fabregat, J. 2003, *Meteorit. Planet. Sci.* 38, 1283

Vinković, D. 2005, *Adv. Space Res.* (in press)

von Zahn, U., Gerding, M., Höffner, J., McNeil, W.J., & Murad, E. 1999, *Meteorit. Planet. Sci.* 34, 1017

von Zahn, U., Höffner, J., & McNeil, W.J. 2002, in: E. Murad, I.P. Williams (eds.), *Meteors in the Earth's Atmosphere* (Cambridge: Cambridge University Press), p. 149

Wasson, J.T. & Kallemeyn, G.W. 1988, *Phil. Trans. R. Soc. Lond. A* 325, 535

Watanabe, J., Tabe, I., Hasegawa, H., Hashimoto, T., Fuse, T., Yoshikawa, M., Abe, S., & Suzuki, B. 2003, *Publ. Astron. Soc. Japan* 55, L23

Zolensky, M.E., Nakamura, K., & Gounelle, M. 2002, *Meteorit. Planet. Sci.* 37, 737

Asteroids, Comets, Meteors
Proceedings IAU Symposium No. 229, 2005
D. Lazzaro, S. Ferraz-Mello & J.A. Fernández, eds.

© 2006 International Astronomical Union
doi:10.1017/S1743921305006794

Asteroid-meteorite links: the Vesta conundrum(s)

C. M. Pieters[1], R. P. Binzel[2], D. Bogard[3], T. Hiroi[1], D.W. Mittlefehldt[3], L. Nyquist[3], A. Rivkin[4] and H. Takeda[5]

[1]Department of Geological Sciences, Brown University, Providence, RI 02912 USA

[2]Department of Earth, Atmosphere, Planetary Science, MIT, Cambridge, Massachusetts, USA

[3]NASA Johnson Space Center, Houston, TX 77058 USA

[4]The Johns Hopkins University Applied Physics Laboratory, Laurel, Maryland, USA

[5]Chiba Institute of Technology, Tsudanuma, Narashino City, Chiba 275, Japan

Abstract. Although a direct link between the HED meteorites and the asteroid 4 Vesta is generally acknowledged, several issues continue to be actively examined that tie Vesta to early processes in the solar system. Vesta is no longer the only basaltic asteroid in the Main belt. In addition to the Vestoids of the Vesta family, the small asteroid Magnya is basaltic but appears to be unrelated to Vesta. Similarly, diversity now identified in the collection of basaltic meteorites requires more than one basaltic parent body, consistent with the abundance of differentiated parent bodies implied by iron meteorites. The timing of the formation of the Vestoids (and presumably the large crater at the south pole of Vesta) is unresolved. Peaks in Ar-Ar dates of eucrites suggest this impact event could be related to a possible late heavy bombardment at least 3.5 Gyr ago. On the other hand, the optically fresh appearance of both Vesta and the Vestoids requires either a relatively recent resurfacing event or that their surfaces do not weather in the same manner thought to occur on other asteroids such as the ordinary chondrite parent body. Diversity across the surface of Vesta has been observed with HST and there are hints of compositional variations (possibly involving minor olivine) in near-infrared spectra.

Keywords.

1. Introduction

The majority of meteorites are believed to be derived from parent bodies in the asteroid belt, even though there are few specific asteroids identified as sources. A principal exception is the association of Vesta with the Howardite, Eucrite, Diogenite class of achondrites (HEDs). Shown in Figure 1 is the original data used in 1970 to argue that the mineralogical properties of the HED meteorites (largely low-Ca pyroxene + plagioclase) are the same as those observed for the asteroid Vesta (McCord *et al.* 1970). At that time, and for several decades later, Vesta was the only asteroid identified to have a basaltic surface. Vesta was therefore believed to be the only large asteroid remaining intact that had melted and differentiated during the early phases of solar system evolution.

Excellent reviews of the character and issues associated with the HED class of meteorites can be found in Takeda (1997), Mittlefehldt *et al.* (1998), Drake (2001), and Keil (2002). Even though the link to Vesta has been scrutinized for decades, a number of rather substantial new issues are actively being discussed as a result of a growing body of new telescopic measurements and detailed analyses of meteorites. A composite spectrum for Vesta is shown in Figure 2 derived from recent surveys. Several key issues are highlighted here in the context of measured properties of Vesta and the HEDs. The

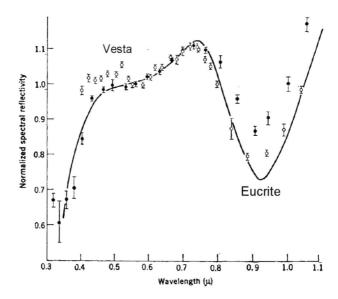

Figure 1. Original telescopic data linking Vesta and the basaltic achondrites. (after McCord *et al.* 1970). The absorption feature near $0.92\,\mu$m for Vesta (points) is an indication of low-Ca pyroxene and is comparable to that seen in laboratory spectra of basaltic achondrites.

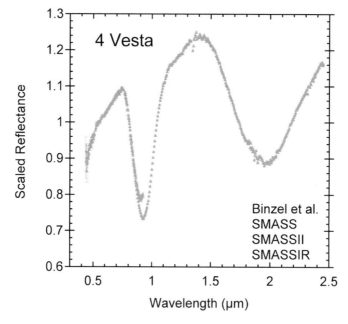

Figure 2. Composite average spectrum of Vesta derived from the SMASS programs of Binzel *et al.* [see: www.smass.mit.edu].

intent is to stimulate discussion that will lead to specific observations and analyses that can be addressed with data acquired during the Dawn rendezvous with Vesta.

Dawn, the 9th mission in NASA's highly successful Discovery Program, will undertake a detailed study of the geophysics, mineralogy, and geochemistry for two of the largest main belt asteroids (Russell *et al.* 2004). Dawn is scheduled to be launched in 2006 and

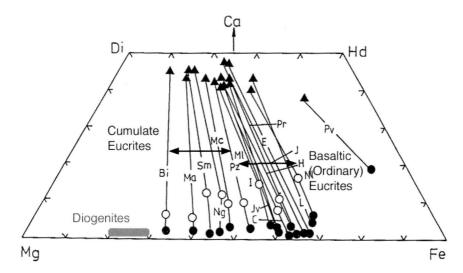

Figure 3. Pyroxene compositions for non-Antarctic Diogenites and Eucrites (after Takeda 1997). The lines connect host pyroxene and exsolution lamellae and the bulk composition is shown as open circles.

will rendezvous with Vesta in 2011 and with Ceres in 2016. The spacecraft will operate for an extended period in nested orbits at each of these minor planets. Dawn science payload includes framing cameras with multi-spectral and stereo capability, a visible to near-infrared imaging spectrometer, a gamma ray/neutron spectrometer, and radio science.

2. The Vesta-HED Link: Is Vesta the Only Basaltic Parent-Body in the Asteroid Belt?

The Vesta-HED link is directly tied to the observed/inferred mineralogy of the asteroid and the measured mineralogy of the meteorites. Pyroxenes are the dominant mafic mineral present in HED meteorites and provide multiple clues about how the parent body evolved (e.g., Takeda 1997). The primary distinctions of pyroxene compositions observed in non-Antarctic HEDs are summarized in Figure 3. For those monomict samples that have not been extensively reprocessed, there appears to be a continuous pyroxene compositional trends across the HED suite (Takada *et al.* 1983; 1997). Diogenites have the most magnesium orthopyroxenes and ordinary (basaltic) eucrites contain the most iron-rich pyroxenes. Although cumulate eucrites are clearly linked to the other eucrites, they have a cumulate texture and their pyroxenes are intermediate in Mg content. Howardites are breccia mixtures of diogenites and eucrites, but individual clasts often fill gaps further illustrating a continuous relation between the components.

Fortunately, remote compositional analyses using visible to near-infrared spectroscopy are very sensitive to pyroxene composition. Ferrous iron in the asymmetric M2 pyroxene site provides highly diagnostic absorptions near 1 and $2\,\mu$ that vary in wavelength with the composition and structure of the pyroxene (e.g., Burns 1993). Typical examples of reflectance spectra for Eucrites, Diogenites, and Howardites are shown in Figure 4. Diogenites with their more Mg-rich pyroxenes have absorption bands at shorter wavelengths than those of the more ferrous eucrites. Howardites, a brecciated and physical mixture,

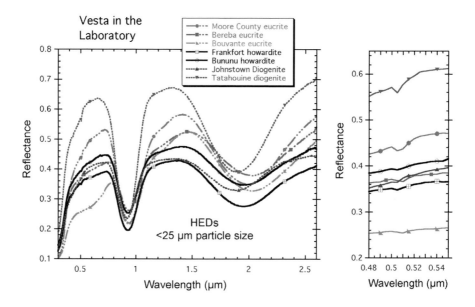

Figure 4. Laboratory bidirectional reflectance spectra of typical examples for Howardites, Eucrites, and Diogenites (HEDs). Right) The same spectra with the visible part of the spectrum expanded. All spectra were acquired at 5 nm spectral resolution.

exhibit spectra with composite absorption band minima intermediate between the two components of the mixture.

The spectrum of each HED meteorite is dominated by the spectral properties of pyroxene. This is illustrated with spectra of mineral separates of a cumulate eucrite, Y-980318, shown in Figure 5. The spectrum of the bulk sample is dominated by the optical properties of the pyroxene. All of the major absorptions are due to electronic transitions of ferrous iron in the M1 and M2 sites of pyroxene (Burns 1993). Ferrous iron in the M1 site of pyroxene produces weaker features than when in the M2 site because the M1 site is more symmetric. Most of the weaker features at shorter wavelengths in the visible are associated with either spin forbidden or charge transfer transitions. On the other hand, iron-poor plagioclase is not absorbing in the near infrared and does not contribute features to the bulk spectrum. The presence of plagioclase provides a relatively transparent medium and enhances the optical path length. A photomicrograph of Y-98318 is shown in Figure 6 illustrating the thick exsolution lamellae characteristic of pyroxene in cumulate eucrites (Takeda 1997).

Reflectance spectra of Vesta acquired using Earth-based telescopes represent an integrated average of the side facing Earth at the time of the measurement. Since Vesta is large (equatorial diameter ∼ 550 km) and bright (IRAS albedo ∼ 0.4), high quality spectra are available (e.g., Gaffey 1997 and Fig. 2). In 1993 detector sensitivity had advanced sufficiently to allow Binzel & Xu (1993) to identify a group of very small asteroids in the vicinity of Vesta that appear to have the same basaltic composition as the larger asteroid. Reflectance spectra of Vesta and representative spectra of these "Vestoids" are shown in Figure 7 and the distribution of currently identified Vestoids are shown in Figure 8. All members of this family of Vestoids are faint, of course, and statistical errors tend to be larger than for Vesta (Xu & Binzel 1993; Burbine *et al.* 2002). Increasing sensitivity of

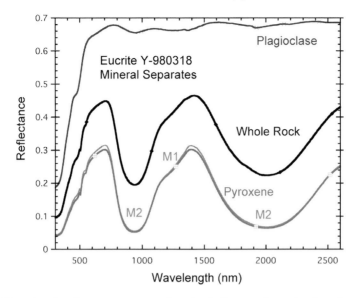

Figure 5. Bidirectional reflectance spectra for the bulk sample and mineral separates from cumulate eucrite Y-980318. The bands near 1 and $2\,\mu$m are due to ferrous iron in the M2 pyroxene site, whereas the weaker band near $1.2\,\mu$m is due to ferrous iron in the more symmetric M1 site of pyroxene.

Figure 6. Photomicrograph of cumulate eucrite Y-980318 showing exsolution lamellae of orthopyroxene inverted from pigeonite. Field of view is 3.3 mm.

modern instruments has provided data of sufficient quality to allow comparison of spectral parameters linked to mineralogy across the Vesta family (Hiroi *et al.* 1995, 1998; Duffard *et al.* 2004; Vernazza *et al.* 2005).

The distribution of basaltic Vestoids extends from Vesta to several key resonances known to act as "escape hatches" to the inner solar system (Wisdom 1985), and thus a mechanism was found to exist for delivery of the HED meteorites from basaltic asteroids in the main belt. This was strengthened when sequential observations with the Hubble Space Telescope allowed a three-dimensional shape model for Vesta to be developed (Thomas *et al.* 1997). The shape model for Vesta revealed a very large crater-like

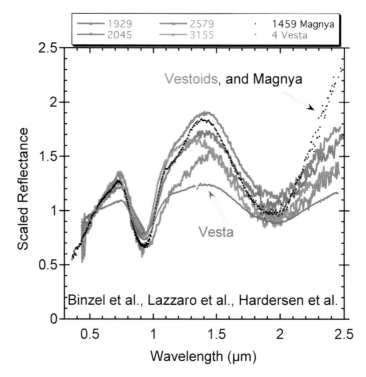

Figure 7. Telescopic reflectance spectra of Vesta, Vestoids, and Magnya. Vesta exhibits a notably flatter continuum and weaker absorption bands.

depression near the south pole as a possible excavation site for the Vestoids, and hence an ultimate source for the HEDs.

Spectra of individual HED meteorites are quite diverse and absorption features are directly linked to meteorite mineralogy (Figures 3 & 4). As might be expected for an impact mixed regolith, the principal spectral properties of Vesta's observed surface are most comparable to those of Howardites. The observed albedo and band strength, however, suggest Vesta's surface materials must also must be relatively fine grained and on the order of $< 25\,\mu$m (Hiroi *et al.* 1994). A comparison of spectra for particle size separates of howardite EET87503 is shown in Figure 9. As is typical for size separates, the finer particles are brighter and the larger particles have a flatter spectrum (larger component of first surface Fresnel reflection). The data are scaled to unity in the visible in Figure 9b. Vesta is seen to be most comparable to the howardite sample with a fine-grained size distribution. Since small particles always coat larger particles in a natural soil, the larger size separates actually have very little analytical use in spectral comparisons and modeling in the near-infrared.

There are a few very important observations, however, that do not readily fit into an all encompassing Vesta-HED story. As can be seen in Figure 7, the Vestoids generally have stronger absorption bands than Vesta and they exhibit a notably steeper continuum from the visible to 1.6 μm. Based on the wavelength of the primary pyroxene absorption bands, the first order pyroxene mineralogy of Vesta and the Vestoids is similar. Both represent basaltic rock types. However, there is no question that the uppermost surface of the large asteroid observed remotely is quite different from that of the smaller relatives (e.g., Hiroi *et al.* 1995, 2001). It is not known yet whether these differences are due to physical properties and/or texture differences, compositional variations implying different source

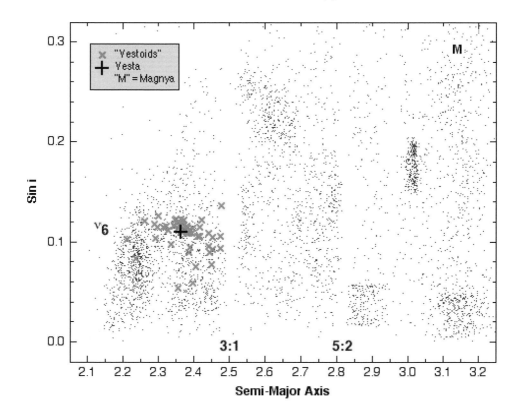

Figure 8. Orbital parameters of Vesta (+), Vestoids (x), and Magnya (M) within in the main asteroid belt (dots). Principal resonances are labeled.

material, or environmental processes that alter surface materials. Examples include: a fine coating of feldspathic dust on a coherent basaltic substrate, a coarse grain basaltic fragment with relatively Fe-rich pyroxenes, a gravity dependent space weathering process, etc. A separate parent body for the Vestoids is not needed, but cannot be eliminated with the current data.

As the character and diversity of small bodies continue to be explored with increasingly sensitive instruments, additional exceptions are found. A small ($\sim$ 30 km diameter) basaltic asteroid, Magnya, was identified in the outer asteroid belt far removed from Vesta (Lazzaro *et al.* 2000), on the other side of several major resonances (see Figure 8). The pyroxene bands for Magnya are at a shorter wavelength than those of Vesta (Hardersen *et al.* 2004) which implies a more Mg bulk pyroxene composition. Independent of actual magnitude, the albedo of Vesta appears to be about twice as high as that of the much smaller Magnya (Hardersen *et al.* 2004). Such a difference could be due to bulk composition, but is more likely a clue to inherent textural differences of small basaltic asteroid fragments. Magnya itself is likely to be a small remnant of a larger body distinct from Vesta. From a dynamical point of view, it is very unlikely that Vesta and Magnya can be related (Lazzaro *et al.* 2000). Thus, Magnya is currently the best evidence for a possible second basaltic parent body having existed in the main asteroid belt.

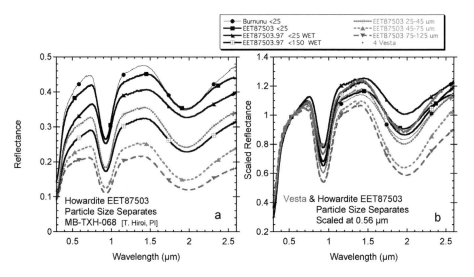

Figure 9. a) Laboratory bidirectional reflectance spectra for bulk powders (lines) and particle size separates (dash, dots) of Howardite EET87503. A spectrum of howardite Bununu is shown for comparison. b) The same as (a), but scaled to unity at 0.56 μm. The composite spectrum of Vesta from Figure 2 is included for comparison in (b). Note that the optical characteristics of the 45–75 and 75–125 μm separates are all outside the range of the < 150 bulk properties even though their actual size is < 150 μm.

3. The Vesta-HED Link: Do All Basaltic Meteorites Share the Same Parent Body?

Oxygen isotopes have long been used to group meteorite classes (Clayton *et al.* 1993) and distinct fields of oxygen isotope values have been interpreted to imply the evolution of different parent bodies. The general presumption is that large differences in oxygen isotopic values implies very different parts of the solar system. The HED meteorites form their own group, another piece of evidence arguing that they have a common parent body, Vesta.

Instrumentation has also improved in terrestrial laboratories and more detailed and higher precision analyses are now possible for small valuable meteorite samples. Two basaltic meteorites classified as eucrites have recently been shown to have oxygen isotopic signatures significantly different from the main family of HEDs: NWA011, which falls far from other HEDs and closer to the CR-chondrites (Yamaguchi *et al.* 2002; Floss *et al.* 2005), and Ibitira which falls between the HED and terrestrial fractionation line (Wiechert *et al.* 2004). Distinction of Ibitira is only made possible with the high precision that can be achieved with modern instruments (Wiechert *et al.* 2004). There are several additional geochemical parameters that suggest the parent body of Ibitira is distinct (Mittlefehldt 2005). A composite figure illustrating the oxygen isotope relations for HEDs, NWA011, and Ibitira is shown in Figure 10.

Minor elements such as Ti, Cr, and Al in the pyroxene structure are known to significantly affect the character of absorption bands as well as the nature of the continuum. However, no systematic analysis documenting the spectroscopic effects of such elements in the pyroxene structure has been undertaken for pyroxenes in the HED range of compositions. Until such effects are better defined, spectra of the two unique basaltic meteorites, Ibitira and NWA011, cannot be readily used to identify their parent body. Both are

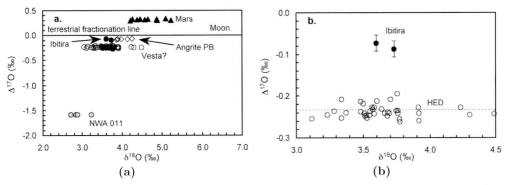

Figure 10. Recent oxygen isotope measurements for the inner planets, HEDs, and other basaltic meteorites. (data taken from Floss *et al.* 2005; Franchi *et al.* 1999; Greenwood *et al.* 2005; Wiechert *et al.* 2001, 2004). Figure (b) contains the same data as (a) at an expanded scale.

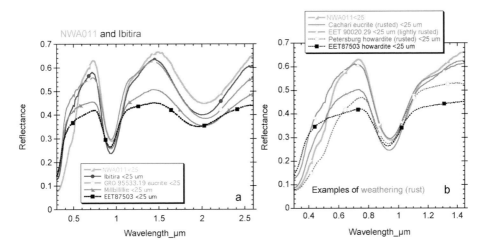

Figure 11. Bidirectional reflectance spectra for NWA011, Ibitira and similar examples of HEDs. All exhibit diagnostic properties indicative of basaltic composition.

basaltic, but neither exhibit spectra that can be distinguished from the diverse range of HEDs measured (see the collection of Takahiro Hiroi [TXH] in the RELAB database at http://www.planetary.brown.edu/relab. Shown in Figure 11 are reflectance spectra of these two unusual meteorites along with examples of eucrites with similar bulk pyroxene composition. NWA011 (sample provided by Akira Yamaguchi) has the unfortunate additional characteristic of terrestrial weathering. Examples of the effects of terrestrial weathering is easily seen in the visible part of the spectrum for NWA011.

4. When Did Major Geological Events Occur on Vesta?

A detailed discussion of the timing and character of the original heating event that produced the basaltic character of Vesta and the HEDs is beyond the scope of this overview. Differentiation occurred only a few million years after the formation of the CAIs and chondrules. As an example, a summary of the early closure dates for cumulate

282 C. M. Pieters *et al.*

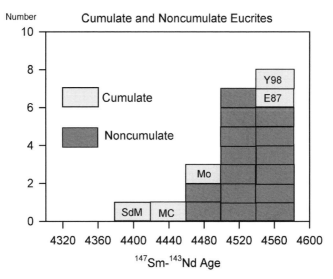

Figure 12. Summary of Sm-Nd ages for cumulate and non-cumulate eucrites [after Nyquist 2005]. The cumulate eurcrites (blue) are labeled: Serra de Mage (SdM), Moore County (MC), Moama (Mo), Yamato 980318, and EET87520. Data from Carlson & Lugmair (2000; EET87520, Moama, Moore County, Serra de Mage) and Nyquist *et al.* (2004; Y980318).

and non-cumulate eucrites as measured by Sm-Nd is shown in Figure 12 (Nyquist 2004). These ancient dates set limits for when Vesta had a solid crust composed of materials that we recognize as HED meteorites.

Discussion of the character of a magma ocean and the relative influence of partial melting and fractional crystallization in the formation of the different types of eucrites and diogenites continues within the meteorite community. The debate focuses largely on whether partial melting or fractional crystallization is more important for the regular trends of compositions observed in the HEDs. Regardless how the components were formed, the resulting sequence is generally believed to be arranged similar to the layered crust model of Takeda (1983, 1997) which is illustrated in Figure 13. Impact craters reworked the surface during and after formation of the primary zones, rearranging and mixing components on a scale that depends on the size of the impact and the depth of the compositional zones. Deposits from the largest, and perhaps earliest, craters also provided sufficient thermal environment for moderate metamorphism.

Argon isotopic analyses are known to be disturbed or reset by major impact events and Ar–Ar dating is thus a good method to track some of these early events (Bogard 1995). Summarized in Figure 14 are Ar-Ar ages for eucrites presented as probability plots (after Bogard 2005). The clustering of data between 4 and 3.5 Gyr is highly suggestive of a period of heavy bombardment similar to the late heavy bombardment suggested for the Moon from lunar samples (Ryder *et al.* 2002). The magnitude of these impacts on the HED parent body (Vesta) would be consistent with energy required to have launched the Vestoids. Since the HEDs seem to share several common peaks in Figure 14, the youngest peak at 3.5 Gyr may indicate that samples destined to become HEDs resided on Vesta until at least as recently as 3.5 Gyr. From a collisional point of view, younger ages for the formation of Vestoids.become increasingly less likely because of the rarity of giant impacts sufficient to form the south pole basin and excavate the Vestoids.

Whenever the Vestoids were formed, the impact event must have been very large and scattered fragmental debris across Vesta itself as well as dusted the surface of the

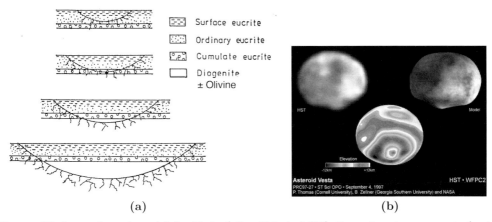

(a) (b)

Figure 13. Layered crust model for Vesta (after Takeda 1997). Impact craters excavate (and mix) different compositional zones. Large craters would produce mixed compositions like the Howardites. The largest crater, such as that observed by HST near Vesta's south pole shown on the right (after Thomas *et al.* 1997), may have formed and distributed the Vestoids.

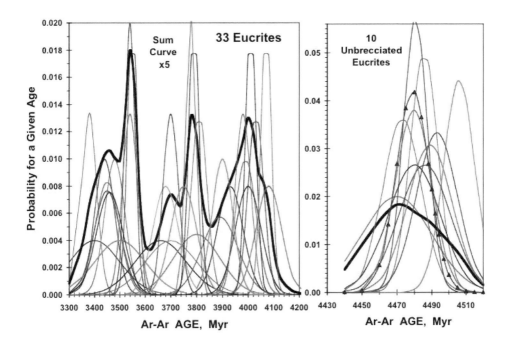

Figure 14. Combined Ar-Ar ages of eucrites (after Bogard, 2005). The probability plot on the left gives impact reset ages of brecciated eucrites. The probability plot on the right gives ages of non-brecciated (basaltic and cumulate) eucrites. The age axes of the two plots are approximately continuous, but they are different in scale. The age peak for the unbrecciated Eucrites is most prominent. The data indicate age resetting events at 4.48, 4.0, 3.8, and 3.5 Gyr.

Vestoids. Both Vesta and the Vestoids exhibit strong diagnostic absorption bands that are quite comparable in wavelength, shape and strength to similar bands found in HED meteorites. This link to the HEDs implies that the surface of Vesta and the Vestoids are directly comparable to particulate samples of HED meteorites and have experienced very little, if any, weathering in the space environment (e.g., Pieters *et al.* 2000).

The optically fresh appearance of the surface of both Vesta and the Vestoids leads to a dilemma, however. If the fresh appearing surface is due to a resurfacing event, it implies that the event happened recently (e.g., Pieters & Binzel 1994). Since both Vesta and all the Vestoids share the same fresh surface property, the event that caused it may have occurred simultaneously – i.e. the formation of the Vestoids. This is inconsistent with the much older Ar-Ar ages suggested for the formation of the large south pole crater and Vestoids.

On the other hand, perhaps the surface geology of both Vesta and Vestoids are indeed as old as suggested by the Ar-Ar data, and the optically fresh appearance is due to a lack of space weathering effects for some other reason. For example, laser induced space weathering experiments indicate pyroxene-rich surfaces are less easily altered than those rich in olivine (Sasaki *et al.* 2002), indicating that the rate of space weathering on the HED parent body would be slower than that of a parent body for olivine-bearing ordinary chondrities. In addition, perhaps the lack of metallic iron for HED-like materials and Vestas location in the main asteroid belt combine to produce an exceptionally low rate of space weathering, and the integrated surface exposure time is insufficient to alter the optical features. Analytical measurement and calculations for the rate of space weathering in the Vesta environment are currently not well bounded.

The dynamical estimates for when the Vestoids of the Vesta family were formed are somewhat intermediate between the Ar-Ar ages and the space weathering predictions, namely ~ 1 Gyr (e.g. Marzari *et al.* 1996). It will be very important for Dawn observations of Vesta's geology and composition to resolve this issue since it relates to the nature and timing of major events in the early solar system.

5. Compositional Variability on Vesta?

Although there is good agreement that the general character of the surface of Vesta is comparable to the basaltic HEDs, the geologic context of specific rock types and/or range of composition is needed to constrain the geologic evolution of the surface. From HST images of Vesta in 4 spectral bands (0.4 to 1.0 μm) we know that the surface is not covered with a uniform debris, but instead exhibits local color variations of 10–20% (Binzel *et al.* 1997). From earth-based telescopes only broad variations can be evaluated as the asteroid rotates. Rotational variations are subtle, but hemispheric correlated spatial differences can be detected (e.g., Gaffey 1997). Two examples provide tantalizing clues of what may be observed at higher resolution when Dawn arrives in 2011.

Water on Vesta. There is a hint that small amounts of water or aqueous fluid existed on Vesta, probably as a contaminant from elsewhere, but possibly linked to early aqueous processes. Evidence comes both from the samples (Treiman *et al.* 2004) as well as telescopic observations (Hasegawa *et al.* 2003, Rivkin *et al.* 2005). Observations across the 3 μm water band are difficult from Earth due to telluric water in Earth's atmosphere. A summary of variations for this part of the spectrum are shown in Figure 15 as a function of rotation (after Rivkin *et al.* 2005). Although some spectral variations appear to be regular with rotational phase, variations due to a hydrous component are not easily distinguished from other compositional variations.

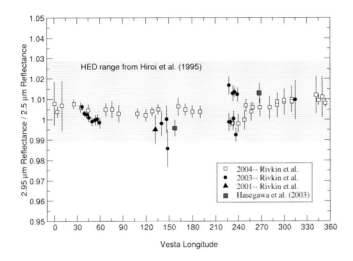

Figure 15. Rotational variations of Vesta spectra near the 3 m water band (after Rivkin 2005). Although the 2.95/2.50 μm ratio should capture variations in water content, small variations in the continuum slope are of the same magnitude.

<u>Olivine on Vesta</u>. The size of the south pole crater on Vesta (Fig. 13) implies material excavated from several 10 s of km in the interior. Most models of a differentiated asteroid include an olivine-rich mantle, but no olivine meteorites occur in the HED suite, although olivine is a minor phase observed in some diogenite meteorites. However, one North Africa olivine diogenite meteorite has recently been found (NWA1877) that contains 45% olivine (Irvine *et al.* 2005). NWA1877 also contains chromite compositions that suggest a primitive mantle source.

The rotationally resolved spectra of Vesta by Gaffey (1997) are all dominated by pyroxene, but one zone on this body is subtly, but distinctly, different. As shown in Figure 16 small variation of band area ratio parameter is observed on Vesta by Gaffey. Band area ratios have been shown to be a useful tool to track mineral ratios in a mixture of olivine and orthopyroxene (Cloutis *et al.* 1986). The technique involves producing a ratio of the area of an absorption near 2 m relative to a continuum (BII) to a comparable area of the absorption near 1 μm (BI). For low-Ca pyroxene common on Vesta (Figure 2) both of these bands are prominent. Olivine, on the other hand, contains no 2 μm band and has a much broader absorption near 1 μm. Thus, in a mixture of orthopyroxene and olivine the band area ratio decreases with increasing olivine content. These rotational data of Vesta have characteristics consistent with a significant olivine component (Gaffey 1997). The data are tantalizing, but without additional data are unfortunately non-unique. Several groups have recently been investigating a range of parameters and find that other variables can also affect the band area ratio such as grain size, temperature, space weathering, and additional mineral diversity. Shown in Figure 17 is a summary of these effects prepared by Ueda *et al.* (2000). Nevertheless, there is no doubt that compositional variations exist on Vesta that are geologically significant (Binzel *et al.* 1997, Gaffey 1997). The instruments on Dawn (Russell *et al.* 2004, 2005) are well poised to characterize the surface composition in the geologic context needed.

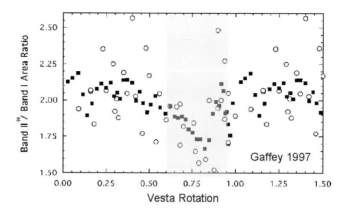

Figure 16. Relative change of the band area ratio (the $2\,\mu$m pyroxene band area divided by the area of the combined ferrous band near $1\,\mu$ m) as a function of rotational position of Vesta (after Gaffey 1977). The solid squares are a 5-point running average. There is one position on Vesta where the $2\,\mu$m band appears relatively weaker (shaded zone near 0.75). This change in the relative strengths of the two bands is consistent with a variety of compositional parameters including a) a minor olivine component since olivine lacks a $2\,\mu$m band, b) finer grain size, or c) variations in space weathering.

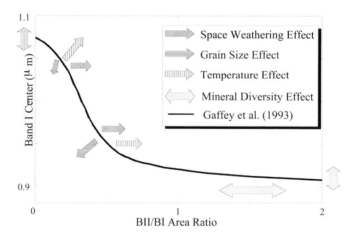

Figure 17. Schematic trends on the Band I center vs. BII/I area ratio plot of Gaffey *et al.* (1993) caused by space weathering effects, grain size, temperature and mineral diversity (after Ueda *et al.* 2000).

6. Summary and Expectations

There has been significant advancement to our understanding of the Vesta - HED link since the original discovery decades ago by McCord *et al.* (1970):

- Is Vesta the ONLY basaltic parent-body in the asteroid belt?

    Vesta remains the only intact basaltic asteroid. Although both the smaller Vestoids and Magnya are clearly basaltic in nature, the Vestoids are probably related to Vesta, but Magnya is not.

- Do ALL basaltic meteorites share the same parent body?

No. HEDs, NWA011, and Ibitira have very similar basaltic mineralogy, but appear to have distinct parent bodies. The plethora of parent bodies required by the suite of iron meteorite implies the existence of many additional differentiated parent bodies in the early solar system (e.g., Mittlefehldt *et al.* 1998).

- When did major geologic events occur on Vesta?

  The date for the event that formed the large unnamed crater at the pole and the Vestoids is unresolved. The timing of the formation of the Vestoids is closely interwoven with understanding the space weathering environment at Vesta.

- What is the compositional range of materials on Vesta?

  The layered crust model for ordinary eucrites – cumulate eucrites – diogenites (top to interior) can be expanded to include olivine diogenites. Although the currently observed surface is dominated by mixtures (howardites), spatial variations are known to exist across Vesta but their inherent compositional distinctions are ambiguous.

When high spatial resolution compositional data for Vesta is acquired by Dawn along with topography and shape, the geologic context will allow many of these issues concerning early evolution of silicate bodies in the solar system to be addressed and resolved. In the process, we will also inevitably find many many more questions about the character and history of small bodies in the solar system that we have not yet even imagined.

Acknowledgements

We thank several additional scientists who generously provided data, insights, or encouragement for this discussion: P. Buchanan, A. Coradini, M. C. De Sanctis, M. Drake, P. Hardersen, T. McCord, T. McCoy, L. McFadden, H. McSween, and C. Russell.

References

Binzel, R.P. & Xu, S. 1993, *Science* 260, 186

Binzel, R.P., Gaffey, M.J., Thomas, P.C., Zellner, B.H., Storrs, A.D., & Wells, E.N. 1997, *Icarus* 128, 95

Binzel, R.P., Bus S.J., & Burbine, T.H. 1999, *Lunar. Planet Sci. XXX* #1216

Bogard, D.D. 1995, *Meteoritics* 30, 244

Bogard, D.D. & Garrison, D.H. 2003, *Meteoritics & Planetary Science* 38 (5), 669

Bogard, D.D. 2005, *Lunar & Planetary Science Conference XXXVI* #1131

Burbine, T.H., Buchanan, P.C., Binzel, R.P., Bus, S.J., Hiroi, T., Hinrichs, J.L., Meibom, A., & McCoy, T.J. 2001, *Meteoritics & Planetary Science* 36, 761

Burns, R.G. 1993, *Mineralogical Application of Crystal Field Theory* (Cambridge: Cambridge University Press) p. 551

Carlson, R.W. & Lugmair, G.W. 2000, in: R.M. Canup & K. Righter (eds.), *Origin of the Earth & Moon* (Tucson: University of Arizona Press), p. 25

Clayton, R.N. 1993, *Annu. Rev. Earth Planet. Sci.* 21, 115

Cloutis, E.A., Gaffey, M.J., Jackowski, T.L., & Reed, K.L. 1986, *Journal Geophysical Research* 91, 11, 641

Cochran, A.L. & Vilas, F. 1998, *Icarus* 134, 207

Cochran, A.L, Vilas, F., Jarvis, K.S., & Kelley, M.S. 2004, *Icarus* 167, 360

Drake, M.J. 2001, *Meteoritics & Planetary Science* 36, 501

Duffard, R., Lazzaro, D., Licandro, J., De Sanctis, M.C., Capria, M.T., & Carvano, J.M. 2004, *Icarus* 171, 120

Floss, C., Taylor, L.A., Promprated, P., & Rumble III, D. 2005, *Meteoritics & Planetary Science* 40, 343

Franchi, I.A., Wright, I.P., Sexton, A.S., & Pillinger, C.T. 1999, *Meteoritics & Planetary Science* 34, 657

Gaffey, M.J. 1977, *Icarus* 127, 130

Gaffey, M.J., Bell, III, J.F., Brown, R.H., Burbine, T.H., Piatek, J.L., Reed, K.L., & Chaky, D.A. 1993, *Icarus* 106, 573

Greenwood, R.C., Franchi, I.A., Jambon, A., & Buchanan, P.C. 2005, *Nature* 435, 916

Hasegawa, S., Murakawa, K., Ishiguro, M., Nonaka, Hi., Takato, N., Davis, C.J., Ueno, M., & Hiroi, T. 2003, *Geophys. Res. Let.* 30 (21), 2123

Hardersen, P.S., Gaffey, M.J., & Abell, P.A. 2004, *Icarus* 167, 170

Hiroi, T., Pieters, C.M., & Takeda, H. 1994, *Meteoritics* 29, 394

Hiroi, T., Binzel, R.P., Sunshine, J.M., Pieters, C.M., & Takeda, H. 1995, *Icarus* 115, 374

Hiroi, T. & Pieters, C.M. 1998, in: T. Harasawa (ed.), *Antarctic Meteorite Research* (Tokyo: National Institute of Polar Research) p. 163

Hiroi, T., Pieters, C.M., Vilas, F., Sasaki, S., Hamabe, Y., & Kurahashi, E. 2001, *Earth Planets Space* 53, 1071

Irving, A.J., Kuehner, S.M., Carlson, R.W., Rumble, D. III, Hupé, A.C., & Hupé, G.M. 2005, *Lunar. Planet Sci. XXXVI* #2188

Keil, K. 2002, in: W.F. Bottke, A. Cellino, P. Paolicchi & R.P.Binzel (eds.), *Asteroids III* (Tucson: University Arizona Press), p. 573

Lazzaro, D., Michtchenko, T.A., Carvano, J.M., Binzel, R.P., Bus, S.J., Burbine, T.H., Moth-Diniz, T., Angeli, C.A., Florczak, M., & Harris, A.W. 2000, *Science* 288, 2033

Marzari, F., Cellino, A., Davis, D.R., Farinella, P., Zappalá, V., & Vanzani, V. 1996, *Astron. Astrophys.* 316, 1996

McCord, T.B., Adams, J.B., & Johnson, T.V. 1970, *Science* 168, 1445

Michtchenko, T.A., Lazzaro, D., Ferraz-Mello, S., & Roig, F. 2002, *Icarus* 158, 343

Mittlefehldt, D.W., McCoy, T.J., Goodrich, C.A., & Kracher, A. 1988, in: J.J. Papike (ed.), *Planetary Materials* (Mineralogical Society of America) 36, 4

Mittlefehldt, D.W. 2005, *Meteoritics. & Planet. Sci.* in press

Nyquist, L., Takeda, H., Shih, C.-Y., & Wiesmann, H. 2004, *Lunar & Planet. Sci XXXV* #1330

Pieters, C.M. & Binzel, R.P. 1994, *Lunar & Planetary Science Conference XXV*, 1083.

Pieters, C.M., Taylor, L.A., Noble, S.K., Keller, L.P., Hapke, B., Morris, R.V., Allen, C.C., McKay, D.S., & Wentworth, S. 2000, *Meteoritics & Planetary Science* 35, 1101

Ryder, G. 2002, *Journal Geophysical Research* 107 (E4, 10.1029/2001JE001583)

Rivkin, A.S., McFadden, L.A., Binzel, R.P., & Sykes, M. 2005, *Icarus* in press

Russell, C.T., Coradini, A., Christensen, U., De Sanctis, M.C., Feldman, W.C., Jaumann, R., Keller, H.U., Konopliv, A., McCord, T.B., McFadden, L.A., McSween Jr., H.Y., Mottola, S., Neukum, G., Pieters, C.M., Prettyman, T.H., Raymond, C.A., Smith, D.E., Sykes, M.V., Williams, B.G., Wise, J., & Zuber, M.T. 2004, *Planetary & Space Science* 52, 465

Sasaki, S., Hiroi, T., Nakamura, K., Hamabe, Y., Kurahashi, E., & Yamada-J, M. 2002, *Adv. Space Rev.* 29 (5), 783

Takeda, H., Mori, H., Delaney, J.S., Prinz, M., Harlow, G.E., & Ishi, T. 1983, *Mem. Natl. Inst. Polar Res. Spec.* 30, 181

Takeda, H. 1997, *Meteoritics & Planet. Sci.* 32, 841

Treiman, A.H., Lanzirotti, A., & Xirouchakis, D. 2004, *Earth & Planetary Science Letters* 219, 189

Thomas, P.C., Binzel, R.P., Gaffey, M.J., Storrs, A.D., Wells, E.N., & Zellner, B.H. 1997, *Science* 277, 1492

Vernazza, P., Mothé-Diniz, T., Barucci, M.A., Birlan, M., Carvano, J.M., Strazzulla, G., Fulchignoni, M., & Migliorini, A. 2005, *Astron. Astrophy.* 436, 1113

Wiechert, U., Halliday, A.N., Lee, D.-C., Snyder, G.A., Taylor, L.A., & Rumble, D. 2001, *Science-Verificar* 294, 345

Wiechert, U.H., Halliday, A.N., Palme, H., & Rumble, D. 2004, *Earth & Planetary Science Letters* 221, 373

Wisdom, J. 1985, *Nature* 315, 731

Yamaguchi, A., Clayton, R.N., Mayeda, T.K., Ebihara, M., Oura, Y., Miura, Y.N., Haramura, H., Misawa, K., Kojima, H., & Nagao, K. 2002, *Science* 296, 334

Asteroids, Comets, Meteors
Proceedings IAU Symposium No. 229, 2005
D. Lazzaro, S. Ferraz-Mello & J.A. Fernández, eds.

Asteroid families

David Nesvorný[1], William F. Bottke[1], David Vokrouhlický[2], Alessandro Morbidelli[3] and Robert Jedicke[4]

[1]Department of Space Studies, Southwest Research Institute, 1050 Walnut St., Suite 400, Boulder, CO 80302, USA, email: davidn@boulder.swri.edu

[2]Institute of Astronomy, Charles University, V Holešovickách 2, CZ-18000 Prague 8, Czech Republic

[3]Observatoire de la Côte D'Azur, Dept. Cassiopee, BP 4224, 06304 Nice Cedex 4, France

[4]Institute for Astronomy, University of Hawaii, 2680 Woodlawn Drive, Honolulu, HI 96822, USA

Abstract. An asteroid family is a group of asteroids with similar orbits and spectra that was produced by a collisional breakup of a large parent body. To identify asteroid families, researchers look for clusters of asteroid positions in the space of proper orbital elements. These elements, being more constant over time than osculating orbital elements, provide a dynamical criterion of whether a group of bodies has a common ancestor. More than fifty asteroid families have been identified to date. Their analysis produced several important insights into the physics of large scale collisions, dynamical processes affecting small bodies in the Solar System, and surface and interior properties of asteroids.

Keywords. minor planets, asteroids

---

1. Introduction

The asteroid belt has collisionally evolved since its formation (see Davis *et al.* 2002 and other chapters in section 4.2 of the *Asteroids III* book). Possibly its most striking feature is the asteroid families that represent remnants of large, collisionally disrupted asteroids (Hirayama 1918; Zappalà *et al.* 1995). In the present asteroid belt most asteroid families can be clearly distinguished from the background population of asteroids.

To analyze these sets separately, we will sort the main-belt asteroids into family and background populations. The analysis of the asteroid families may tell us about things such as asteroid interiors, geological differentiation in the main belt, or about phenomena that alter asteroid colors with time. The analysis of background asteroids is more related to issues such as the primordial temperature gradient in the proto-planetary nebula, subsequent dynamical excitation and mixing of bodies formed at different orbital distances from the Sun.

Here we review the problem of identification of asteroid families in large catalogs of proper elements (section 2). Table 1 lists all asteroid families identified here along with information about the number of family members, size of the parent body, family's age and spectroscopic properties. Sections 3–6 briefly review the methods used to determine these properties and lists relevant references. All families described here are in the main belt and have inclinations $i < 17$ deg; high-inclination main-belt families and families in Jupiter's Trojan swarms are not discussed.

2. Identification

To identify asteroid families, researchers look for clusters of asteroid positions in the space of proper elements: the proper semimajor axis (a_P), proper eccentricity (e_P), and proper inclination (i_P) (Milani & Knežević 1994; Knežević *et al.* 2002). These orbital elements describe the size, shape and tilt of orbits. Proper orbital elements, being more constant over time than instantaneous orbital elements, provide a dynamical criterion of whether or not a group of bodies has a common ancestor.

We used a numerical code that automatically detects a cluster of asteroid positions in the 3-dimensional space of proper elements. We based our code on the Hierarchical Clustering Method (hereafter HCM, Zappalà *et al.* 1990). The HCM requires that members of the identified cluster of asteroid positions in the proper elements space be separated by less than a selected distance (the so-called 'cutoff').

In the first step, we applied the HCM to a catalog of proper asteroid elements (Milani & Knežević 1994; Knežević *et al.* 2002). The catalog we used for this work is already dated. It included 106,284 proper elements known back in 2003. The most recent release of the proper element catalog includes almost 170,000 entries.

The HCM starts with an individual asteroid position in the space of proper elements and identifies bodies in its neighborhood with mutual distances less than a threshold limit (d_{cutoff}). We defined the distance in (a_P, e_P, i_P) space by

$$d = na_P \sqrt{C_a(\delta a_P/a_P)^2 + C_e(\delta e_P)^2 + C_i(\delta \sin i_P)^2}, \qquad (2.1)$$

where na_P is the heliocentric velocity of an asteroid on a circular orbit having the semimajor axis a_P. $\delta a_P = |a_P^{(1)} - a_P^{(2)}|$, $\delta e_P = |e_P^{(1)} - e_P^{(2)}|$, and $\delta \sin i_P = |\sin i_P^{(1)} - \sin i_P^{(2)}|$. The indexes (1) and (2) denote the two bodies in consideration. C_a, C_e and C_i are weighting factors; we adopted $C_a = 5/4$, $C_e = 2$ and $C_i = 2$ (Zappalà *et al.* 1994). Other choices of C_a, C_e and C_i yield similar results.

The cutoff distance d_{cutoff} is a free parameter. With small d_{cutoff} the algorithm identifies tight clusters in the proper element space. With large d_{cutoff} the algorithm detects larger and more loosely connected clusters. For the main belt, the appropriate values of d_{cutoff} are between 1 and 150 m/s. To avoid an a priori choice of d_{cutoff}, we developed software that runs HCM starting with each individual asteroid in our sample and loops over 150 values of d_{cutoff} between 1 and 150 m/s with a 1 m/s step.

In Fig. 1, we illustrate the final product of this algorithm using a 'stalactite' diagram (Zappalà *et al.* 1990, 1994). For each d_{cutoff} on the Y-axis we plot all clusters found by the HCM. For example, with $d_{\mathrm{cutoff}} = 150$ m/s, nearly the whole main belt is linked to a single asteroid, (1) Ceres. We plot a horizontal line segment at $d_{\mathrm{cutoff}} = 150$ m/s with length equal to the total number of members in this cluster. At smaller d_{cutoff} the complex structure of the main asteroid belt emerges. The stalactite diagram is extremely useful when we want to systematically classify this information. In fact, more than fifty significant groups are shown in Fig. 1 – twice the number of robust asteroid families known previously (Bendjoya & Zappalà 2002). We label each stalactite by the lowest numbered asteroid in the group (not all these labels appear in Fig. 1), and proceed to the second step of our algorithm.

In the second step, we select appropriate d_{cutoff} for each individual cluster. Unlike the first step of our algorithm that is fully automated, the second step requires some non-trivial insight into the dynamics of the main-belt asteroids, and cannot be fully automated. To correlate values of d_{cutoff} with structures and processes operating in the main belt, we analyzed projections of families into (a_P, e_P) and (a_P, i_P) planes. Moreover, we used an interactive visualization tool that allows us to work in three dimensions thus

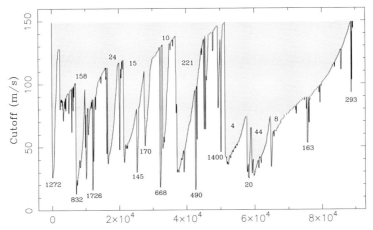

Figure 1. The dynamical structure of the asteroid belt represented in the 'stalactite diagram'. Each stalactite represents an asteroid family and is labeled by the member asteroid that has the lowest designation number. The width of a stalactite at cutoff d shows the number of family members that were identified with d. Large families, such as those associated with (158) Koronis, (24) Themis, (15) Eunomia, (221) Eos, (4) Vesta, and (44) Nysa, appear as thick stalactites that persist over a large range of d_{cutoff}. Smaller families are represented by thin stalactites that are often (but not always, see, e.g., no. 490 corresponding to the Veritas family) vertically short meaning that their determination requires a specific narrow range of d_{cutoff}.

avoiding problems generated by the projection effects into either (a_P, e_P) or (a_P, i_P) planes.

The final product of our algorithm are the asteroid families (Table 1), lists of their members selected at appropriate cutoffs, and the list of background asteroids showing no apparent groups. Figure 2 illustrates this result. From the total of 106,284 main-belt asteroids used here, 38,625 are family members (36.3% of total) and 67,659 are background asteroids (63.7%).

To determine whether our algorithm produced a reasonably complete list of asteroid families, we searched for residual clusters in the background asteroid population using proper elements and asteroid colors simultaneously. Asteroid colors were taken from the Sloan Digital Sky Survey Moving Object Catalog (Ivezić *et al.* 2001; Stoughton *et al.* 2002). We defined the distance in $(a_P, e_P, i_P, PC_1, PC_2)$ space (see section 5 for the definition of principal color components PC_1 and PC_2) by

$$d_2 = \sqrt{d^2 + C_{PC}[(\delta PC_1)^2 + (\delta PC_2)^2]}, \tag{2.2}$$

where d is the distance in (a_P, e_P, i_P) sub-space defined in Eq. (2.1), $\delta PC_1 = |PC_1^{(1)} - PC_1^{(2)}|$, and $\delta PC_2 = |PC_2^{(1)} - PC_2^{(2)}|$. The indexes (1) and (2) denote the two bodies in consideration. C_{PC} is a factor that weights the relative importance of colors in our generalized HCM search. With d in $m\,s^{-1}$, we used typically $C_{PC} = 10^6$ and varied this factor in the 10^4–10^8 range to test the dependence of results. For $C_{PC} < 10^4$ the principal components are given too little weight to be useful. For $C_{PC} > 10^8$ the orbital information, which is essential for correct family identification, is not appropriately used.

We have found no statistically robust concentrations in the extended proper element/color space that would help us to identify new families. This result shows that the list of families in Table 1 based on the available data is (at least nearly) complete. We have also found that the generalized HCM search is useful to identify family 'halos', i.e., populations of peripheric family members that were not joined with the rest of the

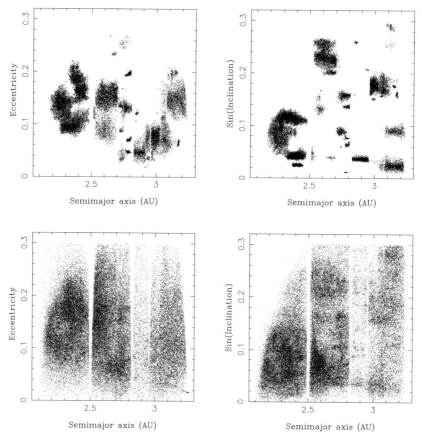

Figure 2. Decomposition of the asteroid belt into family and background asteroids. From top to bottom, the panels show the orbit distribution of the family members and background asteroids, respectively.

family with the standard HCM and cutoffs listed in Table 1. For example, the Koronis family halo located at $a_P = 2.9$ AU and small i_P can be clearly identified. Other 'halos', such as $\sim$30 additional peripheric members of the Nysa-Polana (Cellino *et al.* 2001) family can be identified only in the extended proper element/color space, because their color differs from thet of the local background ($PC_1 \lesssim 0$ for the Polana family).

By choosing cutoff distances d, we compromised (i) to include as many peripheric family members as possible, and (ii) to avoid including too many peripheric interlopers. The selected values of d that are listed in Table 1 are usually restrictive (i.e., at the small end of the acceptable range) because one of the goals of our study is to determine the reliable mean colors for family members. We thus tried to avoid including many peripheric interlopers by using small d.

In total, we have identified $\sim$50 statistically robust asteroid families. We list 42 selected main-belt families in Table 1 (i.e., the Pallas family, high-i or non-main-belt families, and several sub-structures of the prominent main-belt families are not listed). Table 1 includes all most-reliable asteroid families listed in Bendjoya & Zappalà (2002; 25 in total) except a dispersed clump of asteroids around (110) Lydia (Zappalà *et al.* 1994, 1995) which the HCM failed to identify in the new catalog of proper elements. The large overlap between our and previous family classifications shows the consistency of our approach.

By using more proper elements than previous studies we found $\approx$20 new, statistically-robust asteroid families.

3. Size-Frequency Distribution

The Size-Frequency Distribution (hereafter SFD) of observed families is much steeper than that of the local background for absolute magnitudes $H < 13$ (see Zappalà *et al.* 2002 for a review, and Morbidelli *et al.* 2003 for a definition of the local background). If this steep SFD is extrapolated to $H > 13$, small asteroid family members would become more numerous than small background asteroids. That is not what actually happens. Morbidelli *et al.* (2003) have shown that the debiased SFD of asteroid families is shallower than that of the background for $H > 15$ and that the number of family members does not exceed the number of background asteroids down to $H \approx 18$. In fact, the asteroid families probably represent about 30-40% of the main belt population of asteroids down to $D = 1$ km.

4. Size of the parent body

Durda *et al.* (2005) used hydrodynamic modeling of impacts to determine the diameter of the parent body, $D_{\rm PB}$, for 28 families. In essence, they searched for a combination of impact paremeters (including $D_{\rm PB}$) that produced best fits to the observed SFD of large family members (assuming this distribution did not change by secondary fragmentations; Bottke *et al.* 2005). The same method has been recently applied to the Karin cluster yielding $D_{\rm PB} = 31 \pm 3$ km (Nesvorný *et al.* 2005a). The hydrodynamic method avoids problems with the observational incompletness at small sizes and is more rooted in impact physics that the geometrical approch used by Tanga *et al.* (1999).

Table 1 lists $D_{\rm PB}$, $D_{\rm LF}$ and $M_{\rm LF}/M_{\rm PB}$, where $D_{\rm LF}$ is the effective diameter of the largest fragment and $M_{\rm LF}/M_{\rm PB}$ is the largest fragment to parent body mass ratio. $D_{\rm PB}$ that appear in parenthesis in Table 1 were estimated from the observed, observationally-incomplete population of family members. In these cases, true $D_{\rm PB}$ is likely to be larger than the values listed in Table 1. $D_{\rm LF}$ was taken from the SIMPS database (Tedesco *et al.* 2002) or was estimated from the absolute magnitude of the object and albedo appropriate for its taxonomic class. Values estimated via latter method appear in parenthesis in Table 1. Diameter values and $M_{\rm LF}/M_{\rm PB}$ for Bower and Brasilia families are uncertain because 1639 Bower and 293 Brasilia may be interlopers within their own families.

Tanga *et al.* (1999) determined $D_{\rm PB}$ for 14 families using pre-1999 catalog of proper elements and a geometrical approach (see Tanga *et al.* 1999 for a detail description of their method). In general, values listed in Table 1 are similar to those of Tanga *et al.* In many cases, however, the hydrodynamic method combined with our more complete sample of family members produces substantially larger $D_{\rm PB}$. A major disagreement occurs for Eos, Dora, Gefion and Merxia families where hydrodynamic $D_{\rm PB}$ are much larger than Tanga *et al.* ' $D_{\rm PB}$. For Dora and Gefion families, the discrepancy may be (at least partially) understood because Durda *et al.* did not obtain good fits for these families via hydrodynamic modeling. Conversely, Eos and Merxia families for which Durda *et al.* 's fits were good may have larger $D_{\rm PB}$ than thought before.

In total, Table 1 lists estimates of $D_{\rm PB}$ for 35 asteroid families. Out of these, 27 families have $D_{\rm PB} > 100$ km, 24 families correspond to catastrophic breakups ($M_{\rm LF}/M_{\rm PB} \leqslant 0.5$), 4 families were produced by cratering impacts (Juno, Vesta, Massalia, Nemesis; $M_{\rm LF}/M_{\rm PB} \geqslant 0.9$), 10 families correspond to super-catastrophic breakups ($M_{\rm LF}/M_{\rm PB} \leqslant 0.1$), and 19 families have $D_{\rm PB} > 100$ km and $M_{\rm LF}/M_{\rm PB} \leqslant 0.5$. Note that these numbers

Table 1. List of identified, statistically-robust asteroid families. The columns are: lowest-numbered asteroid family member; cutoff limit used (d_{cutoff}); number of family members determined with d_{cutoff}; common taxonomic type(s) of family members; mean PC_1 and PC_2 values from SDSS; age of the family, when available; diameter of the parent body (D_{PB}); diameter of the largest fragment (D_{LF}); and largest fragment to parent body mass ratio (M_{LF}/M_{PB}). Meaning of parenthesis for D_{PB} and D_{LF} values is explained in section 3.2.

Family	d_{cutoff} (m/s)	# of mem.	Tax. Type	PC_1	PC_2	Age (My)	D_{PB} (km)	D_{LF} (km)	M_{LF}/M_{PB} Mass Ratio
			Inner Main Belt, 2.0 < a < 2.5 AU						
4 Vesta	70	5575	V	0.491	-0.288	–	(471)	468	0.98
8 Flora	80	6131	S	–	–	1000 ± 500	203	136	0.3
20 Massalia	50	966	S	0.493	-0.139	150 ± 20	146	146	≈ 1
44 Nysa(Polana)	60	4744	S(F)	–	–	–	–	–	–
163 Erigone	80	410	C/X	0.138	-0.131	280 ± 50	114	73	0.26
			Central Main Belt, 2.5 < a < 2.82 AU						
3 Juno	50	74	S	0.523	-0.150	–	(234)	234	≈ 1
15 Eunomia	80	3830	S	0.624	-0.156	2500 ± 500	≈ 300	255	≈ 0.6
46 Hestia	80	154	S	0.624	-0.151	–	156	124	0.50
128 Nemesis	70	133	C	0.189	-0.196	200 ± 100	195	188	0.90
145 Adeona	60	533	Ch	0.112	-0.189	700 ± 500	184	151	0.55
170 Maria	100	1621	S/L	0.578	-0.107	3000 ± 1000	192	44	0.01
363 Padua	70	303	X/C	0.273	-0.122	300 ± 200	106	(75)	≈ 0.35
396 Aeolia	20	28	–	0.270	-0.187	–	39	34	0.66
410 Chloris	120	135	C	0.241	-0.093	700 ± 400	≈ 175	124	≈ 0.35
569 Misa	80	119	C	0.154	-0.185	500 ± 200	117	73	0.24
606 Brangane	30	30	S	0.441	0.061	50 ± 40	46	36	0.48
668 Dora	70	404	Ch	0.091	-0.190	500 ± 200	≈ 165	27	≈ 0.004
808 Merxia	100	271	S/Sq	0.455	-0.115	240 ± 50	121	33	0.02
847 Agnia	40	252	S/Sq	0.435	-0.169	100 ± 30	60	28	0.10
1128 Astrid	50	65	C	0.221	-0.210	180 ± 80	(41)	35	≈ 0.6
1272 Gefion	80	973	S	0.544	-0.123	1200 ± 400	212	(35)	≈ 0.005
1639 Bower	100	82	–	0.528	0.026	–	(52)	36	≈ 0.3
1644 Rafita	100	382	S	0.538	-0.127	1500 ± 500	63	(42)	≈ 0.3
1726 Hoffmeister	50	235	C	0.058	-0.115	300 ± 200	134	26	0.007
2980 Cameron	60	162	S	0.518	-0.116	–	–	–	–
4652 Iannini	30	18	S	0.324	-0.109	$\lesssim 5$	–	–	–
			Outer Main Belt, 2.82 < a < 3.5 AU						
10 Hygiea	80	1136	C/B	0.081	-0.170	2000 ± 1000	443	407	0.78
24 Themis	90	2398	C/B	0.092	-0.179	2500 ± 1000	448	(225)	≈ 0.13
87 Sylvia	60	19	–	0.137	0.033	–	272	260	0.87
137 Meliboea	120	57	Ch	0.185	-0.161	–	242	145	0.22
158 Koronis	70	2304	S	0.522	-0.111	2500 ± 1000	166	35	0.01
221 Eos	80	4412	K/T/D	0.466	-0.104	1300 ± 200	401	104	0.02
283 Emma	40	76	–	0.129	-0.053	–	185	148	0.51
293 Brasilia	80	95	C/X	0.222	-0.076	50 ± 40	206	55	0.02
490 Veritas	50	284	Ch	0.212	-0.230	8.3 ± 0.5	177	115	0.27
832 Karin	10	84	S	0.387	-0.228	5.8 ± 0.2	31 ± 3	(17)	≈ 0.16
845 Naema	40	64	C	0.135	-0.178	100 ± 50	81	54	0.30
1400 Tirela	70	212	D	0.714	-0.128	–	–	–	–
3556 Lixiaohua	50	97	C/X	0.170	-0.080	300 ± 200	203	30	0.003
9506 Telramund	60	70	S	0.502	-0.166	–	–	–	–
18405 FY12	50	11	X	0.214	0.001	–	–	–	–

do not include Nysa/Polana complex which we did not resolve into two families via HCM. These numbers provide important constraints on the collisional history of the main belt.

5. Spectral properties

Taxonomic classification of asteroid families from visible spectroscopy was recently reviewed by Cellino *et al.* (2002) and Mothé-Diniz *et al.* (2005) (see also Bus & Binzel 2002a, 2002b). The principal result of spectroscopic studies is that members of an asteroid family show similar reflectance spectra. This has been taken as an evidence that main belt asteroids generally have non-differentiated interiors. Nevertheless, important variation of spectral properties can exist among different members of a single dynamical family probably due to somewhat heterogenous composition of the parent asteroid.

Spectroscopy is particularly useful to identify interlopers within families. These objects have the taxonomic type which is typical for background asteroids at the semimajor axis location of a family (S in the inner main belt, C in the outer main belt) and which contrasts with the family's own taxonomic class. For example, asteroids no. 100, 108, 1109, 1209 and 1599 are spectroscopic interlopers in the Hygiea family, asteroids no. 85 and 141 in the Eunomia family, asteroids no. 423 and 507 in the Eos family, asteroids no. 83, 255 and 481 in the Gefion family. It is important to exclude these large interlopers.

Our Table 1 includes 24 families that were not listed in Cellino *et al.* (2002) because their taxonomic type was not known previously. On the other hand, Cellino *et al.* listed 11 families that are not included in our Table 1. Four of these families (14 Bellona, 88 Thisbe, 226 Weringia, 729 Watsonia) are very dispersed asteroid groups in proper element space and have been identified by means of spectroscopy rather than by the analysis of the proper elements. The (2) Pallas family is not included in our list because (2) Pallas has highly inclined orbit and does not appear in the catalog of analytically calculated proper elements that we use here. Three of Cellino *et al.* 's families (125 Liberatrix, 237 Coelestina, 322 Phaeo) were identified by means of the wavelet analysis of the proper elements. These families are statistically less robust because the wavelet analysis is known to impose more relaxed criteria on family membership than the HCM (Zappalà *et al.* 1995). The (2085) Henan and (110) Lydia families are rather dispersed, possibly old families that we have failed to identify as reliable asteroid families by using the most recent proper element catalog. Finally, the HCM fails to identify Cellino *et al.* 's (45) Eugenia family with $d \leqslant 120$ m/s, while all our other families show many members with $d \leqslant 120$ m/s values.

Significant color variation exist between different asteroid families partly because of the varying mineralogy of their parent asteroids and probably also due effects of space weathering (Jedicke *et al.* 2004; Nesvorný *et al.* 2005b). Table 1 lists the color information for 40 asteroid families. Asteroid colors were taken from the Sloan Digital Sky Survey Moving Object Catalog, hereafter SDSS MOC. The second release of SDSS MOC includes five-color CCD photometry for 125,283 moving objects (Ivezić *et al.* 2001; Stoughton *et al.* 2002).

35,401 unique moving objects detected by the survey (i.e., about 28% of the total) have been matched (Jurić *et al.* 2002) to known asteroids listed in the ASTORB file (Bowell *et al.* 1994). The flux reflected by the detected objects was measured almost simultaneously in five bands (measurements in two successive bands were separated in time by 72 seconds) with effective wavelengths 3557 Å (*u* band), 4825 Å (*g* band), 6261 Å (*r* band), 7672 Å (*i* band), and 9097 Å (*z* band), and with 0.1-0.3μm band widths (Fukugita *et al.* 1996). The SDSS photometry is broadly consistent with published spectra of asteroids (Nesvorný *et al.* 2005b).

To make an efficient use of the SDSS MOC, we utilized the Principal Component Analysis (hereafter PCA). The PCA involves a mathematical procedure that transforms a number of possibly correlated variables into a smaller number of uncorrelated variables called principal components. The first principal component accounts for as much of the variability in the data as possible, and each succeeding component accounts for as much of the remaining variability as possible. In the result, the PCA creates linear combinations of the five SDSS colors that maximize the separation between the taxonomic types in the SDSS data.

The first two principal components (i.e., the new uncorrelated variables) that this algorithm yields are given by the following relationships:

$$PC_1 = 0.396(u - g) + 0.553(g - r) + 0.567(g - i) + 0.465(g - z)$$
$$PC_2 = -0.819(u - g) + 0.017(g - r) + 0.09(g - i) + 0.567(g - z), \qquad (5.1)$$

where u, g, r, i, z are the measured fluxes in five bands after correction for solar colors.

We used the subset of entries in the SDSS MOC that were matched to asteroids with known orbit elements (Jurić *et al.* 2002) and that have $\delta PC_1 < 0.1$ and $\delta PC_2 < 0.1$, where δPC_1 and δPC_2 are the measurement errors in the principal components. In total, we studied colors of 7,593 main-belt asteroids of which 3,026 are family members and 4,567 are background main belt objects.

Table 1 lists mean values of PC_1 and PC_2 for individual families. In cases where the taxonomic type of a family was known previously from observations of its large members (see Cellino *et al.* 2002 and references therein) we found that the PC_1 value suggests that this type is also predominant for small asteroid family members observed by the SDSS (see also Ivezić *et al.* 2002).

Figure 3 shows the mean colors of S-, C- and X-complex families. The S-complex families have $PC_1 > 0.35$ while C- and X-complex families have $PC_1 < 0.35$. The V-type Vesta family (denoted in green) differs from the S-type families by small PC_2. The (832) Karin cluster and the (490) Veritas family, the two youngest known asteroid families for which we have good color data, are located on a periphery of regions in (PC_1, PC_2) plane that are populated by the S- and C-type families. There is a significant spread of PC_1 and PC_2 for families of the same taxonomic complex. For example, two C-complex families can differ by ∼0.1-0.2 in PC_1 and/or PC_2. Because measurement errors and errors of the mean colors are ≲0.05, Fig. 3 documents true color differences between families. This result is consistent with studies of spectral variability among families by higher-resolution spectrophotometric measurements (Cellino *et al.* 2002).

6. Age

Accurate determination of the age of an asteroid family can obtained by modeling of orbital dynamics and dispersal of family members via effects of the Yarkovsky force (see Bottke *et al.* 2002 for a review of the Yarkovsky effect). Two different methods have been used to date:

1. Backward Numerical Integration of Orbits. To determine the exact age of a young family, the orbits of the family members can be numerically integrated into the past. The goal is to show that in some previous epoch the orbits of all cluster members were nearly the same (Nesvorný *et al.* 2002). There are two angles that determine the orientation of an orbit in space: the longitude of the ascending node (Ω) and the argument of perihelion (ω). Due to planetary perturbations these angles evolve with different but nearly constant speeds for individual asteroid orbits. Today, the orbits of the family members are oriented

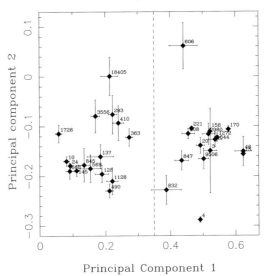

Figure 3. Principal color components PC_1 and PC_2 for asteroid families: PC_1 is a proxy for the spectral slope and PC_2 is related to the spectral curvature (Nesvorný *et al.* 2005b). The bars show 1-σ errors of mean PC_1 and PC_2. Only families with at least two members matching various criteria are shown (see Nesvorný *et al.* 2005b). The C- and S-complex families clearly segregate in PC_1 by the vertical dashed line at $PC_1 = 0.35$. The mean colors of recently-formed Veritas (no. 490) and Karin (no. 832) families are located on a periphery of well defined regions in (PC_1, PC_2) plane that are populated by the C- and S-complex families, respectively. Note the large color difference between the (832) Karin and (158) Koronis families. The Vesta family (no. 4) has the smallest mean PC_2 value among the identified families; the Brangane family (no. 606) has the largest mean PC_2 value.

differently in space because their slightly dissimilar periods of Ω and ω produce slow differential rotation of their orbits with respect to each other. Eventually, this effect allows Ω and ω to obtain nearly uniform distributions in $[0°, 360°]$. For a short time after the parent body breakup, however, the orientations of the fragments' orbits must have been nearly the same. Nesvorný *et al.* (2002, 2003) used this method to determine the ages of the Karin (5.8 ± 0.2 My) and Veritas (8.3 ± 0.5 My) families, and also found that the tight family associated with (4652) Iannini is probably $\lesssim 5$ My old. Unfortunately, Nesvorný *et al.*'s method can not be used to determine ages of asteroid families that are much older than ≈ 10 My, because it is difficult to accurately track the orbits of asteroids on $\gtrsim 10$-My time scales.

 2. *Modeling of Family Spreading via Thermal Forces.* A recent analysis has shown that the asteroid families are subject to slow spreading and dispersal via Yarkovsky thermal effect (Bottke *et al.* 2001). Therefore, old families' orbital parameters in (a_P, e_P, i_P) space do not reflect the immediate outcomes of cratering events or catastrophic disruptions. Instead, they reveal how the family members evolved in (a_P, e_P, i_P) space over long timescales by dynamical diffusion and chaotic resonances. On the other hand, tight clusters in (a_P, e_P, i_P) space should represent young families that have not yet had an opportunity to disperse via dynamical mechanisms. Using theoretical models of the Yarkovsky effect, Nesvorný *et al.* (2005b) and Vokrouhlický *et al.* (2005a,b) have estimated ages for 25 asteroid families. Vokrouhlický *et al.*'s method is more refined and results in smaller error bars if the assumed albedo values, p_v, are correct. The ages listed in Table 1 for Eos, Agnia, Erigone, Massalia, Merxia and Astrid families assume $p_v = 0.13, 0.17, 0.05, 0.21, 0.22, 0.08$, respectively (Vokrouhlický *et al.* 2005a,b).

The ages were not determined for several families listed in Table 1 due to various reasons. (3) Juno and (4) Vesta families correspond to cratering events for which the contribution of ejection speeds to the present spreads of families in a_P is probably very large (Asphaug 1997) making it difficult to separate this contribution from the subsequent spreading by thermal effects. (44) Nysa-Polana clan is a case of two families that overlap in a_P, e_P, i_P space and are difficult to separate (e.g., Cellino *et al.* 2002). (87) Sylvia family is located in a dynamically complicated region at $a_P \sim 3.5$ AU where chaotic resonances (rather than thermal effects) determine the dynamical structure of a family. Similarly, (46) Hestia and (2980) Cameron families (both S-type) were affected by strong chaotic resonances in the past (e.g., the 3:1 mean motion resonance for the Hestia family) that removed large fractions of their original populations. The dynamical structures of these families is complex. (18405) FY12 family has only eleven known members. For this reason, its age cannot be reliably determined by methods described above that are statistical in nature. Finally, (137) Meliboea family is a C-type family in the outer main belt that is probably very old because of its large spread in a_P, e_P, i_P. A large fraction of dynamical interlopers ($\sim 30\%$ based on the SDSS colors) and the difficulty with choosing an adequate d_{cutoff} prevents us from better estimating the age of this family.

References

Asphaug, E. 1997, *Meteoritics and Planetary Science* 32, 965–980.

Bendjoya, P. & Zappalà, V. 2002, in: W. F. Bottke, A. Cellino, P. Paolicchi & R.P. Binzel (eds.), *Asteroids III*, (Tucson: Univ. Arizona Press), p. 613

Bottke, W.F., Vokrouhlický, D. Brož, M., Nesvorný, D., & Morbidelli, A. 2001 *Science* 294, 1693

Bottke, W.F., Vokrouhlický, D. Rubincam, D.P., & Brož, M. 2002, in: W.F. Bottke, A. Cellino, P. Paolicchi & R. Binzel (eds.), *Asteroids III*, (Tucson: Univ. Arizona Press), p. 395

Bottke, W.F., Durda, D.D., Nesvorný, D., Jedicke, R., Morbidelli, A., Vokrouhlický, D., & Levison, H. 2005, *Icarus* 175, 111

Bowell E., Muinonen, K., & Wasserman, L.H. 1994, in: A. Milani (eds.), *Asteroids, Comets and Meteors*, (Dordrecht: Kluwer), p. 477

Bus, S.J. & Binzel, R.P. 2002a., *Icarus* 158, 106

Bus, S.J. & Binzel, R.P. 2002b., *Icarus* 158, 146

Cellino, A., Zappalà, V., Doressoundiram, A., Di Martino, M., Bendjoya, P., Dotto, E., & Migliorini, F. 2001, *Icarus* 152, 225

Cellino, A., Bus, S.J., Doressoundiram, A., & Lazzaro D. 2002, in: W.F. Bottke, A. Cellino, P. Paolicchi & R.P. Binzel (eds.) *Asteroids III*, (Tucson: Univ. Arizona Press), p. 633

Davis, D.R., Durda, D.D., Marzari, F., Campo Bagatin, A., & Gil-Hutton R. 2002, in: W.F. Bottke, A. Cellino, P. Paolicchi & R.P. Binzel (eds.) *Asteroids III*, (Tucson: Univ. Arizona Press), p. 545

Durda, D.D., Bottke, W.F., Nesvorný, D., Enke, B.L., Asphaug, E., & Richardson, D.C. 2005, *Icarus*, to be submitted.

Fukugita, M., Ichikawa, T., Gunn, J.E., Doi, M., Shimasaku, K., & Schneider, D.P. 1996, *Astron. J.* 111, 1748

Hirayama, K. 1918, *Astron. J.* 31, 185

Ivezić, Ž., and 32 colleagues 2001, *Astron. J.* 122, 2749

Ivezić, Ž., Lupton, R.H., Jurić, M., Tabachnik, S., Quinn, T., Gunn, J.E., Knapp, G.R., Rockosi, C.M., & Brinkmann, J. 2002, *Astron. J.* 124, 2943

Jedicke, R., Nesvorný, D., Whiteley, R., Ivezić, Ž., & Jurić M. 2004, *Nature* 429, 275

Jurić, M., and 15 colleagues 2002, *Astron. J.* 124, 1776

Knežević, Z., Lemaître, A., & Milani, A. 2002, in: W.F. Bottke, A. Cellino, P. Paolicchi & R. Binzel (eds.), *Asteroids III*, (Tucson: Univ. Arizona Press), p. 603

Milani, A. & Knežević, Z. 1994, *Icarus* 107, 219

Morbidelli, A., Nesvorný, D., Bottke, W.F., Michel, P., Vokrouhlický, D., & Tanga, P. 2003, *Icarus* 162, 328

Mothé-Diniz, T., Roig, F., & Carvano, J.M. 2005, *Icarus* 174, 54

Nesvorný, D., Bottke, W.F., Dones, L., & Levison, H.F. 2002a, *Nature* 417, 720

Nesvorný, D., Bottke, W.F., Levison, H.F., & Dones, L. 2003, *Astroph. J.* 591, 486

Nesvorný, D., Enke, B.L., Bottke, W.F., Durda, D.D., Asphaug, E., & Richardson, D.C. 2005a, *Icarus*, submitted.

Nesvorný, D., Jedicke, R., Whiteley, R.J., & Ivezić, Ž. 2005b, *Icarus* 173, 132

Stoughton, C., and 191 colleagues 2002, *Astron. J.* 123, 485

Tanga, P., Cellino, A., Michel, P., Zappalà, V., Paolicchi, P., & dell'Oro, A. 1999, *Icarus* 141, 65

Tedesco, E.F., Noah, P.V., Noah, M., & Price, S.D. 2002, *Astron. J.* 123, 1056

Vokrouhlický, D., Brož, M., Morbidelli, A., Bottke, W.F., Nesvorný, D., Lazzaro, D., & Rivkin, A.S. 2005a, *Icarus*, in press.

Vokrouhlický, D., Brož, M., Bottke, W.F., Nesvorný, D., & Morbidelli, A. 2005b, *Icarus*, in press.

Zappalà, V., Cellino, A., Farinella, P., & Knežević, Z. 1990, *Astron. J.* 100, 2030

Zappalà, V., Cellino, A., Farinella, P., & Milani, A. 1994, *Astron. J.* 107, p. 772

Zappalà, V., Bendjoya, P., Cellino, A., Farinella, P., & Froeschle, C. 1995, *Icarus* 116, 291

Zappalà, V., Cellino, A., Dell'Oro, A., & Paolicchi, P. 2002, in: W.F. Bottke, A. Cellino, P. Paolicchi & R.P. Binzel (eds.), *Asteroids III*, (Tucson: Univ. Arizona Press), p. 619

Asteroids, Comets, Meteors 2005
Proceedings IAU Symposium No. 229, 2005
D. Lazzaro, S. Ferraz-Mello & J.A. Fernández, eds.

© 2006 International Astronomical Union
doi:10.1017/S1743921305006812

Solar System binaries

Keith S. Noll[1]

[1]Space Telescope Science Institute, 3700 San Martin Drive, Baltimore, MD 21218, USA
email: noll@stsci.edu

Abstract. The discovery of binaries in each of the major populations of minor bodies in the solar system is propelling a rapid growth of heretofore unattainable physical information. The availability of mass and density constraints for minor bodies opens the door to studies of internal structure, comparisons with meteorite samples, and correlations between bulk- physical and surface-spectral properties. The number of known binaries is now more than 70 and is growing rapidly. A smaller number have had the extensive followup observations needed to derive mass and albedo information, but this list is growing as well. It will soon be the case that we will know more about the physical parameters of objects in the Kuiper Belt than has been known about asteroids in the Main Belt for the last 200 years. Another important aspect of binaries is understanding the mechanisms that lead to their formation and survival. The relative sizes and separations of binaries in the different minor body populations point to more than one mechanism for forming bound pairs. Collisions appear to play a major role in the Main Belt. Rotational and/or tidal fission may be important in the Near Earth population. For the Kuiper Belt, capture in multi-body interactions may be the preferred formation mechanism. However, all of these conclusions remain tentative and limited by observational and theoretical incompleteness. Observational techniques for identifying binaries are equally varied. High angular resolution observations from space and from the ground are critical for detection of the relatively distant binaries in the Main Belt and the Kuiper Belt. Radar has been the most productive method for detection of Near Earth binaries. Lightcurve analysis is an independent technique that is capable of exploring phase space inaccessible to direct observations. Finally, spacecraft flybys have played a crucial paradigm-changing role with discoveries that unlocked this now-burgeoning field.

Keywords. Kuiper Belt, minor planets, asteroids, planets and satellites: formation

1. Two Are Better Than One

From time to time in science there are paradigm-shifting developments that occur at such a rapid pace that it is recognized that an important new subfield of science has emerged. Such is the case for the study of gravitationally bound companions to minor bodies in the solar system, a group of objects that can be referred to as *binary minor planets*, or, more simply, *binaries*. The broad term *binary* used in this context often includes objects that would normally be considered satellites because of the large mass ratio of primary to secondary, others that are true binaries or doubles where the mass ratio of primary to secondary is closer to one, and contact binaries and bilobate objects. Systems consisting of more than two gravitationally bound objects can also exist and one such system has now been observed (Marchis *et al.* 2005a).

For the linguistic purists, it is worth noting that the term *binary* is frequently used to refer to gravitationally bound minor planets regardless of the mutual size of the components. If one cared to do so, a *true* binary could be defined as a system where the

Table 1. Significant Milestones in Solar System Binaries

Date	Object	Significance	Reference
1801	(1) Ceres	search for asteroid satellites begins	
1901	(433) Eros	claim of double from lightcurve	André 1901
1978	Pluto/Charon	first binary TNO	Christy & Harrington 1978
1990	(4769) Castalia	first bilobate NEO imaged	Ostro et $al.$ 1990
1993	(243) Ida	first Main Belt satellite	Belton & Carlson 1994
1997	1994 AW_1	first asynchronous lightcurve NEA	Pravec & Hahn 1997
1998	(45) Eugenia	first groundbased detection of main belt satellite	Merline et $al.$ 1999a,b
2001	2000 DP_{107}	first binary NEA radar detection	Ostro et $al.$ 2000
2001	1998 WW_{31}	second binary TNO	Veillet et $al.$ 2001
2005	(87) Sylvia	first multiple system	Marchis et $al.$ 2005a

barycenter resides outside of either of the two gravitationally bound bodies. By this definition, Pluto and Charon qualify as the first known solar system binary (not counting the Sun-Jupiter binary). The requirement for meeting this definition can be expressed as

$$a > r_p \left(1 + \frac{m_p}{m_s}\right)$$

where a is the semimajor axis of the system, m_p and m_s are the masses of the two components, and r_p is the radius of the primary which can be expressed in terms of the primary mass and density as $r_p = (3m_p/4\pi\rho)^{1/3}$. In practice, however, many systems will not have measured radii or known densities and albedos and thus the location of the barycenter relative to the surfaces will be uncertain. As with many other instances of taxonomical terminology, however, the precise use of the term *binary* is less important than is a detailed knowledge of the objects in question. In this review I will use the term *binary* in the broadest sense to include both true binaries and related classes of objects found among the minor body populations of the solar system.

2. The Discovery of Solar System Binaries

The two-century-long history of searches for and eventual discovery of binary minor planets has been summarized in several earlier reviews, most recently and most thoroughly by Merline et $al.$ (2002) and by Richardson & Walsh (2005). Several milestones in that history are especially noteworthy and are summarized in Table 1.

Over time there have been significant shifts in the prevailing view of the prevalence and nature of binaries in the solar system. Almost immediately after the discovery of (1) Ceres by G. Piazzi in early 1801, searches for satellites began. These searches continued with varying degrees of intensity for nearly two centuries without the definitive detection of a satellite or binary. Reports of possible binaries in the 20th century were numerous, the first being an analysis of the lightcurve of (433) Eros in comparison to the lightcurves of spectroscopic double stars (André 1901). The long series of unsuccessful searches had become so discouraging by the early 1980s that Weidenschilling et $al.$ (1989) titled their review article "Do Asteroids Have Satellites?". It is interesting to note that by that time Pluto's binary companion Charon had already been discovered, but the existence of a large transneptunian population of small bodies was unknown and Pluto was still solidly ensconced as a major planet.

The veil began to be drawn back in 1990 with the discovery of the bilobed nature of the NEA 1989 PB, now known as (4769) Castalia (Ostro et $al.$ 1990). The breakthrough came with the serendipitous discovery of (243) Ida's satellite Dactyl; unlike all previous reports this claim was indisputable. While the reality of bound systems among the minor bodies of the solar system could no longer be questioned, it remained to be seen whether the Ida/Dactyl pair was an anomaly or representative of a significant, but as yet undetected population of bound systems. In 1997–1998 this question was dramatically answered by two major steps forward: the detection of the first asynchronous lightcurve binary (Pravec *et al.* 1997) and the first ground-based detection of an asteroid satellite (Merline *et al.* 1999a). The method of identifying binaries by resolving complex lightcurves into two simple periodic components punctuated by mutual events pioneered by Pravec and colleagues was not universally accepted initially and so the first radar detection of a binary, 2000 DP$_{107}$ by Ostro *et al.* (2000) inaugurated an important new and definitive technique for the identification of NEA binaries. Soon after, this object was also found to have eclipse/occultation events on top of the rotation lightcurve of the primary (Pravec *et al.* 2000a); an important step in proving the reliability of the lightcurve method that is responsible for a large fraction of the known NEA and Main Belt binaries (see Tables 2–3). The detection of the companion of 1998 WW$_{31}$, which we now understand to be the *second* known binary in the TNO population shattered several prevailing assumptions, including the assumed uniqueness of the Pluto/Charon system and the thought that other transneptunian binaries, if they did exist, would have separations and orbits on a scale comparable to the Pluto/Charon system and would thus be difficult to detect around all but the largest TNOs, even with HST.

Taken together, the change in little more than a decade since the discovery of Dactyl can only be described as breathtaking. With such a rapid pace of development, the process of reviewing a field is daunting. In the sections below I have tried to identify aspects of this field that are currently the best developed or where important new shifts in thinking are taking place. This is not the first review of this field and will certainly not be the last. It is my modest hope, however, that it is a fair snapshot of where we are at the time of the IAU Symposium 229, Asteroid, Comets, Meteors in 2005.

3. The Current Inventory

The number of known or suspected binary systems continues to grow rapidly. Table 2 summarizes the inventory of binary minor planets reported as of October 2005. The table includes objects directly observed with imaging instruments or radar and systems identified through lightcurve analysis. Not included are reported instances of possible occultations by companions that are not generally accepted as sufficient evidence for a claim of a binary, although some could be genuine. These more tenuous claims have been summarized by (Weidenschilling *et al.* 1989). The total number of binary systems in Tables 2–4 is 73 including, 24 NEA binaries, 26 Main Belt binaries, 1 Jupiter Trojan, and 22 transneptunian binaries. Despite searches, no binary Centaurs have been found. Neither of the 2 Neptune Trojans are known to be binary. There are insufficient statistics, however, to determine whether these non-detections are of any significance. However, searches for binaries in these populations would certainly be worthwhile both for the potential physical information that binaries can yield and for studies of binary statistics in different populations. With just a few tens of objects spread over many possible groupings of size, spectral type, dynamical class, etc., this list, though impressive, is still inadequate for many of the kinds of questions we would like to ask. It is clear, however, that the known objects are only a tiny fraction of the population of bound systems that are potentially detectable with current and near-future technology.

Table 2. Near Earth Binaries

Object	semimajor axis a (km)	a/r_p	r_s/r_p	Period (hours)	reference[1]
radar[2]					
2000 DP$_{107}$	2.6(2)	6.5	0.41(2)	42.2(1)	[Mg02][P05a]
2000 UG$_{11}$	*0.3*	*2.6*	*0.36(9)*	18.4(1/2)	[No00][P05a]
(66391) 1999 KW$_4$			*0.3-0.4*	17.44(1)	[Be01][PS01][P05a]
1998 ST$_{27}$	<7	<9	*0.15*		[Be03]
2002 BM$_{26}$			*0.2*	<72	[No02a]
2002 KK$_8$			*0.2*		[No02b]
(5381) Sekhmet	1.54(12)	*3.1*	*0.3*	12.5(3)	[No03a][Ne03]
2003 SS$_{84}$			*0.5*	*24*	[No03b]
(69230) Hermes	*0.6*	*2.5-4*	*0.9(1)*	13.894(4)	[Mg03][P05a]
1990 OS	*> 0.6*	*>4*	*0.15*	*18-24*	[Os03]
2003 YT$_1$	*2.7*	*5*	0.19(9)	*30*	[No04a,b]
2002 CE$_{26}$	*5*	*3*	*0.05*	*16*	[Sh04][Sl04]
lightcurve[2]					
1994 AW$_1$			0.49(2)	22.33(1)	[PH97][P05a]
(35107) 1991 VH	*6.5(2.0)*	5.4(6)	0.37(3)	32.66(5)	[P98][P05a]
(3671) Dionysus		4.5(1.0)	0.20(2)	27.74(1)	[Mo97][P05a]
1996 FG$_3$		3.2(4)	0.29(2)	16.135(10)	[P00b][ML00][P05a]
(5407) 1992 AX			0.2(1)	*13.520(1)*	[P00b][P05a]
(31345) 1998 PG			*0.3*	*14.005(1)*	[P00b][P05a]
(88710) 2001 SL$_9$			0.28(2)	16.40(2)	[P01a][P05a]
1999 HF$_1$		4.0(6)	0.22(3)	14.02(1)	[P02][P05a]
(66063) 1998 RO$_1$			0.48(3)	14.54(2)	[P03a][P05a]
(65803) Didymos	*1.1(2)*	2.9(4)	0.22(2)	11.91(1)	[P03b][P05a]
(85938) 1999 DJ$_4$	*0.7*	*3*	0.5(1)	17.73(1)	[P04][Be04][P05a]
2005 AB			⩾0.24	*17.9*	[Rd05]

Published uncertainties in least significant digit(s) shown in parentheses. Estimated quantities or values published without error estimates shown in italics.
[1] Selected references shown. For more complete references see Richardson & Walsh (2005) and Pravec *et al.* (2005a).
[2] Method resulting in discovery. Many objects have been observed by both radar and lightcurve techniques.

[Be01] Benner *et al.* 2001, [Be03] Benner *et al.* 2003, [Be04] Benner *et al.* 2004, [Mg02] Margot *et al.* 2002, [Mg03] Margot *et al.* 2003, [Mo97] Mottola *et al.* 1997, [ML00] Mottola & Lahulla 2000, [Ne03] Neish *et al.* 2003, [No00] Nolan *et al.* 2000, [No02a,b] Nolan *et al.* 2002a,b, [No03] Nolan *et al.* 2003, [No04] Nolan *et al.* 2004, [Os03] Ostro *et al.* 2003, [PH97] Pravec & Hahn 1997, [P98] Pravec *et al.* 1998, [P00b] Pravec *et al.* 2000b, [PS01] Pravec & Sarounova 2001, [P02] Pravec *et al.* 2002, [P03] Pravec *et al.* 2003, [P04] Pravec *et al.* 2004, [P05a] Pravec *et al.* 2005a, [Sh04] Shepard *et al.* 2004, [Sl04] Schlieder *et al.* 2004, [Rd05] Reddy *et al.* 2005, [WPP05] Warner, Pravec, & Pray 2005

3.1. *Population statistics*

The number of binaries is still quite small for any statistical studies, particularly studies related to the binary fraction in different subpopulations. There are, however, interesting hints that are becoming apparent with the current inventory (Table 5). As history has shown with other studies of minor bodies, as the number of discoveries increase, important patterns can be expected to emerge.

Any discussion of the frequency of binaries must take into account observational limits and biases. This is particularly important in the Kuiper Belt where the separations of

Table 3. Main Belt Binaries

Object	semimajor axis a (km)	a/r_p	r_s/r_p	Period (days)	reference[1]
imaging					
(243) Ida	*108*	*7.0*	*0.045*	*1.54*	[BC94][Me02]
(45) Eugenia	1196(4)	*11.1*	*0.06*	4.7244(10)	[Me99b][Me02][Ma04]
(90) Antiope	170(1)	*3.1(5)*	*0.99*	0.68862(5)	[Me00a][De05]
(762) Pulcova	*810*	*11.6*	*0.14*	*4.0*	[Me00b][Me02]
(87) Sylvia[2]	1356(5)	*17.6(7)*	0.12(2)	3.6496(7)	[Br01][St01a][Ma05a]
	706(5)	*9.2(4)*	0.045(15)	1.3788(7)	[Ma05a]
(107) Camilla	1235(16)	*11*	0.040(4)	3.710(1)	[St01b][Ma05b]
(22) Kalliope	1065(8)	*11.8(4)*	0.22(2)	3.590(1)	[Me01a][MB01][Ma03]
(3749) Balam	310(20)	*52*	*0.22*	110(25)	[Me02a][Ma05b]
(121) Hermione	768(11)	*7.4(3)*	*0.06*	2.582(2)	[Me02b][Ma05c]
(1509) Escalonga			*0.33*		[Me03a]
(283) Emma	596(3)	*8*	*0.08*	3.360(1)	[Me03b][Ma05b]
(379) Huenna	3400(11)	*74*	*0.08*	80.8(4)	[Mg03b][Ma05b]
(130) Elektra	1252(30)	*13.5*	*0.02*	3.92(3)	[Me03c][Ma05b]
(22899) 1999 TO$_{14}$			*0.3*		[Me03d]
(17246) 2000 GL$_{74}$			*0.4*		[Tm04]
(4674) Pauling			*0.3*		[Me04a]
lightcurve					
(3782) Celle			0.42(2)	1.5238(13)	[Ry03]
(1089) Tama			*0.7*	0.6852(2)	[Bh04a]
(1313) Berna				1.061(5)	[Bh04b]
(4492) Debussy				*1.108*	[Bh04c]
(854) Frostia			*0.4*	*1.565*	[Bh04d]
(5905) Johnson			0.40(4)	0.907708(2)	[W05a,c]
(76818) 2000 RG$_{79}$			0.37(3)	0.5885(4)	[W05b]
(3982) Kastel					[P05b]
(809) Lundia			*1*	*0.64*	[Kr05]
(9069) Hovland			*0.3*		[W05c]
Trojan					
(617) Patroclus)	685(40)	*11(1)*	0.92(5)	2.391(3) *or* 4.287(2)	[Me01b][Ma05d]

Published uncertainties in least significant digit(s) shown in parentheses. Estimated quantities or other values published without error estimates shown in italics.
[1] Selected references shown. For additional references see Merline *et al.* (2002) or Richardson & Walsh (2005).
[2] Average of long and short ellipsoid axes used for r.
[BC94] Belton & Carlson 1994, [Br01] Brown *et al.* 2001, [De05] Descamps *et al.* 2005, [Kr05] Kryszczynska *et al.* 2005, [Ma03] Marchis *et al.* 2003, [Ma04] Marchis *et al.* 2004, [Ma05a,b,c,d] Marchis *et al.* 2005a,b,c,d, [MB01] Margot & Brown 2001, [Me99] Merline *et al.* 1999, [Me00a,b] Merline *et al.* 2000a,b, [Me01a,b] Merline *et al.* 2001a,b, [Me02a,b] Merline *et al.* 2002a,b, [Me03a,b,c,d] Merline *et al.* 2003a,b,c,d, [Me04a] Merline *et al.* 2004a, [P05b] Pravec *et al.* 2005b, [Ry03] Ryan *et al.* 2003, [St01a,b] Storrs *et al.* 2001a,b, [Tm04] Tamblyn *et al.* 2004, [W05a,b,c] Warner *et al.* 2005a,b,c

the majority of binaries are not resolved by ground-based observations and among the smaller radii families in the Main Belt where potential binaries are extremely faint.

Gross population statistics are most frequently cited for each of the major small-body populations, although, as discussed below this is probably an oversimplification. In the most recent and extensive work on the subject, Pravec *et al.* (2005a) find $15 \pm 4\%$ of NEAs

Table 4. Transneptunian Binaries

Object	semimajor axis a (km)	a/r_p	r_s/r_p	e	Period (days)	reference[1]
classical						
(88611) 2001 QT$_{297}$	27,300(340)	*410*	*0.72*	0.240(3)	825(3)	[Op03] [K05]
1998 WW$_{31}$	22,300(800)	*300*	*0.83*	0.82(5)	574(10)	[V02]
(58534) 1997 CQ$_{29}$	8,010(80)	*200*	*0.91*	0.45(3)	312(3)	[N04a]
(66652) 1999 RZ$_{253}$	4,660(170)	*56*	*1.0*	0.460(13)	46.263(6/74)	[N04b]
2001 QW$_{322}$			*1.0*			[Kv01]
2000 CF$_{105}$			*0.73*			[N02a][N02b]
2000 CQ$_{114}$			*0.81*			[SNG04]
2003 UN$_{284}$			*0.76*			[MC03]
2003 QY$_{90}$			*0.95*			[EKC03]
2005 EO$_{304}$			*0.58*			[KE05]
(80806) 2000 CM$_{105}$			*0.58*			[SN05]
2000 OJ$_{67}$			*0.69*			[SN05]
(79360) 1997 CS$_{29}$			*0.95*			[SN05]
1999 OJ$_4$			*0.54*			[SN05]
scattered						
2003 EL$_{61}$	49,500(400)		*0.22*	0.050(3)	49.12(3)	[Br05a,b]
2001 QC$_{298}$	3,690(70)		*0.79*		19.2(2)	[Mg04]
(82075) 2000 YW$_{134}$			*0.55*			[SN05]
(48639) 1995 TL$_8$			*0.46*			[SN05]
2003 UB$_{313}$			*0.13*			[Br05c,d]
resonant						
Pluto/Charon	19,636(8)	16.7(4)	0.53(3)	0.0076(5)	6.38722(2)	[TB97]
(26308) 1998 SM$_{165}$	11,310(110)		*0.42*		130(1)	[BT02][Mg04]
(47171) 1999 TC$_{36}$	7,640(460)		*0.39*		50.4(5)	[TB02][Ma02][Mg04]

Published uncertainties in least significant digit(s) shown in parentheses. Estimated quantities or other values published without error estimates shown in italics.
[1] Selected references shown.
[BT02] Brown & Trujillo 2002, [Br05a,b,c,d] Brown *et al.* 2005a,b,c,d, [EKC03] Elliot, Kern & Clancy 2003, [Kv01] Kavelaars *et al.* 2001, [KE05] Kern & Elliot 2005, [K05] Kern 2005 thesis, [Ma02] Marchis *et al.* 2002, [Mg04] Margot *et al.* 2004, [MC03] Millis & Clancy 2003, [N02a,b] Noll *et al.* 2002a,b, [N04a,b] Noll *et al.* 2004a,b, [Op03] Osip *et al.* 2003, [SNG04] Stephens, Noll & Grundy 2004, [SN05] Stephens & Noll 2005, [TB97] Tholen & Buie 1997, [V02] Veillet *et al.* 2002

are binary (11 binaries) with their lightcurve survey sensitive to primaries with $r_p > 0.15$km and $r_s/r_p > 0.18$. Margot *et al.* (2002) found 5 binary companions of primaries larger than 0.2 km in diameter in their sample of "∼50" NEAs studied by radar. The authors quote a fraction of near 16% which implies that 31 of the 50 meet the diameter criterion. Though not stated in Margot *et al.* the small size of the sample implies a relatively large uncertainty of $+9/-5$%. Of the remaining 19 smaller primaries, none were found to have binaries which I use to estimate an upper limit of < 9%, although this is likely subject to observational bias. Interestingly, the frequency of detected NEA binaries agrees with the number estimated from terrestrial double craters (Bottke & Melosh 1996). Despite this apparent agreement, however, there is clearly room for improvement in the statistics and for studies of the fraction of NEA binaries as a function of size, spin, and orbit characteristics. It will be especially interesting to be able to compare binary frequency in NEAs with the frequency in comparably sized Main Belt populations.

In the Main Belt, Merline *et al.* (2002) report ∼2% of the 300 objects in their Main Belt survey have relatively large ("tens of km") binaries. Of the 6 Trojans searched by Merline *et al.* (2002), one, Patroclus, was found to be binary. The sample size is insufficient to

Table 5. Fraction of Binaries in Minor Planet Populations

Population	binaries (%)	reference
NEA		
$d > 0.3$ km	15 ± 4	Pravec *et al.* 2005a
$d > 0.3$ km, $2.2 <P< 2.8$ hr	$66\pm^{10}_{12}$	"
$d > 0.2$ km	$16\pm^{9}_{5}{}^{*}$	Margot *et al.* 2002
$d < 0.2$ km	$< 9^{*}$	"
Main Belt		
"Average"	~2	Merline *et al.* 2002
10km $< d < 50$km	$\sim 10 \pm^{7}_{3}{}^{*}$	Colas *et al.* 2005
Koronis	$22\pm^{18}_{9}{}^{*}$	Merline *et al.* 2004b
Karin	$< 10^{*}$	"
Veritas	$< 9^{*}$	"
Vestoids		Ryan *et al.* 2004
Hungarias		Warner *et al.* 2005
Transneptunian		
"Average"	$11\pm^{5}_{2}$	Stephens & Noll 2005
Cold Classical	$22\pm^{10}_{5}$	"
All other	$5.5\pm^{4}_{2}$	"
"Bright" TNOs	$75\pm^{10}_{30}{}^{*}$	Brown *et al.* 2005d

* Uncertainties not reported, calculated in this work based on available information.

determine the binary frequency, but ongoing search programs (e.g. Merline *et al.* HST cycle 14 program 10512) may remedy this shortcoming.

A bewildering variety of frequencies are cited for the transneptunian population, some based on gross statistics of the number of known binaries divided by the number of known TNOs. However, this is a particularly inaccurate method for estimating the frequency of TNO binaries since many are undetectable by typical ground-based observations. Of the 22 known TNBs, 13 have been found with the Hubble Space Telescope. The other 8 have been found at ground based telescopes all with separations greater than 0.3 arcsec. The fraction of objects with widely separated, relatively bright companions is clearly small. Schaller & Brown (2003) failed to find any companions in a sample of 150 TNOs observed with Keck. This fact can also be deduced from the relatively small fraction of objects found by groundbased telescopes, 8, out of the more than 1000 separate TNOs so far observed at least once. However, the inhomogeneity of these ground based observations precludes any quantitative analysis of these statistics. A subset of the HST-discovered objects come from a uniform sample observed with the NIC2 camera and can provide a better estimate of the global frequency of binaries of 11% $\pm^{5}_{2}$ (Stephens & Noll 2005).

Neither Kuiper Belt, Main Belt, nor NEAs are homogeneous populations. It is, therefore, naive to expect that the frequency of binaries in each of the distinct subpopulations will be identical. If they are not, then the gross statistics in the previous paragraphs have marginal utility because they depend on the makeup of the sample. In the case of transneptunians, the populations are identified by their orbital dynamics. Distinct populations identified are the classical Kuiper belt (nonresonant, cubewanos), the resonant objects (including, but not limited to the Plutinos in the 3:2 resonance), and the scattered disk objects. The classical belt is further divided into two overlapping populations, one, the cold classical disk, have inclinations relative to the Kuiper Belt plane of less than 5 degrees. The hot classicals are identified by their inclinations in excess of this cutoff. The scattered disk consists of both near and extended scattered groups. In the Main Belt, the subpopulations of most interest are collisional families, but there may also be differences in binary frequency with taxonomical class and size. For the NEAs there may also be differences as a function of size and orbital dynamics.

An analysis of 84 TNOs observed with NICMOS has revealed a total of 9 previously unknown binaries (Stephens & Noll 2005). This sample is large enough that it is possible, for the first time, to identify a factor of four higher fraction of binaries in the cold classical disk than in the combined dynamically excited populations. The statistics for other dynamical classes of TNOs remains too small for further discrimination, but ongoing observations with HST may result in rapid progress. Reporting on the detection of the companion to 2003 UB_{313}, Brown *et al.* (2005d) note that 3 of 4 of the brightest TNOs (Pluto, 2005 FY_9, 2003 EL_{61}, and 2003 UB_{313}) have satellites. However, because satellites 3.2-4.2 magnitudes fainter than their primary (companions of 2003 EL_{61}, and 2003 UB_{313} respectively) would not have been detected in the large NICMOS survey (Stephens & Noll 2005) it is premature to conclude that large TNOs may have a higher fraction of binaries. Deeper observations of a substantial number of smaller, fainter, TNOs will be needed to test the hypothesis of a size-dependence in binary frequency.

Several studies have now begun to focus on the search for binaries within specific collisional families within the Main Belt. Merline *et al.* (2004b) reported the discovery of 2 Koronis binaries out of a sample of 9 while finding no binaries among 17 Karin cluster targets and 18 Veritas family targets, all observed with HST. This is a significantly larger fraction than would be expected in a population with only 2% binaries, though it remains subject to small number uncertainties. Merline *et al.* (2004b) speculated that the fraction of binaries might depend on the size of the parent body and the ability of the subsequent ejecta to form bound pairs. An alternative explanation is that the binary frequency in a collisional family increases as the size relative to the largest fragment decreases. In this particular case, the searches in the Koronis family are sampling relatively smaller members of the family. The apparent higher proportion of binaries in the Koronis family appears to contradict another hypothesis that newer collisional families would have more binaries than old families. Colas *et al.* (2005) report the detection of 4 binaries among a sample of ~40 asteroids with diameters between 10 and 50 km. This is suggestive of a difference in binary frequency as a function of size, but, as with the other families, the statistics are still weak.

4. Physical Properties: mass, albedo, and density

One of the most practical benefits of binaries is the ability to derive the system mass from observation of the orbit. With observations at multiple epochs and the centuries-old Kepler's laws, the determination of mass is conceptually simple. In practice, solutions are found through iterative numerical methods. Because of the difficulties in acquiring large blocks of observing time on major facilities, in many cases data are very sparse. Specialized methods for constraining orbit solutions with limited data have been developed (Herstroffer & Vachier 2005). Orbits, and hence masses, have been obtained for only a subset of the known binaries. For systems found through lightcurve analysis, the period and relative sizes of the components are determined, but the semimajor axis, a, of the mutual orbit is not. Lacking a, the mass cannot be directly determined, though in some cases with modelling, reasonable guesses can be made (e.g. Pravec *et al.*

2005a). The values listed in Table 6 are limited to those objects with directly measured orbits.

An important question is whether brightness can be used as a proxy for mass. This depends on the uniformity of the albedo of the objects in the minor body population in question. These are not known a priori. For NEAs and Main Belt asteroids, it is usually possible to directly measure the diameter of the objects. In that case, it is a relatively simple step to determine an albedo from the measured apparent magnitude and known distance. Complications arise, particularly for NEAs and smaller Main Belt asteroids due to the effects of non-spherical shape, phase function, and complex viewing geometries. Spectral taxonomy can used to estimate albedo for objects with unknown diameters. Transneptunian objects are too distant to be resolved. Measurement of thermally emitted flux coupled to observed reflected flux and models of rotation and thermal inertia can be used to determine albedo. Only a few TNOs have had their albedos measured in this way from the ground and from the Spitzer Space Telescope (Grundy *et al.* 2005a,b). Interestingly, the determination of a mass for binaries allows one to constrain the albedo as a function of assumed density. This approach can provide significant constraints on albedo because it varies as the 2/3 power of assumed density; over a "reasonable" range of density (500–2000 kg/m^3) albedos vary by a factor of 2.5 (Noll *et al.* 2004a). So far, there are no apparent correlations of TNO albedo with any other observable (Grundy *et al.* 2005a), so estimates by proxy are currently not possible.

The determination of density has the same limitations as does the determination of albedos. If direct size determinations are available, density is calculable once the mass is known. Sizes are best determined for NEAs and Main Belt asteroids since these can be resolved by radar or direct imaging. Non-spherical shape dominates the uncertainties for NEAs to the degree that measured densities are frequently so uncertain as to be unconstraining (Table 6). Main Belt binaries result in the best density constraints because good diameters can be measured for the primary and because many of the systems are very asymmetric with most of the mass in the primary so that assumptions about the relative albedo of the primary and secondary introduce negligible uncertainty. For both the Main Belt and the NEA populations the densities that have been determined so far are consistent with expected densities based on meteorite samples and reasonable bulk porosities (references). The lack of measured diameters impedes the determination of densities in the transneptunian population. Aside from the well-constrained Pluto/Charon binary, only one object has an estimated density, (47171) 1999 TC$_{36}$ (Stansberry *et al.* 2005). Interestingly, the density of (47171) 1999 TC$_{36}$ appears to be extremely low, $\rho = 500$–900 kg/m^3 implying a significant mismatch between surface area and filled volume.

5. Formation and Destruction of Binaries

An important area of investigation is the effort to understand when, how, and under what conditions binaries and multiples are formed and how long they are able to survive. Theoretical work to date has identified three main formation mechanisms: capture, collisions, and rotational fission. Currently, the leading models for binary formation differ for each of the major populations. Richardson & Walsh (2005) have thoroughly reviewed this topic and I include here only a few of the most salient points.

TNBs have had the greatest variety of models proposed for their formation. Stern (2002) proposed a collisional model but noted that the then assumed fraction of 1% binaries could not be met unless he assumed that the mean albedo was higher than the formerly assumed value of $p = 0.04$ adopted from comet Halley. While it is now apparent

Table 6. Mass and Density of Solar System Binaries

Object	system mass	density (g/cm^3)	references
NEAs	(10^9kg = Gg)		
(66391) 1999 KW$_4$	2330(230)	2.6(1.6)	[Me02]
2000 DP$_{107}$	460(50)	1.7(1.1)	[Me02]
2000 UG$_{11}$	5.1(5)	0.8(0.6)	[Me02]
(5381) Sekhmet		2.0(0.7)	[Ne03]
(3671) Dionysus		*1.0-1.6(3/2)*	[P05a]
(65803) Didymos		*1.7-2.1(3/2)*	[P05a]
1991 VH		*1.4-1.6(5)*	[P05a]
1996 FG$_3$		*1.3(6)*	[P05a]
Main Belt	(10^{18}kg = Zg)		
(87) Sylvia	14.78(6)	1.2(1)	[Ma05a]
(107) Camilla	10.8(4)[1]	*1.88*	[Ma05b]
(130) Elektra	10.1(9)[1]	3.8(3)	[Ma05b]
(22) Kalliope	7.3(2)[1]	2.03(16)	[Ma03]
(45) Eugenia	5.8(2)[1]	1.1(3)	[Ma05b]
(121) Hermione	5.4(2)[1]	1.1(3)	[Ma05c]
762 Pulcova	*2.6*	1.8(8)	[Me02]
(283) Emma	1.49(2)	0.87(2)	[Ma05b]
(90) Antiope	0.82(1)	0.6(2)	[De05]
(379) Huenna	0.477(5)	1.16(13)	[Ma05b]
243 Ida	*0.042*	2.6(5)	[Me02]
Trojan	(10^{18}kg = Zg)		
617 Patroclus	*1.5*	1.3(5)	[Me02]
	1.4(2)*or*	0.8(1)*or*	[Ma05d]
	4.3(2)	2.6(1)	"
Transneptunian	(10^{18} kg = Zg)		
Pluto/Charon	14,710(20)	1.99(7)/1.66(15)	[TB97]
2003 EL$_{61}$	4,200(100)		[Br05b]
(47171) 1999 TC$_{36}$	13.9(2.5)	*0.5-0.9*	[Mg04][Mg05][Sb05]
2001 QC$_{298}$	10.8(7)		[Mg04]
(26308) 1998 SM$_{165}$	6.78(24)		[Mg04]
(66652) 1999 RZ$_{253}$	3.7(4)		[N04b]
1998 WW$_{31}$	2.7(4)		[V02]
(88611) 2001 QT$_{297}$	2.3(1)		[Op03][K05]
(58534) 1997 CQ$_{29}$	0.42(2)		[N04a]

[1] Mass calculated from measured period and semimajor axis.
[Br05b] Brown *et al.* 2005, [De05] Descamps *et al.* 2005, [K05] Kern 2005, [Ma03] Marchis *et al.* 2003, [Ma04] Marchis *et al.* 2004, [Ma05a,b,c,d] Marchis *et al.* 2005a,b,c,d, [Mg04] Margot *et al.* 2004, [Mg05] Margot *et al.* 2005, [Me02] Merline *et al.* 2002, [Ne03] Neish *et al.* 2003, [N04a,b] Noll *et al.* 2004a,b, [P05a] Pravec *et al.* 2005a, [Sb05] Stansberry *et al.* 2005, [Op03] Osip *et al.* 2003, [V02] Veillet *et al.* 2002

that albedos of many TNOs are indeed higher than 4%, the collisional model cannot produce the much higher fraction of binaries that are now being found. A collision model has also been proposed for the Pluto/Charon binary (Canup 2005) but stochastic models such as this have little applicability to larger populations of binaries. Capture models have evolved in complexity with the most recent work by Astakhov *et al.* (2005) using chaos-assisted capture giving a reasonable distribution of orbital parameters compared to the observed sample. The hybrid collision-exchange capture model of Funato *et al.* (2004) can now be ruled out because the eccentricity of observed orbits does not meet the predictions of this model.

In the Main Belt, collisions appear to be the dominant mode of forming binaries. The evidence for this is the small secondary to primary size ratio for most of the objects (Table 3) and the apparent higher incidence of binaries in collision families (Merline *et al.* 2004b). Numerical models of formation of binaries (e.g. Durda *et al.* 2004) identify two main classes of post-collision binaries: fragments that are captured or reaccreted around a remnant primary and pairs (or more) of fragments that are mutually captured.

NEA binaries are distinctive in the rapid rotation of the primaries, typically less than 3 hours (Pravec *et al.* 2005a). This is taken to be an indication that these objects may have formed through spin-up (from the YORP effect or collisions) and subsequent fission. The role of tidal forces during close encounters with Earth and Mars has usually been assumed (Richardson *et al.* 1998, Walsh & Richardson 2005). However, the recent evidence for a similarly large fraction of binaries among rapidly-rotating small asteroids populations in the Main Belt opens the question of whether NEA binaries are formed in the Main Belt and survive the orbital perturbations that bring them into NEA orbits or whether binaries are efficiently formed during the relatively brief 10 Myr mean lifetime of NEAs.

Survival of binaries, once they are formed, is an important, and so far largely-neglected question. Petit & Mousis (2004) have investigated the survival of TNBs and have found that some of the most widely separated may have lifetimes shorter than the age of the solar system. If this is correct, the initial inventory of binaries may have been even higher than the already high fraction we see today. An important question, in addition to lifetime against collisional erosion, is the survival of binaries during scattering events. The fact that most binaries orbit within 10% of the Hill radius or less (Figure 1), regardless of population, may be the signature of tidal disruption of the most loosely bound systems.

Given the large size range and potentially different histories of these populations, the variety of formation mechanisms may not entirely surprising, but from an aesthetic point of view, it is clearly less than satisfactory. A question for the future is whether or not there are underlying and unifying modes for formation that apply across all the classes of minor bodies in the solar system. At present, collisions appear to be the most universal component of any such model, even if, in some cases, they are only part of the formation scenario. The apparent prevalence of binaries in the Kuiper Belt raises the possibility that bound systems are commonplace in accreting dust disks and that further evolution is dominated by the destruction (or lack thereof) of preexisting binaries. The most promising case for consolidation is for NEA binaries which may turn out to be survivors of a preexisting binary population in the Main Belt.

6. Binaries Beyond

The Main Belt asteroid (87) Sylvia has now been confirmed as the first known minor planet triple (Marchis *et al.* 2005a,b). The discovery of triples was not entirely unexpected as numerical simulations of collisions produce triples and higher multiples as well as binaries, at least in the early post-collision time scale (Leinhardt & Richardson 2005). Stability of multiples is an important issue that will have to be addressed, but as demonstrated by the existence of stellar triples and higher-order multiples, stable configurations can be found. Ultimately, the incidence of triples relative to binaries may help constrain models of formation, especially collisional models.

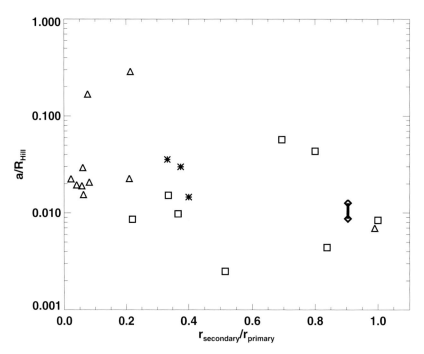

Figure 1. The characteristics of binaries in different solar system populations are shown in this figure. (asterisks=NEAs, triangles=MainBelt, diamond=Trojan, squares=TNOs) The separation, in terms of the fraction of the Hill radius is plotted against the size ratio of the binary pair. We have plotted only objects for which a system mass determination has been made from observed orbital data (Tables 2–4,6). All of the binary populations have most objects within a similar small fraction of the Hill radius. The main difference between populations is the large number of objects in the Kuiper Belt with large size ratios. The lack of TNBs with small secondary to primary size ratios is heavily influenced by observing bias and the number of such systems remains to be determined. It is clear that large primary to secondary ratio systems are rare in the Main Belt and among NEAs.

No multiples have been identified among NEAs or TNOs as of mid-October 2005. However, (47171) 1999 TC_{36} has been proposed as a potential triple as a way of understanding the apparent low density derived from Spitzer observations (Margot *et al.* 2005, Stansberry *et al.* 2005). Interestingly, an unexplained large change in brightness on one night has been reported for this object (Ivanova *et al.* 2005). Further monitoring of the lightcurve is called for. The TNO 1997 CQ_{29} also has reported large variations in one of the two components (Noll *et al.* 2004, Margot *et al.* 2005). This may be the signature of a close binary or contact system, or may simply represent large albedo and shape effects. The largest TNOs have larger Hill radii (for objects of the same density r_H scales with the radius of the primary) and a larger fraction of their Hill spheres are searchable with instruments of a given resolution. Stern (2003) reviewed searches for additional satellites of Pluto and prospects for improved searches using the Hubble Space Telescope and the New Horizons spacecraft. With a high fraction of Pluto-sized objects now known to have one satellite, it seems more likely that multiples will eventually be found.

The discovery of numerous binary systems has stimulated creative and unconventional thinking. Ćuk & Burns (2005) propose a variant of the YORP effect which would apply to binaries. For NEAs, which are small enough and close enough to the Sun to be affected, this effect can explain the nearly circular and synchronous orbits of secondaries. The timescale for orbital evolution under the BYORP effect is surprisingly short, generally less than 10^5 years. Because of the short evolutional time scale, Ćuk & Burns speculate that most NEA binaries may be in stable states where YORP and BYORP cancel and are small. This prediction is testable with accurate observations of orbital elements for NEA binary pairs, a test that should be possible within the next five years. On a much grander scale, Cintala *et al.* (2005) suggest an "experiment" aimed at deorbiting a binary companion with multiple explosive charges. They list candidates in order of Δv needed to deorbit the secondary and crash it into the primary. The authors identified 2000 UG$_{11}$ as the best candidate with a 230 m diameter primary and a 140 m secondary orbiting with a semimajor axis of 410 m.

Comet nuclei are thought to originate from the same pool of primordial bodies that populate the outer solar system and are thus likely to share a similar propensity for binaries. Study of comet nuclei is a notoriously difficult problem because of the small size of the objects and because of the persistent presence of coma, even at large heliocentric distances. Disruption of comet nuclei is observed, with perhaps the best known example being comet Shoemaker-Levy 9 (c.f. Weaver *et al.* 1995). There are, however, no conclusive observations of a gravitationally bound cometary binary. Marchis *et al.* (1999) speculated that some aspects of comet Hale-Bopp might be more easily explained if that object were a bound binary, but this was not a unique interpretation. It seems likely, however, that some fraction of comets, particularly "new" Oort-cloud comets could be binary systems. Observations of nuclei at large heliocentric distances with high-angular resolution could address this question.

The immediate future of the study of binaries in minor planets is very promising. Discovery will continue, probably at an increased pace, due both to the heightened awareness of the existence and detectability of these bodies and to the availability of instrumentation capable of finding faint companions at small separations. Binaries will continue to provide critical physical data that is not obtainable in other way. The growing number of detected binaries will enable the nascent study of binaries fraction as a function of dynamical history, composition, size, rotation rate, and other conceivable correlations. There are many possible avenues of investigation. Among the most promising are apparently high fraction of binaries in the dynamically unperturbed classical Kuiper Belt (Stephens & Noll 2005) and the apparently high fraction of binaries in some Main Belt collisional families (Merline *et al.* 2005) because both have the potential to yield clues on the origin and survival of binaries.

We have now reached the point where it is possible to state that binaries must be a common feature in the evolution of protoplanetary disks like the one that formed our solar system. Debris disks that we now observe around other stars probably share an affinity for forming pairs. It remains to be seen if the great utility of binaries as mass-measuring tools and in teasing out the early history of the solar system will result in insights that extend beyond to inform our understanding of planetary systems in general. If nothing else, however, the triumph of the detection and utilization of minor planet binaries must stand as an object lesson in the value of persistence and as an especially poignant reminder of the unexpected complexity of planetary systems.

7. Online Resources

In a rapidly developing field such as this, online compilations are invaluable resources for researchers and students. While online resources are typically not referenceable, because of their ephemeral nature, it is worth mentioning several resources that are currently available. A comprehensive compilation of solar system binaries is regularly updated by W. R. Johnston at his website http://www.johnstonsarchive.net/astro/asteroid-moons.html. The Ondrejov Asteroid Photometry Project led by Petr Pravec maintains an online resource at http://sunkl.asu.cas.cz/~ppravec/ that includes prepublication data and an updated list of binary NEAs. Joel Parker maintains the Distant EKOs (Edgeworth-Kuiper Belt Objects) web pages that include a compilation of transneptunian binaries. Franck Marchis has an extensive online collection of binary asteroid orbital information, much of which has not been published at the time of this review. Updated information is available at http://astron.berkeley.edu/~fmarchis/ and was used in the tables in this review; references, however, were to published materials.

Acknowledgements

I would like to acknowledge ongoing and productive collaborations with my colleagues, particularly W. Grundy, D. Stephens, D. Osip, J. Spencer, and M. Buie. A tip of the hat also to A. Storrs and B. Zellner, early collaborators who helped me appreciate the excitement of objects smaller than Saturn. A special note of thanks goes to A. Lubenow who was the key enabler for complex solar system programs on the Hubble Space Telescope for two decades and who did so with his unique mixture of deep technical expertise and intolerance for nonsense. This work was supported in part by grant GO 10514 from the Space Telescope Science Institute which is operated by AURA under contract from NASA.

References

André, C.L.F. 1901, *Astronomische Nachtrichen* 155, 27

Astakhov, S.A., Lee, E.A. & Farrelly, D. 2005, *MNRAS* 360, 401

Behrend, R. *et al.* 2004a, *IAUC* 8265

Behrend, R., Roy, R., Sposetti, S., Waelchli, N., Pray, D., Berger, N., Demeautis, C., Matter, D., Durkeee, R., Klotz, A. Starkey, D., & Cotrez, V. 2004b, *IAUC* 8292

Behrend, R. 2004c, *IAUC* 8354

Behrend, R., Bernasconi, L., Klotz, A., & Durkee, R. 2004, *IAUC* 8389

Belton, M. & Carlson, R. 1994, *IAUC* 5948, 2

Benner, L.A.M., Ostro, S.J., Giorgini, J.D., Jurgens, R.F., Margot, J.L., & Nolan, M.C. 2001, *IAUC* 7632

Benner, L.A.M., Nolan, M.C., Margot, J.L., Ostro, S.J., & Giorgini, J.D. 2003, *AAS/Division for Planetary Sciences Meeting* Abstracts, 35, 24.01

Benner, L.A.M., Nolan, M.C., Ostro, S.J., Giorgini, J.D., Margot, J.L., & Magri, C. 2004, *IAUC* 8329

Bottke, W.F. & Melosh, H.J. 1996, *Icarus* 124, 372

Brown, M.E., Margot, J.L., de Pater, I., & Roe, H. 2001, *IAUC* 7588

Brown, M.E., Trujillo, C.A. & Rabinowitz, D. 2005a, *IAUC* 8577

Brown, M.E., Bouchez, A.H., Rabinowitz, D., Sari, R., Trujillo, C.A., van Dam, M., Campbell, R., Chin, J., Hartman, S., Johansson, E., Lafon, R., Le Mignant, D., Stomski, P., Summers, D., & Wizinowich, P. 2005b, *ApJ* 632, L45

Brown, M.E. 2005c, *IAUC* 8610

Brown, M.E., *et al.* 2005d, *ApJ* submitted

Canup, R.M. 2005, *Science* 307, 546

Christy, J.W. & Harrington, R.S. 1978, *AJ* 83, 1005

Cintala, M.J., Durda, D.D. & Housen, K.R. 2005, *36th Annual Lunar and Planetary Science Conference* 36, 2160

Colas, F., Behrend, R., Klotz, A., Roy, R., & Bernasconi, L. *IAU Symp. 229* abstract

Ćuk, M. & Burns, J.A. 2005, *Icarus* 176, 418

Descamps, P., Marchis, F., Michalowski, T., Berthier, J., Hestroffer, D., Vachier, F., Colas, F., & Birlan, M. 2005, *IAU Symp. 229* abstract

Durda, D.D., Bottke, W.F., Enke, B.L., Merline, W.J., Asphaug, E., Richardson, D.C., & Leinhardt, Z.M. 2004, *Icarus* 170, 243

Elliot, J.L., Kern, S.D. & Clancy, K.B. 2003, *IAUC* 8235

Durda, D.D., Bottke, W.F., Enke, B.L., Merline, W.J., Asphaug, E., Richardson, D.C., & Leinhardt, Z.M. 2004, *Icarus* 170, 243

Funato, Y., Makino, J., Hut, P., Kokubo, E., & Kinoshita, D. 2004, *Nature* 427, 518

Goldreich, P., Lithwick, Y., & Sari, R. 2002, *Nature* 420, 643

Grundy, W.M., Noll, K.S. & Stephens, D.C. 2005a, *Icarus* 176, 184

Grundy, W.M., Spencer, J.R., Stansberry, J.A., Buie, M.W., Chiang, E.I., Cruikshank, D.P., Millis, R.L., & Wasserman, L.H. 2005b, *AAS/Division for Planetary Sciences Meeting Abstracts*, 37, 52.07

Herstroffer, D. & Vachier, F. *IAU Symp. 229* abstract

Ivanova, V., Borisov, G. & Belskaya, I. 2005, *IAU Symp. 229* abstract

Kavelaars, J.J., Petit, J.-M., Gladman, B., & Holman, M. 2001, *IAUC* 7749

Kern, S.D. & Elliot, J.L. 2005, *IAUC* 8526

Kern, S.D. 2005, PhD thesis, MIT

Kryszczynska, A., Kwiatkowski, T., Hirsch, R., Polinska, M., Kaminski, K., & Marciniak, A. 2005, *CBET* 239

Leinhardt, Z.M. & Richardson, D.C. 2005, *Icarus* 176, 432

Marchis, F. Boehnhardt, H., Hainaut, O.R., & Le Mignant, D. 1999, *Astron. & Astrophys.* 349, 432

Marchis, F., Descamps, P., Hestroffer, D., Berthier, J., Vachier, F., Boccaletti, A., de Pater, I., & Gavel, D. 2003, *Icarus* 165, 112

Marchis, F., Descamps, P., Hestroffer, D., Berthier, J., & de Pater, I. 2004, *AAS/Division for Planetary Sciences Meeting* Abstracts, 36, 46.02

Marchis, F., Descamps, P., Hestroffer, D., & Berthier, J. 2005a, *Nature* 436, 822

Marchis, F., Berthier, J., Clergeon, C., Descamps, P., Hestroffer, D., de Pater, I., & Vachier, F. 2005b, *IAU Symp. 229* Abstract

Marchis, F., Hestroffer, D., Descamps, P., Berthier, J., Laver, C., & de Pater, I. 2005c, *Icarus* in press

Marchis, F., Hestroffer, D., Descamps, P., Berthier, J., Bouchez, A.H., Campbell, R.D., Chin, J.C.Y., van Dam, M.A., hartman, S.K., Johansson, E.M., Lafon, R.E., Le Mignant, D., de pater, I., Stomski, P.J., Summers, D.M., & Wizinovitch, P.L. 2005d, *AAS/Division for Planetary Sciences Meeting* Abstracts, 37, 14.07

Margot, J.-L. & Brown, M.E. 2001, *IAUC* 7703

Margot, J.L., Nolan, M.C., Benner, L.A.M., Ostro, S.J., Jurgens, R.F., Giorgini, J.D., Slade, M.A., & Campbell, D.B. 2002, *Science* 296, 1445

Margot, J.L. & Keck, W.M. 2003b, *IAUC* 8182

Margot, J.L., Nolan, M.C., Negron, V., Hine, A.A., Campbell, D.B., Howell, E.S., Benner, L.A.M., Ostro, S.J., Giorgini, J.D., & Marsden, B.G. 2003a, *IAUC* 8227

Margot, J.L., Brown, M.E., Trujillo, C.A., & Sari, R. 2004, *AAS/Division for Planetary Sciences Meeting* Abstracts, 36, 08.03

Margot, J.L., Brown, M.E., Trujillo, C.A., Sari, R., & Stansberry, J.A. 2005, *AAS/Division for Planetary Sciences Meeting* Abstracts, 37, 52.04

Merline, W. J., Close, L.M., Dumas, C., Chapman, C.R., Roddier, F., Menard, F., Slater, D.C., Duvert, G., Shelton, C., Morgan, T., & Dunham, D.W. 1999a, *IAUC* 7129

Merline, W.J., Close, L.M., Dumas, C., Chapman, C.R., Roddier, F., Menard, F., Slater, D.C., Duvert, G., Shelton, C., & Morgan, T. 1999b, *Nature* 401, 565

Merline, W.J., Close, L.M., Shelton, J.C., Dumas, C., Menard, F., Chapman, C.R., & Slater, D.C. 2000a, *IAUC* 7503

Merline, W.J., Close, L.M., Dumas, C., Shelton, J.C., Menard, F., Chapman, C.R., & Slater, D.C. 2000b, *Bulletin of the American Astronomical Society* 32, 1017

Merline, W.J., Menard, F., Close, L., Dumas, C., Chapman, C.R., & Slater, D.C. 2001a, *IAUC* 7703

Merline, W.J., Close, L.M., Siegler, N., Potter, D., Chapman, C.R., Dumas, C., Menard, F., Slater, D.C., Baker, A.C., Edmunds, M.G., Mathlin, G., Guyon, O., & Roth, K. 2001b, *IAUC* 7741

Merline, W.J., Weidenschilling, S.J., Durda, D.D., Margot, J.L., Pravec, P., & Storrs, A.D. 2002, in: W.F. Bottke, A. Cellino, P. Paolicchi & R.P. Binzel (eds.), *Asteroids III* (Tucson: University of Arizona Press), p. 289

Merline, W.J., Close, L.M., Siegler, N., Dumas, C., Chapman, C.R., Rigaut, F., Menard, F., Owen, W.M., & Slater, D.C. 2002a, *IAUC* 7827

Merline, W.J., Tamblyn, P.M., Dumas, C., Close, L.M., Chapman, C.R., Menard, F., Owen, W.M., Slater, D.C., & Pepin, J. 2002b, *IAUC* 7980

Merline, W.J., Close, L.M., Tamblyn, P.M., Menard, F., Chapman, C.R., Dumas, C., Duvert, G., Owen, W.M., Slater, D.C., & Sterzik, M.F. 2003a, *IAUC* 8075

Merline, W.J., Dumas, C., Siegler, N., Close, L.M., Chapman, C.R., Tamblyn, P.M., & Terrell, D. 2003b, *IAUC* 8165

Merline, W.J., Tamblyn, P.M., Dumas, C., Close, L.M., Chapman, C.R., & Menard, F. 2003, *IAUC* 8183

Merline, W.J., Tamblyn, P.M., Chapman, C.R., Nesvorny, D., Durda, D.D., Dumas, C., Storrs, A.D., Close, L.M., & Menard, F. 2003d, *IAUC* 8232

Merline, W.J., Tamblyn, P.M., Dumas, C., Menard, F., Close, L.M., Chapman, C.R., Duvert, G., & Ageorges, N. 2004a, *IAUC* 8297

Merline, W.J., Tamblyn, P.M., Nesvorny, D., Durda, D.D., Chapman, C.R., Dumas, C., Storrs, A.D., Feldman, B., Owen, W.M., Close, L.M., & Menard, F. 2004b, *AAS/Division for Planetary Sciences Meeting* Abstracts, 36, 46.01

Merline, W.J., Tamblyn, P.M., Nesvorny, D., Durda, D.D., Chapman, C.R., Dumas, C., Owen, W.M., Storrs, A.D., Close, L.M., & Menard, F. 2005, *AAS/Division for Planetary Sciences Meeting* Abstracts, 37, 03.07

Mottola, S., Hahn, G., Pravec, P., & Sarounova, L. 1997, *IAUC* 6680

Mottola, S. & Lahulla, F. 2000, *Icarus* 146, 556

Neish, C.D., Nolan, M.C., Howell, E.S., & Rivkin, A.S. 2003, *American Astronomical Society Meeting* Abstracts, 203, 134.02

Nolan, M.C., Margot, J.-L., Howell, E.S., Benner, L.A.M., Ostro, S.J., Jurgens, R.F., Giorgini, J.D., & Campbell, D.B. 2000, *IAUC* 7518

Nolan, M.C., Howell, E.S., Magri, C., Beeney, B., Campbell, D.B., Benner, L.A.M., Ostro, S.J., Giorgini, J.D. & Margot, J.-L. 2002a, *IAUC* 7824

Nolan, M.C., Howell, E.S., Ostro, S.J., Benner, L.A.M., Giorgini, J.D., Margot, J.-L., & Campbell, D.B. 2002b, *IAUC* 7921

Nolan, M.C., Howell, E.S., Rivkin, A.S., & Neish, C.D. 2003a, *IAUC* 8163

Nolan, M.C., Hine, A.A., Howell, E.S., Benner, L.A.M., & Giorgini, J.D. 2003b, *IAUC* 8220

Nolan, M.C., Howell, E.S. & Hine, A.A. 2004a, *IAUC* 8336

Nolan, M.C., Howell, E.S. & Miranda, G. 2004b, *AAS/Division for Planetary Sciences Meeting* Abstracts, 36, 28.08

Noll, K. S., Stephens, D.C., Grundy, W.M., Millis, R.L., Spencer, J., Buie, M.W., Tegler, S.C., Romanishin, W., & Cruikshank, D.P. 2002a, *AJ* 124, 3424

Noll, K., Stephens, D., Grundy, W., Spencer, J., Millis, R., Buie, M., Cruikshank, D., Tegler, S., & Romanishin, W. 2002b, *IAUC* 7857

Noll, K.S. 2003, *Earth Moon and Planets* 92, 395

Noll, K.S., Stephens, D.C., Grundy, W.M., Osip, D.J., & Griffin, I. 2004a, *AJ* 128, 2547

Noll, K.S., Stephens, D.C., Grundy, W.M., & Griffin, I. 2004b, *Icarus* 172, 402

Ostro, S.J., Chandler, J.F., Hine, A.A., Rosema, K.D., Shapiro, I.I., & Yeomans, D.K. 1990, *Science* 248, 1523

Ostro, S.J., Margot, J.-L., Nolan, M.C., Benner, L.A.M., Jurgens, R.F., & Giorgini, J.D. 2000, *IAUC* 7496

Ostro, S.J., Nolan, M.C., Benner, L.A.M., Giorgini, J.D., Margot, J.L., & Magri, C. 2003, *IAUC* 8237

Petit, J.-M. & Mousis, O. 2004, *Icarus* 168, 409

Pravec, P. & Hahn, G. 1997, *Icarus* 127, 431

Pravec, P., Wolf, M. & Sarounova, L. 1998, *Icarus* 133, 79

Pravec, P., Kusnirak, P., Hicks, M., Holliday, B., & Warner, B. 2000a, *IAUC* 7504

Pravec, P., Sarounová, L., Rabinowitz, D.L., Hicks, M.D., Wolf. M., Krugly, Y.N., Velichko, F.P., Shevchenko, V.G., Chiorny, V.G., Gaftonyuk, N.M., & Genvier, G. 2000b, *Icarus* 146, 190

Pravec, P., Kusnirak, P. & Warner, B. 2001, *IAUC* 7742

Pravec, P. & Sarounova, L. 2001, *IAUC* 7633

Pravec, P., Šarounová, L., Hicks, M.D., Rabinowitz, D.L., Wolf, M., Scheirich, P., & Krugly, Y.N. 2002, *Icarus* 158, 276

Pravec, P., Kusnirak, P., Sarounova, L., Brown, P., Esquerdo, G., Pray, D., Benner, L.A.M., Nolan, M.C., Giorgini, J.D., Ostro, S.J., & Margot, J.-L. 2003a, *IAUC* 8216

Pravec, P., Benner, L.A.M., Nolan, M.C., Kusnirak, P., Pray, D., Giorgini, J.D., Jurgens, R.F., Ostro, S.J., Margot, J.-L., Magri, C., Grauer, A., & Larson, S. 2003b, *IAUC* 8244

Pravec, P., Kusnirak, P., Sarounova, L., Brown, P., Kaiser, N., Masi, G., & Mallia, F. 2004, *IAUC* 8316

Pravec, P., *et al.* 2005a, *Icarus* in press

Pravec, P., Kusnirak, P., Kornos, L. Vigli, J., Pray, D., Durkee, R., Cooney, W., Gross, J., & Terrell, D. 2005b, *IAUC* 8609

Reddy, V., Dyvig, R., Pravec, P., & Kusnirak, P. 2005, *IAUC* 8483

Richardson, D.C., Bottke, W.F. & Love, S.G. 1998, *Icarus* 134, 47

Richardson, D.C. & Walsh, K.J. 2005, *Ann. Rev. Earth & Planet. Sci.* in press

Ryan, W.H., Ryan, E.V., Martinez, C.T., & Stewart, L. 2003, *IAUC* 8128

Ryan, W.H., Ryan, E.V. & Martinez, C.T. 2004a, *Planet. Space Sci.* 52, 1093

Ryan, W.H., Ryan, E.V. & Martinez, C.T. 2004b, *AAS/Division for Planetary Sciences Meeting Abstracts*, 36, 46.09

Schaller, E.L. & Brown, M.E. 2003, *AAS/Division for Planetary Sciences Meeting Abstracts*, 35, 39.20

Schlieder, J.E., Shepard, M.K., Nolan, M., Benner, L.A.M., Ostro, S.J., Giorgini, J.D., & Margot, J.L. 2004, *AAS/Division for Planetary Sciences Meeting Abstracts*, 36, 32.30

Shepard, M.K., Schlieder, J., Nolan, M.C., Hine, A.A., Benner, L.A.M., Ostro, S.J., & Giorgini, J.D. 2004, *IAUC* 8397

Stansberry, J.A., Cruikshank, D.P., Grundy, W.G., Margot, J.L., Emery, J.P., Fernandez, Y.R., & Rieke, G.H. 2005, *AAS/Division for Planetary Sciences Meeting Abstracts*, 37, 52.05

Stern, S.A. 2002, *AJ* 124, 2300

Stern, S.A. 2003, *Lunar and Planetary Institute Conference Abstracts*, 34, 1106

Stephens, D.C., Noll, K.S. & Grundy, W. 2004, *IAUC* 8289

Storrs, A., Vilas, F., Landis, R., Wells, E., Woods, C., Zellner, B., & Gaffey, M. 2001a, *IAUC* 7590

Storrs, A., Vilas, F., Landis, R., Wells, E., Woods, C., Zellner, B., & Gaffey, M. 2001b, *IAUC* 7599

Tamblyn, P.M., Marline, W.J., Chapman, C.R., Nesvorny, D., Durda, D.D., Dumas, C., Storrs, A.D., Close, L.M., & Menard, F. 2004, textitIAUC 8293

Tholen, D.J. & Buie, M.W. 1997, in: S.A. Stern & D.J. Tholen (eds.), *Pluto and Charon* (Tucson: University of Arizona Press), p. 193

Veillet, C., Parker, J.W., Griffin, I., Marsden, B., Doressoundiram, A., Buie, M., Tholen, D.J., Connelley, M., & Holman, M.J. 2002, *Nature* 416, 711

Walsh, K.J. & Richardson, D.C. 2005, *AAS/Division for Planetary Sciences Meeting Abstracts*, 37, 14.11

Warner, B., Pravec, P., Kusnirak, P., Pray, D., Galad, A., Gajdos, S., Brown, P., & Krzeminski, Z. 2005a, *IAUC* 8511

Warner, B.D., Pravec, P. & Pray, D. 2005b, *IAUC* 8592

Warner, B., Pravec, P., Harris, A.W., Galad, A., Kusnirak, P., Pray, D.P., Brown, P., Krzeminski, Z. and 10 more authors 2005, *IAU Symp. 229* Abstract

Weaver, H.A., A'Hearn, M.F., Arpigny, C., Boice, D.C., Feldman, P.D., Larson, S.M., Lamy, P., Levy, D.H., Marsden, B.G., Meech, K.J., Noll, K.S., Scotti, J.V., Sekanina, Z., Shoemaker, C.S., Shoemaker, E.M., Smith, T.E., Stern, S.A., Storrs, A.D., Truger, J.T., Yeomans, D.K., & Zellner, B. 1995, *Science* 267, 1282

Weidenschilling, S.J., Paolicchi, P. & Zappalá, V. 1989, in:R.P. Binzel, T. Gehrels & M.S. Matthews (eds.), *Asteroids II* (Tucson: University of Arizona Press) p. 643

Weidenschilling, S.J. 2002, *Icarus* 160, 212

Asteroids, Comets, Meteors
Proceedings IAU Symposium No. 229, 2005
D. Lazzaro, S. Ferraz-Mello & J.A. Fernández, eds.

Outer irregular satellites of the planets and their relationship with asteroids, comets and Kuiper Belt objects

Scott S. Sheppard[1]

[1]Department of Terrestrial Magnetism, Carnegie Institution of Washington, Washington, DC 20015, USA
email: sheppard@dtm.ciw.edu

Abstract. Outer satellites of the planets have distant, eccentric orbits that can be highly inclined or even retrograde relative to the equatorial planes of their planets. These irregular orbits cannot have formed by circumplanetary accretion and are likely products of early capture from heliocentric orbit. The irregular satellites may be the only small bodies remaining which are still relatively near their formation locations within the giant planet region. The study of the irregular satellites provides a unique window on processes operating in the young solar system and allows us to probe possible planet formation mechanisms and the composition of the solar nebula between the rocky objects in the main asteroid belt and the very volatile rich objects in the Kuiper Belt. The gas and ice giant planets all appear to have very similar irregular satellite systems irrespective of their mass or formation timescales and mechanisms. Water ice has been detected on some of the outer satellites of Saturn and Neptune whereas none has been observed on Jupiter's outer satellites.

Keywords. planets and satellites: general, Kuiper Belt, minor planets, asteroids, comets: general, solar system: formation

1. Introduction

Satellites are stable in the region called the Hill sphere in which the planet, rather than the sun, dominates the motion of the object (Henon 1970). The Hill sphere radius of a planet is defined as

$$r_H = a_p \left[\frac{m_p}{3M_\odot} \right]^{1/3} \tag{1.1}$$

where a_p and m_p are the semi-major axis and mass of the planet and $M_\odot$ is the mass of the sun. Table 1 shows the sizes of each giant planet's Hill sphere as seen from the Earth at opposition.

Most planetary satellites can be classified into one of two categories: regular or irregular (Kuiper (1956); Peale (1999)).

The regular satellites are within about $0.05r_H$ and have nearly circular, prograde orbits with low inclinations near the equator of the planet. These satellites are thought to have formed around their respective planets through circumplanetary accretion, similar to how the planets formed in the circumstellar disk around the sun. The regular satellites can be subdivided into two types: classical regulars and collisional shards (Burns 1986).

The classical regular satellites are large (several hundred to thousands of kilometers in size) and have evenly spaced orbits. The regular collisional shards are small (less than a few hundred kilometers) and are believed to have once been larger satellites but have been shattered or tidally disrupted over their lifetimes. These shards are usually very near

Table 1. Irregular Satellites of the Planets

Planet	Irr[a] (#)	m_p (10^{25}kg)	r_{min}[b] (km)	a_{crit} (10^6km)	r_H[c] (deg)	r_H (10^7km)
Mars	0	0.06	0.1		0.8	0.1
Jupiter	55	190	1.5	6.6	4.7	5.1
Saturn	26	57	3	5.7	3.0	6.9
Uranus	9	9	7	2.9	1.5	7.3
Neptune	6(7)	10	16	3.8	1.5	11.6

a) The number of known irregular satellites.
b) Minimum radius that current outer satellite searches would have detected to date.
c) Size of the Hill sphere as seen from Earth at opposition.

the planet where tidal forces and meteor fluxes are very high. Many collisional shards are associated with known planetary rings.

In contrast, the irregular satellites have semi-major axes $> 0.05r_H$ with apocenters up to $0.65r_H$ (Figure 1). Irregular satellites have eccentric orbits that are usually highly inclined and distant from the planet. They can have both prograde and retrograde orbits. The irregular satellites can not have formed around their respective planet with their current orbits and are likely the product of early capture from heliocentric orbits (Kuiper 1956).

Orbital characteristics displaying strong capture signatures instead of in situ formation around the planet is often used to define an irregular satellite. Throughout this work we will use a more strict definition. We follow others and define irregular satellites as those satellites which are far enough from their parent planet that the precession of their orbital plane is primarily controlled by the sun instead of the planet's oblateness. In other words, the satellite's inclination is fixed relative to the planet's orbit plane instead of the planet's equator. In practice this means any satellite with a semi-major axis more distant than the critical semi-major axis (Burns 1986), $a_{crit} \sim (2J_2 r_p^2 a_p^3 m_p/M_\odot)^{1/5}$, is an irregular satellite (Table 1). Here J_2 is the planet's second gravitational harmonic coefficient and r_p is the planet's equatorial radius. Figures 1 and 2 show the orbital characteristics of the known irregular satellites. In these figures all known regular satellites would fall very near the origin.

Almost all known planetary satellites fall into one of the three types mentioned above. A few exceptions do exist. The formation of the Earth's Moon is best explained through a collision between a Mars sized body and the young Earth (see Canup & Asphaug 2001 and references therein). Mars' two small satellites Phobos and Deimos resemble regular collisional shards, but some have suggested that they may be captured bodies similar to the irregular satellites of the giant planets (Burns 1992). No outer irregular satellites of Mars are known to exist (Sheppard & Jewitt 2004). Both Neptune's Triton and Saturn's Iapetus have at times been considered irregular satellites. These two objects stand out because both Triton and Iapetus are about ten times larger than any other known irregular satellite. Triton has all the characteristics of a regular satellite except that its orbit is retrograde. The best explanation for a retrograde orbit is through capture. Iapetus also has all the characteristics of a regular satellite but its inclination of 7 degrees is significantly larger than any other known regular satellite. Even so, this inclination is not as high as the vast majority of irregular satellites. Iapetus' relatively large inclination is probably because it is the most distant regular satellite of Saturn. At this distance the circumSaturnian nebula was probably of low density which dissipated quickly. These factors would have significantly slowed or stopped the process of orbital evolution.

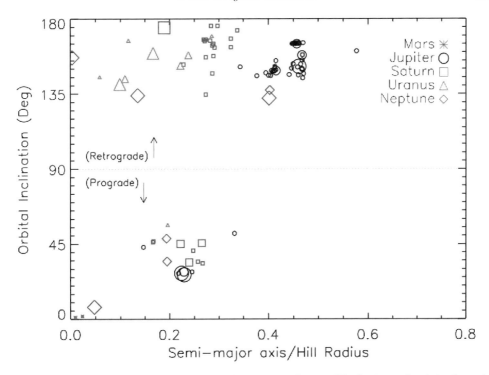

Figure 1. All 96 Known irregular satellites of the giant planets. The horizontal axis is the ratio of the satellites semi-major axis to the respective planet's Hill radius. The vertical axis is the inclination of the satellite to the orbital plane of the planet. The size of the symbol represents the radius of the object: Large symbol $r > 25$ km, medium symbol $25 > r > 10$ km, and small symbol $r < 10$ km. Neptune's Triton can be seen in the upper left of the figure while Nereid is near the lower left. Mars' two satellites are plotted for comparison. All 53 known regular satellites would fall near the origin of this plot. [Modified from Sheppard *et al.* 2005]

Finally, Neptune's outer satellite Nereid is usually considered an irregular satellite but its relatively small semi-major axis and low inclination yet exceptionally large eccentricity suggest it may be a perturbed regular satellite, perhaps from Triton's capture (Goldreich *et al.* 1989; Cuk & Gladman 2005; Sheppard *et al.* 2006).

2. Irregular Satellite Discovery

Irregular satellite discovery requires large fields of view because of the large planetary Hill spheres. Sensitivity is needed because the majority of irregular satellites are small (radii < 50 km) and therefore faint. With the use of large field-of-view photographic plates around the end of the 1800's the first distinctive irregular satellites were discovered (Figure 3). In 1898 the largest irregular satellite of Saturn, Phoebe (radius ~ 60 km), was discovered and in 1904 the largest irregular of Jupiter, Himalia (radius ~ 92 km), was discovered (see Kuiper 1961 for a review of early photographic surveys).

Until 1997 only ten or eleven irregular satellites were known and the last discovered irregular satellite was in 1975 on photographic plates (Kowal *et al.* 1975). Since 1997 eighty-six irregular satellites have been discovered around the giant planets (Gladman et al. 1998;2000;2001; Sheppard & Jewitt 2003; Holman *et al.* 2004; Kavelaars *et al.* 2004; Sheppard *et al.* 2005;2006). Jupiter's retinue of irregular satellites has increased from 8 to 55, Saturn's from 1 to 26, Uranus' from 0 to 9 and Neptune's from 1 to 6 (or seven if

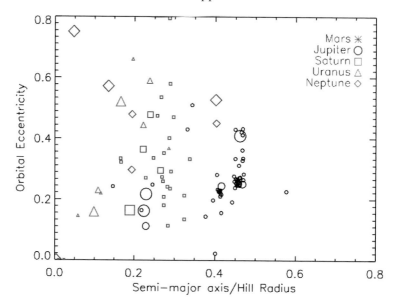

Figure 2. All 96 Known irregular satellites of the giant planets. The horizontal axis is the ratio of the satellites semi-major axis to the respective planet's Hill radius. The vertical axis is the orbital eccentricity. The size of the symbol represents the radius of the object: Large symbol $r > 25$ km, medium symbol $25 > r > 10$ km, and small symbol $r < 10$ km. Again, all 53 known regular satellites would fall near the origin of this plot, where Triton and Mars' satellites are located. [Modified from Sheppard *et al.* 2005]

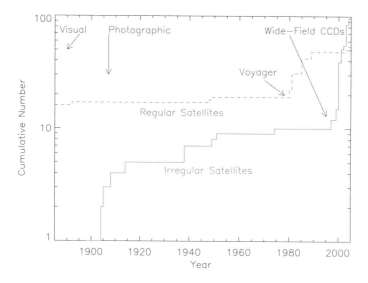

Figure 3. The number of irregular and regular satellites discovered since the late 1800's. Key technological advances which resulted in a jump in discoveries are listed.

including Triton). Table 1 shows information about the current irregular satellite systems around the giant planets.

The number of known irregular satellites (96 as of November 2005) have recently surpassed the number of known regular satellites (Figure 3). The main reason is new

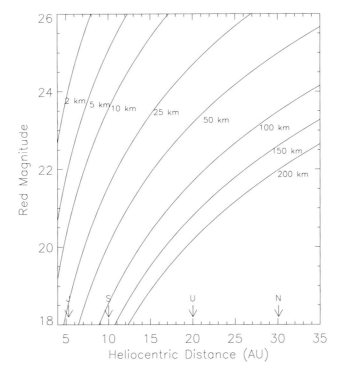

Figure 4. The distances of the planets versus the observable small body population diameter for a given red magnitude assuming an albedo of 0.04. Jupiter's closer proximity allows us to probe the smallest satellites.

technology. The recent development of sensitive, large scale CCD detectors has allowed these faint outer planetary satellites to be discovered. Because of the proximity of Jupiter (Figure 4), it currently has the largest irregular satellite population (Sheppard and Jewitt 2003).

3. Capture of Irregular Satellites

Only the four giant planets have known irregular satellite populations (Figure 1). The likely reason is that the capture process requires something that the terrestrial planets did not have. Capture of a heliocentric orbiting object is likely only if the object approaches the planet near its Lagrangian points and has an orbital velocity within about 1% that of the planet. Objects may temporarily orbit a planet (i.e. Shoemaker-Levy 9) but because of the reversibility of Newton's equations of motion some form of energy dissipation is required to permanently capture a body. Without dissipation the object will be lost in less than a few hundred years (Everhart 1973; Heppenheimer & Porco 1977). In the present epoch a planet has no known efficient mechanism to permanently capture satellites (Figure 5).

Kuiper (1956) first suggested that the irregular satellites were originally regular satellites which escaped from the planet's Hill sphere to heliocentric orbit because of the decreasing mass of the planet. These "lost" satellites would have similar orbits as the parent planet. Eventually the satellite would pass near the planet and be slowed down from the mass escaping from the planet. The satellite would thus be captured in an

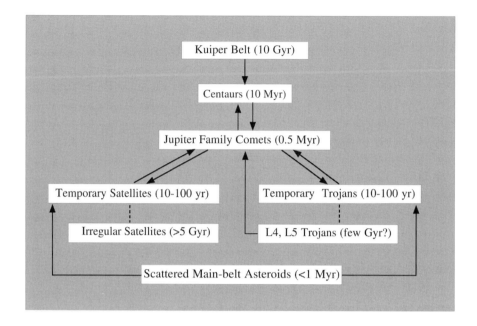

Figure 5. Interrelations among the small body populations in the solar system. Solid arrows denote established dynamical pathways. Dashed lines show pathways which currently have no known energy dissipation source and thus can not lead to permanent capture but only temporary capture. During the planet formation epoch such pathways may have existed. Numbers in parentheses indicate the approximate dynamical lifetimes of the different populations. (Figure from Jewitt et al. (2004))

"irregular" type orbit. It is now believed that the giant planets never lost significant amounts of mass and thus irregular satellites are unlikely to be escaped regular satellites.

The dissipation of energy through tidal interactions between the planet and irregular satellites is not significant for such small objects at such large distances (Pollack *et al.* 1979). The creation of irregular satellites from explosions of the outer portions of the massive ice envelopes of the large regular satellites from saturation by electrolysis seems unlikely and no observational evidence supports such explosions on the regular satellites (Agafonova & Drobyshevski 1984).

Three viable mechanisms have been proposed for irregular satellite capture. Satellite capture could have occurred efficiently towards the end of the planet formation epoch due to gas drag from an extended planetary atmosphere (Kuiper 1956; Pollack, Burns & Tauber 1979), the enlargement of the Hill sphere caused by the planet's mass growth (Heppenheimer & Porco 1977) and/or higher collisional or collisionless interaction probabilities with nearby small bodies (Colombo & Franklin 1971; Tsui 2000). Below we discuss each of these in more detail.

3.1. *Capture by Gas Drag*

During early planet formation the giant planets likely had primordial circumplanetary nebulae (Pollack *et al.* 1979; Cuk & Burns 2004). An object passing through this gas and dust near a planet would have experienced gas drag. In order to significantly slow an object for capture it would need to encounter about its own mass within the nebula. Conditions at the distances of the irregular satellites are unknown, but rough estimates suggest that if the object was larger than a few hundred kilometers it would not have been significantly affected. If the object was very small it would have been highly slowed and would have spiraled into the planet. If the object was just the right size (a few km to a few hundred kilometers) is would have experienced just enough gas drag to be captured (Pollack et al. 1979). Hydrodynamical collapse of the primordial planetary nebula would have to occur within a few thousand years of capture in order for the satellites to not experience significant orbital evolution and eventually spiral into the planet from gas drag. In this scenario the current irregular satellites are only the last few captured bodies which did not have time to spiral into the planet. Retrograde objects would have experienced larger gas drag during their time within the nebula and thus their orbits should be more modified toward smaller eccentricities, inclinations and semi-major axes. Observations currently show that both the progrades and retrogrades have similar modification. Gas drag would also allow for larger objects to be captured closer to the planet since the nebula would be more dense there. In the action of gas drag smaller irregular satellites should have their orbits evolve faster and should have been preferentially removed. No size versus orbital characteristics are observed for any of the irregular satellites of the planets.

3.2. *Pull-down Capture*

Another way an object can become permanently captured is if the planet's mass increased or the Sun's mass decreased while the object was temporarily captured, called pull-down capture (Heppenheimer & Porco 1977). Either of these scenarios would cause the Hill sphere of the planet to increase making it impossible for the object to escape with its current energy. Again, the enlargement of the Hill sphere would have to happen over a short timescale. Likely mass changes of the Sun or the planet would need to be greater than about 40% over a few thousand years (Pollack *et al.* 1976). The Hill sphere of the planet would also increase if the planet migrated significantly away from the sun (Brunini 1995). This mechanism is not a likely cause of permanent capture because the large migrations required to make temporary capture permanent within a few thousand years would severely disrupt any satellite systems (Beauge *et al.* 2002).

3.3. *Capture Through Collisional or Collisionless Interactions*

Finally, a third well identified mechanism of capture could be from the collision or collisionless interaction of two small bodies within the Hill sphere of the planet (Colombo & Franklin; Tsui 2000; Astakhov *et al.* 2003; Funato *et al.* 2004; Agnor & Hamilton 2004). This could occur as asteroid-asteroid or asteroid-satellite encounters. These encounters could dissipate the required amount of energy from one or both of the objects for permanent capture. This mechanism for capture would operate much more efficiently during the early solar system when many more small bodies where passing near the planets. An interesting point of this capture mechanism is that it would be fairly independent of the mass or formation scenario of the planet and mostly depend on the size of the Hill sphere and number of passing bodies.

4. Dynamics of Irregular Satellites

The known irregular satellites are stable over the age of the Solar System though strongly influenced by solar and planetary perturbations (Henon 1970; Carruba *et al.* 2002; Nesvorny *et al.* 2003). The perturbations are most intense when the satellite is near apoapsis. High inclination orbits have been found through numerical simulations to be unstable due to solar perturbations (Carruba *et al.* 2002; Nesvorny *et al.* 2003). Satellites with inclinations between $50 < i < 130$ degrees slowly have their orbits stretched making them obtain very high eccentricities. The high eccentricities are obtained in $10^7 - 10^9$ years and cause the satellite to eventually be lost from the system either through exiting the Hill sphere or colliding with a regular satellite or the planet.

A number of irregular satellites have been found to be in orbital resonances with their planet. These resonances protect the satellites from strong solar perturbations. The two main types of resonance found to date are Kozai resonances and secular resonances (Kozai 1962; Carruba *et al.* 2002; Nesvorny *et al.* 2003). The irregular satellites known or suspected of being in resonances are Jupiter's irregular satellites Sinope, Pasiphae, Euporie (S/2001 J10), S/2003 J18 and Carpo (S/2003 J20) and Saturn's irregular satellites Siarnaq (S/2000 S3), Kiviuq (S/2000 S5) and Ijiraq (S/2000 S6) and Uranus' Stephano (Saha & Tremaine 1993; Whipple & Shelus 1993; Nesvorny *et al.* 2003; Cuk & Burns 2004; R. Jacobson person communication). These resonances occupy a very small amount of orbital parameter space. The evolution of satellites into these resonances implies some sort of slow dissipation mechanism which allowed the satellites to acquire the resonances and not jump over them. This could be obtained from weak gas drag, a small increase in the planet's mass or a slow migration of the planet.

From numerical and analytical work it has been found that retrograde orbits are more stable than prograde orbits over large time-scales (Moulton 1914; Hunter 1967; Henon 1970; Hamilton & Krivov 1997). Analytically the retrogrades may be stable up to distances of $\sim 0.7 r_H$ while progrades are only stable up to $0.5 r_H$ (Hamilton & Krivov 1997). This is consistent with known orbits of retrogrades and progrades to date. Known retrograde (prograde) irregular satellites have semi-major axes out to $\sim 0.47 r_H$ ($\sim 0.33 r_H$) and have apocenters up to $\sim 0.65 r_H$ ($\sim 0.47 r_H$).

Many of the irregular satellites have been found to show dynamical groupings (Gladman *et al.* 2001; Sheppard & Jewitt 2003; Nesvorny et al. 2003). At Jupiter the dynamical groupings are well observed in semi-major axis and inclination phase space (Figure 1) and are probably similar to families found in the main belt asteroids which are created when a larger parent body is disrupted into several smaller daughter fragments. The irregular satellites at the other giant planets are mostly grouped in inclination phase space and not in semi-major axis phase space. It would be unlikely that a fragmented body would create daughter bodies with such significantly different semi-major axes. This inclination clustering may just be because of resonance effects or that these particular inclinations are more stable. Still, there do appear to be some irregular satellites at Saturn, Uranus and Neptune that do cluster in semi-major axis and inclination phase space like those seen at Jupiter but these groups are not well populated. Further satellites in these putative dynamical families may be observed when smaller satellites are able to be discovered in the future.

Fragmentation of the parent satellites could be caused by impact with interplanetary projectiles (principally comets) or by collision with other satellites. Collisions with comets are improbable in the current solar system but during the heavy bombardment era nearly 4.5 billion years ago they would have been highly probable (Sheppard & Jewitt 2003). Large populations of now defunct satellites could also have been a collisional source in creating the observed satellite groupings (Nesvorny *et al.* 2004). No size versus orbital property correlations are seen in the groupings which suggest breakup occurred after any significant amounts of gas were left.

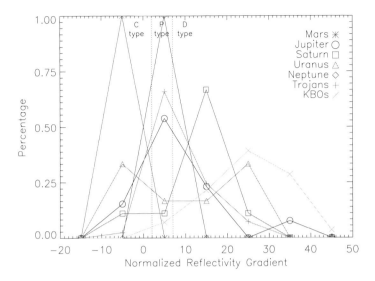

Figure 6. The colors of the irregular satellites of Jupiter, Saturn, Uranus and Neptune compared to the KBOs, Trojans and Martian satellites. The Jupiter irregular satellites are fairly neutral in color and very similar to the nearby Jupiter Trojans. Saturn's irregulars are significantly redder than Jupiter's but do not reach the extreme red colors seen in the KBOs. Uranus' irregular satellites are very diverse in color with some being the bluest known while others are the reddest known irregular satellites. Only two of Neptune's irregulars have measured colors and not much can yet be said except they don't show the very red colors seen in the Kuiper Belt. The general linear colors of the C, P and D-type asteroids are shown for reference (Dahlgren & Lagerkvist 1995). Irregular satellite colors are from Grav *et al.* 2003; 2004a.

The detection of dust in bound orbits about Jupiter in the outer Jupiter system from the Galileo spacecraft is attributed to high velocity impacts of interplanetary microme-teoroids into the atmosphereless outer satellites (Krivov *et al.* 2002). The micron sized dust is in prograde and retrograde orbits with a number density (10 km^{-3}) about ten times larger than in the local interplanetary medium.

5. Physical Properties of Irregular Satellites

Most of the space in the giant planet region of the solar system is devoid of objects which make irregular satellites one of the only dynamical clues as to what affected most of the mass in the solar system 4.5 billion years ago. Irregular satellites were likely asteroids or comets in heliocentric orbit which did not get ejected into the Oort cloud or incorporated in the planets. They may be some of the only small bodies remaining which are still relatively near their formation locations within the giant planet region. The irregular satellite reservoirs lie between the main belt of asteroids and Kuiper Belt which makes them a key to showing us the complex transition between rocky objects in the main asteroid belt and the expected very volatile rich objects in the Kuiper Belt.

5.1. *Visible and Infrared Colors*

Colors of the irregular satellites are neutral to moderately red (Tholen & Zellner 1984; Luu 1991; Rettig *et al.* 2001; Maris et al. 2001; Grav *et al.* 2003; 2004a). Most do not show the very red material found in the distant Kuiper Belt (Figures 6 and 7). The Jupiter irregular satellite colors are very similar to the C, P and D-type carbonaceous outer main

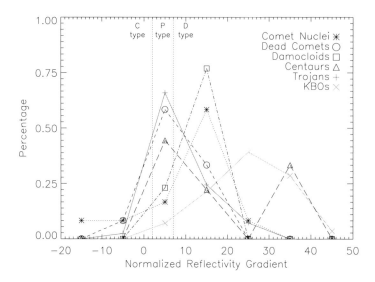

Figure 7. Same as Figure 6 except this plot shows the colors of the comet nuclei, dead comet candidates, Damocloids, Centaurs, Trojans and KBOs. It is plotted as a separate graph from Figure 6 to avoid confusion between the many different types of objects. Jupiter's irregulars are similar in color to the dead comets and some of the Centaurs. Saturn's irregulars are similar in color to the Damocloids and active comet nuclei. Comet nuclei and dead comet colors are from Jewitt (2002) and references therein. Centaur and KBO colors are from Barucci *et al.* (2001); Peixinho *et al.* (2001); Jewitt & Luu (2001) and references therein. Damocloid colors are from Jewitt (2005) and references therein.

belt asteroids (Degewij *et al.* 1980) as well as to the Jupiter Trojans and dead comets. Colors of the Jupiter irregular satellite dynamical groupings are consistent with, but do not prove, the notion that each group originated from a single undifferentiated parent body. Optical colors of the 8 brightest outer satellites of Jupiter show that the prograde group appears redder and more tightly clustered in color space than the retrograde irregulars (Rettig *et al.* 2001; Grav et al. 2003). Near-infrared colors recently obtained of the brighter satellites agree with this scenario and that the Jupiter irregular's colors are consistent with D and C-type asteroids (Sykes *et al.* 2000; Grav *et al.* 2004b).

The Saturn irregular satellites are redder than Jupiter's but still do not show the very red material observed in the Kuiper Belt. The colors are more similar to the active cometary nuclei and damocloids. Buratti *et al.* (2005) show that the color of the dark side of Iapetus is consistent with dust from the small outer satellites of Saturn. Buratti *et al.* also find that none of Saturn's irregular satellites have similar spectrophotometry as Phoebe. The irregular satellites of Uranus have a wide range of colors from the bluest to the reddest. These satellites may show the extreme red colors observed in the Kuiper Belt and have a distribution similar to the Centaurs. Neptune's irregulars have limited observational data but to date they don't show the extreme red colors seen in the Kuiper Belt.

5.2. *Spectra and Albedos*

Near-Infrared and optical spectra of the brightest Jupiter satellites are mostly linear and featureless (Luu 1991; Brown 2000; Jarvis et al. 2000; Chamberlain & Brown 2004; Geballe *et al.* 2002). Jarvis et al. (2000) finds a possible 0.7 micron absorption feature in Jupiter's Himalia and attributes this to oxidized iron in phyllosilicates which is typically

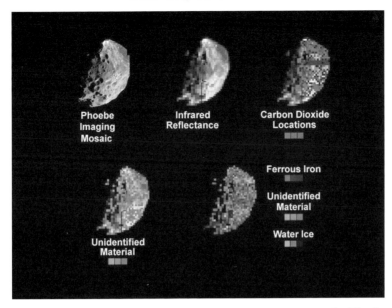

Figure 8. Phoebe's mineral distribution as seen by the Cassini spacecraft. Phoebe appears to have a very volatile rich surface which is unlike the irregular satellites at Jupiter. (Produced by NASA/JPL/University of Arizona/LPL using data from the Cassini Imager and VIMS; see Porco *et al.* 2005 and Clark *et al.* 2005).

produced by aqueous alteration. The spectra of Jupiter's irregular satellites are consistent with C-type asteroids. The irregular satellites at Saturn and Neptune appear to be remarkably different with rich volatile surfaces.. The largest Saturn irregular, Phoebe, has been found to have water ice (Owen *et al.* 1999) as has the large Neptune irregular satellite Nereid (Brown, Koresko & Blake 1998).

Jupiter's irregular satellites have very low albedos of about 0.04 and 0.05 which again along with their colors are consistent with dark C, P and D-type Carbon rich asteroids in the outer main belt (Cruikshank 1977) and very similar to the Jovian Trojans (Fernandez *et al.* 2003). Saturn's Phoebe has an average albedo of about 0.07 (Simonelli *et al.* 1999) while Neptune's Nereid was found to have an albedo of 0.16 from Voyager data (Thomas *et al.* 1991). These albedos are more similar to the higher albedos found in the Kuiper Belt (Grundy et al. 2005; Cruikshank *et al.* 2005). These are in comparison to the average albedos of comet nuclei 0.03, extinct comets 0.03, and Jovian Trojans 0.06 (Fernandez *et al.* 2003).

The Cassini spacecraft obtained resolved images of Himalia and showed it to be an elongated shaped object with axes of 150 x 120 km with an albedo of about 0.05 (Porco *et al.* 2003). Cassini obtained a mostly featureless near-infrared spectrum of Jupiter's JVI Himalia (Chamberlain & Brown 2004).

Cassini obtained much higher resolution images of Saturn's irregular satellite Phoebe (Figure 8) with a flyby of 2071 km on June 11, 2004. The images showed Phoebe to be intensively cratered with many high albedo patches near crater walls (Porco *et al.* 2005). Phoebe's density was found to be 1630 ± 33 kg m^{-3} (Porco et al. 2005). The spectra showed lots of water ice as well as ferrous-iron-bearing minerals, bound water, trapped CO_2, phyllosilicates, organics, nitriles and cyanide compounds on the surface (Clark *et al.* 2005). Phoebe's volatile rich surface and many compounds infer the object was formed beyond the rocky main belt of asteroids and maybe very similar to the composition of comets. Its relatively high density compared to that observed for comets and inferred for

Kuiper Belt objects makes it a good candidate to have formed near its current location where the highly volatile materials are still unstable to evaporation.

6. Comparison of the Giant Planet Irregular Satellite Systems

6.1. *Giant Planet Formation*

Irregular satellites are believed to have been captured around the time of the formation of the giant planets. Thus their dynamical and physical properties are valuable clues as to what happened during the planet formation process. Because of the massive hydrogen and helium envelopes of the gas giants Jupiter and Saturn, they presumably formed quickly in the solar nebula before the gas had time to significantly dissipate. The less massive and deficient in hydrogen and helium ice giants Uranus and Neptune appear to have taken a drastically different route of evolution.

There are two main models for giant planet formation. The standard model of core accretion assumes the cores of the giant planets were formed through oligarchic growth for about 10^6 to 10^8 years. Once they obtained a core of about ten Earth masses they quickly accreted their massive gaseous envelopes (Pollack *et al.* 1996). The disadvantage of this model is that the protoplanetary disk likely dissipated within a few million years while the core accretion model requires long timescales to form the planets. Because of the lower surface density and larger collisional timescales for more distant planets the core accretion model can not adequately form Uranus and Neptune in the age of the solar system.

The second giant planet formation mechanism is through disk instabilities. This model suggests parts of the solar nebula became unstable to gravitational collapse (Boss 2001). In this model the planets would form on timescales of only about 10^4 years. The disadvantages are it doesn't allow for massive cores and does not appear to be applicable to the small masses of Uranus and Neptune.

Both giant planet formation models have trouble forming Uranus and Neptune (Bodenheimer & Pollack 1986; Pollack *et al.* 1996). Any theory on the different formation scenarios of Uranus and Neptune to that of Jupiter and Saturn should take into account the irregular satellite systems of each. The recent theory that Uranus and Neptune lost their hydrogen and helium envelopes by photoevaporation from nearby OB stars (Boss, Wetherill, & Haghighipour 2002) would have caused all their irregular satellites to be lost because of the significant decrease in the planet's mass (Sheppard and Jewitt (2003); Jewitt & Sheppard (2005)). Another recent theory is that Uranus and Neptune formed in the Jupiter-Saturn region with subsequent scattering to their current locations (Thommes, Duncan, & Levison 2002). Any large migration by the planets would have disrupted any outer satellite orbits (Beauge *et al.* 2002).

6.2. *Population and Size Distributions of the Irregular Satellites*

When measured to a given size the population and size distributions of the irregular satellites of each of the giant planets appears to be very similar (Figure 9) (Sheppard and Jewitt 2005; Jewitt and Sheppard 2005). In order to model the irregular satellite size distribution we use a differential power-law radius distribution of the form $n(r)dr = \Gamma r^{-q}dr$, where Γ and q are constants, r is the radius of the satellite, and $n(r)dr$ is the number of satellites with radii in the range r to $r+dr$. All giant planet irregular satellite systems appear to have shallow power law distribution of $q \sim 2$. If we don't include Triton the largest irregular satellite of each planet is of the 150 km scale with about one hundred irregular satellites expected around each planet with radii larger than about

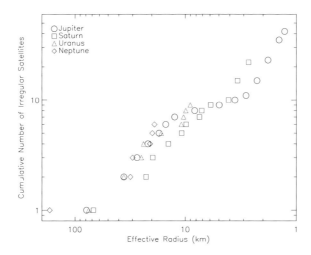

Figure 9. Cumulative radius function for the irregular satellites of Jupiter, Saturn, Uranus and Neptune. This figure directly compares the sizes of the satellites of the giant planets assuming all satellite populations have similar low albedos. The planets have statistically similar shallow size distributions of irregular satellites. Neptune's irregular satellite size distribution is plotted without including Triton. [Modified from Sheppard et al. 2005]

1 km. This is unexpected considering the different formation scenarios envisioned for the gas giants versus the ice giants.

7. Discussion and Conclusions

The irregular satellites of each planet are a distinct group of bodies not necessarily linked to the two prominent reservoirs of the main asteroid belt or the Kuiper Belt. These satellites may have formed relatively near their current locations and were subsequently captured by their respective planet near the end of the planet formation epoch. With the development of large, sensitive, digital detectors on large class telescopes in the late 1990's the discovery and characterization of the irregular satellites improved dramatically. We find that the gas giants Jupiter and Saturn and the ice giants Uranus and Neptune all have a system of irregular satellites which have similar sizes, populations and dynamics.

Current observations favor the capture mechanism of collisional or collisionless interactions within the Hill spheres of the planets. This capture mechanism is fairly independent of the planets formation scenario and mass unlike gas drag or pull-down capture (Jewitt & Sheppard 2005). Because the less massive ice giants are more distant from the Sun their Hill spheres are actually larger than the gas giants. These increased Hill spheres may compensate for the lower density of small bodies in the outer solar nebula and thus allow all the giant planets to capture similar irregular satellite systems. Recent discoveries of binaries in the Kuiper Belt show that such objects may be quite common in the outer solar system. These binary pairs would be ideal for creating irregular satellites of the giant planets through three body interactions as has been shown for the capture of Triton (Agnor & Hamilton 2004). In fact, the equally sized binary pairs in the Kuiper Belt may have formed in a similar manner (Funato *et al.* 2004).

Three body interactions would have been much more probable in the early solar system just after planet formation when leftover debris was still abundant. This capture process would allow for the possible scattering predicted for Uranus and Neptune unlike gas drag and pull-down capture since capture by three body interactions would still operate after any scattering. Three body capture also agrees with the results of Beauge *et al.* (2002) in which they find the irregular satellites would have to have formed after any significant planetary migration as well as Brunini *et al.* (2002) who find that Uranus' irregular satellites would have to be captured after any impact which would have tilted the planet's rotation axis. Also, Triton may have disrupted the outer satellites of Neptune and capture of these irregulars may have occurred after Triton was captured (Cuk & Gladman 2005). These scenarios all point to satellite capture happening just after the planet formation process.

If three body interactions were the main capture mechanism then one may expect the terrestrial planets to have irregular satellites. The terrestrial planets had very small Hill spheres compared to the giant planets because of their low mass and proximity to the Sun. In addition, the terrestrial planets had no population of regular satellites for passing objects to possibly interact with. This may explain why Mars and the other terrestrial planets have no outer satellites, though Mars' two inner satellites may have been capture through three body interactions. Perhaps Phobos and/or Deimos were once binary asteroids.

The observed irregular satellite dynamical families were probably created after capture. In order to have a high probability of impact for the creation of families either the captured had to occur very early on when collisions were much more probable than now or there must have been a much larger population of now defunct satellites around each planet.

The non-detection of volatiles on Jupiter's irregular satellites whereas volatiles are seen on Saturn's and Neptune's bodes well for the objects to have formed near their current location. The currently limited data on the albedos, colors and densities of the irregular satellites appear to show that each planet's irregular satellites are physically distinct. Jupiter's irregulars are remarkably similar to the Jovian Trojans and dead comets. Saturn's are significantly redder but neither Jupiter's or Saturn's show the very red material observed in the Kuiper Belt. Uranus' irregulars have a wide range of colors with some being the bluest and others being the reddest.

Acknowledgements

Support for this work was provided by NASA through Hubble Fellowship grant # awarded by the Space Telescope Science Institute, which is operated by the Association of Universities for Research in Astronomy, Inc., for NASA, under contract NAS 5-26555.

References

Agafonova, I. & Drobyshevski, E. 1985, *EM&P* 33, 1
Agnor, C. & Hamilton, D. 2004, *BAAS* 36, 40.14
Astakhov, S., Burbanks, A., Wiggins, S., & Farrelly, D. 2003, *Nature* 423, 264
Barucci, M., Fulchignoni, M., Birlan, M., Doressoundiram, A., Romon, J., & Boehnhardt, H. 2001, *AA* 371, 1150
Beauge, C., Roig, F., & Nesvorny, D. 2002, *Icarus* 158, 483
Bodenheimer, P. & Pollack, J. 1986, *Icarus* 67, 391
Boss, A. 2001, *ApJ* 563, 367
Boss, A. 2002, *ApJ* 599, 577
Boss, A., Wetherill, G., & Haghighipour, N. 2002, *Icarus* 156, 291
Brown, M., Koresko, C., & Blake, G. 1998, *ApJ* 508, 175
Brown, M. 2000, *AJ* 119, 977
Brunini, A. & Conicet, P. 1995, *EM&P* 71, 281
Brunini, A., Parisi, M., & Tancredi, G. 2002, *Icarus* 159, 166

Buratti, B., Hicks, M., & Davies, A. 2005, *Icarus* 175, 490

Burns, J. 1986, *in Satellites ed. by J. Burns and M. Matthews (University of Arizona Press, Tucson)* pg. 117

Burns, J. 1992, *in Mars ed. by Kieffer, Jakosky, Snyder, Matthews (University of Arizona Press, Tucson)* pg. 1283

Canup, R. & Asphaug, E. 2001, *Nature* 412, 708

Carruba, V., Burns, J., Nicholson, P., & Gladman, B. 2002, *Icarus* 158, 434

Colombo, G. & Franklin, F. 1971, *Icarus* 15, 186

Chamberlain, M. & Brown, R. 2004, *Icarus* 172, 163

Clark, R., Brown, R., Jaumann, R. *et al.* 2005, *Nature* 435, 66

Cruikshank, D. 1977, *Icarus* 30, 224

Cruikshank, D., Stansberry, J., Emery, J., Fernandez, Y., Werner, M., Trilling, D., & Rieke, G. 2005, *ApJ* 624, 53

Cuk, M. & Burns, J. 2004, *Icarus* 167, 369

Cuk, M. & Gladman, B. 2005, *ApJ* 626, 113

Dahlgren, M. & Lagerkvist, C. 1995, *AA* 302, 907

Degewij, J., Zellner, B., & Andersson, L. 1980, *Icarus* 44, 520

Everhart, E. 1973, *AJ* 78, 316

Fernandez, Y., Sheppard, S., & Jewitt, D. 2003, *AJ* 126, 1563

Funato, Y., Makino, J., Hut, P., Kokubo, E., & Kinoshita, D. 2004, *Nature* 427, 518

Geballe, T., Dalle, O., Cruikshank, D., & Owen, T. 2002, *Icarus* 159, 542

Gladman, B., Nicholson, P., Burns, J., Kavelaars, J., Marsden, B., Williams, G., & Offutt, W. 1998, *Nature* 392, 897

Gladman, B., Kavelaars, J., Holman, M., Petit, J., Scholl, H., Nicholson, P., & Burns, J. 2000, *Icarus* 147, 320

Gladman, B., Kavelaars, J., Holman, M., Nicholson, P., Burns, J. *et al.* 2001, *Nature* 412, 163

Goldreich, P., Murray, N., Longaretti, P., & Banfield, D. 1989, *Science* 245, 500

Grav, T., Holman, M., Gladman, B., & Aksnes, K. 2003, *Icarus* 166, 33

Grav, T., Holman, M., & Fraser, W. 2004a, *ApJ* 613, 77

Grav, T. & Holman, M. 2004b, *ApJ* 605, 141

Grundy, W., Noll, K., & Stephens, D. 2005, *Icarus* 176, 184

Hamilton, D. & Krivov, A. 1997, *Icarus* 128, 241

Henon, M. 1970, *AA* 9, 24

Heppenheimer, T. & Porco, C. 1977, *Icarus* 30, 385

Holman, M., Kavelaars, J., Grav, T., Gladman, B. *et al.* 2004, *Nature* 430, 865

Hunter, R. 1967, *MNRAS* 135, 245

Jarvis, K., Vilas, F., Larson, S., & Gaffey, M. 2000, *Icarus* 145, 445

Jewitt, D. & Luu, J. 2001, *AJ* 122, 2099

Jewitt, D. 2002, *AJ* 123, 1039

Jewitt, D., Sheppard, S., & Porco, C. 2004, in: F. Bagenal (ed.), *Jupiter: the planet, satellites and magnetosphere* (Cambridge: Cambridge University Press), p. 263

Jewitt, D. & Sheppard, S. 2005, *Space Sci Rev* 116, 441

Jewitt, D. 2005, *AJ* 129, 530

Kavelaars, J., Holman, M., Grav, T., Milisavljevic, D., Fraser, W., Gladman, B. J., Petit, J.-M., Rousselot, P., Mousis, O., & Nicholson, P. 2004, *Icarus* 169, 474

Kessler, D. 1981, *Icarus* 48, 39

Kowal, C., Aksnes, K., Marsden, B., & Roemer, E. 1975, *AJ* 80, 460

Kozai, Y. 1962, *AJ* 67, 591

Krivov, A., Wardinski, I., Spahn, F., Kruger, H., & Grun, E. 2002, *Icarus* 157, 436

Kuiper, G. 1956, *Vistas Astron.* 2, 1631

Kuiper, G. 1961, in: G.P. Kuiper and B.M. Middlehurst (eds.), *Planets and Satellites)* (Chicago: University of Chicago Press) p. 575

Luu, J. 1991, *AJ* 102, 1213

Maris, M., Carraro, G., Cremonese, G., & Fulle, M. 2001, *AJ* 121, 2800

Moulton, F. 1914, *MNRAS* 75, 40

Nesvorny, D., Alvarellos, J., Dones, L., & Levison, H. 2003, *AJ* 126, 398

Nesvorny, D., Beauge, C., & Dones, L. 2004, *AJ* 127, 1768

Owen, T., Cruikshank, D., Dalle, O., Geballe, T., Roush, T., & de Bergh, C. 1999, *Icarus* 139, 379

Peale, S. 1999, *ARA&A* 37, 533

Peixinho, N., Lacerda, P., Ortiz, J., Doressoundiram, A., Roos-Serote, M., & Gutierrez, P. 2001, *AA* 371, 753

Pollack, J., Burns, J., & Tauber, M. 1979, *Icarus* 37, 587

Pollack, J., Hubickyj, O., Bodenheimer, P., Lissauer, J., Podolak, M., & Greenzweig, Y. 1996, *Icarus* 124, 62

Porco, C., West, R., McEwen, A. *et al.* 2003, *Sci* 229, 1541

Porco, C., Baker, E., Barbara, J. *et al.* 2005, *Sci* 307, 1243

Rettig, T., Walsh, K., & Consolmagno, G. 2001, *Icarus* 154, 313

Saha, P. & Tremaine, S. 1993, *Icarus* 106, 549

Sheppard, S. & Jewitt, D. 2003, *Nature* 423, 261

Sheppard, S., Jewitt, D., & Kleyna, J. 2004, *AJ* 128, 2542

Sheppard, S., Jewitt, D., & Kleyna, J. 2005, *AJ* 129, 518

Sheppard, S., Jewitt, D., & Kleyna, J. 2006, *AJ* submitted

Simonelli, D., Kay, J., Adinolfi, D., Veverka, J., Thomas, P., & Helfenstein, P. 1999, *Icarus* 138, 249

Sykes, M., Cutri, R., Fowler, J., Tholen, D., Skrutskie, M, Price, S., & Tedesco, E. 2000, *Icarus* 143, 371

Tholen, D. & Zellner, B. 1984, *Icarus* 58, 246

Thomas, P., Veverka, J., & Helfenstein, P. 1991, *JGR* 96, 19253

Thommes, E., Duncan, M., & Levison, H. 2002, *AJ* 123, 2862

Tsui, K. 2000, *Icarus* 148, 139

Whipple, A. & Shelus, P. 1993, *Icarus* 101, 265

Asteroids, Comets, Meteors
Proceedings IAU Symposium No. 229, 2005
D. Lazzaro, S. Ferraz-Mello & J.A. Fernández, eds.

© 2006 International Astronomical Union
doi:10.1017/S1743921305006836

Collisional evolution of asteroids and Trans–Neptunian objects

Adriano Campo Bagatin[1]

[1]Departmento de Fisica, Ingenieria de Sistemas y Teoria de la Señal
Escuela Politecnica Superior, Universidad de Alicante, Alicante, Spain
email: adriano@dfists.ua.es

Abstract. Since Pietrowsky's first analytical study of collisional systems of asteroids (1953), through Dohnanyi's comprehensive theory (1969), to the analytical and numerical studies of the last two decades, the collisional evolution of populations of asteroids —and to a less extent, of Trojans and TNOs— has been investigated by many researchers.

The study of such systems is an intrinsically delicate mathematical problem, as their evolution in time is properly described in terms of systems of first–order, non–linear differential equations. Physically, the limited knowledge of some of the collisional properties, rotations and internal structure of bodies, and the complex interplay with dust, non–gravitational effects and dynamical interactions with planets, make the study of the collisional evolution a hard multi–parametric problem. Nevertheless, the task is worth the effort, in fact the understanding of evolutionary processes in the solar system's small body belts provides the main tools to discriminate between the many different theoretical scenarios proposed to explain the formation of the solar system itself.

This review tries to give an updated overall view of the research done in this field, and to show the connections between apparently independent phenomena that may affect the evolution of collisional systems of asteroids and TNOs.

Keywords. Minor planets, Asteroids, Kuiper Belt

---

1. Introduction

In 1918, when Hirayama firstly pointed out that groups of asteroids —that he was observing to have almost identical orbital elements— may have a common origin, the idea that catastrophic collisions among them could sometimes occur began to raise among astronomers. The pioneering work of Pietrowsky (1953) on the frecuency of collisions, in which he estimated relative velocities to be around 5 km/s, and concluded about an expected stationary distribution for asteroids, was the next step into considering the asteroid belt as a collisional system. Dohnanyi (1969), Dohnanyi (1971) and Hellyer (1970), Hellyer (1971) finally studied analytically the evolution of the cascades of fragments coming from asteroid breakups.

The increasing number of discoveries in the two following decades confirmed the sensation that mutual collisions have been shaping the size distribution of the main belt asteroids during and after the 'heavy bombardment' phase occured some 4500 millions of years ago.

In the meantime, the development of computer science made possible that the sets of differential equations describing the evolution in time of the populations of a whole range of sizes could be discretised and numerically handled. This allowed the first numerical models (Davis *et al.* 1985) of the collisional evolution of the asteorid belt to appear and make us understand new features of the main asteroid belt. By that time, another population of small bodies had been confirmed: more than hundred Trojan asteroids had been already found at the stable Lagrangian points of the restricted three–body problem that considers the Sun and planet Jupiter as primary bodies. Soon it was recognised to be a collisional system as well.

The last decade of the XX century put an end to the two–century era of speculations about how asteroids actually look like, and was spotted by the first ever images of *true* asteroids. They offered new opportunities for calculating collision rates and to shed some light into collisional evolution and internal structures.

The origin of short period comets with periods smaller than some puzzled astronomers of the first half of the past century. Edgeworth (1943) and 1949 published his conjecture on the existence of a disk of planetesimals beyond the orbit of planet Neptune, a potential source for comets. During the same period, Kuiper was developing (1951) his own ideas about the same topic. Fernandez (1980) finally showed dynamically the necessity for the existence of such a region. The Kuiper Belt —as the predicted region was finally called at that time— began to be unveiled in 1992, when Luu & Jewitt discovered the first of a long list of objects with semimajor axis beyond Neptune, nowadays known as Trans–Neptunian Objects (TNOs). As more and more TNOs were discovered, it was an evidence that they formed just another population of bodies in a collisional system. Today we see that the structure is complicated by the existende of at least three dynamically different populations: the 'Plutinos', around the 3:2 mean motion resonance with Neptune at 39.4 AU, the Classical Disk, from 40 to some 50 AU, and the Scattered Disk, with large eccentricities and inclinations (Gladman *et al.* 2001, Gladman *et al.* 2002). Beyond that region, a population of large objects ($D > 500km$) in a wide variety of eccentricities and inclinations is being discovered at present.

The present review tries to update and complement the comprehensive review by Davis *et al.* (2002), concentrating on the two major collisional systems in the solar system, the Main Asteroid Belt and the Trans-Neptunian region.

Section 2 is dedicated to summarize the available observational evidences, Section 3 will review the theoretical work developed up to the present, while the main numerical models and their conclusions will be revisited in Section 4. Section 5 will finally try and draw some conclusions, and especially formulate questions yet to be answered in the future.

Amidst all the devoted researchers who have intensely contributed to the knowledge and the understanding of small bodies in the solar system, Paolo Farinella (1953–2000) owns a special place in the recent past of this field. His peculiar skill for interconnecting different areas to explain actual phenomena was only equal to his special hability for seeding and keeping fruitful connections between researchers across the world. To the memory of him this review is dedicated.

2. Observational constraints

Any model of the collisional evolution of asteroids and TNOs has to be checked by trying to match the observed characteristics, those features of the belt that are the product of 4.5 billion years of collisional history. Various authors have attempted to infer the primordial population of asteroids by integrating backwards in time from the present belt distribution, but the inherent instability of the collisional problem prevents such results from converging, as expected from earliest Dohnanyi's result, which showed that under specific circumstances the asteroid population was collisionally relaxed and independent of the starting population.

Direct observables are the size distributions, the existence of Asteroids Vesta and Psyche, and the characteristics of asteroid families. Other observables may help constraining

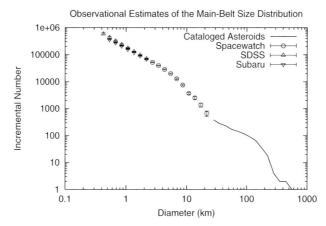

Figure 1. Estimates of the main-belt size distribution, as extrapolated from different observational surveys.

the collisional evolution of asteroids and TNOs, but they may be affected by modelling and non parameter–free assumptions: the distribution of craters on asteroids and on the satellites of planets satellites, the cosmic–ray exposure ages, the NEO populations, and the dust production rates, among others.

2.1. *Size distributions*

The first constraint on models of collional evolution of small bodies populations is the size-frequency distribution.

2.1.1. *Size distribution of the Main Asteroid Belt*

It is widely believed that at least asteroids larger than some $300\ km$ are primordial objects, while the rest of the population is collisionally evolved. The bump in the mainbelt asteroid size distribution centered around $100\ km$ has been explained in a variety of different ways: i) Davis *et al.* (1979), Davis *et al.* (1984) found that this feature marked the transition from the gravity strength regime to the strength–dominated regime and hence was a product of collisional physics. ii) Durda *et al.* (1998) found that this feature was a secondary bump produced by the wave from the strength-gravity regime which occurs at much smaller sizes. iii) Campo Bagatin *et al.* (2002) argued that this shows the effect of non–self–similarity in the physics of fragmentation, that produce a multi–fractal structure in the mass distribution of km–size objects.

Several lines of evidence point to the existence of a variable index in the size distribution of the small asteroid population. Cellino *et al.* (1991) analyzed IRAS data for different zones of the asteroid belt and different size ranges and found incremental size distribution indexes ranging from -2 to -4 for asteroids larger than a few tens of km. Present estimates assume a -4 exponent for the power–law of the distribution in the $5\ km$ to $20\ km$ size range. The Sloan Digital Sky Survey (SDSS) (Ivezic *et al.* 2001) and the Subaru Sub-km Main Belt Asteroid Survey (SMBAS) (Yoshida *et al.* 2003) find similar power–laws for the distributions of objects below $5\ km$ in the size range from $\sim 400\ m$ to $\sim 5\ km$. The SDSS finds an exponent of -2.30 ± 0.05 in the $\sim 400\ m$ to $\sim 5\ km$ size range, and the SMBAS finds -2.19 ± 0.02 for the $\sim 500\ m$ to $\sim 1\ km$ size range. For smaller sizes, Cheng (2004) assumes a -3.5 exponent based on the analysis of the size distribution of boulders on the surface of asteroid 433 Eros, visited by the NEAR probe.

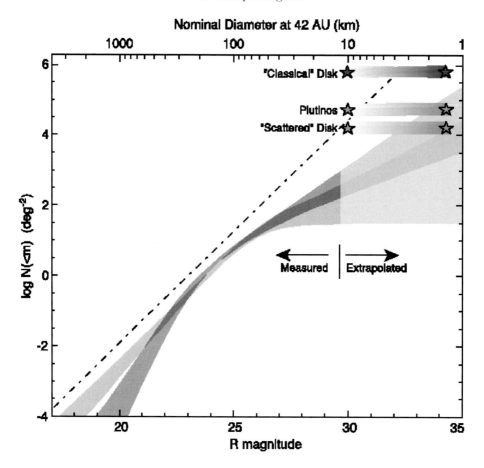

Figure 2. Bernstein *et al* (2004)'s estimates for the distribution of objects – both in R–magnitude and in related size—in the Trans–Neptunian region.

2.1.2. *Size distribution of TNOs*

Since the discovery of the first TNOs in 1992, many other objects have been discovered beyond the orbit of Neptune in a more or less serendipitous way. Sistematic surveys of the Trans-Neptunian populations have been started in the last decade (Gladman *et al.* 1998; Gladman *et al.* 2001, Chiang & Brown, 1999; Luu & Jewitt, 1998; Petit & Gladman, 2003) that have allowed to state the existence of three dynamically different populations for the approximately 1000 objects observed to date. Un-biased samples are needed in order to get reliable orbital elements for TNOs and be able to correctly characterise them dynamically. Nevertheless, it is not straightforward to determine size distributions for those populations, as strong uncertainties remain – among others – about the albedos of TNOs.

Bernstein *et al.* (2004) provided size-bands distributions for TNOs, extrapolating the result of a 0.02 deg^2 survey using the Advanced Camera for Surveys aboard the Hubble Space Telescope (HST), discovering objects down to $m = 28.3$. They show a clear pattern consisting in a steep power-law distibution for the large size end of TNOs, followed by a shallower (including values below the Dohnanyi's exponent) size distributions below a transition size around 70–100 km.

It can easily seen that the situation is much more fuzzy that in the case of asteroids, and it is hard to say that unambiguous observables are available for TNOs at all.

2.2. *Vesta and Psyche*

A powerful constraint on the asteroid collisional history is the existence of the basaltic crust of Vesta, which dates back to the earliest era of the solar system. Any collisional model must preserve this thin crust during the collisional bombardment (Davis *et al.* 1984). HST observations of Vesta revealed the existence of a ~ 450 *km* diameter basin caused by the impact of a ~ 40 *km* diameter projectile (Marzari *et al.* 1996; Asphaug 1997). This impact, presumably the largest since Vesta's crust formed, provides a very specific constraint on Vesta's collisional history.

On the other hand, asteroid 16 Psyche appears to be the collisionally exposed core of a parent body which was virtually identical to Vesta. The possibility to disrupt the Psyche parent body without the formation of a family associated with it, while preserving the crust of Vesta, was investigated by Davis *et al.* (1999). They concluded that it was very difficult, but not impossible —based on collisional modelling— to create Psyche and preserve Vesta's crust, but other explanations for Psyche should be sought. Furthermore, recent work suggests a mean density for Psyche of $1.8 \pm 0.6 g/cm^3$ (Viateau 2000), which is inconceivably low for the iron core of a differentiated body.

2.3. *Asteroid families*

Asteroid families are an observable which is a direct product of collisions. There are over 60 statistically significant clusters identified in asteroid proper elements (Benjoya & Zappalà 2002). Marzari *et al.* (1999) used this number of recognized families in the present belt as a constraint on the overall asteroid collisional history, and found that a small mass initial belt best reproduced the observed number and type of families originated by disruption of parent bodies larger than 100 km diameter. Non–gravitational forces like Yarkovsky effect together with chaotic resonances can push family members to slowly disperse over time (Nesvorný *et al.* 2002).

3. Theoretical studies

3.1. *Asinthotic collisional evolution*

The task of analytic studies on the collisional evolution of the mass distribution of asteroids has been tackled in the past by various researchers: Pietrowski (1953), Hellyer (1970, 1971) and Dohnanyi (1969, 1971).

Dohnanyi's theory is especially important as it also includes and develops the studies made by the other quoted authors. The most important result of this theory is that, as the collisional process in the main asteroid belt gives raise to a cascade of fragments shifting mass toward smaller and smaller sizes, a simple power–law equilibrium mass distribution is approached under well defined assumptions on the collisional response parameters and on the size range. This equilibrium distribution extends over all the size range of the population, except near its high–mass end. It corresponds to a number of bodies dN in the mass interval $(m, m + dm)$, or in the diameter interval $(D, D + dD)$, proportional to $m^{-11/6}dm$ or to $D^{-3.5}dD$, respectively, with the proportionality coefficients decreasing with time as the disruptive process goes on. Relaxation to this equilibrium mass distribution may be sometimes fast: i.e., in the asteroid belt it occurs over a time span much shorter than the age of the solar system.

Two critical assumptions of the Dohnanyi model are: (i) all the collisional response parameters are size–independent, implying that the transition from cratering to fragmentation outcomes occurs for a fixed projectile-to-target mass ratio, and no gravitational reaccumulation of fragments is taken into account. (ii) The population has an upper cutoff in mass, but no lower cutoff.

The $-11/6$ value of the mass distribution index has been recovered by Paolicchi (1994), and in a general way by Tanaka *et al.* (1996) who showed that the equilibrium exponent is independent on the details of collisional outcomes as long as the fragmentation model is self-similar. Williams & Wetherill (1994) have shown that the $-11/6$ exponent changes less than 10^{-4} when Dohnanyi's collisional physical assumptions — e.g., the relative importance of cratering and catastrophic breakup events, the mass distribution of fragments from a single impact, etc. — are varied in a substantial way. Martins (1999) found that for a non-stationaty state, the size distribution can be expressed by a power series of $m^{-5/3}$.

3.2. *Releasing Dohnanyi's assumptions*

The basic assumptions of Dohnanyi's theory are not fulfilled in real planetary collisional systems. In contraddiction with assumption (i), collisional response parameters are not size–independent, and they are often expressed as scaling laws describing how the collisional outcomes vary with target and impactor size. In its most general form, a scaling law extrapolates the outcomes of collisions —for which we have direct, physical experimental experience— to the outcomes of collisional events at size scales much larger (or smaller) than the ones that can be handled under laboratory conditions. One important component of scaling laws is determining the shattering impact specific energy, that is given in terms of the energy per unit target mass required for the catastrophic shattering of the target, Q_S^*, such that the largest fragment produced has a mass of one-half that of the original target. Dimensional analyses (Farinella *et al.* 1982; Housen & Holsapple 1990; Housen *et al.* 1991), impact experiments (Ryan 1992; Martelli 1994; Holsapple 1993; Housen & Holsapple 1999; Holsapple *et al.* 2002), and hydrocode studies (Benz & Asphaug 1999) show that Q_S^* scales as roughly $D^{-0.24}$ to $D^{-0.61}$ for strength-dominated targets. Recent results (Housen 2004) show that rotation may have an effect on the results of catastrophic fragmentation making it easier to shatter objects in rotation state.

The size dependence of Q_S^* in the gravity-dominated regime shows a dependence on D^α, with $\alpha \sim 2$, and it is due to the effect of the self–compression exerted by gravity on the body's interior at increasing sizes.

Another way to see the scaling problem is considering the critical impact specific energy, Q_D^*, the energy per unit target mass required for catastrophic disruption of the target, i.e., such that the largest resulting object, formed by partial reaccumulation of fragmente, has a mass one-half that of the original body (e.g., Davis *et al.* 1985; Love & Ahrens 1996; Melosh & Ryan 1997; Benz & Asphaug 1999). Estimates of Q_D^* scale as $D^{1.13}$ to $D^2.00$. Hydrocode studies (Benz & Asphaug 1999), numerical collisional models (Durda *et al.* 1998), and observations of asteroid spin rates (Pravec & Harris 2001; Pravec *et al.* 2002) suggest that the transition from the strength-dominated to the gravity-dominated regime occurs at target diameters about 100–500 meters, while the other quoted authors generally place the transition in the 1–10 kilometer size range.

Both self–gravity and strain–rate effects on the scaling of Q_S^* influence the resulting mass distribution in every single collision and they are expected to affect the collisional cascade in a noticeable way, as will be illustrated in sec. 4 (Fujiwara 1989; Davis *et al.* 1989; Davis *et al.* 1994; Campo Bagatin *et al.* 1994a). The self–gravity of celestial

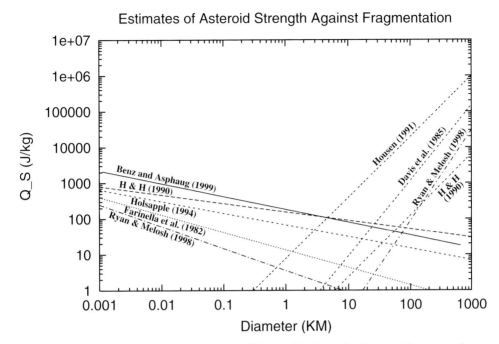

Figure 3. Different published estimates of the scaling laws for the specific energy for fragmentation, Q_S^*.

bodies grows at increasing sizes, its effect being —apart from the self–compression quoted above— to directly affect the escape velocity of the produced fragments remarkably, especially in targets greater than a few–km size. This effect combines with the experimental results found by many authors (Davis & Ryan, 1990; Nakamura & Fujiwara 1991; Giblin et al. 1994; Giblin 1998), indicating a shallow mass–velocity dependence in the velocity distribution of the ejected fragments of a fragmentation event as a function of mass, and they are yet another source for non–self–similarity.

The large amount of data available today for asteroids (about 200,000 catalogued) seem to confirm that the real distribution departs from a single exponent power–law, at least for objects larger that a few km. The fact that the size distribution of multi– km objects is not represented by a single exponent power–law is not surprising from a theoretical point of view. As a matter of fact, the Dohnanyi's result may be interpreted in terms of fractal distributions, that are the natural outcomes of self–similar multiplicative cascades. When the cascade is non self–similar the whole process is rather described by multifractal distributions, and they may be conveniently fitted in this way, as shown by Campo Bagatin *et al.* (2001). Deviations from Dohnanyi's exponent may occur when size– dependent specific energy for fragmentation (or disruption) holds either in the strenght or in the gravity regimes: O'Brien & Greenberg (2003) showed that if $Q*_S \propto D^s$, then $d^N (D, D + dD) \propto D^{-p}$, with $p = (7/2 + s)/(3/2 + s)$, and that wavy behaviour may be triggered due to transition between the two scaling regimes.

Assumption (ii) in Dohnanyi's theory is not fulfilled either: in fact, it is not clear what the size range of the asteroidal and TNO populations are. In some case the greatest body is well known (e.g.: Ceres, for the asteroid belt), but this is not the case for the smallest end of the distribution, and we do not know if there is a more or less sharp size cutoff due to physical constraints or to the effect of non–gravitational forces. The consequences of the existence of such a cutoff at small sizes were explained and discussed by Campo Bagatin *et al.* (1994a), resulting in deviations from a single power–law, as wavy patterns appear in the numerically simulated distributions of equilibrium. Cheng (2004) reexamined analytically the colisional evolution in the asteroid belt and calculated the relative importance of various collisional processes versus asteroid size, confirming that self–similar size distributions do not apply to asteroids due to size–dependent collisional processes, and underlines that as destruction and creation rates do not balance, a collisional model cannot explain why the asteroid size distribution is close to the Dohnanyi's slope.

Notwithstanding the enumerated 'emendaments' to the Dohnanyi's theory, his analytical finding is still a very useful result, as it is valid for the average distribution of collisional systems in stationary conditions.

3.3. *Relative velocities and collision probabilities*

In order to understand the actual amount of impacts that may occur within a collisional system, a useful magnitude is the intrinsic collisional probability, P_i, that gives the collision rate per unit cross-section area. The actual average number of collisions onto a target of radius R_T by projectiles of radius R_P within a time ΔT is then given by:

$$N_{col} = P_i (R_T + R_P)^2 N_T N_P \Delta T$$

where P_i depends on the average relative encounter velocity and on the available volume. N_T and N_P are the number of targets and projectiles, respectively. The basic theory for calculating collision rates and speeds was developed by Öpik (1951) and Wetherill (1967) and applied by various researchers in subsequent years, but there has been recent work to refine such calculations and to extend them to both Trojans, Hildas and TNOs in recent years. The main corrections are related to specific features of the orbital distribution of the population under study or to statistical mechanics (Greenberg 1982; Farinella & Davis 1992; Bottke & Greenberg 1993; Bottke *et al.* 1994; Vedder 1998; Dell'Oro & Paolicchi 1998; Dell'Oro *et al.* 1998, Dell'Oro *et al.* 2001). Alternatively, different authors have preferred a direct numerical approach based on the integration of the orbits of asteroids over a sufficiently long timespan. The derived distribution of close encounters and mutual speeds recorded during the integration can be extrapolated to infer the collision probability and characteristic impact speed (Yoshikawa & Nakamura 1994; Marzari *et al.* 1996; Dahlgren 1998).

4. Collisional evolution models

4.1. *The asteroid belt*

To trace the detailed time history of collisional evolution and to include realistic collisional physics requires numerical models. As the asteroid belt is the best known between the solar system collisional systems, most of the numerical research concentrated on its study.

Even if not all models use the same physical characterisations, they may depend on a number of poorly known critical parameters that govern the mass distribution and the re-accumulation of fragments after their formation:

1) The scaling laws for Q_S^* or Q_D^* may vary widely, as outlined in section 2.

2) The inelasticity parameter, f_{KE}, that determines what fraction of the relative kinetic energy of the collision goes into kinetic energy of the fragments. Its value may well

depend on the composition, internal structure, and size of the bodies, and it is often taken between1% and 10%. f_{KE} is also implicitly embedded into the specific energy for disruption (Q_D^*) (Campo Bagatin *et al.* 2001).

3) A relationship of the form $V(m) = Cm^{-r}$ may aplly for the mass and velocity of ejected fragments, like argued experimentally (Nakamura *et al.* 1992; Giblin 1998), with mass–velocity dependence ranging from none to $r = 1/6$, as found by Giblin (1998) and Giblin *et al.* (2004). This relationship is important because even a shallow dependence does make a significant difference in the amount of mass that can be re–accumulated by objects. The effects of this dependence have been studied by Petit & Farinella (1993) and Campo Bagatin *et al.* (1994b).

Davis *et al.* (1985) pioneered the era of collisional evolution models based on realistic collisional physics, including the effect of self–compression, and provided the first attempt to fit current observed distributions with a self–consistent numerical model.

Davis *et al.* (1994) and Campo Bagatin *et al.* (1994a) introduced new algorithms for the collisional outcomes of single asteroidal collisions, taking into account experimental results as well as scaling–laws based on dimensional analysis. Campo Bagatin *et al.* (1994a) also studied the effect of releasing (ii) assumption in Dohnanyi's theory, including a sudden break in the size distribution of the studied population which introduced a wavy pattern in the final distribution.

Durda (1993) and Durda & Dermott (1997) examined the influence of Q_D^* on the shape of an evolving size distribution by showing that the power–law exponent of a population in collisional equilibrium is a function of the size dependence of Q_D^*. When Q_D^* decreases with target size, as is the case in the strength-scaling regime, the slope index of the equilibrium size distribution is steeper than Dohnanyi's; instead, when Q_D^* increases with size, as in the gravity-scaling regime, the resulting slope index is shallower than, -3.5. Durda *et al.* (1998) presented results from numerical experiments illustrating in a more systematic fashion the sensitivity of an evolved size distribution on the shape of the strength scaling law. They found that if there is non-linearity in the relationship between $logQ_S^*$ and $logD$ then a structure is introduced into the evolved size distribution due to non-linearity in collisional lifetimes of the colliding objects, a result also shown in Campo Bagatin (1998).

Gil–Hutton & Brunini (1999) included the effect of early collisions by scattered comets from the Uranus–Neptune zone.

One of the most interesting issues about small bodies in the solar system is their internal structure (Asphaug *et al.* 2002; Britt *et al.* 2002). Campo Bagatin *et al.* (2001) modelled asteroid collisional evolution considering two different kinds of objects: monoliths (simulated by high f_{KE}), and gravitational aggregates (low f_{KE}) (often called rubble–piles) (Richardson *et al.* 2002). The relative number of gravitational aggregates can be estimated to be be between 50%to 100% in the $10 - 200 \ km$ range, depending on the adopted scaling law and on the choice of other collisional parameters, especially the inelasticity parameter, f_{KE}. The size scale at which a significant ratio of gravitational aggregates is expected seems to depend strongly on the reaccumulation model. For instance: assuming a mass-velocity relationship with $r = 1/6$ provides some 10% aggregate asteroids for 2 km-size bodies, while ignoring that dependence avoids reaccumulated objects up to some 10 km. Even if reaccumulation is treated in a simple way in the model, it shows that it is necessary to make further efforts in understanding the dependence of ejection speeds of fragments on their masses.

Vokhroulický & Farinella (1998) revived the effect predicted by the russian engineer Ivan O. Yarkovsky, who noted (1900) that the diurnal and seasonal heating of a rotating object in space would cause it to experience a radial force that —even if small— could lead

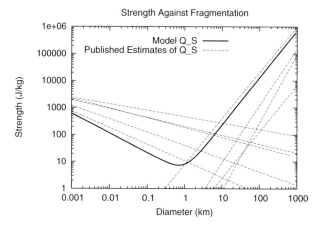

Figure 4. O'Brien & Greenberg (2005) recent best fit scaling law for Q_S^*.

to large secular effects in the orbit of small bodies. The Yarkovsky effect is a potential mechanism for removal of asteroids from the main belt, and further calculations have demonstrated that it may help to explain the mechanisms of refilling the Near Earth Asteroids populations (Bottke *et al.* 2000, and Bottke *et al.* 2002)). Penco *et al.* (2004) have included this effect in their model by updating the Campo Bagatin *et al.* (1994a,b, and 2001) algorithms, and found that the Yarkovsky non–gravitational force may affect significantly the absolute number of objects at given size ranges, but that it is not efficient enough to produce unambiguous overall effects in the final size distribution, that is. A wave pattern driven by this effect alone may arise, but it can be completely overruled by other effects (i.e., a low–mass cutoff, non–self–similarity in Q_S^*, etc.). In any case the effect on the size distribution cannot probably be ignored if one wants to compute a reliable Q_S^* scaling law from the observed size distributions.

Bottke *et al.* (2005) collisional evolution model includes an approximate method to estimate the effect of early dynamical depletion evolution governed by Jupiter's accretion. They find that the best fit to the observed asteroid belt distribution and to other observables correspond to an evolution which rapidly reached the wavy pattern recently observed, including the bump at the 100–km size. They conclude that this should be a fossil signature of the initial evolution of the belt, settled at the end of the Late Heavy Bombardement period, and that the distribution of objects larger than this size is probably unchanged since that time. The scaling law for Q_D^* that allows to achieve such results is very similar to results produced by numerical hydrocode simulations of asteroid impacts (Benz & Asphaug 1999).

O'Brien & Greenberg (2005) tackled the issue of collisional evolution including an evaluation of the Yarkovsky non–gravitational force and considering the removal of objects from the asteroid belt by dynamical effects. The authors explored the parameter space as to fulfill the available observables, making the model reproduce fairly well the observed main asteroid belt distribution and the estimated size distribution of NEAs by means of a best fit scaling–law for Q_S^* and Q_D^* (Figure 4).

4.2. *The trans–Neptunian region*

The physical properties of TNOs are likely to be different from those of asteroids. TNOs seem to be made of different kinds of ices with none to some extent of mixture with silicates, and the recent Deep Impact experiment on comet Temple I shall hopefully throw some light on this issue. A few set of experiments have been run impacting pure and porous water ice at some hundred meters per second (Ryan *et al.* 1999; Giblin *et al.* 2004), or silicate–ice porous mixtures (Arakawa *et al.* 2002). Authors agree in estimating the energy density for shattering (Q_S^*) in the $10^5 - 10^6 \, erg/g^3$ range, while f_{KE} varies around 5%.

The first numerical model of the TNO populations was performed by Davis & Farinella (1997), based on Davis *et al.* (1994)'s former model for the collisional evolution of asteroids. They concluded that the population of TNOs larger than 'break' size 100 *km* are not significantly altered by collisions, while objects with size below that seem to be distributed close to the steady–state. They also argue that the population of Centaurs can hardly be considered to be originated by collisions in the TNO region, while short–period comets might well be collisional fragments that experienced some type of alteration in the inteior of the parent bodies.

Kenyon & Bromley (2004) modelled the TNO region by means of a multi–annulus coagulation code; their results show a 'break' size in the 1–30 *km* range, a shallow power–law index (-3.5) in the size distribution of TNO populations at large sizes and an even shallower one ($b \simeq -3.0$, that is below the steady state distribution) for bodies below the break size. These results are not in agreement with estimations from observations and with the results of models by other authors. However, their simulations seem to be affected by the presence of a sharp cutoff at the low–size end, resulting in the standard pattern for the final distribution shown by Campo Bagatin *et al.* (1994a)

Krivov *et al.* (2005) developed a detailed model for the collisional evolution of a disk of particles evolving through collisions by considering mass and orbital elements as independent variables of a phase space, instead of the classical mass–semimajor axis binning. They derive a kinetic equation for a distribution function that contains information on the combined mass, spatial and velocity distributions of particles. When applied to the TNO population, the model provides qualitatively similar results to Davis & Farinella (1997) and to Bernstein *et al.* (2004), as far as the two–slope power–law for the size distribution is concerned. They also investigated the dependence of collisional lifetime on size, and the mass density as a function of size at given heliocentric distances, as well as the collisional mass loss with varying initial masses.

Pan & Sari (2005) semi–analytical model calculates the size dependence distribution for TNOs considering them as gravity–dominated objects with negligible material strength. In this way they evaluate the size distribution and find that the largest objects are distributed according to an exponent -5 down to a transition size located around 40 *km*, somewhat below numerical and present observational estimates. Smaller object are found to follow a much shallower power–law distribution, with an exponent close to -3.

Campo Bagatin & Benavidez (2005) modelled the TNO region within a classical particle–in–a–box scheme, by dividing it into three different –but potentially interacting– populations, according to the dynamical characterisation presently accepted. They follow the evolution for each of the three populations and find the common pattern shown by other authors: two power–law distributions linked at a transition size. The three populations have different transition sizes, with an average around 100 *km* in agreement with Davis & Farinella (1997), Krivov *et al.* (2005) and Bernstein (2004). The model also shows that the final power–law ditribution for objects larger than the transition size is basically the same that at the beginning of the simulations, whatever the initial conditions were, implying that this population suffered little collisional evolution and that it should be mostly primordial. On the other hand, the power–law size distribution of objects below the transition size does not seem to be affected by different initial conditions, and it seems to be in a relaxed state: different initial conditions give raise to different absolute numbers of objects in the size intervals, but they produce the same exponent for the power–law size distribution.

5. Open questions and conclusions

Remarkable improvements to the knowledge of the collisional evolution of populations of asteroids and TNOs have been achieved in the last two decades, especially due to the increased volume of data on the asteroid size distribution and the TNO populations; the close encounters with asteroids and comets have provided important clues about the internal structure and collisional rates. As a matter of fact, the modelling of the collisional evolution of asteroid populations is a very complex problem, due to the fact that a number of poorly known free parameters are embedded in the modelling of the involved physical phenomena. More observational data —on one hand— and more refined modelling techniques —on the other hand— are needed in order to get a comprehensive and self-consistent picture of fragmentation and collisional evolution of small bodies populations. Even if advances in the correct description of fragmentation physics has been possible, a better understanding of what is the outcome of energetic collisions certainly remains a challenge for the future.

The development of smooth particle hydrocode (SPH) calculations of the fragmentation process coupled with N-body integrators to follow the trajectories of fragments once the material interactions have ceased (Michel *et al.* 2002; Michel *et al.* 2004) have introduced a new paradigm for collisional outcomes. The new models argue that gravitational reaccumulation is the fundamental mechanism for forming individual bodies during disruptive collisions. Hence, the velocity field established by the collision would determine the number and size of fragments, not the propagation of cracks that fracture material bonds. Further work is certainly needed in this area, but this approach suggests a very different physical basis for understanding the outcomes of disruptive collisions. A better understanding of the response of gravitational aggregates to collisions is indeed necessary as well. Once these techniques will be reliable enough to be sistematically run, it will be desirable that in the future they may provide an alternative way of feeding the algorithms for the solution of the kinetic equations typical of the collisonal evolution problem.

Between the very many interesting issues regarding the collisional evolution of asteroids, a number of open questions are especially waiting to be answered:

• Collisions have modified asteroid rotation rates over solar system history, so information about the collisional history of the belt is embedded in the spin rates of asteroids of all sizes. How do rotations evolve as part of the collisional evolution? As discussed by Davis *et al.* (1989), the uncertainties in modelling how collisions alter rotation rates are so large that any definitive work on this topic is yet to be done and there have been no publications in the past decade on the collisional alteration of asteroid rotation rates. Do rotations have some effect on fragmentation, like suggested by Housen (2004)? Does this signicantly affect the final size distribution of asteroids?

• The somehow wavy distribution observed in the actual asteroid size distribution may be the result of the interplay between the existence of scaling–laws —with the effect of propagating wavy patterns induced by non uniform Q_S^*— and the potential existence of a cutoff in the small–size end, abrupt enough to trigger its own wavy pattern. Then, how is the small-particle end of the actual asteroid population, and how does it affect the overall evolution? The answer to this question is not obvious, but many non-gravitational forces do indeed act on interplanetary matter and are efficient at different sizes (Burns, 1979): the solar wind, Poynting-Robertson drag, and the Yarkovsky effect. Clearly, more work is needed in this area to explicitly treat within collisional models the various non-gravitational forces that act to remove especially dust–size partticles from the asteroid population, and especially to determine how strong and abrupt a small–mass cut-off must be in order for wave–like features to appear in the evolved size distribution. One of the key matters in this issue is the way sub-cm particles do fragment when they undergo shattering events. Do scaling–laws still make sense for collisions among dust particles? Can the collisional physics at this size range rather be explored with more detail?

• The lack of self–similarity implies a multifractal structure for multi–km objects — rather than the single exponent fractal behaviour characteristic of self–similarity— that may be used to match their distribution. Starting from observed collisional populations, and establishing the driving parameters of the corresponding multifractal distributions, is it possible to relate them with critical parameters governing the collisional cascade, such as Q_D^*? In other words, could we recover potatoes back from mushed potatoes?

As for the populations of Trans–Neptunian Objects, the state of knowledge is indeed, at present, at a lower level respect to the asteroid belt; therefore there are many issues pending in order to allow reliable comparisons of collisional evolution studies with the observed populations.

The first, and more urgent issue is to achieve a large reliable list of orbital parameters which are not affected by any bias. Establishing the dynamical characterisation of these objects is the cornerstone on which to build further understanding. Related to this is the recent discovery of popolations of massive objects well beyond $50\,AU$ that may, once again, change our view and understanding of the outer solar system, as well as of the mechanisms of its formation.

A second issue is certainly to acquire more data on the physical characteristics of TNOs, in particular on their albedos and surface compositions, that are still not univocally determined.

Some of the main questions that the study of collisional evolution of TNOs may help to answer are:

• How did the Scattered Disk, the Centaurs populations and the outer part of the Trans–Neptunian region form and evolve in time?

• What was the initial mass of this region?

• Are TNOs larger than some size mostly pristine bodies? And what fraction of km–size populations are gravitational aggregates?

More and more questions will soon add, hopefully, as answers will be found to the above ones: this is the main rule of the game of knowledge.

References

Arakawa, M., Lelyva-Kopystinski. J., & Maeno. N. 2002, *Icarus* 158, 516

Asphaug, E. 1997, *Meteor. Planet. Sci.* 32, 965

Asphaug, E., Ryan, E.V. & Zuber, M.T. 2002, in: W.F. Bottke, A. Cellino, P. Paolicchi & R.P. Binzel (eds.), *Asteroids III* (Tucson: Arizona University Press), p. 423

Bendjoya, P. & Zappalà, V. 2002, in: W.F. Bottke, A. Cellino, P. Paolicchi & R.P. Binzel (eds.), *Asteroids III* (Tucson: Arizona University Press), p. 613.

Benz, W. & Asphaug, E. 1999, *Icarus* 142, 5

Bernstein, G.M., Trilling, D.E., Allen, R.L., Brown, M.E., Holman, M., & Malhotra, R. 2004, *Astron. J.* 128, 1364.

Bottke, W.F. & Greenberg, R. 1993, *Geophys. Res. Let.* 20, 879

Bottke, W.F., Nolan, M. C., Greenberg, R., & Kolvoord, R.A. 1994, *Icarus* 107, 255

Bottke, W.F., Morbidelli, A., Jedicke, R., Petit, J.-M., & Gladman, B. 2000, *Science* 288, 2190

Bottke, W.F., Durda D.D., Nesvorný, D., Jedicke, R., Morbidelli, A., Vokrouhlický, & Levison, H.F. 2005, *Icarus* 175, 111

Bottke, W.F., Vokrouhlický, D., Rubincam, D.P., & Broz, M. 2002, in: W.F. Bottke, A. Cellino, P. Paolicchi & R.P. Binzel (eds.), *Asteroids III* (Tucson: Arizona University Press), p. 379

Britt, D.T., Yeomans, D., Housen, K., & Consolmagno, G. 2002, in: W.F. Bottke, A. Cellino, P. Paolicchi & R.P. Binzel (eds.), *Asteroids III* (Tucson: Arizona University Press), p. 485

Burns, J.A., Lamy, P.L., & Soter, S. 1979, *Icarus* 40, 1

Campo Bagatin, A. 1998, *Ph.D. thesis, Servivio de Publicaciones de la Universidad de Valencia*, Spain

Campo Bagatin, P., Cellino, A., Davis, D.R., Farinella, P., & Paolicchi, P. 1994, *Planet. Space Sci.* 42, 1079

Campo Bagatin, A., Farinella, P. & Petit, J.-M. 1994, *Planet. Space Sci* 42, 1099

Campo Bagatin, A., Petit, J.-M. & Farinella, P. 2001, *Icarus* 149, 198

Campo Bagatin, A., Martinez, V. & Paredes, S. 2002, *Icarus* 157, 549

Campo Bagatin, A. & Benavidez, P.G. 2005, *IAU Symp. 229, Asteroids, Comets and Meteors* Abstract

Cellino, A., Zappalá, V. & Farinella, P. 1991, *Mon. Not. Royal Astr. Soc.* 253, 561

Cheng, A. 2004, *Icarus* 169, 357.

Chiang, E.I. & Brown, M.E. 1999, *Astron. J.* 118, 1411.

Dahlgren, M. 1998, *Astron. Astrophys.* 336, 1056

Davis, D. R., Chapman, C. R., Weidenschilling, S. J., & Greenberg, R. 1984, *Lunar Planet. Sci. XV* 192.

Davis, D.R., Chapman, C.R., Greenberg, R., Weidenschilling, S.J., & Harris, A.W. 1979, in: T. Gehrels (ed.), *Asteroids* (Tucson: Univ. of Arizona Press), p. 528

Davis, D. R., Chapman, C. R., Weidenschilling, S. J., & Greenberg, R. 1985, *Icarus* 62, 30

Davis, D.R., Durda, D.D., Marzari, F., Campo Bagatin, A., & Gil-Hutton, R. 2002, in: W.F. Bottke, A. Cellino, P. Paolicchi & R.P. Binzel (eds.), *Asteroids III* (Tucson: Arizona University Press), p. 545

Davis, D.R. & Ryan, E.V. 1990, *Icarus* 83, 156

Davis, D.R., Weidenschilling, S.J., Farinella, P., Paolicchi, P., & Binzel, R.P. 1989, in: R.P. Binzel, T. Gehrels & M.S. Matthews (eds.), *Asteroids II* (Tucson: Univ. of Arizona Press), p. 805

Davis, D.R. & Farinella, P. 1997, *Icarus* 125, 50

Davis, D.R., Farinella, P. & Marzari, F. 1999, *Icarus* 137, 140

Davis, D.R., Ryan, E.V. & Farinella, P. 1994, *Planet. Space. Sci.* 42, 599

Dell'Oro, A.& Paolicchi, P. 1998, *Icarus* 136, 328

Dell'Oro, A., Marzari, P. Paolicchi F., Dotto, E., & Vanzani, V. 1998, *Astron. Astrophys.* 339, 272

Dell'Oro, A., Marzari, F., Paolicchi, P., & Vanzani, V. 2001, *Astron. Astrophys.* 366, 1053

Dohnanyi, J. W. 1969, *J. Geophys. Res.* 74, 2531

Dohnanyi, J. W. 1971, in: T. Gehrels (ed.) *NASA-SP 267.* p. 263

Durda, D.D. 1993, Ph.D. thesis, Univ. of Florida.

Durda, D.D. & Dermott, S.F. 1997, *Icarus* 130, 140

Durda, D.D., Greenberg, R. & Jedicke, R. 1998, *Icarus* 135, 431

Edgeworth, K.E. 1943, *J.B.Astron.Assoc.* 20, 181

Edgeworth, K.E. 1949, *Mon Not. R. Astron. Soc.* 109, 600

Farinella, P. & Davis, D.R. 1992, *Icarus* 97, 111

Farinella P., Paolicchi, P. & Zappalá, V. 1982, *Icarus* 52, 409

Fernandez, J. 1980, *Mon. Not. R. Astr. Soc.* 192, 481

Fujiwara, A. 1989, in: R.P. Binzel, T. Gehrels & M.S. Matthews (eds.), *Asteroids II* (Tucson: Univ. of Arizona Press), p. 240

Giblin, I., Martelli, G., Smith, P.N., & Di Martino, M. 1994, *Pl. Space Sci.* 42, 1027.

Giblin, I. 1998, *Planet. Space. Sci.* 46, 921

Giblin, I., Davis, D.R. & Ryan, E.V. 2004, *Icarus* 171, 487.

Gil-Hutton, R. & Brunini, A. 1999, *Planet. Space Sci.* 47, 331

Gladman, B., Kavelaars, J.J., Nicholson, P., Loredo, T., & Burns, J.A. 1998, *Astron. J.* 116, 2042

Gladman, B., Kavelaars, J.J., Petit, J.-M., Morbidelli, A., Holman, M., & Loredo, T. 2001, *Astron.J.* 122, 1051

Gladman, B., Holman, M., Grav, T., Kavelaars, J., Nicholson, P., Aksnes, K., & Petit, J.-M. 2002, *Icarus* 157, 269

Greenberg, R. 1982, *Astron. J.* 87, 184

Hellyer, B. 1970, *Mon. Not. Royal Astr. Soc.* 148, 383

Hellyer, B. 1971, *Mon. Not. Royal Astr. Soc.* 154, 279

Holsapple, K.A. 1993, *Ann. Rev. Earth Planet. Sci. XXI* 333

Holsapple, K.A., Giblin, I., Housen, K., Nakamura., A., & Ryan, E.V. 2002, in: W.F. Bottke, A. Cellino, P. Paolicchi & R.P. Binzel (eds.), *Asteroids III* (Tucson: Arizona University Press), p. 443

Housen, K., Schmidt, R.M. & Holsapple, K.A. 1991, *Icarus* 184, 180

Housen, K. 2004, *Proc. 35th Lun. Plan. SCi. Conf.*

Housen, K. & Holsapple, K. 1990, *Icarus* 84, 226

Housen, K.R. & Holsapple, K.A. 1999, *Icarus* 142, 21

Ivezić, Z. and 30 more authors 2001, *Astron. J.* 122, 2749

Kenyon, S.J. & Bromley, B.J. 2004, *Astron. J.* 128, 1916

Krivov, A. V., Sremcević, M. & Spahn, F. 2005, *Icarus* 174, 105

) Kuiper, G.P. 1951, in: J.A. Hynek (ed.), *Proceedings of a topical symposium, commemorating the 50th anniversary of the Yerkes Observatory and half a century of progress in astrophysics*, (New York: McGraw-Hill), p. 357

Luu, J. & Jewitt, D. 1992, *Nature* 362, 730.

Luu, J.X. & Jewitt, D.C. 1998, *Astrophys. J.* 502, L91

Love, S.G. & Ahrens, T.J. 1996, *Proc. Lunar Planet. Sci. 27th* 777

Martelli, G., Ryan, E.V., Nakamura, A.M., & Giblin. I. 1994, *Icarus* 42, 1013

Martins, V. 1999, *Pl. Space Sci.* 5 , 687

Marzari, F., Farinella, P. & Davis, D.R. 1999, *Icarus* 142, 63

Marzari, F., Scholl, H. & Farinella, P. 1996, *Icarus* 19, 192

Melosh, H.J. & Ryan, E.V. 1997, *Icarus* 129, 562

Michel, P., Benz, W., Tanga, P., Benz, W., & Richardson, D.C. 2002, *Icarus* 161, 198

Michel, P., Benz, W. & Richardson, D.C. 2004, *Icarus* 168, 420

Nakamura A. & Fujiwara. A. 1991, *Icarus* 92, 132

Nakamura A., Seguiyama, K., & Fujiwara. A. 1992, *Icarus* 100, 127

Nesvorný, D., Morbidelli, A., Vokrouhlický, D., Bottke, W.F., & Broz, M. 2002, *Icarus* 157, 155

O'Brien, D.P. & Greenberg, R. 2003, *Icarus* 164, 334

O'Brien, D.P. & Greenberg, R. 2005, *Icarus* 178, 179

Öpik, E.J. 1951. *Proc. Roy. Irish Acad.* 54, 165

Pan, M. & Sari, R. 2004, *Icarus* 173, 342

Paolicchi, P. 1994, *Planet. Space Sci.* 42, 1093

Petit, J.-M. & Farinella, P. 1993, *Celest. Mech.* 57, 1

Pietrowski, S. 1953, *Acta Astronomica* A5, 115.

Penco, U., Dell'Oro, A., Paolicchi, P., Campo Bagatin, A., La Spina, A., & Cellino, A. 2004, *Pl. Space. Sci.* 52, 1087

Petit, J.-M. & Gladman, B. 2003, *C.R. Physique* 4, 743

Pravec, P. & Harris, A.W. 2001, *Icarus* 148, 12

Pravec, P., Harris, A.W. & Michalowski, T. 2002, in: W.F. Bottke, A. Cellino, P. Paolicchi & R.P. Binzel (eds.), *Asteroids III* (Tucson: Arizona University Press), p. 113

Richardson, D.C., Leinhardt, Z.M., Melosh, H.J., Bottke, W.F., & Asphaug, E. 2002, in: W.F. Bottke, A. Cellino, P. Paolicchi & R.P. Binzel (eds.), *Asteroids III* (Tucson: Arizona University Press), p. 501

Ryan, E.V. 1992, Ph.D. dissertation, University of Arizona, Tucson.

Ryan, E.V., Davis, D.R. & Giblin, I. 1999, *Icarus* 142, 56

Tanaka, H., Inaba, S. & Nakazawa, K. 1996, *Icarus* 123, 450

Vedder, J.D. 1998, *Icarus* 131, 283

Viateau, B. 2000, *Astron. Astrophys.* 354, 725

Vokrouhlický, D. & Farinella , P. 1998, *Astron. J.* 116, 2032

Wetherill, G.W. 1967, *J. Geophys. Res.*, 72, 2429

Williams, D.R. & Wetherill, G.W. 1994, *Icarus* 107, 117

Yoshida, F., Nakamura, T., Watanabe, J., Kinoshita, D., Yamamoto, N., & Fuse, T. 2003, *Publ. Astron. Soc. Jpn.* 55, 701

Yoshikawa, M. & Nakamura, T. 1994, *Icarus* 108, 298

Asteroids, Comets, Meteors
Proceedings IAU Symposium No. 229, 2005
D. Lazzaro, S. Ferraz-Mello & J.A. Fernández, eds.

© 2006 International Astronomical Union
doi:10.1017/S1743921305006848

Non-gravitational forces acting on small bodies

Miroslav Brož[1], D. Vokrouhlický[1], W.F. Bottke[2], D. Nesvorný[2], A. Morbidelli[3] and D. Čapek[1]

[1]Institute of Astronomy, Charles University, Prague, V Holešovičkách 2, 18000 Prague 8,
Czech Republic
email: vokrouhl@mbox.cesnet.cz, mira@sirrah.troja.mff.cuni.cz, capek@sirrah.troja.mff.cuni.cz

[2]Southwest Research Institute, 1050, Walnut St., Suite 400, Boulder, CO-80302, USA
email: bottke@boulder.swri.edu, davidn@boulder.swri.edu

[3]Observatoire de Nice, Dept. Cassiopee, BP 4229, 06304 Nice Cedex 4, France
email: morby@obs-nice.fr

Abstract. Non-gravitational perturbations, regardless being many orders of magnitude weaker than gravity, hold keys to fully understand the evolution of small Solar System bodies. This is because individual bodies, or their entire groups, manifest traces of a long-term accumulated changes by these effects.

For meteoroids and small asteroids in the 10 cm–10 km size range, the principal non-gravitational force and torque arise from an anisotropic thermal emission of the absorbed solar radiation. Related perturbations of the orbital and rotational motion are called the Yarkovsky and YORP effects.

We review the most important Yarkovsky- and YORP-driven processes, in the Main Asteroid Belt. These include: steady and size-dependent semimajor axis drift, secular changes of rotational period and obliquity, efficient transport towards low-order resonances, interaction with weaker higher-order resonances, captures in secular and spin-orbit resonances.

Many independent observations can be naturally interpreted in the framework of Yarkovsky/YORP models, like cosmic ray exposure ages of meteorites, current population and size-distribution of near-Earth objects, the existence of unstable resonant asteroids or the structure of asteroid families.

Keywords. minor planets, asteroids; interplanetary medium; radiation mechanisms: thermal

1. Introduction

Current observations of small Solar System bodies provide many important constraints for dynamical studies. Laboratory analyses of collected meteorite samples, astrometric and photometric observations of small asteroids in the Earth's neighbourhood or relatively larger asteroids orbiting in the Main Asteroid Belt allowed us to recognise, during the last ten years, the importance of non-gravitational phenomena affecting their orbital evolution.

In this review, we are going to focus on small asteroidal bodies in the size-range from 10 cm up to 10 km, which do not exhibit any outgassing and cometary activity. The principal accelerations affecting the motion of these small bodies are listed in Table 1.

The largest non-gravitational accelerations caused by the interaction with the solar radiation field – like the Yarkovsky/YORP effect, the radiation pressure or the Poynting-Robertson drag – are, roughly speaking, 10 orders of magnitude weaker than solar gravity. At a first glimpse, they seem to be too subtle phenomena, but we have to take into

Table 1. The approximate values of radial and transversal accelerations affecting bodies in the size-range 10 cm to 10 km. The solar gravity is scaled to unity. For comparison, typical gravitational perturbations by planets and large asteroids are $GM_{pl} \simeq 10^{-3}$ and $GM_{ast} \lesssim 10^{-9}$.

acceleration	radial	transversal
gravity	$GM_\odot \simeq 1$	
Yarkovsky/YORP effect	10^{-7} to 10^{-11}	10^{-8} to 10^{-12}
radiation pressure	10^{-6} to 10^{-11}	
Poynting-Robertson drag		10^{-10} to 10^{-15}
solar wind, Lorentz force, plasma drag	$< 10^{-15}$	

account also the direction of the acceleration vector and the effect of its eventual long-term accumulation.

Of course, a small radial acceleration, not exceeding the solar gravity, does not have significant orbital effects (it only slightly decreases or increases the orbital velocity), while a transversal acceleration may cause a secular change of energy (and hence the semimajor axis of the orbit). Some types of accelerations also tend to average-out along the orbit, while others can accumulate over millions or even billions of years. If we take into the account these two issues, the Yarkovsky/YORP effect is by far the strongest non-gravitational force in the size-range 10 cm to 10 km and, hereinafter, we will focus on the Yarkovsky/YORP only.

How much a body can change its orbit? What are the secular effects? Typically, the Yarkovsky/YORP force can push a 10-m meteoroid's semimajor axis by 0.1–0.2 AU, before being disrupted by a random collision with another body. Similarly, a small 1-km Main-Belt asteroid can move by 0.05 AU (within its collisional lifetime). These are certainly significant shifts, comparable to the distances between major resonances or to the sizes of asteroid families (i.e., the prominent concentrations of asteroids in the proper-element space). They give a hint that the Yarkovsky/YORP effect plays an important role in the evolution of small Solar-System bodies.

We present a brief overview of Yarkovsky and YORP effects principles in Section 2 and the most direct observational evidences for these phenomena in Section 3. Section 4 is devoted to various unstable populations, which the Yarkovsky/YORP helps to sustain, and Section 5 to evolutionary processes shaping asteroid families.

2. The Yarkovsky/YORP effect principles

The basic principle of the Yarkovsky/YORP thermal effect is the absorption of solar radiation by a body and its anisotropic thermal reemission. The temperature differences on the surface, together with an uneven shape of the body, then lead to a recoil force and torque (Figure 1). (A detailed discussion on the mathematical theory describing the Yarkovsky/YORP effect can be found in Bottke *et al.* (2002) and references therein.) Contrary to the direct radiation pressure and its relativistic counterpart, the Poynting-Robertson effect, the radiation is absorbed and thermally reprocessed here. Due to a finite thermal conductivity of the material, there is some "thermal lag" between the absorption and the emission. This is also the reason, why the Yarkovsky/YORP effect sensitively depends on the rotational state (obliquity γ and period P).

The Yarkovsky/YORP effect is negligible in case of very small and very large bodies: the upper limit for size D is a natural consequence of the fact, that the force is approximately proportional to the surface area (D^2), the mass $\propto D^3$ and thus the resulting acceleration $\propto 1/D$. The lower limit is given by the conduction of heat across the whole

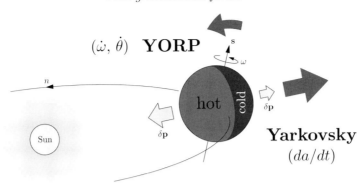

Figure 1. An illustration of the Yarkovsky/YORP effect principle. As an asteroid absorbs the solar radiation, its part facing the Sun becomes hotter than the reverse one. The infrared emission from the surface is then anisotropic, what gives rise to the Yarkovsky force, affecting the orbital motion of the asteroid, and the YORP torque, modifying the spin state.

small body, which effectively diminishes temperature differences on the surface and the corresponding infrared emission is then almost isotropic.

In the next sections, we will need to know the principal secular effects of the force and torque on the orbital and rotational dynamics. The Yarkovsky force is related to the orbital dynamics (Rubincam 1995; Vokrouhlický 1998, 1999). Its diurnal variant, driven by the rotational frequency, dominates for bodies with low thermal conductivity (e.g., with regolith on the surface). It can either increase or decrease semimajor axis a and the change Δa is proportional to the cosine of the obliquity γ. In case of the seasonal variant, the changes of temperature on the surface are mainly driven by the orbital frequency. It is a usual situation for bodies with higher thermal conductivity (regolith-free surface). The semimajor axis a steadily decreases and $\Delta a \propto -\sin^2 \gamma$.

The YORP torque (Rubincam 2000; Vokrouhlický & Čapek 2002) works for non-spherical bodies only. It has an asymptotic behaviour – it pushes the obliquity towards 0 or 180° and the rotation period towards 0 or ∞. (We note, however, that the behaviour of the YORP and collisional evolution close to these asymptotic spin states is poorly understood today and it will certainly be a subject of forthcoming studies.) Because of the dependence of the Yarkovsky force on the obliquity we can expect a complicated interplay between the Yarkovsky and YORP effects.

Of course all variants of the Yarkovsky forces and the YORP torque are caused by a single temperature distribution on the surface of the body – they are actually a single phenomenon. Nevertheless, we find the above division conceptually useful.

What do we need to calculate the Yarkovsky/YORP? To properly calculate the temperature distribution on the surface of an asteroid (and then straightforwardly the corresponding IR emission, force and torque) we need to know its orbit (i.e., the position of the radiation source), size and shape, spin axis orientation and period, mass, density of surface layers, albedo, thermal conductivity, capacity and IR emissivity of the material.

These are many a priori unknown parameters. In the "worst" case (and for vast majority of asteroids), we know only the orbit and broad-band photometry results (from which we can "guess" an approximate albedo, size and thermal parameters). How to overcome this lack of physical parameters? One possibility is to study only asteroids known very well, like (6489) Golevka (Figure 2). However, we can also use a collective dynamics approach – study whole groups of bodies (like asteroid families) and treat the unknown

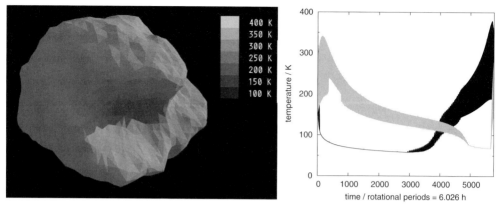

Figure 2. The temperature distribution on the surface of the asteroid (6489) Golevka, calculated by a numerical solution of the 1-dimensional heat diffusion equation, individually for all 4092 surface elements of the shape model. For two selected surface elements, located on roughly opposite sides of the body, we plot the time evolution of the temperature (the time is counted as the number of rotations and covers one complete orbit). Both seasonal and diurnal variations of the temperature, due to the changing distance from the Sun, illumination geometry and shadowing, are clearly visible. Adapted from Chesley *et al.* (2003).

thermal parameters as statistical quantities, it means to select a reasonable probability distribution and assign them randomly to the individual bodies.

3. The Yarkovsky and YORP: the most direct observational evidence

Following a previous prediction by Vokrouhlický *et al.* (2000), Chesley *et al.* (2003) were the first to directly detect the non-gravitational semimajor axis drift due to the Yarkovsky effect. Vokrouhlický *et al.* (2000) computed the position of (6489) Golevka during its 2003 close approach to the Earth using all previous radar and optical astrometry data and two models of Golevka's motion: (i) purely gravitational only and (ii) with the Yarkovsky acceleration included (Figure 3).

The respective radar ranging to Golevka, reported by Chesley *et al.* (2003), confirmed the 15 km $O-C$ difference in the distance from the dish, what is outside 3-σ error interval of the purely gravitational model, but it fits very well with the Yarkovsky model. Because the latter involves a non-gravitational acceleration, they were also able to constrain the bulk density of Golevka to $2.7^{+0.4}_{-0.6}\,\mathrm{g/cm^3}$.

The current state-of-the-art model by Čapek & Vokrouhlický (2005) assumes Golevka consists of two layers: low conductivity surface and high conductivity core. It enables to put a lower limit for the surface thermal conductivity K, which should be at least 10^{-2} or $10^{-1}\,\mathrm{W/m/K}$, (i.e., substantially larger than the laboratory-measured conductivity of the lunar regolith $10^{-3}\,\mathrm{W/m/K}$). This is in a rough agreement with thermophysical models, which Delbó *et al.* (2003) use to interpret observed infrared fluxes coming from Near-Earth asteroids. The average value of K for all observed NEA's seems to be of the same order.

Unfortunately, we do not have any direct measurement of the YORP effect yet. However, a strong evidence of the ongoing YORP evolution comes from the analysis of a group of Koronis-family asteroids, which has a bimodal obliquity distribution (Slivan 2002; Slivan *et al.* 2003). The prograde group has periods 7.5–9.5 h, obliquities 42°–50° and even similar ecliptic longitudes of the poles within 40°. The values for the retrograde group are $P < 5\,\mathrm{h}$ or $> 13\,\mathrm{h}$ and $\gamma \in (154°, 169°)$ (Figure 4). This observational result

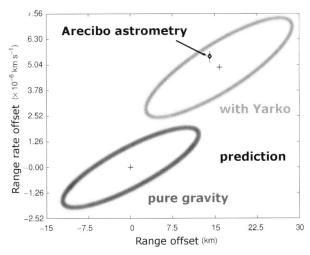

Figure 3. Range vs. range rate (i.e., the quantities measured by radar) for the close approach of (6489) Golevka in May 2003. The predictions of the two theoretical models of Golevka's motion, purely gravitational and with Yarkovsky, are plotted with their 90% confidence ellipses. The astrometric observation by the Arecibo radar is denoted by the black point and arrow. Adapted from Chesley *et al.* (2003).

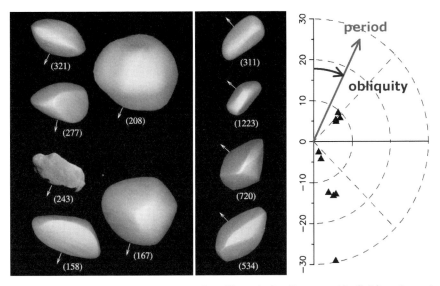

Figure 4. Shape models and spin vectors of 11 Koronis family asteroids (left) and a polar plot period vs. obliquity for the same group (right). Adapted from Slivan *et al.* (2003).

was very surprising, because collisions should produce a random distribution of rotational states, surely not the bimodal.

Vokrouhlický *et al.* (2003) thus constructed a model of spin state evolution, which included solar torques and the YORP thermal torque. Let's take the prograde-rotating asteroids as an example (Figure 5). They analysed the evolution of asteroids, which initially had periods $P = 4$–5 h and obliquities γ evenly distributed in the interval $(0°, 90°)$. They found the evolution is firstly driven by the YORP effect toward an asymptotic state (γ decreases and P increases). After some 1 Gy, when the precession rate reaches the value $\simeq 26''/$y, the spin is captured in the s_6 spin-orbit resonance and it pushes γ to

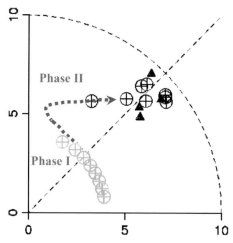

Figure 5. Period vs. obliquity polar plot depicting Slivan's prograde-rotating group. The observed asteroids are denoted by triangles, the initial state of the numerical model by gray circles and the final state after 2.5 Gy by black circles. The dashed line with an arrow shows an evolutionary path and two phases: (i) the YORP driven and (ii) the resonance capture. Adapted from Vokrouhlický *et al.* (2003).

$\sim 50°$, P to ~ 8 h and also forces the spin axes to be really parallel in space. Around the time 2.5 Gy, what is an approximate age of the Koronis family, the match of the model with the observations is perfect. Similarly, it is possible to explain the existence of the retrograde-rotating group; there is no significant spin-orbit resonance in this case and the spin axes of the retrograde-rotating asteroids are let to evolve freely toward the YORP asymptotic states.

Generally, thermal torques seem to be more important than collisions for asteroids smaller than 40 km, because today we can still clearly see the traces of the YORP-driven evolution and the collisions have not been able to randomise the spin states during several past Gy.

4. Delivery into unstable regions

Various unstable populations, like meteoroids hitting the Earth, Near-Earth asteroids, or Main-Belt asteroids located inside major mean motion resonances, have dynamical lifetimes shorter than the age of the Solar System and provide a nice opportunity for dynamicists to look for sources and transport mechanisms.

4.1. *Meteorite transport from the Main Belt*

Meteorite transport from the Main Belt is the eldest application of the Yarkovsky effect (Öpik 1951; Peterson 1976; Farinella *et al.* 1998; Vokrouhlický & Farinella 2000; Bottke *et al.* 2000). The meteorites reach the Earth in two stages: (i) a Yarkovsky-driven change of the semimajor axis spanning ~ 10 My, and (ii) a capture in a powerful gravitational resonance, which increases eccentricity of the orbit up to 1 in a mere ~ 1 My (Figure 6). Approximately 1% of meteoroids then collide with the Earth (and can be found as meteorites), but most of them fall directly to the Sun.

The main motivation for the introduction of the above Yarkovsky model were the observed cosmic ray exposure (CRE) ages of meteorites, which measure, how long time the meteorite spent in the interplanetary space as a small fragment. The model naturally explains that (i) the CRE ages are much longer than resonance residence times alone;

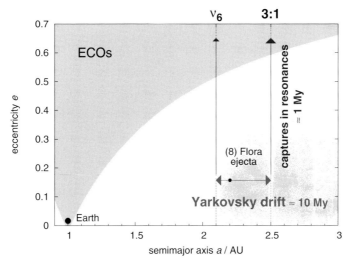

Figure 6. A schematic semimajor axis vs. eccentricity plot of the Yarkovsky-enabled model for the meteorite transport from the Main Belt. In the first stage, spanning typically $\sim 10\,\mathrm{My}$, the Yarkovsky effect pushes the semimajor axes of meteoroids toward principal gravitational resonances (like ν_6 secular resonance with Saturn and $3/1$ mean motion resonance with Jupiter). In the second stage, the resonances pump the eccentricities quickly and thus in $\sim 1\,\mathrm{My}$ the orbit reaches Earth-crossing space.

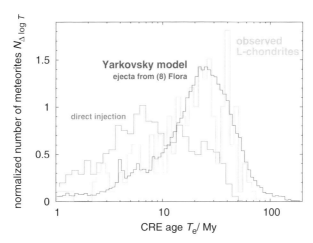

Figure 7. The observed distribution of cosmic ray exposure ages of L-chondrites (thick gray line), compared with the model distribution of Yarkovsky-driven ejecta from (8) Flora (bold line) and with an old model (thin gray line), which assumed only a direct injection of fragments into resonances. The non-random peaks on the observed CRE distribution, which were not possible to fit within a steady-state model, are most probably stochastic events, i.e., large craterings or disruptions, which produced many fragments at once. Adapted from Vokrouhlický & Farinella (2000).

(ii) there is a strong dependence of the CRE's on the material – namely the CRE's of iron meteorites are $10\times$ longer than of stones; (iii) the most stony meteorites have the CRE's of the order $10\,\mathrm{My}$ (see Figure 7). The Yarkovsky drift is able to supply meteoroids from a wide range of parent bodies (not only from the vicinity of resonances); it is effective enough to explain the observed meteorite flux of the order $3 \times 10^5\,\mathrm{kg/y}$. Moreover, petrologic and mineralogical studies (Burbine *et al.* 2003) show the number

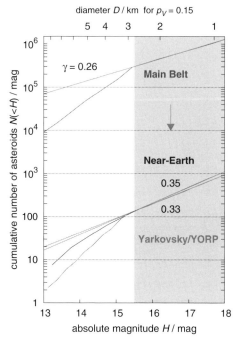

Figure 8. Cumulative distribution of absolute magnitudes H for the three populations: observed Main-Belt Asteroids, observed Near-Earth Asteroids and the Yarkovsky/YORP model population, which assumes transport from the Main Belt to the near-Earth space. The slopes γ of the distributions ($N(<H) \sim 10^{\gamma H}$) were all fitted in the interval $H \in (15.5, 18)$ mag. Adapted from Morbidelli & Vokrouhlický (2003).

of parent bodies of iron meteorites is larger than of stones. This is because hard irons are more resistant to collisions, their total semimajor-axis drift (within the collisional lifetime) is larger and thus they can effectively sample larger volume of the Main Asteroid Belt.

4.2. *Delivery of Near-Earth Asteroids from the Main Belt*

Observations of the Near-Earth Asteroids provide two important constraints: (i) the cumulative distribution of their absolute magnitudes has a slope $\gamma = 0.35$ ($N(<H) \sim 10^{\gamma H}$ in the magnitude range 15.5 to 18; Figure 8), and (ii) their removal rate by planetary scattering is ~ 200 bodies larger than 1 km per My.

Morbidelli & Vokrouhlický (2003) assumed the same basic scenario as for meteorites and constructed a Yarkovsky/YORP model of the transport from the Main Asteroid Belt (this source has the slope $\gamma = 0.26$, again in the interval $H \in (15.5, 18)$ mag). Their model yield a flux of 150–200 bodies (>1 km) into the main J3/1 and ν_6 resonances (which then quickly became NEA's) and the slope of the resulting model NEA population is $\gamma = 0.33$. So, the Yarkovsky/YORP effect is efficient enough to keep the current NEA population in steady state and it also explains, why the observed slope of NEA's is moderately shallower than that of MBA's.

4.3. *Resonant populations resupplied from the Main Belt*

Low-order mean motion resonances with Jupiter usually harbour small populations of objects with dynamically unstable orbits (and sometimes also stable ones). We consider here the J7/3 resonance at approximately 2.96 AU heliocentric distance and the J2/1

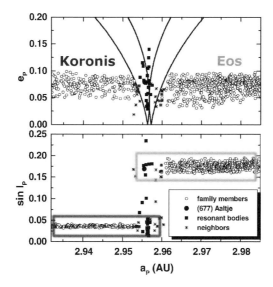

Figure 9. Proper semimajor axis vs. proper eccentricity and inclination in the surroundings of 7/3 mean motion resonance with Jupiter. The resonant asteroids and two adjacent asteroid families, the Koronis and Eos, are plotted. From Tsiganis *et al.* (2003).

resonance at approximately 3.25 AU as two examples, which were previously studied in some detail.

There are 22 observed unstable asteroids in the J7/3 resonance. Tsiganis *et al.* (2003) proved, that the Yarkovsky drift may keep the resonant population in steady state, as it pushes members of the neighbouring Koronis and Eos families towards the resonance. An independent confirmation, that the resonant bodies are truly related to the families is the observed confinement of inclinations – the mean inclinations of the two resonant groups, 2° and 10° respectively, correspond to the mean inclinations of the Koronis and the Eos family (Figure 9).

The J2/1 resonance harbours some 150 asteroids and 50 of them are on dynamically unstable orbits. Brož *et al.* (2005) simulated the evolution of neighbouring Main-Belt asteroids pushed by the Yarkovsky effect towards the J2/1 resonance. They verified this flux of Main-Belt bodies keeps the unstable resonant population in steady state. Moreover, the orbital evolutionary tracks of the Main-Belt asteroids, their dynamical lifetimes inside the J2/1 resonance and also size distribution are consistent with the actual observed unstable resonant asteroids. A few observed unstable objects, which escape from the J2/1 in less then 2 My, are most probably inactive Jupiter-Family comets.

The long-lived asteroids, confined to stable island of the J2/1 resonance, cannot be explained within the Yarkovsky model and the problem of their origin remains open.

5. Processes shaping asteroid families

Asteroid families are prominent clusters of asteroids, which are located close to each other in the space of proper elements a_p, e_p and $\sin I_p$ and usually also exhibit some spectral similarities. Families are thought to be remnants of large collisions producing fragments, which then has been evolving due to the Yarkovsky/YORP effect, gravitational resonances and further secondary collisions. The primary collisions can scale from large catastrophic disruptions of parent bodies to smaller cratering events (Michel *et al.*

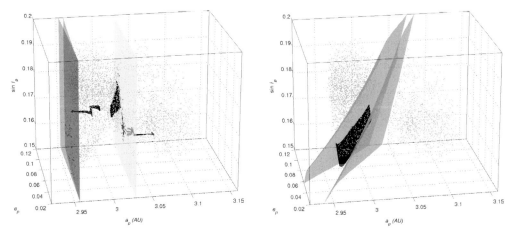

Figure 10. The Eos family in the 3-dimensional space of proper elements a_p, e_p and $\sin I_p$. The three resonances, J7/3 and J9/4 (left) and z_1 (right) are plotted together with examples of bodies drifting by the Yarkovsky effect and interacting with these resonances. Adapted from Vokrouhlický *et al.* (2005a).

2001, Durda *et al.* 2005). Typical velocities, which fragments gain with respect to the parent body, are of the order of a few tens of m/s.

Bottke *et al.* (2001) and Vokrouhlický *et al.* (2005a) demonstrated the post-impact evolution of asteroid families using two examples: the Koronis and the Eos family. They reported three general processes, how the Yarkovsky drift together with gravitational resonances can dramatically affect the overall shape of the families, i.e., the distribution of their members in the space of proper orbital elements. We can call these processes "bracketing", "crossing" and "trapping".

At first, notice the shape of the Eos family (Figure 10): it is sharply cut at a low value of proper semimajor axis a_p, there is a evident paucity of asteroids, especially the bigger ones, at large-a_p's and the family is also somewhat distorted or elongated towards low-a_p, low-e_p and low-$\sin I_p$. These observed features nicely coincide with analytically computed borders of resonances, namely with the 7/3 mean motion resonance with Jupiter at 2.955 AU, the J9/4 resonance at 3.03 AU and the $z_1 = g - g_6 + s - s_6$ secular resonance.

We explain the observations this way: initially, just after the parent body disruption, the family was more compact; asteroids drifting due to the Yarkovsky/YORP effect towards smaller semimajor axis meet the powerful J7/3 resonance, which scatters their eccentricities and inclinations, or pumps them up to planet crossing orbits, and consequently no asteroids are visible behind. The J7/3 resonance thus brackets the Eos family (Figure 10, left).

The asteroids drifting in the opposite direction, towards larger semimajor axis, meet the weaker J9/4 resonance. Some of them are able to cross it, but the rest is scattered. This crossing explains, why there is less asteroids behind the J9/4, and why the paucity is size-dependent – the smaller asteroids drift faster and typically cross the J9/4 resonance at low eccentricity and inclination (Figure 10, left).

Many Eos-family members are trapped in the z_1 secular resonance; they drift in semimajor axis by the Yarkovsky effect and they are also forced to follow the libration centre of the resonance, which position, however, depends on all three orbital elements a_p, e_p and $\sin I_p$. Thus, not only the semimajor axis changes, but also eccentricity and inclination and the stream of asteroids forms at small values of a_p, e_p and $\sin I_p$, i.e., the elongated shape of the family (Figure 10, right).

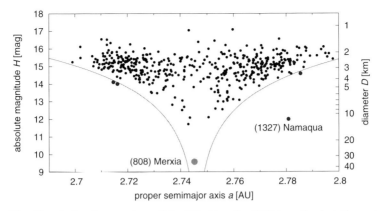

Figure 11. The Merxia family members (identified by the HCM method at the cut-off velocity 80 m/s) in the semimajor axis–absolute magnitude plot. The gray dots outside the 'V'-shape are probable interlopers.

In case of the Koronis family the situation is slightly different. This family is split in two parts, each of which has a different mean value of proper e_p (but the same mean $\sin I_p$). Their division correlates with the position of the secular resonance $g - 2g_5 + 3g_6$. A detailed study shows that, unlike in the Eos case, long-lasting captures in this resonance are not possible and drifting orbits necessarily jump over it. During this process their e_p is always lifted by ~ 0.025, right the observed difference between the mean e_p values of the two parts of the Koronis family. Because the resonance does not involve s-frequencies, the inclinations are not affected at all.

To conclude, if one assumes an initially compact impact-generated family (with a reasonable ejection velocity field compatible with hydrocode models), and takes into account the above evolutionary processes, it is possible to understand the currently observed extent of the family and its overall shape.

5.1. *"Eared" families and a new method of family-age determination*

The age of an asteroid family, i.e., the time of the collision which generated the family, is a very important parameter, not only for dynamical studies, but also for physical ones, space-weathering models, etc. One indication of the family age seems to be a typical 'V'-shape, which many families exhibit in the proper semimajor axis a_p–absolute magnitude H plane; see Figure 11 for an example of the Merxia family. This shape is a natural consequence of two phenomena: (i) the initial impact, because smaller fragments (with higher H's) gain higher velocities with respect to the parent body and fall farther from the centre, and (ii) the Yarkovsky/YORP effect, because the smaller fragments drift faster in semimajor axis and subsequently move farther from the centre.

There are several outliers visible at the (a_p, H) plot, which do not fit to this scheme. Most probably, they are interlopers, which are not related to the Merxia family. Indeed, the big asteroid (1327) Namaqua is an X-type, which is spectrally incompatible with the S-type Merxia family asteroids.

The problem is, that we do not know the initial spread, just after the impact and we cannot calculate the age simply from the current extent of the family, since the Yarkovsky drift is only responsible for an unknown part of it. Luckily, there is more information hidden in the (a_p, H) plot – notice the depletion of small asteroids in the centre and the overdensity at extreme values of the semimajor axis. Sometimes we call this funny feature an "eared" family. Might this be a YORP effect fingerprint? The YORP effect tilts the

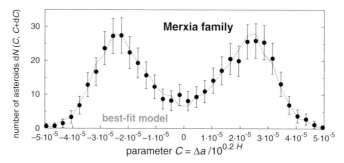

Figure 12. The distribution of the Merxia family members in the C-parameter and the comparison with the best fit model by Vokrouhlický *et al.* (2005b).

Table 2. List of asteroid families and their ages estimated by the method of Vokrouhlický *et al.* (2005b).

family	age/My	family	age/My
Agnia	100^{+30}_{-20}	Erigone	280^{+30}_{-50}
Astrid	180^{+80}_{-40}	Massalia	152^{+18}_{-18}
Eos	1300^{+150}_{-200}	Merxia	238^{-23}_{+52}

spin axes of asteroids directly up or down what enhances the Yarkovsky semimajor-axis drift and can drive the smaller asteroids towards the edges of the family. Possibly, it can allow us to resolve the ambiguity and to determine the age more precisely.

To check it, Vokrouhlický *et al.* (2005b) constructed a family evolution model, which accounts for: (i) an isotropic ejection of fragments (and random periods P and obliquities γ at the beginning), (ii) the Yarkovsky drift, (iii) the YORP effect, and (iv) collisional reorientations. There are four free parameters in the model: (i) the initial velocity dispersion V of 5-km fragments (for a size D, $V(D) = V\frac{5\,\mathrm{km}}{D}$), (ii) the YORP "strength" c_{YORP} (iii) the family age T, and (iv) the surface thermal conductivity K.

They fit this model with observations using a 1-dimensional C-parameter, which is closely related to the semimajor axis a_p and the absolute magnitude H: $C = \Delta a_p/10^{0.2H}$, where Δa_p is the distance from the family centre. The best fit for the Merxia family (Figure 12) yields the following results: the initial dispersion in semimajor axis was roughly one half of the currently observed one (what is in agreement with a statistical argument of Dell'Oro *et al.*, 2004); the initial velocity was small ($V = 24^{+6}_{-12}$ m/s), what is in agreement with impact models (Michel *et al.* 2001); the YORP effect is important ($c_{\mathrm{YORP}} = 0.6^{+1.4}_{-0.4}$); asteroids are probably covered with a low-conductivity layer ($K = 0.005$ W/m/K); and the family is of the young age ($T = 238^{+52}_{-23}$ My). See Table 2 for results concerning other asteroid families.

Up to now, the analysis of the Merxia family was done in the (a_p, H) plane only. We can, however, use also information hidden in the proper eccentricity e_p and inclination $\sin I_p$. The distribution of the Merxia members is clearly uneven in the (a_p, e_p) plane – the spread of e_p increases abruptly at $a_p \doteq 2.75$ AU. Vokrouhlický *et al.* (2005b) successfully explain it as a Yarkovsky transport across the three-body mean motion resonance with Jupiter and Saturn 3J−1S−1. It is actually an independent confirmation that the Yarkovsky semimajor-axis drift is calculated correctly, because the smaller spread of e_p before the resonance is increased by the resonance crossing and then matches the observed spread of the family members behind the resonance.

The chronology method mentioned in this section does not work for "too young" or "too old" families. The former have not had enough time to evolve by the Yarkovsky/YORP and to exhibit the "ears". The latter are much older than the typical time-scale of the YORP-driven evolution and the model does not account for the evolution of totally spun-up or spun-down asteroids.

Let us finally mention, that the freshest clusters, like Karin, Iannini or Veritas, were precisely dated by a direct backward N-body integration, which revealed a convergence of orbital perihelia and nodes corresponding to the time of the impact (Nesvorný *et al.* 2002, 2003; Nesvorný & Bottke 2004; see also the review by Nesvorný *et al.*, this volume).

6. Conclusions and future work

The non-gravitational forces, namely the Yarkovsky/YORP effect relevant for small asteroidal bodies in the size-range 10 cm to 10 km, are now inevitable ingredients of dynamical models. Today, there is a dozen of "big" applications of the Yarkovsky/YORP models; we mentioned some of them in this brief review.

The precise measurement of Golevka's non-gravitational drift was only a first step. Within the next decade, we expect a dozen of similar Yarkovsky detections by precise radar astrometry (Vokrouhlický *et al.* 2005c, 2005d) or future optical astrometry with GAIA.

Yarkovsky semimajor-axis drift of the order ~ 10 km per 10 years becomes crucial for an accurate orbit determination and even for estimates of an impact hazard (Giorgini *et al.* 2002). Especially, when the calculation of an impact probability depends on the fact, if the asteroid misses or hits a phase-space "keyhole", which is much smaller then the diameter of the Earth.

Further step forward might be a thorough combination of dynamical models with infrared observations of NEA's and their thermophysical models (Delbó *et al.* 2003) – they supply independent constraints (with different correlations) on Yarkovsky/YORP-related parameters, like the thermal conductivity.

We can await the first direct detection of the YORP effect in the forthcoming years, either from ground-based photometric measurements and corresponding lightcurve modelling, or from the space-borne mission Hayabusa, which now orbits the asteroid (25143) Itokawa (e.g.,Vokrouhlický *et al.* 2004).

The dynamical studies of asteroid families provide also predictions of physical properties and rotational states of individual asteroids, which can serve as good opportunities for further observational tests (similar to Vokrouhlický *et al.* (2005e) who photometrically observed (2953) Vysheslavia and confirmed its retrograde rotation predicted by Vokrouhlický *et al.* (2001)). For example, the small members of the families with intermediate ages (discussed in Section 5.1) should exhibit preferential values of obliquities due to the YORP torque and Yarkovsky drift: the asteroids located far from the family-centre at lower/larger values of semimajor axis should have retrograde/prograde rotations. The most suitable families for such survey seem to be the Massalia or the Erigone, located in the inner Main Belt, what makes them more easily observable.

An appealing project would be to determine systematically the ages of all asteroid families, including large and old ones. However, we have to face several obstacles: (i) we still lack the direct measurements of basic physical parameters (albedos, masses, shapes, spectra) for most family members and we cannot expect the situation dramatically improves in the next few years; (ii) a modelling of several subsequent YORP cycles have not been developed yet.

There is already a number of examples, how the YORP torque affects rotational states of asteroids (we discussed some in Sections 3 and 5.1). Moreover, there are further indications: (i) the distribution of rotational periods of all $\sim$1500 asteroids, we have lightcurves for, reveals an excess of very slow and very fast rotators (Pravec & Harris 2000); (ii) small NEA's have a non-Maxwellian distribution of periods; and (iii) there seems to exist a preference of retrograde-rotating asteroids among NEA's (La Spina *et al.* 2004), what is in concert with the positions of Main-Belt escape routes, fed by the obliquity-dependent Yarkovsky drift. A detailed model for a long-term YORP-driven period and obliquity evolution, concerning the entire Main-Belt and NEA's, does not exist yet. Also a possible YORP origin of binaries created by asteroid fission have not been studied in detail.

Acknowledgements

The work of DV, DČ and MB was supported by the Grant Agency of the Czech Republic by grant 205/05/2737.

References

Bottke, W.F., Rubincam, D.P., & Burns, J.A. 2000, *Icarus* 145, 301

Bottke, W.F., Vokrouhlický, D., Brož, M., Nesvorný, D., & Morbidelli, A. 2001, *Science* 294, 1693

Bottke, W.F., Vokrouhlický, D., Rubincam, D.P., & Brož, M. 2002, in: W.F. Bottke, A. Cellino, P. Paolicchi & R.P. Binzel (eds.), *Asteroids III* (Tucson: The University of Arizona Press), p. 395

Brož, M., Vokrouhlický, D., Roig, F., Nesvorný, D., Bottke, W.F., & Morbidelli, A. 2005, *Mon. Not. R. Astron. Soc.* 359, 1437

Burbine, T.H., McCoy, T.J., Meibom, A., Gladman, B., & Keil, K. 2003, in: W.F. Bottke, A. Cellino, P. Paolicchi & R.P. Binzel (eds.), *Asteroids III* (Tucson: The University of Arizona Press), p. 653

Chesley, S.R., Ostro, S.J., Vokrouhlický, D., Čapek, D., Giorgini, J.D., Nolan, M.C., Margot, J.-L., Hine, A.A., Benner, L.A.M., & Chamberlin, A.B. 2003, *Science* 302, 1739

Čapek, D. & Vokrouhlický, D. 2005, *Icarus*, to be submitted

Delbó, M., Harris, A.W., Binzel, R.P., Pravec, P., & Davies, J.K. 2003, *Icarus* 166, 116

Dell'Oro, A., Bigongiari, G., Paolicchi, P., & Cellino, A. 2004, *Icarus* 169, 341

Durda, D.D., Bottke, W.F., Nesvorný, D., Asphaug, E., & Richardson, D.C. 2005, *Icarus*, submitted

Farinella, P., Vokrouhlický, D., & Hartmann, W.K. 1998, *Icarus* 132, 378

Giorgini, J.D., Ostro, S.J., Benner, L.A.M., Chodas, P.W., Chesley, S.R., Hudson, R.S., Nolan, M.C., Klemola, A.R., Standish, E.M., Jurgens, R.F., Rose, R., Chamberlin, A.B., Yeomans, D.K., & Margot, J.-L. 2002, *Science* 296, 132

La Spina, A., Paolicchi, P., Kryszczynska, A., & Pravec, P. 2004, *Nature* 428, 400

Michel, P., Benz, W., Tanga, P., & Richardson, D.C. 2001, *Science* 294, 1696

Morbidelli, A. & Vokrouhlický, D. 2003, *Icarus* 163, 120

Nesvorný, D., Bottke, W.F., Dones, L., & Levison, H.F. 2002, *Nature* 417, 720

Nesvorný, D., Bottke, W.F., Levison, H.F., & Dones, L. 2002, *Ap.J.* 591, 486

Nesvorný, D. & Bottke, W.F. 2004, *Icarus* 170, 324

Öpik, E.J. 1951, *Proc. R. Irish Acad.* 54, 165

Peterson, C. 1976, *Icarus* 29, 91

Pravec, P. & Harris, A.W. 2000, *Icarus* 148, 12

Rubincam, D.P. 1995, *J. Geophys. Res.* 100, 1585

Rubincam, D.P. 2000, *Icarus* 148, 2

Slivan, S.M. 2002, *Nature* 419, 49

Slivan, S.M., Binzel, R.P., Crespo da Silva, L.D., Kaasalainen, M., Lyndaker, M.M., & Krčo, M. 2003, *Icarus* 162, 285

Tsiganis, K., Varvoglis, H., & Morbidelli, A. 2003, *Icarus* 166, 131

Vokrouhlický, D. 1998, *Astron. Astrophys.* 335, 1093

Vokrouhlický, D. 1999, *Astron. Astrophys.* 344, 362

Vokrouhlický, D., Brož, M., Farinella, P., & Knežević, Z. 2001, *Icarus* 150, 78

Vokrouhlický, D. & Čapek, D. 2002, *Icarus* 159, 449

Vokrouhlický, D. & Farinella, P. 2000, *Nature* 407, 606

Vokrouhlický, D., Milani, A., & Chesley, S.R. 2000, *Icarus* 148, 118

Vokrouhlický, D., Nesvorný, D., & Bottke, W.F. 2003, *Nature* 425, 147

Vokrouhlický, D., Čapek, D., Kaasalainen, M., & Ostro, S.J. 2004, *Astron. Astrophys.* 414, L21

Vokrouhlický, D., Brož, M., Morbidelli, A., Bottke, W.F., Nesvorný, D., Lazzaro, D., & Rivkin, A.S. 2005a, *Icarus*, in press

Vokrouhlický, D., Brož, M., Bottke, W.F., Nesvorný, D., & Morbidelli, A. 2005b, *Icarus*, submitted

Vokrouhlický, D., Čapek, D., Chesley, S.R., & Ostro, S.J. 2005c, *Icarus*, 173, 176

Vokrouhlický, D., Čapek, D., Chesley, S.R., & Ostro, S.J. 2005d, *Icarus*, in press

Vokrouhlický, D., Brož, M., Michałowski, T., Slivan, S.M., Colas F., Šarounová, L., & Velichko, F. 2005e, *Icarus*, in press

Asteroids Comets Meteors 2005
Proceedings IAU Symposium No. 229, 2005
D. Lazzaro, S. Ferraz-Mello & J.A. Fernández, eds.

© 2006 International Astronomical Union
doi:10.1017/S174392130500685X

Unbiased orbit determination for the next generation asteroid/comet surveys

A. Milani[1], G. F. Gronchi[1], Z. Knežević[2], M. E. Sansaturio[3], O. Arratia[3], L. Denneau[4], T. Grav[4], J. Heasley[4], R. Jedicke[4] and J. Kubica[5]

[1]Department of Mathematics, University of Pisa, Piazza Pontecorvo 5, 56127 Pisa, Italy
email: milani@dm.unipi.it
[2]Astronomical Observatory, Volgina 7, 11160 Belgrade 74, Serbia and Montenegro
[3]E.T.S. de Ingenieros Industriales, University of Valladolid Paseo del Cauce s/n 47011
Valladolid, Spain
[4]Institute for Astronomy, University of Hawaii, Honolulu, HI, 96822
[5]Carnegie Mellon University, Robotics Institute, Pittsburgh, PA, 15213

Abstract. In the next generation surveys, the discovery of moving objects can be successful only if an observation strategy and the identification/orbit determination procedure are appropriate for the diverse apparent motions of the target sub-populations. The observations must accurately measure the displacement over a short interval of time; observations believed to belong to the same object have to be connected into *tracklets*. Information contained in tracklets is in most cases not sufficient to compute an orbit: two or more of them must be *identified* to provide an orbit. We have developed a method for recursive identification of tracklets allowing an unbiased orbit determination for all sub-populations and efficient enough to cope with the data flow expected from the next generation surveys. The success of the new algorithms can be easily measured only in a simulation, by consulting a posteriori some "ground truth".

We present here the results of a simulation of the orbit determination for one month of operations of the future Pan-STARRS survey, based upon a Solar System Model with a downsized population of Main Belt asteroids and a full size populations of Trojans, NEO, Centaurs, Comets and TNO. The results indicate that the method already developed and tested to find identifications of NEO and Main Belt asteroids are directly applicable to Trojans. The more distant objects often require modified algorithms, fitting orbits with only 4 parameters in a coordinate system specially adapted to handle very short arcs of observations. These orbits are mostly used as intermediate results, allowing to find full solutions as more tracklets are identified.

When the number density of detections is as large as expected from the next generation surveys, both joining observations into tracklets and identifying tracklets can produce some false results. The only reliable way to remove them is a procedure of tracklet/identification management. It compares the tracklets and the identifications with a complex logic, allowing to discard almost all the false tracklets and all the false identifications. However, the distant objects still present a challenge for orbit determination: they require three tracklets in separate nights. If this requirement is met we have found no problem in achieving an unbiased orbit determination for all populations. Further work will lead to more advanced simulations, in particular by introducing a realistic model for astrometric and photometric errors.

Keywords. surveys, orbit determination, identification, population models

---

1. Introduction

Most of the next generation astronomical surveys are going to be "general surveys", with the goal to discover and catalog as many astronomically significant objects as possible with the available resources, rather than "specialized surveys" dedicated to one specific population. That is, with the same telescope and camera, if possible on the same images, they will try to discover and measure whatever can be detected, without restricting the discoveries to the interests of a specific astronomical sub-community. This approach can have great advantages with respect to restricted interest surveys, which are too often forced to use marginal resources. However, is it really possible to share the resources with satisfactory efficiency for all the subprojects?

In this paper we will address the problems arising in the discovery of moving solar system objects†. The target populations could be asteroids (Near Earth, Main Belt, Jupiter Trojans), comets (short and long period), Trans Neptunian Objects (regular, Plutinos, scattered disk), Centaurs, Trojans of other planets, satellites of the outer planets, and even classes of objects not discovered yet. Can all these populations be the target of the same survey? For this there are two main critical issues: the observation strategy and the identification/orbit determination procedure.

To observe an object moving with respect to the fixed stars the survey needs to take a sequence of images of the same field: *detection* can take place by having just $m = 2$ images with a time interval Δt long enough to be able to measure the displacement, and short enough to be able to connect into *tracklets*‡ the individual observations believed to belong to the same object; the procedure to form tracklets from individual observations is discussed in Section 4. The optimal value for Δt depends upon the *number density* of the detectable moving objects per unit area on the images and upon the *proper motion* (angular velocity on the celestial sphere), with typical values between 15 minutes and 1 hour. The first critical issue can thus be summarized as follows: can we select m, Δt and the scanned portion of the sky in a way optimal for all moving objects populations? Of course not, but it appears to be possible to find a reasonable compromise solution. We shall show in Section 3 that an interesting mix of populations can be observed on the same images, and discuss in Section 6 the problems arising from this compromise in the orbit determination of the fastest moving and the slowest moving objects.

The second critical issue arises from the fact that a tracklet in most cases does not allow to determine an orbit and to establish to which population the detected object belongs. Even for $m > 2$, if Δt is short, a tracklet contains only information which can be summarized in an *attributable*, a 4-dimensional vector containing two angles and two angular rates (a vector tangent to the celestial sphere): orbital elements containing 6 independent parameters cannot be uniquely determined. A tracklet with this property is called a *Too Short Arc* (TSA). The only way to compute a unique orbit is to *identify* two TSAs, separated by a time interval much longer than Δt, as belonging to one and the same object. Then there are more equations than unknowns, and a least squares solution is well defined. The classical methods for identification and orbit determination are not suitable to cope with the data flow expected from the next generation surveys. Thus it has been necessary to develop entirely new algorithms, to code them in efficient software and to perform extensive tests without waiting for the real data. So far, the interest has been concentrated on the specific problems of orbit determination for NEO, both because it can be more difficult and because of the immediate interest in impact monitoring. However, distant objects are a serious challenge for orbit determination because the angular rate relative accuracy is intrinsically very poor. The specific problems arising for distant objects and the new algorithms required are discussed in Section 5.

The success of the algorithms to find identifications can be assessed in terms of completeness and reliability. These can be easily measured only in a simulation (such as the

---

† The problem can be even more general, e.g., sharing resources with searches for supernovae and candidate gamma ray burst sources.

‡ In Milani *et al.* 2004 we use Very Short Arc instead of tracklet, for the same concept.

one presented in Section 3), by consulting a posteriori some "ground truth" specifying which observation belongs to which object. In this case, completeness is the ratio between the objects for which identifications and orbits (of a specified quality) have been obtained and the total number observed, (lack of) reliability is the ratio between *false identifications* and *true* ones. An identification is false if it contains observations belonging to different objects; also a tracklet can be false. As the survey goes deeper in apparent magnitude the number density increases and the problem of false tracklets and identifications becomes more serious. Section 6 discusses how high levels of completeness and reliability can be achieved, and Section 7 assesses the quality of the resulting orbits.

The studies described in this paper are intended as preparation for one of the next generation survey, the Panoramic Survey Telescope and Rapid Response System (Pan-STARRS), an all-sky survey telescope under development at the University of Hawaii. Pan-STARRS is composed of 4 individual 1.8 m telescopes observing the same region of sky simultaneously. Each telescope will have a $\simeq 7$ deg^2 field of view and will be equipped with $\simeq 1$ billion pixel CCDs in the focal plane, resulting in a spatial sampling of $\simeq 0.3$ arcseconds per pixel. With exposure times of ~ 30 s it is estimated that the system will reach a limiting magnitude of $V \sim 24.5$. The design of Pan-STARRS is weighted toward its primary purpose, which is to detect potentially hazardous Solar System objects. However, the wide-field, repetitive nature of the system make it ideal to detect a host of other astronomical phenomena, ranging from Solar System to cosmology.

A single 1.8 meter telescope prototype, essentially a one quarter part of the full system, is currently being built on Haleakala in Maui, Hawaii. This prototype will allow for testing all the technology that is being developed for Pan-STARRS and will be used to make a full-sky survey to provide an astrometric and photometric calibration data set that will be used for the full system. First light on this prototype is scheduled for early 2006.

One of the features of this project is that the survey will be fully simulated to develop and test the observing strategy and the data processing chain, from raw images to orbits, before real data are processed. This has the purpose of achieving optimum performance from the very beginning of the operational life rather than having to solve the problems a posteriori, when it might be late for some corrections.

The simulation requires an assumed solar system model. If the goal is to show that the survey can achieve a satisfactory discovery rate on different populations, then the model must include a representative sample of each target population. Although population models were available for some populations, a global model of all the objects observable (in a survey much deeper than the current ones) had to be build specifically for this purpose, see Section 2. This model was used in an *observation simulation* to show that even less numerous and dimmer populations of moving objects can nevertheless be observed and successfully pass through the procedure ending in a reliable orbit catalog.

2. The Pan-STARRS Solar System model

After ten years of operations Pan-STARRS should have detected roughly 10 million small objects of the Solar System. The Moving Object Processing System (MOPS) churns through these detections and identifies those corresponding to known and new objects resulting in a final database of orbits for all of them. To test and compare old and new algorithms for these tasks we are developing a realistic model of the small body populations of the Solar System that could be observed during the lifetime of Pan-STARRS. The use of a realistic model will ensure that the MOPS efficiency at finding tracklets/identifications and at computing orbital parameters is tested well before the system becomes operational.

It will eventually also allow the MOPS to monitor its efficiency while it is operating, by processing synthetic data in parallel with the real data.

To our knowledge this is the first attempt to create a model of the solar system for all populations of small bodies, including those recently discovered. As an example, the Statistical Asteroid Model (Tedesco *et al.* 2005) provides only a model of the Main Belt. Unfortunately, we found it impossible to create models for each of the solar system's small body populations using a single algorithm. Our knowledge and the distribution of the members for each population varies greatly and this required a different technique for generating each synthetic sub-population. Furthermore, the model will incorporate objects for which very few members are known (e.g. objects interior to the Earth's orbit) or populations that are entirely unknown (e.g. interstellar objects, distant major planets). It will be released on the Internet and described in detail in another paper soon.

Since there exist no directly measured size distributions of small bodies we chose to create the SSM based on measured absolute (H) or apparent magnitude distributions.

• **Near Earth Objects (NEO) and objects Interior to Earth's Orbit (IEO).** The NEO model is based on the orbit element and absolute magnitude distributions of Bottke *et al.* 2002 (for $H < 24$) and the H distribution of Rabinowitz *et al.* (2000) for $H \geqslant 24$. IEOs are included as a consequence of the transport model generating the known NEO population. Very small objects could be detected by Pan-STARRS when passing very close to Earth, the lower limit being set by the *trailing loss*†. We chose to arbitrarily set the limit at $H = 25$, corresponding to asteroids of diameter 35m/70m (depending upon the composition), unlikely to result in damage if they impact on Earth. This is below the recommended diameter limit of 100m for the next series of NEO surveys as promoted by Stokes *et al.* (2003). According to the size distributions cited above, this choice results in a model population of about 250,000 NEOs.

• **Main Belt Objects (MBO).** We expect that there are $\sim 10^7$ small objects within reach of Pan-STARRS. It is believed (Jedicke *et al.* 2002) that the known MB sample is nearly complete for $H < 14.5$ (diameters greater than about 6.5 km) based on the lack of new discoveries of asteroids in that range. Our synthetic model was generated by randomly selecting known MB asteroids with $H < 14.5$ and cloning them until we reached 10 million objects. The cloning process 'smears' each orbital element and creates a 'fuzzy' MB. This synthetic model relies on the assumption that there is no or little dependence of the orbit distribution of MB objects with their size so that the distribution of orbital elements for objects with $H < 14.5$ should be representative of the unbiased MB orbit population. The H distribution for these objects is as specified in Jedicke *et al.* (2002) who recommend that the known H-distribution be used for objects larger than the completeness limit and the Sloan Digital Sky Survey H-distribution (Ivezić *et al.* 2001) for smaller objects. To reduce the number of objects in the model it includes only those objects that can achieve $V < 24.5$ at opposition.

• **Trojans (TRO) of outer planets.** The Pan-STARRS SSM currently contains synthetic Jupiter Trojans whose orbit elements were generated without correcting for observational selection effects in the known population. Briefly, the observed population distribution in semi-major axis, eccentricity, inclination, and mean longitude difference with Jupiter's were fit to reasonably shaped analytic functions. The synthetic population was then generated randomly from those functional forms, using the size distribution from Jewitt and Luu (2000)‡, resulting in 320,000 objects. We have generated 20,000

---

† A very close object has a large proper motion, thus its image spreads on many pixels.
‡ Jewitt and Luu (2000) provided results only for L4. We assumed a L5 cloud identical to the L4 cloud. This implies the L5 population might be overestimated with respect to the L4 one.

synthetic Trojans of Mars, $40,000$ of Saturn, $20,000$ of Uranus and $20,000$ of Neptune in the same manner as for the Jupiter Trojans but then rotating and scaling the orbital elements to the appropriate position and distance of each outer planet.

• **Centaurs (CEN).** The Centaur model is based on the observational analysis of Jedicke and Herron (1997) who fit the dynamical 'residence-time' distribution of the Centaurs in (a, e, i) space from Duncan *et al.* (1995) to reasonable analytic functions. The differential H distribution was modelled as $n(H) \propto 10^{0.61 \times H}$ (Jedicke & Herron 1997, Sheppard *et al.* 2000). Over $60,000$ synthetic Centaurs were generated according to the analytic distributions with minimum opposition magnitude $V \leqslant 24.5$.

• **Trans-Neptunian Objects (TNO).** In the Pan-STARRS SSM the TNOs encompass both the classical and resonant types. Following Levison & Morbidelli (2003) and Gomes *et al.* (2004), the synthetic population was generated using a symplectic integrator (Wisdom & Holman 1991) to model the migration of the outer solar system into a near planar disk of small bodies. The migration causes a large number of objects to be scattered inwards and later to be ejected completely due to close encounters with Jupiter and Saturn. Other objects get caught in resonances with Neptune that sweep through the disk creating a large population of resonant objects. The synthetic objects for our SSM were derived from the $7,159$ stable orbits from the simulation which were sampled at thousand year intervals. The position and velocity of the objects were rotated and scaled such that Neptune at the time the orbital elements were extracted was close to the position of Neptune at some chosen epoch. The new position and velocity were used to compute the osculating orbital elements: $72,000$ simulated objects have been generated. The absolute magnitude was then chosen from an apparent magnitude distribution consistent with Bernstein *et al.* (2004).

• **Scattered Disk Objects (SDO).** With only ~ 80 SDO currently known it is exceedingly difficult to determine the unbiased orbit distribution of these objects. We assumed that the a, e and i distributions were independent and applied a reasonable but *ad hoc* bias correction to each and then fit a common functional form to the corrected distribution. The synthetic orbits for the SDOs were randomly selected from those fits. The apparent magnitude distribution of the SDOs was taken from Elliot *et al.* (2005). Requiring that each synthetic object be brighter than $V = 24.5$ at some time during Pan-STARR's operational lifetime we generated over $20,000$ SDOs.

• **Short Period Comets (SPC) and Long Period Comets (LPC).** As was done for the SDO, the synthetic SPC and LPC populations for this preliminary SSM is based only on the observed orbit distribution for each class of object. Analytical functional forms were fit to the observed distributions and synthetic orbit elements were then generated randomly according to those distributions. The apparent magnitude distribution of objects at perihelion was chosen to be $N(> R) \sim e^{0.04(R-24.5)}$. Only objects that reached $V \leqslant 24.5$ during the ten years of Pan-STARRS operation were included in the model, for a total of $10,000$ SPC and $10,000$ LPC.

3. Survey Simulation

The Pan-STARRS Solar System Survey Simulation (S^4) used in this work is a single lunation from a multi-year S^4; the details will be described in a future paper.

In an ecliptic reference frame centered on opposition (λ', β) the S^4 is defined by two near-quadrature areas with $|\beta| < 10°$, $-120° < \lambda' < -90°$ or $+90° < \lambda' < +120°$ (~ 550 deg^2 each) and also the opposition region with $|\lambda'| < 30°$ and $|\beta| < 40°$ (~ 4550 deg^2). A single Pan-STARRS field covers about 7 deg^2 so that each near-quadrature area needs at least 84 fields to be covered, while the opposition region

Single Lunation Sky Plane Density

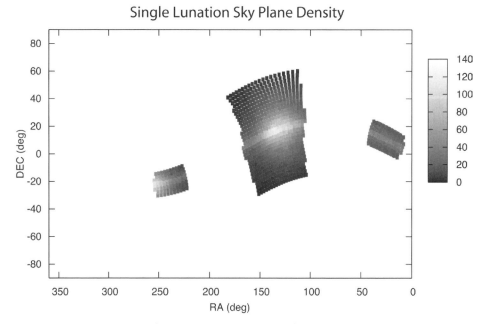

Figure 1. Sky plane density (number per square degree) of all types of objects in the survey simulation used in this work. Note the peak density corresponding to the center of the L5 Trojan cloud in the opposition region, and the signature of the L4 cloud in the quadrature area on the left.

requires at least 660 pointings. The nominal survey requires about 5 nights of clear sky per lunation and provides 3 visits to each field within that time† Each field is visited twice with about a 15 minute time separation and we required that the minimum altitude for observations be 20°. We used Paulo Holvorcem's TAO (http://pan-starrs.ifa.hawaii.edu/project/MOPS/tao.html) field scheduler to handle the scheduling within a night and wrote our own wrapper routine to handle multi-night observing including a crude approximation of the weather. The efficiency of scheduling the selected fields in all the regions was close to 100%.

To generate synthetic data for this work we determined the astrometric position and apparent magnitude for those objects with $V \leqslant 24.5$ in the SSM that appear in the simulated survey fields. This simulation used only $1/100^{th}$ of the MBO but the full density of all other types of objects. The synthetic detections have no astrometric or photometric error: we have assumed, to weight the observations to be fitted, a standard deviation of 0.1 arcsec in each angular coordinate. There were no false detections generated within each simulated Pan-STARRS field. Still, the sky-plane density of objects tops at $122/\text{deg}^2$ in the center of the Jupiter Trojan cloud as shown in Fig. 1.

4. Tracklets

We define a *tracklet* as a set of observations which *could* belong to the same moving object. In the simulation described in the previous Section, most tracklets are composed of only 2 observations (with about a 15 minute separation); tracklets with up to 4 observations can result from the superposition near the edge of two fields. *A posteriori*,

† To maximize the number of discoveries with a given telescope time and performance, exactly the same windows should be visited in three separate nights. Thus the windows should be defined with respect to the opposition at some reference epoch during the lunation.

after the inventory of detections in one frame has been extracted, the only information available to suggest that 2 observations belong to the same object is their proximity on the sky at slightly different times.

At the sky-plane densities anticipated for the final Pan-STARRS system of $\sim 250/$ deg^2, with a comparable density of false detections at the 5 σ level, we will need to join observations using some *a priori* knowledge or other information supplementing the positional one (e.g. trail orientation, trail length, apparent brightness).

Even at the reduced sky-plane densities of this work the task of identifying all possible pairs of detections that are closely spaced on the sky within the 7 deg^2 Pan-STARRS field-of-view is not trivial. Calculating the distance between every possible pair has a computational complexity $O(N^2)$ for N observations, thus it is inefficient. We have implemented an approach using kd-trees (Kubica *et al.* 2005a, 2005b) to search for the close pairs. This technique uses a tree structure (like a binary search generalized into multiple dimensions) to quickly eliminate large groups of detections in the search process: thus the search time for close sets of detections is $O(N \log N)$.

For instances where tracklets of more than 2 observations are possible these are created by fitting observations to a linear function of time within some error, with the distance between the first and last observation limited by a maximum proper motion. Thus if four observations are within the allowable error off a linear fit, but only combinations of three satisfy the maximum proper motion cut, two tracklets will be created with two observations in common. When the tracklet determination is done, tracklets that are entirely included in another tracklet are removed.

In fact, a tracklet composed as discussed above may be *false*, that is it may contain observations belonging to different moving objects†. A tracklet quality control can be performed by fitting degree 2 polynomials of time to the individual observations, separately for right ascension and declination (Milani et al. 2005b, Section 2). If there are 4 (or more) observations, high residuals with respect to the quadratic fits is a good diagnostic for false tracklets. By using RMS > 0.3 arcsec as a control, 47% of the false tracklets with 4 observations have been discarded. For the tracklets with 3 or more observations the second derivatives of right ascension and declination as estimated in the quadratic fits could be used as diagnostics of false tracklets. However, significant second derivatives can be interpreted as an indication that the observed object is very near. To discard a tracklet on the basis of these metrics would introduce the risk of discarding the discoveries of some NEO. For the tracklets with 2 observations there is no way to remove the false ones at this stage. This implies that the task of deleting the 0.6% of the tracklets which are false needs to be included in the identification process; see Section 6.

5. Identification

Given a tracklet, in most cases the information contained allows to compute only a 4-dimensional attributable (that is the tracklet is a TSA). An attributable does not contain enough information to produce a full set of orbital elements: in fact the range r and the range rate $\dot{r}$ of the corresponding moving object are left completely undetermined. In Milani et al. (2004) we have introduced some dynamical and physical constraints to these undetermined variables, confining them to a bounded region of the $(r, \dot{r})$ plane, the *Admissible Region*: each point of this 2-dimensional region corresponds to a Virtual Asteroid (VA), with a full orbit. In the same paper we have also discussed a method for

---

† It may also contain some observation not belonging to any moving object, but to a variable star etc.; this does not occur in the present simulation.

sampling the Admissible Region by triangulation, to obtain a finite (and not too large) set of VAs that can be propagated in time till the epoch of another attributable.

In Milani et al. (2005b) we have considered the problem of joining two attributables, referring to different times, to produce preliminary orbits to be used as first guess in a differential corrections procedure fitting the whole set of observations generating the two attributables. We call this kind of identification a *linkage*. The VAs defined by the triangulation of the Admissible Region of one attributable are propagated to the epoch of the other one, then the *identification penalty* technique described in Milani et al. (2000) is used to assess whether the second attributable can belong to the same object.

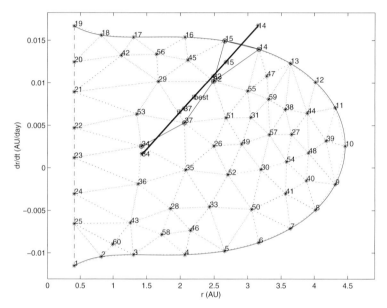

Figure 2. The triangulated Admissible Region for $2003BH_{84}$ corresponding to the attributable from the discovery night and the identification with an attributable of 12 days later. The nodes of the triangulation with moderate identification penalty are joined with solid lines; the bold line shows a part of the LOV. The nominal solution is marked "best" and the true solution by a crossed square sign.

Even if in this way we can obtain some preliminary orbits, the differential corrections algorithm may not converge: to avoid this problem, we use a *constrained differential correction* algorithm, providing solutions along the Line Of Variation (LOV solutions with only 5 free parameters, see Milani *et al.* 2005a). These solutions are meant as a sampling of the part of the LOV (a 1-dimensional set) lying inside the Admissible Region, that form a *second generation* set of VAs. The next step is testing these orbits against the available attributables in a third night of observations to see if they can belong to the same object: this kind of identification is called *attribution*. In most cases this allows to fit a full least squares orbit of good quality (see Section 7). This procedure is illustrated by Figure 2 in which the triangulated Admissible Region is shown for the asteroid $2003BH_{84}$; the VAs of the triangulation that have moderate identification penalty served as starting points for constrained differential corrections, which resulted in convergent LOV solutions. The true solution, computed using additional observations, is close to the LOV.

The procedure above is very effective in obtaining identifications and good orbits for NEO and MB (Milani *et al.* 2005b). In the present simulation we have found that the same algorithms can straightforwardly be applied to Jupiter and Mars Trojans and to

Centaurs, but for distant objects (including TNO, SDO, Trojans of the other outer planets) we found additional problems to be solved. A distant object has a proper motion less than 100 arcsec per day near opposition, even less in the near-quadrature areas. Within $\Delta t = 15$ minutes, the motion is less than 1 arcsec, thus the angular rates are determined with very poor relative accuracy.

Another related problem occurs when two tracklets, belonging to distant objects, from different nights are proposed for identification. The two sets of observations together might still be a TSA: the projection of the orbit on the celestial sphere might be a great circle within the observational errors: because the observations of the same night are very close, the great circle through the nightly average points fits all the observations in a satisfactory way. Then the attempt to compute an orbit might fail even if the identification is true. After this failure, the recursive chain of identifications is interrupted and even if the same object has been observed in a third night, this will not be found.

The solution we adopted is to accept a lower quality orbit when it is found that the joined observations from 2 tracklets are still a TSA. The procedure has three steps. First, we test whether the set of all observations from both nights is a TSA or not: this is done by computing the two components of curvature tangent to the celestial sphere, namely *acceleration* (along track component) and *geodetic curvature* (cross track), together with their uncertainty as deduced from the covariance matrix. If the curvature components are in absolute value less than their standard deviations then the set of observations is still a TSA, and differential corrections starting from a rough preliminary orbit are likely to diverge. In this case the second step is to compute a 4-parameter orbit, with values of range and range-rate fixed at the preliminary orbit values and the 4-dimensional attributable coordinates fitted: essentially, we compute a single attributable for the two nights together. The third step is to use the output of this fit, converted to some other coordinates (e.g., cartesian coordinates) as the first guess for constrained differential correction. This succeeded for $\simeq 82\%$ of the distant objects; however, for the remaining 18% we kept the 4-parameter orbits as a VA to be used in the next recursive step.

These 4-parameter orbits, even when it is not possible to upgrade them to 5-parameter LOV orbits, are used to compute predictions for a third observing night: by using the identification penalty we select the candidate 3-night identifications. When the data from three distinct nights (with an interval of at least 4 days between each couple of nights) are joined together, they have significant curvature, even for distant objects: the angular rates are estimated very roughly, but the average angles for each night are determined to sub-arcsec accuracy, and the three couples of average angles are very unlikely to be found on a great circle with such a precision. Thus, if the proposed identification is true the differential corrections always converge to a LOV orbit, independently from the quality of the 2-night orbit used to provide a first guess. Starting from the LOV orbit a full least squares orbit can be computed in $\simeq 87\%$ of the cases. With this method, the distant objects are not more difficult than the other ones from the point of view of finding 3-night identifications confirmed by a least squares orbit (with either 5 or 6 parameters).

The whole procedure is somewhat more complicated than the description above, because to achieve top completeness in a computationally efficient way the procedure must be organized in a sequence of iterations. The first iteration is based on a "smart triangulation", that is the Admissible Region is triangulated only for those tracklets which are likely to belong either to a NEO or to a distant object (this choice is based on the value of the proper motion and on the overall size and number of disconnected components of the Admissible Region). For most tracklets, likely to belong to MB or Jupiter Trojans, only 2 VAs are computed (one in the MB and one in the Trojan region).

The second iteration operates on all tracklets left unidentified in the first, triangulating the admissible region for all with a metric optimized to find NEO; indeed, all NEO identifications are found. The third iteration solves the few remaining cases, mostly Jupiter Trojans and distant objects: one especially difficult object was the only long period comet observed over 3 nights in this simulation, which required loosening the controls on maximum eccentricity in the differential correction iterations: the eccentricity of the final orbit is 0.988 ± 0.02, but an eccentricity > 1.5 was reached during the iterations.

In conclusion, all the identifications, both for objects observed in 2 nights and for those observed in 3 and more nights, can be found. The problem, as dicussed in the next Section, are the false identifications found along with the true ones.

6. Identification management

The properties of our identification and orbit determination procedure we want to measure are *completeness* and *reliability*. Completeness is measured by the ratio between the true identifications found and the ones hidden in the data (known by means of the "ground truth"). Reliability is measured by the fraction of false identifications among those proposed. The overall completeness depends upon the fraction of simulated objects for which all tracklets have been used in least square orbital fit and the fraction of objects which have been "lost", that is all their tracklets remained unlinked. Moreover, we need to take into account the fraction of true but incomplete identifications, in which not all the tracklets belonging to the same object have been identified. To compute these metrics we need first to join all the identifications with 2, 3 (and possibly more) tracklets.

The final stage of this identification management procedure is the *normalization* of the identification database (Milani et al. 2005b, Section 7.1). The purpose is to remove all duplications and contradictions accumulated in the identification process. We first sort all the identifications found by "quality", that is, an identification is *superior* if either it contains more nights or has the same number of nights and lower RMS of residuals.

The normalization procedure uses the following binary relations among identifications: *compatible* (all the tracklets belonging to the first are among the tracklets of the second), *independent* (none of the tracklets belonging to the first are among the ones of the second) and *discordant* (neither compatible nor independent).

Then we scan this sorted list from the top to reduce it to a *normalized* list of identifications. The first one is kept in the normalized list. Each of the others is compared with all the ones already included and it is kept in the normalized list if it is independent from all the others. It is removed if it is compatible with some of the previous ones. It is also removed if it is discordant with a previously included one with more tracklets.

Discordant identifications appear as contradictions, unless they have been removed by an identification containing all the tracklets of both: e.g., A=B=C and B=C=D are discordant unless A=B=C=D is in the list. Thus the discordant identifications with the same number of nights are both removed from the normalized list at the end of the procedure. This improves the reliability, because it is very likely that the tracklets used in a false identification belong to an object which has been observed in other nights: thus, if the true identifications are found, the false identifications will be discordant with them and will not appear in the normalized list. However, the price to be paid for this is the loss of some true identifications, resulting in lower completeness.

In conclusion, there is a trade-off between reliability and completeness: the normalization procedure defined above ensures top reliability. Moreover, this procedure needs to be applied in batch, working on all the observations of several observing nights, not one by one, not even night by night. After completing the search for all possible identifications, we can normalize the list and then remove the tracklets belonging to normalized identifications, leaving the unidentified ones for another iteration of the procedure.

The special attention we pay to reliability is due to the quadratic growth of the false tracklets/identifications number with the number density of observable objects (per unit area on the sky, see Figure 1). In the area corresponding, in the simulation we are using, to the center of the L5 Trojan cloud, the number density is already such that both false tracklets and false identifications are common. In a full simulation with all the MB objects the number of false tracklets/identifications should grow by an order of magnitude.

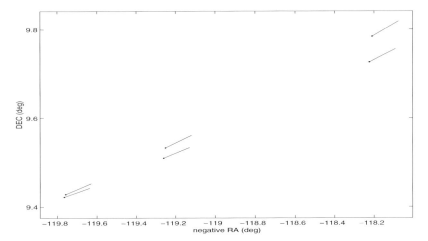

Figure 3. The attributables of 6 tracklets from the peak number density region at the center of the L5 Trojan cloud; the points represent the average positions, the segment the angular motion in one day. The corresponding tracklets can be identified in 4 ways, two of which are false.

A surprising result of this simulation is that false identifications with good least squares fits of all the observations can be found even without simulating astrometric noise. As an example, we have found $\simeq 50$ false identifications with 3 nights of data: the exact number depends upon the value of the controls. All these false identifications, without exceptions, are removed by the normalization procedure, but each one of them can "kill" one or two true identifications which are discordant. In Figure 3 we show one example of "Trojan twins", so close on the celestial sphere and with so similar angular rates that the tracklets may be identified in 4 different ways, all with RMS below 0.06 arcsec.

Thus what matters for false identifications is the number density of detected objects with similar angular rates: e.g., either Jupiter Trojans or distant objects. If either a NEO or a MB happened to be close in position to one of these, it would not be confused.

Table 1. Results for three-nighters

Class	no.tot	complete	incomp	lost	comp%	inc%	lost%
All	74904	74855	0	49	99.9	0.0	0.1
Tro J	55147	55098	0	49	99.9	0.0	0.1
Main B	6326	6326	0	0	100.0	0.0	0.0
Neo	1086	1086	0	0	100.0	0.0	0.0
Com/cent	711	711	0	0	100.0	0.0	0.0
Distant	11634	11634	0	0	100.0	0.0	0.0

The results of the identification procedure applied to this simulation are summarized in Table 1 for 3-nighters and Table 2 for 2-nighters. There were also some objects observed

Table 2. Results for two-nighters

Class	no.tot	complete	lost	comp%	lost%
All	24191	24171	20	99.9	0.1
Tro J	15989	15980	9	99.9	0.1
Main B	2202	2200	2	99.9	0.1
Neo	596	596	0	100.0	0.0
Com/cent	245	245	0	100.0	0.0
Distant	5159	5150	9	99.8	0.2

in more than 3 nights ($1,013$ in 4, 295 in 5 and 60 in 6 nights), and for all of them complete identifications were found in 100% of the cases. Note that of the very few lost objects not even one is due to the difficulty of finding the identification: they are all due to discordance with false identifications.

Of course without astrometric and photometric errors the results should be "perfect", and as shown in the Tables essentially they are, regardless of the orbit class. This means the algorithms are effective, it does not mean the same results can be obtained under more realistic conditions, e.g., with astrometric and photometric noise.

The identification management described in this Section turns out to be also effective as "tracklet management", to remove the false tracklets and join tracklets of the same night belonging to the same object. In fact, we have found not a single case of a false tracklet included in an identification (confirmed by a least squares solution). We found no incomplete identifications (see the third column in Table 1), implying that all the tracklets of the same night belonging to the same object have been included. Thus we can discard all the tracklets discordant (with some but not all observations in common) with the tracklets included in identifications: no true tracklet is wrongly removed.

The result is that 94.6% of the false tracklets are discarded. Moreover, most cases of multiple tracklets in the same night for the same objects are "solved" by merging in an identification together with others. Thus the leftover database of unidentified tracklets contains few false and few couples of true tracklets belonging to the same object in the same night. That is, tracklet management is mostly done. The question is whether the result would be that good in a more realistic simulation (with noise) and with real data.

7. Orbit determination

The definition of identification we use requires that an orbit can be fit to all the observations included in the identified tracklets. However, the quality of the identification is not the same as the quality of the orbit obtained with it. Indeed, one orbit could fit very well all the available observations of a given object, and still be quite different from the "true" orbit (in this context, the orbit used for generating the simulated observations).

Three metrics have to be considered in the quality of the orbit determination resulting from a true identification:

(1) The number of parameters which have been fit. An orbit can be a full least squares solution, with 6 parameters, a constrained LOV solution with 5 fit parameters, a 4-parameter fit with two parameters (range and range rate) kept fixed.

(2) The distance between the "true" orbit and the one determined, measured taking into account the covariance of the latter. If X_0 is the true orbit, X the one determined with normal matrix C_X, we use the 6-dimensional norm of the orbit error

$$D_6 = ||X - X_0||_6 = \sqrt{(X - X_0)^T\, C_X\, (X - X_0)/6}.$$

(3) The distance between the attributable predicted for an epoch one month later based on the true orbit, and the one based on the determined orbit, taking into account the expected uncertainty of the prediction. If A_0 is the attributable predicted one month later based on X_0, and C_{obs} is the normal matrix for such an attributable resulting from observations at that later date†, A is the attributable predicted with X and $C_A = \Gamma_A^{-1}$ is its normal matrix obtained by propagating the covariance Γ_X, we use the 4-dimensional norm of the prediction error

$$K_4 = ||A - A_0||_4 = \sqrt{(A - A_0)^T \, C \, (A - A_0)/4} \text{ where } C = C_A - C_A \, (C_A + C_{obs})^{-1} \, C_A.$$

The value of these metrics depends upon the purpose, that is how we are going to use these orbits. The norm D_6 is useful in the hypothesis that the same object is rediscovered under similar conditions (especially with the same number of observed nights) after some comparatively long time, years later. Then, if D_6 is small, the orbit identification between the two independent discoveries is easy. The norm K_4 is useful in the hypothesis that the same object is detected one month later, even in a single night. K_4 is the same metric used in the search for attributions of single tracklets: if it is small, the attribution is easy.

The results obtained for the identifications found, subdivided by number of nights of observations, are as follows. For 3-nights identifications, the norm D_6 is more than 3 in 0.4% of the cases with 6-parameter full orbits, in $\simeq 10\%$ of the cases with 5-parameter orbits (which are only 2% of the total); there are no 4-parameter orbits. The K_4 norm is less than 1 in all cases. In conclusion, almost all the 3-nights orbits are good enough both for next month attributions and for next apparition orbit identifications; this confirms the results by Spahr et al. (2004). 4-nights orbits are of course even better.

The results for 2-nighters are very different: the norm D_6 is more than 10 in 0.4% and more than 3 in 4% of the 6-parameter orbits. For 5-parameter orbits, 13% of the cases have $D_6 > 10$ and 58% have $D_6 > 3$. For 4-parameter orbits, 0.2% of the cases have $D_6 > 10$, 17% have $D_6 > 3$; however, the normal matrix of such orbits has rank 4, and the norm D_6 may not be a good diagnostic. From this test we can conclude not only that 2-nights identifications provide only inaccurate orbits (in agreement with Spahr et al. 2004), but also that the linear approximation to estimate the uncertainty (by means of the normal and covariance matrix) can fail, especially for 5-parameter LOV orbits. This indicates that orbit identification with rediscoveries in another apparition would be difficult, but not necessarily impossible, given the extreme robustness of the methods to identify 2-nights LOV solutions discussed in Milani *et al.* (2005a).

The K_4 norm for 2-nighters tells a different story. For 5- and 6-parameter orbits the value is < 1 for almost all cases (exceptions are only 0.02%). For the 4-parameter orbits, which are needed only for distant objects (see Section 5), K_4 is larger than 10 in 91.5% of the cases, even > 100 in 1.6% of the cases. This implies that, if a distant object is observed in only two nights in one lunation, and in only 1 night in the next one, it could be difficult to find a 3-nights identification with the algorithms used in this work‡.

Further progress in the algorithms for identification is possible (and we are working on them). Moreover, it is necessary to run a multi-lunation simulation to try to achieve good orbit determination for distant objects. However, the conclusion for now is that for the distant objects (TNO, scattered disk, Trojans of S-U-N) the survey observation planning should guarantee three tracklets in three separate nights in each lunation, otherwise a

† The uncertainty is computed under the same conditions of the present simulation, e.g. assuming two observations with an interval of 15 minutes.

‡ It is possible that other methods would work in such a case, e.g., a procedure using Gauss' preliminary orbit method with some smart trick to avoid the $O(N^3)$ computational complexity.

non negligible fraction of the discoveries for this population could be lost. If this requirement is satisfied, we see no obstacle from the identification and orbit determination to a *unbiased* survey discovering all sub-populations of solar system objects. The problem will have to be reassessed with more realistic simulations, e.g., introducing astrometric and photometric errors and a probabilistic detection model, before this conclusion can be reliably applied to the real data.

Acknowledgements

This research has been funded by: the Italian *Ministero dell'Università e della Ricerca Scientifica e Tecnologica*, PRIN 2004 project "The Near Earth Objects as an opportunity to understand physical and dynamical properties of all the solar system small bodies", *Ministry of Science and Environmental Protection of Serbia* through project 1238 "Positions and motion of small Solar System bodies", the *Observatorio de Mallorca* (OAM), the Spanish *Ministerio de Ciencia y Tecnología* and the European funds *FEDER* through the grant AYA2001-1784. The design and construction of the Panoramic Survey Telescope and Rapid Response System by the University of Hawaii Institute for Astronomy are funded by the United States Air Force Research Laboratory (AFRL, Albuquerque, NM) through grant number F29601-02-1-0268.

References

Bernstein, G. M., Trilling, D. E., Allen, R. L., Brown, M. E., Holman, M., & Malhotra, R. 2004, *AJ* 128, 1364

Bottke, W. F., Morbidelli, A., Jedicke, R., Petit, J., Levison, H. F., Michel, P., & Metcalfe, T. S. 2002, *Icarus*, 156, 399

Duncan, M. J., Levison, H. F., & Budd, S. M. 1995, *AJ* 110, 3073

Elliot, J.L., Kern, S.D., Clancy, K.B., Gulbis, A.A.S., Millis, R.L., Buie, M.W., Wasserman, L.H., Chiang, E.I., Jordan, A.B., Trilling, D.E., & Meech, K.J. 2005, *AJ* 129, 1117

Gomes, R.S., Morbidelli,A., & Levison, H.F. 2004, *Icarus* 170, 492

Ivezić, Ž., and 32 colleagues 2001, *AJ* 122, 2749

Jedicke, R. & Herron, J. D. 1997, *Icarus* 127, 494

Jedicke, R., Larsen, J., & Spahr, T. 2002, in: A. Cellino, B. Bottke, P. Paolicchi & R.P. Binzel (eds.), *Asteroids III* (Tucson: University of Arizona), p. 71.

Jewitt, D. C. & Luu, J. X. 2000, *AJ* 120, 1140

Kubica, J., Moore, A., Connolly, A., & Jedicke, R., 2005 *Signal and Data Processing of Small Targets*, in press

Kubica, J., Moore, A., Connolly, A., & Jedicke, R. 2005, *The Eleventh ACM SIGKDD International Conference on Knowledge Discovery and Data Mining*, in press.

Levison, H.F. & Morbidelli, A. 2003, *Nature* 426, 419

Milani, A., La Spina, A., Sansaturio, M.E., & Chesley, S.R. 2000, *Icarus*, 144, 39

Milani, A., Gronchi, G.F., de' Michieli Vitturi, M., & Knežević, Z. 2004, *CMDA* 90, 59

Milani, A., Sansaturio, M.E., Tommei, G., Arratia, O., & Chesley, S.R. 2005a, *A & A* 431, 729

Milani, A., Gronchi, G.F., Knežević, Z., Sansaturio, M.E., & Arratia, O. 2005b, *Icarus*, in press

Rabinowitz, D.L., Helin, E., Lawrence, K., & Pravdo, S. 2000, *Nature* 403, 165

Sheppard, S. S., Jewitt, D. C., Trujillo, C. A., Brown, M. J. I., & Ashley, M. C. B. 2000, *AJ* 120, 2687

Spahr, T., Chesley, S., Heasley, J., & Jedicke, R. 2004, *AAS, DPS meeting #36, #32.19*

Stokes, G. H. and 11 co-authors. 2003, Report of the Near-Earth Object Science Definition Team. August 22, 2003 (available at `http://neo.jpl.nasa.gov/neo/report.html`).

Tedesco, E. F., Cellino, A., & Zappalá, V. 2005, *AJ* 129, 2869

Wisdom, J. & Holman, M. 1991, *AJ* 102, 1528

Asteroid, Comets, Meteors
Proceedings IAU Symposium No. 229, 2005
D. Lazzaro, S. Ferraz-Mello & J.A. Fernández, eds.

Origin of water on the terrestial planets

Michael J. Drake[1] and Humberto Campins[2]

[1]Lunar and Planetary Laboratory, University of Arizona, Tucson, AZ 85721, USA
email: drake@lpl.arizona.edu

[2]Physics Department, University of Central Florida, Orlando, FL 32816, USA
email: campins@physics.ucf.edu

Abstract. We examine the origin of water in the terrestrial planets. We list various geochemical measurements that may be used to discriminate between different endogenous and exogenous sources of water. Late stage delivery of significant quantities of water from asteroidal and cometary sources appears to be ruled out by isotopic and molecular ratio considerations, unless either comets and asteroids currently sampled spectroscopically and by meteorites are unlike those falling to Earth 4.5 Ga ago or our measurements are not representative of those bodies. The dust in the accretion disk from which terrestrial planets formed was bathed in a gas of H, He and O. The dominant gas phase species were H_2, He, H_2O, and CO. Thus grains in the accretion disk must have been exposed to and adsorbed H_2 and water. We examine the efficacy of nebular gas adsorption as a mechanism by which the terrestrial planets accreted "wet". A simple model suggests that grains accreted to Earth could have adsorbed 1 - 3 Earth oceans of water. The fraction of this water retained during accretion is unknown, but these results suggest that at least some of the water in the terrestrial planets may have originated by adsorption.

Keywords. Earth, planets and satellites: formation

1. Introduction

Water is a common chemical compound in our solar system. In addition to Earth, it has been identified in asteroids, comets, meteorites, Mars, in the atmospheres, rings and moons of giant planets, and there is evidence for it in the poles of our Moon and of Mercury. The high deuterium to hydrogen (D/H) ratio of Venus atmosphere has been interpreted as evidence for Venus once having had far more water in the past than is present today. Here we address the origin of water in the terrestrial planets.

There is no agreement on the origin of water in the terrestrial planets. Possible sources of water can be divided into endogenous and exogenous. Endogenous sources include direct adsorption of water from gas onto grains in the accretion disk and accretion of hydrous minerals forming in the inner solar system. Exogenous sources include comets, hydrous asteroids, and phyllosilicates migrating from the asteroid belt. Recent reviews discuss this topic (e.g., Drake 2005; Campins *et al.* 2004; Lunine *et al.* 2003; Drake & Righter 2002; Pepin & Porcelli 2002; Robert, Gautier & Dubrulle 2000).

We know virtually nothing about water in Mercury and Venus. We do have information on Martian water through studies of the Martian meteorites. However, Mars presents unique challenges because compositional information is more limited than for Earth. An important example is the global value of the D/H ratio for Martian water. It is possible that we do not know Mars' intrinsic D/H isotopic ratio. Unlike Earth, Mars lacks plate tectonics and, hence, has no means of cycling water between mantle and crust. The Martian meteorites we measure on Earth may simply be sampling water delivered to Mars by cometary and asteroidal impacts subsequent to planetary formation. Note, however, that the contrary has been argued (Watson *et al.* 1994), i.e., that Martian

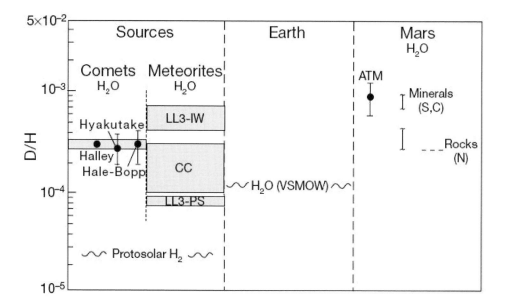

Figure 1. The D/H ratios in H_2O in three comets, meteorites, Earth (Vienna standard mean ocean water - VSMOW), protosolar H_2, and Mars. "CC" = carbonaceous chondrites, "LL3-IW" = interstellar water in Semarkona, "LL3-PS" = protostellar water in Semarkona. After Drake & Righter (2002).

meteorites provide evidence for an intrinsic Martian mantle D/H ratio up to 50% higher than Earth's mantle (see Figure 1).

It seems likely that the origin of water on the other terrestrial planets cannot be divorced from the origin of water on Earth. Because we know most about water on Earth, the rest of this paper will be spent examining the origin of such water.

2. Water on Earth

2.1. *Accretion History*

The accretion of the Earth was dominated by a violent series of events. Dynamical theory of planetary accretion points inexorably to a hierarchy of accreting planetesimals, with the largest object accreting at any given time during the growth of a planet being one tenth to one third of the mass of the growing planet (Wetherill 1985; Chambers 2001). During the later stages of accretion, these collisions deposited enough energy to at least partly melt the Earth and, possibly, to have completely melted the Earth. The Earth probably experienced serial magma ocean events. The final, most massive impact ejected material into Earth orbit. This disk of material subsequently accreted to form the Moon (Canup & Asphaug 2001).

Metal delivered by accreting planetesimals sank through these serial magma oceans and ponded at their bases for some period of time before transiting diapirically through the lower mantle to the center of the planet. Metal appears to have equilibrated with silicate at the base of the magma ocean. The mean depth calculated corresponds to at least the depth of the current upper mantle/lower mantle boundary of the Earth (Drake

Table 1. Ratios of ^{129}Xe to ^{132}Xe in Earth and Mars reservoirs. OIB refers to ocean island basalts. MORB refers to mid-ocean ridge basalts After Musselwhite *et al.* (1991).

	Mars	Earth
Atmosphere	2.4	0.985
Basalts	1.0–1.5	0.988 OIB* 1.00–1.14 MORB*

2000). That depth probably represents some ensemble average memory of metal-silicate equilibrium in a series of magma oceans and should not be taken as the literal depth of the last magma ocean. Comparable conclusions can be drawn for the Earth's Moon, Mars, and Vesta (Righter & Drake 1996).

As discussed below, the primitive atmosphere and ocean appear to have formed very early in Earth's history. Core formation, magma ocean solidification, ocean and atmospheric outgassing were essentially complete by 4.45 Ga ago.

2.2. *Evidence for an Early Water Ocean*

It had long been thought that the accretion disk at 1 AU was too hot for hydrous phases to be stable (e.g., Boss 1998), and that water was delivered to Earth after it formed by bombardment from asteroids and/or comets. However, recent geochemical evidence increasingly argues against asteroids and comets being the main sources of Earths water and points to the Earth accreting "wet" throughout its growth (Drake & Righter 2002; Drake 2005). In either case the Earth may have had water oceans very early in its history.

Evidence for the existence of an early water ocean on Earth comes from the different reservoirs of ^{129}Xe/^{132}Xe in the atmosphere, mid-ocean ridge basalt (MORB) source, and ocean island basalt (OIB) source (Table 1). ^{129}I has a half-life of about 16 Ma and is produced only during nucleosynthesis in a precursor astrophysical environment. All ^{129}I would have decayed to ^{129}Xe within 5 – 7 half-lives, or about 100 Ma following nucleosynthesis. Thus distinct ^{129}Xe/^{132}Xe reservoirs on Earth (and Mars) must have formed within 100 Ma of nucleosynthesis of ^{129}I.

It is difficult, although not impossible, to fractionate I from Xe by purely magmatic processes (Musselwhite & Drake 2000). The problem is that I and Xe are both volatile and incompatible (they both have low vaporization temperatures and both prefer magmas to solid mantle), although Xe is a little less incompatible. However, water is extremely effective at fractionating I from Xe (Musselwhite, Drake & Swindle 1991) because I dissolves in liquid water (it is a halogen) while Xe bubbles through (it is a gas).

If accretion ceased while ^{129}I still existed and any magma ocean solidified, liquid water could become stable at the Earths surface. Musslewhite (1995) showed that outgassed ^{129}I could be recycled hydrothermally into the oceanic crust and subducted back into the mantle, which was by now partially degassed of Xe. Subsequent decay of ^{129}I would give a MORB source with an elevated ^{129}Xe/^{132}Xe ratio in the mantle relative to the earlier outgassed atmosphere (Fig 2). A related conclusion was drawn for Mars (Musselwhite, Drake & Swindle 1991). It appears that the Earth had a primitive atmosphere and large bodies of water by 4.45 Ga ago.

Further evidence for the existence of an early water ocean on Earth comes from detrital zircons. Wilde *et al.* (2001) and Mojzsis, Harrison & Pidgeon (2001) independently reported zircons of 4.4 Ga age and 4.3 Ga age respectively. On the basis of magmatic

Iodine Recycling

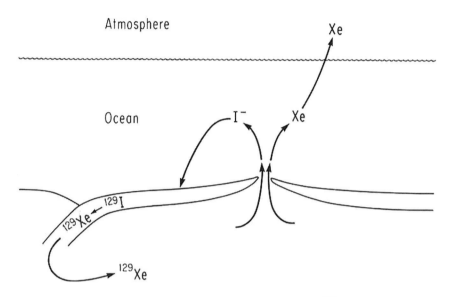

Figure 2. Cartoon illustrating outgassing of I and Xe while ^{129}I was still available. Xenon bubbles through the water ocean into the atmosphere. ^{129}I dissolves in water and is recycled hydrothermally into the crust and subducted back into the mantle. Subsequent decay of ^{129}I leads to mid-ocean ridge basalts (MORB) being erupted with elevated ^{129}Xe/^{132}Xe compared to the atmosphere. Separate Xe reservoirs imply fractionation of I from Xe within 100 Ma after nucleosynthesis of ^{129}I. From Musselwhite (1995).

oxygen isotope ratios and micro-inclusions of SiO_2, these authors concluded that the zircons formed from magmas involving re-worked continental crust that had undergone low temperature interaction with a liquid hydrosphere. Wilde *et al.* (2001) specifically conclude that the 4.4 Ga zircons represent the earliest evidence of both continental crust and water oceans on Earth.

Support for this conclusion was provided by Watson & Harrison (2005), who used a geothermometer based on the Ti-content of zircons to conclude that the zircons crystallized at ~700°C, a temperature indistinguishable from granitoid zircon growth today. In other words, water-bearing evolved magmas were present within 200 Ma of solar system formation. The implication is that modern patterns of crust formation, erosion, and sediment formation had already been established prior to 4.4 Ga, implying liquid water oceans already existed.

It is not possible to estimate the amount of water outgassed by 4.4 Ga ago. Water has been outgassed from the terrestrial mantle over the age of the solar system, but it has also been recycled back into the interior through subduction of oceanic plates at plate margins. Because the downgoing slab gets outgassed during subduction, not all water is transferred back to the mantle. The storage potential for water of silicates located primarily in the transition zone between lower and upper mantle is about 5 Earth oceans (Ohtani 2005). The time-dependent fluxes of water from the interior of the Earth to the surface and back cannot at present be deciphered.

3. Proposed Sources of Water and Methods of Discrimination

3.1. *Discriminators*

In principle, it should be possible to determine the main sources of and relative contributions to Earth's water if they have distinct chemical and isotopic signatures. Signatures that are used as discriminators include:

a) the D/H ratio of water in Earth, Mars, comets, meteorites, and the solar nebula

b) the relative abundances and isotopic ratios for noble gases on Earth, Mars meteorites, comets and the solar nebula

c) the ratio of noble gases to water on Earth, meteorites and comets

d) the isotopic composition of the highly siderophile (very strongly metal-seeking) element Os in Earth's primitive upper mantle (PUM), in the Martian mantle, and in meteorites

3.1.1. *D/H Ratios*

All deuterium is believed to have formed in the early universe. Stellar nuclear reactions convert D into He, thus lowering the D/H ratio of the universe with time. However, fractionation processes can produce local enhancements in the D/H ratio. For example, low-temperature ion-molecule reactions in the cores of molecular clouds can enhance the D/H ratio in icy grains by as much as two orders of magnitude above that observed in the interstellar medium (e.g., Gensheimer, Mauersberger & Wilson 1996). In our solar system, there is evidence for more than one reservoir of hydrogen (Drouart, Dubrulle, Gautier, *et al.* 1999; Mousis *et al.* 2000; Robert 2001; Hersant, Gautier & Hure 2001 and references therein). The solar nebula gas D/H ratio is estimated from observations of CH_4 in Jupiter and Saturn to be a low $2.1 \pm 0.4 \times 10^{-5}$ (Lellouch *et al.* 2001). Jupiter and Saturn likely obtained most of their hydrogen directly from solar nebula gas (this estimate is also consistent with protosolar D/H value inferred from the solar wind implanted into lunar soils; Geiss & Gloecker 1998). A second reservoir, enriched in D compared with the solar nebula gas, contributed to bodies that accreted from solid grains, including comets and meteorites.

We have D/H ratios from water in three comets (all from the Oort cloud), Halley $(3.2 \pm 0.1 \times 10^{-4}$, Eberhardt *et al.* 1995); Hyakutake $(2.9 \pm 1.0 \times 10^{-4}$, Bockele-Morvan *et al.* 1998); and Hale-Bopp $(3.3 \pm 0.8 \times 10^{-4}$, Meier *et al.* 1998). These are all about twice the value for terrestrial water $(1.49 \times 10^{-4}$, Lecuyer, Gillet & Robert 1998), about fifteen times the value for the solar nebula gas $(2.1 \pm 0.4 \times 10^{-5}$, Lellouch *et al.* 2001), and consistent with the range of values for hot cores of dense molecular clouds $(2$ to 6×10^{-4}, Gensheimer, Mauersberger & Wilson 1996). Carbonaceous chondrites have the highest water abundance of all meteorites (up to 17 wt%; Jarosewich 1990) and their D/H ratios range from 1.20×10^{-4} to 3.2×10^{-4} (Lecuyer, Gillet & Robert 1998). The largest D enrichment in a water-bearing mineral in a meteorite was measured at $7.3 \pm 1.2 \times 10^{-4}$ in the LL3 chondrite Semarkona (Deloule & Robert 1998). These results are illustrated in Figure 1.

The measurement of the D/H ratio of water in three comets is a significant development. Different authors interpret these ratios in very different ways. Some (e.g., Dauphas, Robert & Marty 2000; Morbidelli *et al.* 2000; Drake & Righter 2002; Robert 2001) consider the high D/H ratio in these comets as evidence against a cometary origin of most of the terrestrial water. Others (e.g., Delsemme 2000, Owen & Bar-Nun 2001) argue that comets are the main reservoir of deuterium-rich water that raised the terrestrial D/H a factor of six above the protosolar value. Complicating the matter further, recent laboratory measurements of the D/H ratio in sublimating ices have shown that fractionation

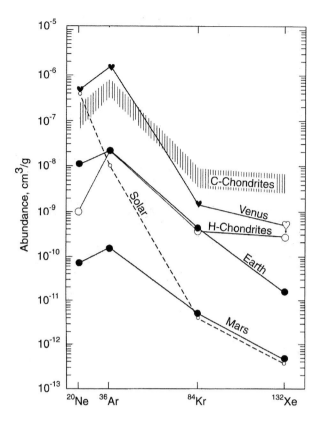

Figure 3. Noble gases in Venus, Earth, Mars, and meteorites. After Owen & Bar-Nun (1995).

can occur during sublimation (Weirich, Brown & Lauretta 2004; Moores, Brown, Lauretta, *et al.* 2005). During sublimation of water ice samples in a vacuum the D/H ratio of the evolved gas varies with time and can increase or decrease relative to the initial D/H ratio, depending on the nature of the sample. The root cause is interpreted to be differential diffusion and sublimation of HDO and H_2O. This result means that the measured D/H ratios in cometary comae may not be representative of the bulk cometary values.

3.1.2. *Noble Gases*

Noble gases are chemically inert and very volatile. Hence they probably arrive at a planet along with other volatiles, quickly move to the planet's atmosphere, and avoid the chemical complications of planetary evolution. Thus the noble gas characteristics of a planetary atmosphere can be tracers of the source of the planet's volatiles.

Figure 3 shows the abundances of noble gases in Venus, Earth and Mars compared with solar abundances and those in two kinds of meteorites. Note that the proportions of Ar, Kr, and Xe in the atmospheres of Earth and Mars are remarkably similar, and completely different from the abundances found in meteorites or the Sun (solar wind). We concentrate here on the possibility of cometary and meteoritic contributions to Earth and Mars.

We do not know much about noble gases in comets, and some of the measurements we do have appear contradictory. Krasnopolsky *et al.* (1997) reported an upper limit in comet Hale-Bopp of 0.5% of the solar Ne/O ratio. Stern *et al.* (2000) reported a tentative

Ar detection in comet Hale-Bopp and a roughly solar Ar/O ratio. Weaver *et al.* (2002) reported upper limits for Ar/O of < 10% and < 8% of the solar value in comets LINEAR 2001 A2 and LINEAR 2000 WM1, respectively. All these observations have been made of comets from the Oort cloud. The more sensitive upper limits for Ar in the LINEAR comets are not consistent with the detection reported by Stern *et al.* (2000) in comet Hale-Bopp. At this point it is not clear if comet Hale-Bopp was unusually rich in Ar or if the tentative detection is somehow flawed.

3.1.3. *Siderophile Elements in Earths Mantle*

The relatively high abundances of highly siderophile (very strongly metal-seeking) elements (HSEs) at 0.003× CI in Earth's primitive upper mantle (PUM) and their roughly chondritic element ratios suggest that these elements arrived after Earth's core formation had ceased (e.g., Drake & Righter 2002). Had these elements arrived sooner they would have been quantitatively extracted into Earth's core. This material is commonly termed the "late veneer". Drake & Righter (2002) argue that Earth-building materials shared some but not all the properties with extant meteorites, i.e., no primitive material similar to Earth's mantle is currently in our meteorite collections. More specifically, Re and Os are two HSEs that are linked by beta decay, $^{187}Re = {}^{187}Os + \beta$ thus Os isotopes can be used to constrain the origin of the "late veneer". Carbonaceous chondrites, the only abundant water-bearing meteorites, have a significantly lower $^{187}Os/^{188}Os$ ratio of 0.1265 than Earth's primitive upper mantle (PUM) value of 0.1295, effectively ruling them out as the source of the "late veneer" (Fig. 4). [Note that external ±2 sigma precision on the $^{187}Os/^{188}Os$ ratio is within the size of each individual data point in the histogram (Brandon *et al.* 2005).] Mars primitive upper mantle appears to have an even higher $^{187}Os/^{188}Os$ ratio of 0.132 (Brandon *et al.* 2005). The Earth's mantle $^{187}Os/^{188}Os$ ratio overlaps anhydrous ordinary chondrites and is distinctly higher than anhydrous enstatite chondrites, while the Martian mantle is higher than all common meteorite types.

3.2. *Sources*

In this section we examine the principal proposed sources of water in the terrestrial planets in the light of these discriminators. We divide these sources into two categories, endogenous and exogenous and we evaluate their possible contributions using the discriminators discussed in section 3.1.

3.2.1. *Endogenous Sources*

3.2.1.1. *Primordial Gas Captured from the Solar Nebula*

It has been argued that a primordial atmosphere could not have been captured directly from the solar nebula, principally because of Earth's D/H ratio in water and noble gas abundances. However, Campins *et al.* (2004) point out that the processes involved in planetary accretion, degassing, and the evolution of a hydrosphere and atmosphere are complex and may have fractionated the chemical and isotopic signatures of the source(s) of water. Hydrogen, for example, may be an important constituent in the outer core and possibly inner core of Earth (e.g., Okuchi 1997, 1998). If H and D were fractionated in that process, the residual D/H ratio in the hydrosphere may not reflect that of the original source. Following accretion, the surface of the Earth continued to be modified by large impacting asteroids and comets (Section 3.2.2). Large impact events have the capacity to completely volatilize any oceans (Zahnle & Sleep 1997) and blow off portions of Earth's atmosphere (Melosh & Vickery 1989). Fractionation of D/H and noble gas/water ratios may occur as a consequence of impact processes, which may also mask the signatures of the original source material.

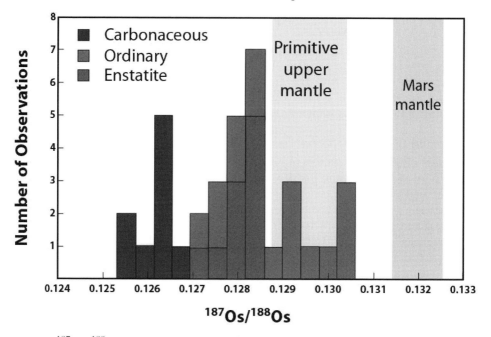

Figure 4. $^{187}Os/^{188}Os$ ratios in carbonaceous, ordinary, and enstatite chondrites, and in the Earth's primitive upper mantle (PUM), are distinct and are diagnostic of the nature of the Earth's "late veneer". Mars is not plotted because the uncertainty in its initial $^{187}Os/^{188}Os$ ratio is larger than range of the X-axis. After Drake and Righter (2002).

3.2.1.2. *Adsorption of Water onto Grains in the Accretion Disk*

The terrestrial planets grew in an accretion disk of gas and dust grains. Hydrogen, He, and O_2 dominated the gas in which the dust was bathed. Some of that H_2 and O_2 combined to make water vapor. If thermodynamic equilibrium was attained, there were about two Earth masses of water vapor in the accretion disk inside of 3 AU (Drake 2005). The mass of the Earth is 5×10^{27} g. The mass of one Earth ocean is 1.4×10^{24} g. The extreme maximum estimate for the amount of water in the Earth is about 50 Earth oceans (Abe *et al.* 2000), with most estimates being 10 Earth oceans or less. For example, an estimate based on the water storage potential of minerals in the silicate Earth is about 5 - 6 Earth oceans (Ohtani 2005). Thus the mass of water vapor available in the region of the terrestrial planets far exceeded the mass of water accreted.

Could water vapor be adsorbed onto grains before the gas in the inner solar system was dissipated? Stimpfl, Lauretta & Drake (2004) have examined the role of physisorption by modeling the adsorption of water on to grains at 1000°K, 700°K, and 500°K using a Monte Carlo simulation. Stimpfl, Lauretta & Drake (2004) "exploded" the Earth into $0.1\mu m$ spheres of volume equal to Earth, recognizing that grains in the accretion disk are not spherical and would be fractal in nature. If the surface area of the fractal grain was 100 times that of a sphere of corresponding volume, then one quarter of an ocean of water could be adsorbed at 1000°K, one Earth ocean could be adsorbed at 700°K, and three Earth oceans could be adsorbed at 500°K. Obviously in the accretion disk there would be a size-frequency distribution of grain sizes and variable fractal dimensions, but the exercise shows that the concept deserves further exploration. This work is discussed in more detail in Drake (2005).

Of course, there are issues of retention of water as the grains collide and grow to make planets. However, it is clear that volatiles are not completely outgassed even in planetary scale collisions. For example, primordial ^{3}He, far more volatile than water, is still outgassing from Earth's mantle 4.5 Ga after an almost grown Earth collided with a Mars-sized body to make the Moon.

Stimpfl, Lauretta & Drake (2004) showed that the efficiency of adsorption of water increases as temperature decreases, that is, the process should have been more efficient further from the Sun than closer to the Sun. Thus, it is likely that Mars, Earth, and Venus all accreted some water by adsorption, with Mars accreting the most both because of its greater distance from the Sun and the lower energy of collisions during accretion because of its smaller final mass. The current differences in the apparent water abundances among the terrestrial planets are probably the result of both different initial inventories and subsequent geologic and atmospheric processing.

There is an interesting consequence for the evolution of planetary redox states if the terrestrial planets accreted "wet". Okuchi (1997) showed that when Fe-metal and water were compressed to 30–100 kbars and heated to 1200°C - 1500°C, iron hydride formed. In a magma ocean environment, metal sinking to form planetary cores should contain H, and OH should be left behind in the molten silicate. As more metal was delivered to the planet as it accreted, more H would be extracted into the core and more OH liberated in the silicate. Thus planetary mantles should become progressively more oxidized with time, perhaps explaining the high redox states relative to the iron - wüstite buffer. This process might explain the correlation of the degree of oxidation of silicate mantles with planet mass (Righter, Drake & Scott 2006).

3.2.2. *Exogenous Sources*

3.2.2.1. *Comets*

Comets were long considered the most likely source of water in the terrestrial planets. A cometary source was attractive because it is widely believed that the inner solar system was too hot for hydrous phases to be thermodynamically stable (Boss 1998). Thus an exogenous source of water was needed.

There are elemental and isotopic reasons why at best 50% and, most probably, a very small percentage of water accreted to Earth from cometary impacts (Drake & Righter, 2002). Figure 1 compares the isotopic composition of hydrogen in Earth, Mars, three Oort Cloud comets, and various early solar system estimates. It is clear that 100% of Earth's water did not come from Oort Cloud comets with D/H ratios like the three comets measured so far. D/H ratios in Martian meteorites do agree with the three cometary values (Figure 1). That may reflect the impact of comets onto the Martian surface in a non-plate tectonics environment that precludes recycling of surface material into the Martian mantle (Drake 2005). Conversely, there are models (Lunine *et al.* 2003) that have asteroids and comets from beyond 2.5 AU as the main source of Mars' water.

So what limits the cometary contribution to Earth's water? Consider, for example, that perhaps Earth accreted some hydrous phases or adsorbed water, and some amount of additional water came from comets. Indigenous Earth water could have had D/H ratios representative of the inner solar system, i.e., low values because of relatively high nebular temperatures, perhaps like protosolar hydrogen ($2 - 3 \times 10^{-5}$, Lecluse & Robert 1994) in which case a cometary contribution of up to 50% is possible. Alternatively, indigenous Earth water could have had D/H ratios representative of a protosolar water component identified in meteorites ($\sim 9 \times 10^{-5}$, Deloule & Robert 1995), in which case there could be as little as a $10 - 15\%$ cometary contribution (Owen & Bar-Nun 2000).

There are caveats to using cometary D/H ratios to limit the delivery of cometary water to Earth. First, we do not know that Oort Cloud comets Halley, Hale-Bopp, and Hyakutake are representative of all comets. Certainly they are unlikely to be representative of

Kuiper Belt objects, the source of Jupiter family comets, as Oort Cloud comets formed in the region of the giant planets and were ejected while Kuiper Belt objects have always resided beyond the orbit of Neptune. Second, D/H measurements are not made of the solid nucleus, but of gases emitted during sublimation. As mentioned in Section 3.1, differential diffusion and sublimation of HDO and H_2O may make such measurements unrepresentative of the bulk comet. The D/H ratio would be expected to rise in diffusion and sublimation, as has been confirmed in preliminary laboratory experiments on pure water ice (Weirich, Brown & Lauretta 2004). Lower bulk D/H ratios would increase the allowable amount of cometary water. Intriguing experiments on mixtures of water ice and TiO_2 grains by Moores et al. (2005) suggest that D/H ratios could be lowered in sublimates. Third, the D/H ratios of organics and hydrated silicates in comets are unknown, although that situation may be rectified by analysis of samples returned by the Stardust mission. Note, however, that D/H ratios up to $50 \times$ Vienna Standard Mean Ocean Water (VSMOW) have been measured in some chondritic porous interplanetary dust particles (CP-IDPs) which may have cometary origins (Messenger 2000), and higher aggregate D/H ratios of comets would decrease the allowable cometary contribution to Earth's water.

Delivery of water from comets can also be evaluated in light of other cometary geochemical data. For an assumed Ar/H_2O ratio of 1.2×10^{-7} in the bulk Earth, comets like Hale-Bopp with an approximately solar ratio of Ar/H_2O (Stern, Slanter, Festou, et al. 2000) would bring in 2×10^4 more Ar than is presently in the Earths atmosphere (Swindle & Kring 2001), if 50% of Earth's water, the maximum amount permitted by D/H ratios, was derived from comets. It is unclear if this measurement of comet Hale-Bopp is applicable to all comets. However, the true Ar/O ratios in comets would have to be at least three orders of magnitude below solar in order to be consistent with the Ar abundance of the Earth's atmosphere.

Another approach to estimating the contribution of cometary materials to Earth's water budget can be made by considering the implications for the abundances of noble metals and noble gases. Dauphas & Marty (2002) show that the total mass of cometary and asteroidal material accreting to Earth after core formation is $0.7 - 2.7 \times 10^{25}$g and that comets contribute < 0.001 by mass or $< 0.7 - 2.7 \times 10^{22}$g. Given that the minimum mass of water in the Earth, one Earth ocean, is 1.4×10^{24}g, comets can contribute less than 1% of Earth's water.

3.2.2.2. Asteroids

Asteroids are a plausible source of water based on dynamical arguments. Morbidelli et al. (2000) have shown that up to 15% of the mass of the Earth could be accreted late in Earths growth by collision of one or a few asteroids originating in the Main Belt. However, there are strong geochemical arguments against a significant contribution of water from asteroids, unless one postulates that Earth was hit by a hydrous asteroid unlike any falling to Earth today as sampled by meteorites. One cannot prove this hypothesis wrong, as it could involve a single, unique event.

However, if asteroidal material falling to Earth 4.5 Ga ago was the same as that which falls today, one can effectively rule out asteroids as a source of water. The reason involves the Os isotopic composition of the so-called "late veneer", the material that may have contributed the highly siderophile elements (HSEs) that are present to within 4% of chondritic proportions at about 0.003 of chondritic absolute abundances (Fig. 5).

Of current meteorite falls, only carbonaceous chondrites have significant amounts of water (Jarosewich 1990). Enstatite and ordinary chondrites are essentially anhydrous. The late addition of water must be accompanied by other chemical elements such as Re

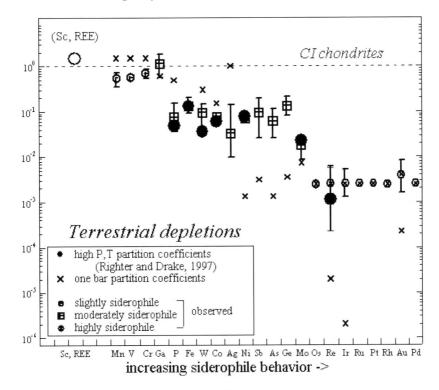

Figure 5. Comparison of the observed siderophile element depletions in Earth's upper mantle (open symbols - abundances normalized to CI chondrites and refractory elements), with those calculated using one bar partition coefficients (crosses) and high pressure / high temperature calculated partition coefficients (solid circles). Calculated depletions using the latter partition coefficients overlap the observed depletions, consistent with metal-silicate equilibrium and homogeneous accretion. After Drake (2000).

and Os. Earth's Primitive Upper Mantle (PUM) has a significantly higher $^{187}Os/^{188}Os$ ratio than carbonaceous chondrites, effectively ruling them out as the source of the "late veneer" (Fig. 4). The PUM $^{187}Os/^{188}Os$ ratio overlaps anhydrous ordinary chondrites and is distinctly higher than anhydrous enstatite chondrites. Mars' PUM has an even higher $^{187}Os/^{188}Os$ ratio, making an asteroidal source of Martian water even more unlikely.

The identification of anhydrous meteorites with the "late veneer" (Walker *et al.* 2002) effectively rules out asteroidal material being a late source of Earth's and Mars' water. There is one caveat, however; thermal processing of asteroids was occurring 4.5 Ga ago. One cannot exclude the possibility that ordinary chondrites once contained water and those falling to Earth today have lost it by metamorphism, even though it was still present 4.5 Ga ago. However, the preservation of aqueous alteration products in some carbonaceous chondrites (McSween 1979) suggests that loss of water from initially hydrous asteroids is unlikely to proceed to the anhydrous limit.

3.2.2.3. *Early Accretion of Water from Inward Migration of Hydrated Phyllosilicates*

Most solar nebula models suggested that the growth zones of the terrestrial planets were too hot for hydrous minerals to form (e.g., Cyr, Sears & Lunine 1998; Delsemme 2000; Cuzzi & Zahnle 2004). Ciesla & Lauretta (2004) suggest that hydrous minerals were formed in the outer asteroid belt region of the solar nebula and were then transported to the hotter regions of the nebula (i.e., Earth and Mars) by gas drag, where they were incorporated into the planetesimals that formed there. These hydrated minerals were able to survive for long periods in hotter regions due to sluggish dehydration kinetics. Note that this mechanism differs from the delivery of water by stochastic impacts with large planetary embryos originating in the outer asteroid belt region (Section 3.2.2.2). Drake (2005) points out that it seems unlikely that phyllosilicates could be decoupled from other minerals and transported into the inner solar system. Thus the proposed radial migration of hydrated minerals appears subject to the same objection involving Os isotopes discussed in Section 3.2.2.2.

3.2.2.4. *Water and Organics*

It has been postulated that water and organics were delivered from the same cometary and/or asteroidal source (e.g., Chyba 1993; Delsemme 2000). In light of the new evidence, such combined delivery seems less likely. In addition to comets and asteroids being inconsistent with some geochemical properties of the Earth, it is unlikely that a complex organic material would have survived the magma ocean accompanying the formation of the Moon at the end of accretion of the Earth (Drake 2000). However, it seems possible that complex organic material may have been delivered to Earth after it formed and liquid water oceans became stable.

Comets, which are known to be rich in organic molecules, have been postulated to be the principal source of terrestrial amino acids (Pierazzo & Chyba 1999; Chyba 1993). Some meteorites are also rich in carbon and organic compounds, hence, asteroids may have also contributed significantly to Earths organic inventory. In fact, Kring & Cohen (2002) point out that during late heavy bombardment, asteroidal material probably delivered a large mass of organic material to Earths surface, as much as 160 times larger than that in the total land biomass today. Even if some of the organic molecules were dissociated during the impacts, Kring & Cohen (2002) propose the formation of impact-generated hydrothermal regions with lifetimes up to 10^6 years, where complex organic molecules might reassemble.

3.3. *Outstanding Challenges*

Geochemical data for the Earth, Mars, meteorites, and comets cast increasing doubt on either asteroids or comets being the principal source of water in the terrestrial planets. The idea that the terrestrial planets could have obtained much or all of their water by adsorption on to grains directly from gas in the accretion disk is relatively new (Stimpfl, Lauretta & Drake 2004; Drake 2005). Many major issues remain to be resolved in order to make further progress. Below we list some of the most important.

1. The D/H ratio of the nebular gas is inferred from spectroscopic measurements of CH_4 in the atmospheres of Jupiter and Saturn to be $2.1 \pm 0.4 \times 10^{-5}$ (Lellouch *et al.* 2001), much lower than VSMOW. Spectroscopic or direct measurements of solar D/H cannot be made because practically all of the Suns deuterium has been burned to make He. If the nebular D/H ratio really is as low as implied by the Jovian and Saturnian atmospheres, a mechanism to raise the D/H ratio of nebular gas from solar to VSMOW is needed.

2. It is likely that the D/H and Ar/O ratios measured in cometary comas and tails are not truly representative of cometary interiors. Reconnaissance experiments have been shown that D/H ratios in laboratory experiments can increase or decrease with time due to differential diffusion and sublimation, depending on the physical nature of the starting material (Weirich, Brown & Lauretta 2004; Moores *et al.* 2005). Further, measurements of Ar/O ratios in comets are either upper limits or 3 sigma detection limits and are somewhat uncertain. Depending on the siting of Ar in comets, Ar/O ratios may also be unrepresentative of cometary interiors. The Deep Impact mission will be the first attempt to expose fresh cometary interior material for spectral analysis with ground-based and

space-based high spectral resolution spectrometers. It is unclear if improvement in our understanding of cometary D/H ratios and Ar/O ratios will be made.

3. The key argument against an asteroidal source of Earths water is that the Os isotopic composition of Earths primitive upper mantle matches that of anhydrous ordinary chondrites, not hydrous carbonaceous chondrites. But are the parent bodies of the ordinary chondrites anhydrous? Could ordinary chondritic meteorites be derived from the metamorphosed outer parts of hydrous asteroids, in which case impact of a bulk asteroid could deliver water? It is probable that spacecraft spectral examination of very deep impact basins in S-type asteroids will be needed to address this question.

4. A related question is why there are any anhydrous primordial bodies, such as the parent bodies of anhydrous meteorites, in the solar system if adsorption of water from gas in the accretion disk was an efficient process, as preliminary calculations suggest it might have been.

5. The timing of loss of gas from the accretion disk in the region of the terrestrial planets is unknown. For adsorption to be efficient, nebular gas must persist long enough for grains to adsorb water. Radial migration of phyllosilicates also depends on the presence of gas. The timing of loss of gas from the accretion disk will be intimately connected to the currently unknown mechanism of loss.

References

Abe, Y., Ohtani, E., Okuchi, T., Righter, K., & Drake, M.J. 2000, in: R.M. Canup & R. Righter (eds.), *Origin of the Earth and Moon* (Tucson: Univ. Arizona Press), p. 413

Bockelee-Morvan, D., Gautier, D., Lis, D.C., Young, K., Keene, J., Phillips, T., Owen, T., Crovisier, J., Goldsmith, P.F., Bergin, E.A., Despois, D., & Wooten, A. 1998, *Icarus* 133, 147

Boss, A.P. 1998, *Ann. Rev. Earth Planet. Sci.* 26, 53

Brandon, A.D., Humayun, M., Puchtel, I.S., & Zolensky, M.E. 2005, *Geochim. Cosmochim. Acta* 69, 1619

Campins, H., Swindle, T.D., & Kring, D.A. 2004, in: J. Seckbach (ed.) *Origin, Evolution and Biodiversity of Microbial Life in the Universe*, p. 569

Canup, R.M. & Asphaug, E. 2001, *Nature* 412, 708

Chambers, J.E. 2001, *Icarus* 152, 205

Chyba 1993, *Geochimica et Cosmochimica Acta* 57, 3351

Ciesla, F.J., Lauretta D.S., & Hood L.L. 2004, *Lunar Planet. Sci. XXXV*, abstract 1219.

Cuzzi, J.N. & Zahnle, K.J. 2004, *Ap. J.* 614, 490

Cyr, K.E., Sears, W.D., & Lunine , J.I. 1998, *Icarus* 135, 537

Dauphas, N. & Marty B. 2002, *Jour. Geophys. Res.* 107, E12-1

Dauphas, N., Robert, F., & Marty, B. 2000, *Icarus* 148, 508

Deloule, E. & Robert F. 1995, *Geochim. Cosmochim. Acta* 59, 4695

Delsemme, A.H. 2000, *Icarus* 146, 313

Drake, M.J. 2000, *Geochim. Cosmochim. Acta* 64, 2363

Drake, M.J. 2005, *Meteoritics and Planetary Science* 40, 519

Drake, M.J. & Righter, K. 2002, *Nature* 416, 39

Drouart, A., Dubrulle, B., Gautier, D., & Robert, F. 1999, *Icarus* 140, 129

Eberhardt, P., Reber, M., Krankowsky, D., & Hodges, R.R. 1995, *Astron. & Astrophys.* 302, 301

Geiss, J. & Gloeckler, G. 1998, *Space Science Reviews* 84, 239

Gensheimer, P.D., Mauersberger, R., & Wilson, T.L. 1996, *Astron. & Astrophys.* 314, 281

Hersant, F., Gautier, D., & Huré, J.-M. 2001, *Ap.J.* 554, 391

Jarosewich, E. 1990, *Meteoritics* 25, 323

Krasnopolsky, V.A., Mumma, M.J., Abbott, M., Flynn, B.C., Meech, K.J., Yeomans, D.K., Feldman, P.D., & Cosmovici, C.B. 1997, *Science* 277, 1488

Kring, D.A. & Cohen B.A. 2002, *J. Geophys. Res.* 107, 4-1

Lecluse, C. & Robert, F. 1994, *Geochim. Cosmochim. Acta* 58, 2927

Lecuyer, C., Gillet, P., & Robert, F. 1998, *Chem. Geol.* 145, 249

Lellouch, E., Bézard B., Fouchet T., Feuchtgruber H., Encrenaz T., & de Graauw T. 2001, *Astron. & Astrophys.* 370, 610

Lunine, J.I., Chambers, J., Morbidelli, A., & Leshin, L.A. 2003, *Icarus* 165, 1

McSween, H.Y. 1979, *Revs. Geophys. Space Phys.* 17, 1059

Meier, R., Owen, T.C., Matthews, H.E., Jewitt, D.C., Bockelee-Morvan, D., Biver, N., Crovisier, & Gautier, D. 1998, *Science* 279, 842

Melosh, H.J. & Vickery, A.M. 1989, *Nature* 338, 487

Messenger, S. 2000, *Nature* 404, 968

Mojzsis, S.J., Harrison, T.M., & Pidgeon, R.T. 2001, *Nature* 409, 178.

Moores, J.E., Brown, R.H., Lauretta, D.S., & Smith, P.H. 2005, *Lunar and Planetary Science XXXVI*, abstract 1973

Morbidelli, A., Chambers J., Lunine J.I., Petit J.M., Robert F., Valsecchi G.B., & Cyr K.E. 2000, *Meteoritics and Planetary Science* 35, 1309

Mousis, O., Gautier, D., Bockelée-Morvan, D., Robert, F., Dubrulle, B., & Drouart, A. 2000, *Icarus* 148, 513

Musselwhite, D.S. & Drake, M.J. 2000, *Icarus* 148, 160

Musselwhite, D.S., Drake, M.J., & Swindle, T.D. 1991, *Nature* 352, 697

Musselwhite, D.S. 1995, Ph.D. thesis, University of Arizona, Tucson.

Ohtani, E. 2005, *Elements* 1, 25

Okuchi, T. 1997, *Science* 278, 1781

Okuchi, T. 1998, *J. Phys. Condensed Matter* 10, 11595

Owen, T. & Bar-Nun, A. 2001, *Origins Life Evol. Biosphere* 31, 435

Owen, T. & Bar-Nun, A. 2000, in: R.M. Canup & K. Righter (eds.) *Origin of the Earth and Moon* (Tucson: Univ. of Arizona Press), p. 459

Owen, T. & Bar-Nun, A. 1995, *Icarus* 116, 215

Pepin, R.O. & Porcelli D. 2002, *Reviews in Mineralogy and Geochemistry* 47, 191

Pierazzo & Chyba 1999, *Meteoritics and Planetary Science* 34, 909

Righter, K. & Drake, M.J. 1996, *Icarus* 124, 513

Righter, K., Drake, M.J., & Scott E. 2006, in: L. Leshin & D. Lauretta (eds.), *Meteorites and the Earth Solar System II* (Tucson: Univ. of Arizona Press), (submitted).

Robert, F. 2001, *Science* 293, 105

Robert, F., Gautier, D., & Dubrulle, B. 2000, *Space Sci. Rev.* 92, 201

Stern, S.A., Slater D.C., Festou M.C., Parker J.W., Gladstone G.R., A'Hearn M.F., & Wilkinson E. 2000, *ApJ.* 544, L169

Stimpfl, M., Lauretta, D.S., & Drake, M.J. 2004, *Meteoritics and Planetary Science* 39, A99

Swindle, T.D. & Kring D.A. 2001, in: Eleventh Annual V.M. Goldschmidt Conference, LPI Contribution No. 1088, Lunar and Planetary Institute, Abstract 3785

Walker, R.J., Horan M.F., Morgan J.W., Becker H., Grossman J.N., & Rubin A.E. 2002, *Geochim. Cosmochim. Acta* 66, 4187

Watson, L.L., Hutcheon I.D., Epstein S., & Stolper E.M. 1994, *Science* 265, 86

Watson, E.B. & Harrison, T.M. 2005, *Science* 308, 841

Weaver, H.A., Feldman P.D., Combi M.R., Krasnopolsky V., Lisse, C.M., & Shermansky, D.E. 2002, *ApJ.* 576, L95

Weirich, J.R., Brown R.H., & Lauretta D.S. 2004, *Bull. Amer. Astron Assoc.* 36, 1143.

Wetherill, G.W. 1985, *Science* 228, 877

Wilde, S.A., Valley, J.W., Peck, W.H., & Graham, C.M. 2001, *Nature* 409, 175

Zahnle, K.J. & Sleep, N.H. 1997, in: P.J. Thomas, *et al.* (eds.) *Comets and the Origin and Evolution of Life*, (New York: Springer Velag), P. 194

Asteroids, Comets, Meteors
Proceedings IAU Symposium No. 229, 2005
D. Lazzaro, S. Ferraz-Mello & J.A. Fernández, eds.

The interior of outer Solar System bodies

M. T. Capria[1] and A. Coradini[2]

[1]INAF - IASF, Area Ricerca Tor Vergata, 00133 Roma, Italy
email: mariateresa.capria@rm.iasf.cnr.it
[2]INAF - IFSI, Area Ricerca Tor Vergata, 00133 Roma, Italy
email: angioletta.coradini@rm.iasf.cnr.it

Abstract. The population of small bodies of the outer Solar System is composed by objects of different kind and size, such as comets, Kuiper Belt objects and Centaurs, all sharing however a common characteristic, that is to be rich in ices and other volatiles. The knowledge of the composition and properties of these bodies would help in better understanding the processes that shaped the solar nebula at large heliocentric distances and determined the formation and evolution of the planets. A large number of observational results are now available on these bodies, due to successful space missions and increasingly powerful telescopes, but all our instruments are unable to probe the interiors. However, we are beginning to see how these seemingly different populations are related to each other by dynamical and genetic relationships. In this paper we try to see what could be their thermal evolution and how and when it brings to their internal differentiation. In fact, in this way we can try to foresee what should be the surface expression of their differentiation and evolution and try to link the surface properties, as probed by instruments, with the interior properties. One thing to note about the cometary activity is that it is well interpreted when assuming that the comets are small, fragile, volatile-rich and low-density objects. This view, despite of the strong differences noted in the few comet nuclei observed in situ, has not been disproved. On the other side, the observations of the Kuiper belt objects are possibly indicating that they are large, probably collisionally evolved objects (Farinella & Davis 1996), maybe with larger densities. We are now facing a kind of paradox: we have from one side the comets, and from the other side a population of much denser and larger objects; we know that a dynamical link exists between them, but how can we go from one type of population to another? In this paper the current status of our knowledge on the subject is reviewed, taking into account the results of thermal modeling and the results of observations.

Keywords. comets: general, Kuiper Belt, planets and satellites: formation

1. Introduction

In this paper we analyze the internal structure of the small objects that originated in the outer solar system, beyond Neptune: Kuiper Belt objects (KBOs), Centaurs and short period comets.

All these bodies accreted from the protoplanetary nebula extending beyond the region where planets formed, but we do not know exactly where: in fact many of them were displaced and scattered from the accreting proto-Uranus and Neptune. In particular, the bodies that are now present in the Oort cloud are probably belonging to this category. It is also commonly accepted that the bodies populating the Kuiper Belt formed in a region roughly overlapping their present position. Moreover it has been suggested that the Kuiper Belt zone could be the source of most short period comets (Fernandez 1980). In this frame, Centaurs, with their instable orbits, represent bodies caught "on the way".

From a physical point of view all these bodies, having been originated in the same place, should be closely related and should have the same intrinsic physical nature.

Obviously there are effects, related to the size of these bodies, that can finally lead to a different evolution, but we should be able to decipher their main evolutionary path and establish common genetic relationships. The small bodies present in the outer solar system are characterized by a high content of volatiles elements, that can lead under certain thermodynamic conditions to the development of an intrinsic activity, giving rise to the sublimation and loss of water ice and high-volatility carbon compounds. The properties of these bodies can be the result of the physical and chemical conditions prevailing in the solar nebula to the moment of their accretion and of the processes acting on them during the subsequent evolution. We know also that this region was characterized by a high degree of "mobility", as shown by Morbidelli (2004), that strongly influenced the subsequent history and hence the present structure and composition of many objects.

Lets now try to identify the region that we want to take into account: while the outer edge is obviously the Oort cloud, the in-bound edge is more difficult to define due to the fact that comets undergo a multistage capture process, and that they can be even scattered into the inner solar system eventually contributing to the Near Earth Objects population. For sake of simplicity we can consider the asteroid belt as the deeper inner edge.

The present structure and appearance of these bodies has been affected by their dynamical history, by the surface aging (reddening of surfaces due to irradiation), by their activity (when present, as in the case of comets) and by their collisional evolution. This last process, in particular, could have heavily shaped them: the comets could even be collisional fragments directly ejected from the Kuiper belt.

As far as the sizes are concerned, it is very difficult, after the discovery of several "minor bodies" of very large size like for example Varuna (Jewitt & Sheppard 2002) and 2003 UB313 (Brown 2005), to define a clear limit between them and the "real" planets. However, in this review we will not deal with planet-sized bodies, that probably underwent processes closer to the ones that shaped the planets than to the ones affecting small bodies.

The interior of these bodies cannot be directly probed, however we can at least infer their composition from the theories of protosolar nebula chemical evolution. Cosmogonical theories usually predict that the first condensates grow through different accumulation processes, that include low velocity mutual collisions. In the process of adhesion different parameters play important roles, affecting both the velocity and the mass distribution of grains. Among those affecting mainly the velocity distribution of particles, we have to mention gas turbulence and gas-dust drag forces, while the mass distribution depends not only on the relative particle velocity but also on their sticking efficiencies. The relative importance of gravitational instability with respect to collisional coagulation can have consequences on the final structure of the cometesimals and on the porosity of the resulting bodies. For this reason, density (porosity) is an observationally derived property having cosmogonical significance. We can obtain hints on the internal structure also from the modeling of the thermal evolution of ice-rich bodies, and from the new data collected both by planetary missions and by ground based observations of the different objects belonging to this category. In this paper we will try to combine the different sources of data and to see how they can be used to improve our theoretical approaches in order to create a general scheme.

After the previous, very general definitions, we will briefly discuss the objects included in our review from the point of view of what it is known about their interiors from observation; after that, we will discuss the hints that we can obtain from formation models and from thermal evolution models; then, we will try to reach some conclusions.

2. Kuiper Belt objects

After the discovery of the first Kuiper Belt object, the number of KBOs directly detected has greatly increased, as well as the area of the solar system in which they are found. The orbits of the so-called classical Kuiper Belt objects fall into two main categories (Luu & Jewitt 2002): objects with $a < 41$ AU and $e > 0.1$ (like Pluto and Charon) that are in mean motion resonances with Neptune and objects with $41 < a < 50$ AU and $e < 0.1$ (like 1992 QB1) that are not in resonant orbits. Another component of the trans-neptunian region, the so-called scattered Kuiper Belt, has been added in the last years: these objects are characterized by highly eccentric orbits extending to ~ 130 AU and could be planetesimals that were scattered out of the Uranus-Neptune region into eccentric orbits. It is probable that the Kuiper Belt could extend much farther than we know, and that some objects can be found up the Oort Cloud. In addition, many if not all of the so-called short-period comets (orbital periods < 200 years) likely had their origin in the Kuiper Belt, and represent the detectable (via proximity) subclass of what may be an enormous number of such bodies in the 30-100 AU region: the Kuiper Belt region should contain enough bodies to explain the observed rate of short period comets.

Jewitt (1999) argued that the inclination distribution of the trans-Neptunian objects is important because it controls the velocity dispersion of these objects and hence determines whether the collisional regime is erosive or agglomerative. The size distribution of KBOs is not well known, however it should be related to the primordial phases of solar system evolution, even if it was changed by the resonances and the progressive erosion due to different phenomena. The size distribution of a population of objects can be considered a useful diagnostic for understanding the processes that lead to the erosion and/or accretion of planetary bodies. From recent observations and theoretical studies, it is emerging that objects in the trans-neptunian region probably follow a complex size distribution. All the sizes are derived from brightness values and assume that the objects have an albedo of 0.09 (see Minor Planet Center, "Minor Planet Lists", on line, http://cfa-www.harvard.edu/iau/lists/MPLists.html, updated 19 August 2005). Thus, there are at least factor-of-two uncertainties in the estimates of the sizes of KBOs. On 28 July 2005, Santos-Sanz *et al.* (2005) announced the discovery of an object nearly as large as Pluto, now designated 2003 EL61. Brown, Trujillo & Rabinowicz (2005) reported, almost at the same time, the discovery of two large trans-neptunian bodies: 2003 UB313 is clearly larger than Pluto, based on its observed brightness, the other, designated 2005 FY9, is over half the diameter of Pluto.

Due to the distance and the consequent faintness of these objects, the observation is difficult: we have only lightcurves and low resolution spectra, from which we can have an idea of the surfaces. These surfaces are surprisingly different between them (Luu & Sheppard 2002), and probably heterogeneous. The observed color distribution could be attributed to the reddening due to irradiation and subsequent polymerization of the surface carbonized compounds and erosion activity/resurfacing. In this case, a transport mechanism able to bring fresh material on the surface is required.

Nothing we know about the interior of KBOs (and Centaurs), if not by extrapolation on what we know about the comets, which should originate from them. It is highly possible that the largest ones between the KBOs be in some way internally differentiated, due to the high content of refractory and hence of radioactive nuclei releasing heat in the whole nucleus, but the dividing line between objects large enough to be differentiated and those that should not is unknown. Collisions too could have had a role in thermally processing the interiors.

As far as the masses (and densities) are concerned we do not have data, with the exception of Varuna for which a density has been estimated (Jewitt & Sheppard 2002). Following the analysis of the lightcurve, the body could be a rotationally distorted rubble-pile, so it would be porous at an unknown scale and low density (~ 1000 kg/m^3). On the contrary, surprisingly, Jewitt & Luu (2004) discovered that Quaoar spectrum seems to indicate the presence on the surface of crystalline ice. Crystalline ice is formed only at temperatures above 110 K, well above the present temperature of Quaoar that, according with the previously cited authors, is about 50 K. The interpretation of this observation could be a strong indication of an interior activity of this large object leading to the generation of ice volcanism, similar to the one presently observed on Enceladus, or to the recent(?) exposition of the underneath layers of crystalline ice, being the first layers of amorphous ice removed by impact. Another possibility is that the ice on the surface could have been heated above 110 K by micrometeorite impacts. In the first case, the crystallinity could be an indication of the differentiation which the object undergoes, probably due to the combined effects of radioactive decay, primordial bombardment and compaction due to the body self-gravity. Therefore we cannot exclude that in some cases the large KBOs could resemble Triton more than a dead icy body.

However, the impact hypothesis cannot be excluded since the role of impacts in the Kuiper Belt has been relevant, as supported by the observation that the number of binaries is surprisingly high. The formation of binaries is explained by two competing theories. One entails the physical collision of bodies (Weidenschilling 2002) while the other utilizes dynamical friction or a third body to dissipate excess momentum and energy from the system (Goldreich, Lithwick & Sari 2002). In both cases the formation of multiple systems asks for a higher density of the KBO disk, that allowed the formation of binary and multiple bodies (Nazzario & Hyde 2005). This implies that the probability of collisions was higher than the present one. As noted by Funato, Makino, Hut *et al.* (2004) this will also allow direct mass determination in a near future.

It is to be stressed again that KBOs observation is indicating a contradictory situation: from one side the analysis of Varuna (Jewitt & Sheppard 2002) seems to indicate a porous interior, while the presence of crystalline ice on Quaoar spectrum seems to indicate a differentiation process, leading to compaction and differentiation.

3. Centaurs

A growing number of bodies, referred to as Centaurs, have been identified with orbits crossing those of Saturn, Uranus, and Neptune. These bodies can be seen as transition bodies between the KBOs and the comets (Levison & Duncan 1994, Hahn & Bailey 1990): the fact that their orbits, on the basis of dynamical calculations, are not stable over the lifetime of the solar system, suggests that the Centaurs were formerly residing in the Kuiper Belt and only recently have been delivered into their current orbits. The over-all appearance, based on photometrical measurements, is also not in contrast with this hypothesis: the colour diversity and redness of the Centaurs match those of KBOs. This common origin with KBO makes the Centaurs very interesting, because they could pro-vide compositional information on the more distant Kuiper Belt objects and information about their subsequent processing.

The estimated diameters are between 20 and 200 km, from Earth-based telescopic ob-servations at thermal wavelengths and assuming low visual albedo. Three of these objects (Chiron, Pholus, and 1993 HA2) are large and bright enough for ground-based telescopic spectral observations at visual and near-infrared wavelengths (0.4 to 1.0 microns). Chiron

(Hartmann *et al.* (1990); Meech & Belton 1990) and 5145 Pholus (Tegler, Romanishin, Consolmagno *et al.* (2005)) have independent assessments of their diameters. Chiron is a uniquely complex body, apparently undergoing sporadic outbursts of activity that may be driven by the sublimation of CO. Chiron has a flat reflectance spectrum similar to that of the C-type asteroids. Thermal-infrared observations suggest that its visual albedo appears to be low (0.04 to 0.1), further indicating a similarity to the C-type asteroids.

In marked contrast to Chiron's bluish colour, the visible reflectance of 5145 Pholus and 1993 HA2 is extremely red and exhibits steep upward slopes toward longer wavelengths. Unfortunately, the red slope alone is insufficient to identify a specific solid material; the presence of several distinctive absorption features has led to the hypothesis that the surface of Pholus is composed of a mixture of H_2O ice and a variety of organic materials. The organic materials suggested to date include light hydrocarbons or methanol ice, tholins similar to those hypothesized on Titan, polymeric HCN and carbon black. The presence of light hydrocarbons suggests that Pholus has been less chemically processed than comets and asteroids. Intriguingly, extrapolations of orbital calculations suggest that the current orbit is dynamically new and that Pholus may have arrived recently from the Kuiper Belt.

This is in agreement with the data recently collected by the VIMS Sperctrometer on the Cassini mission, that have found a complex chemistry on the surface of Phoebe. The origin of Phoebe, which is the outermost large satellite of Saturn, is of particular interest because its inclined, retrograde orbit suggests that it was gravitationally captured by Saturn, having accreted outside the region of the solar nebula in which Saturn formed. By contrast, Saturn's regular satellites (with prograde, low-inclination, circular orbits) probably accreted within the sub-nebula in which Saturn itself formed. The imaging spectroscopy of Phoebe shows ferrous-iron-bearing minerals, bound water, trapped CO_2, probable phyllosilicates, organics, nitriles and cyanide compounds. The detection of these compounds on Phoebe makes it one of the most compositionally diverse objects yet observed in our solar system. It is likely that Phoebe's surface contains primitive materials from the outer solar system, indicating a surface of cometary/KBO origin (Clark *et al.* (2004)).

An origin in the Kuiper Belt can probably also be assumed for a body such as Triton, and the identification of Pluto and Charon as the two largest known Kuiper Belt objects has been discussed (Cruikshank 2005). The identification of these large outer solar system bodies as KBOs provides a simple and consistent framework for their formation in the outer solar system: in this picture, Pluto and Triton formed in the trans-Neptunian region and became two of its largest members. Suggested capture mechanisms include a collision with an original satellite (Goldreich *et al.* (1989), or gas drag in a proto-Neptunian nebula (McKinnon & Leith 1995).

4. Comets

In the last years space missions such as Deep Space 1, Stardust and very recently Deep Impact gave us wonderful pictures of comet surfaces, but unfortunately they did not give us information on their interior structure. Most of our knowledge about comets is still based on the results of Giotto that studied the comet 1P/Halley in 1986, and on the data from two unusually active recent comets: C/1995 O1 (Hale-Bopp) and C/1996 B2 (Hyakutake). The imaged surfaces of 19P/Borrelly, 81P/Wild 2 and 9P/Tempel 1, all short period comets, are strikingly different each other, supporting the idea that, when looked from nearby, comets are diverse, probably due to the several paths that their internal evolution can follow only changing a few key parameters, such as the amount of volatiles and the presence or not of a refractory crust. Moreover there is a strict dependence of the comet evolution on the orbital path that it follows, and this evolution can deeply change its surface appearance (Coradini, Capaccioni, Capria *et al.* (1997a); Coradini, Capaccioni, Capria *et al.* (1997b)). We shall expect an even larger variability when we will be able to image the surface of a long period comet. Deep Impact, that successfully hit with a cupper projectile the nucleus of 9P/Tempel1, will possibly give us some hints at least about the composition of the layers close to the surface, when the analysis of the data will be completed. To the moment of this writing, there are no firm conclusions about the composition and density of these layers.

We have many ideas but not many constraints on the interior structure of cometary nuclei. What little we know about the interior of comet nuclei is inferred from observations and from our limited knowledge of the way in which comets accreted in the protoplanetary nebula, although we should keep in mind that the behavior could be history dependent and that the abundance ratio of ejected volatiles in the coma does not represent the nucleus abundances (Huebner & Benkhoff 1999). The Jupiter Family comets originated in the zone of the solar system presently known as Kuiper Belt and were successively scattered closer to the Sun, so they should have been formed in a colder environment with respect to the long period comets, but it is still unclear if this can bring any difference in their composition and properties. Whereas the cometary surface and the layers immediately behind are somewhat processed during each perihelion passage, the interior could be much more primordial. It is well possible that most nuclei are not homogeneous bodies: chemical and physical inhomogeneities should be present as a consequence of formation conditions and differentiation processes. Compositional differences could exist between the various populations of comets depending on their formation zone and subsequent processing. Even if a comet is born or find itself, at some stage of its existence, relatively homogeneous from a compositional point of view, thermal evolution will soon change this situation: diurnal heat wave penetrates for few centimeters under the surface, but seasonal and orbital heat waves penetrate much more deeply, depending on the thermal properties of the local matter.

As already written, we know something more about comets interior, in particular about density, than about the interior of the other bodies that are the subject of this review. In what follows we will review what we know about the porosity and density of comet nuclei.

4.1. *Comets are porous and low density objects*

It is usually assumed, both in interpreting observations and in modeling the evolution of nuclei, that comet nuclei are porous, low-density objects. The reasons for which we assume that comet matter is porous have been listed by Benz & Asphaug (1999): 1) It would be impossible to sustain the observed sublimation of volatile ices if only the volatiles present on a surface layer would be sublimating; 2) The analysis of non-gravitational forces acting on comet 1P/Halley indicated that the mass of the comet nucleus must be less than the mass of that body if it consisted of compacted material. The resulting density indicates that the nucleus is porous. 3) The material created during KOSI experiments (Laemmerzahl, Gebhard, Gruen *et al.* (1995)) was very porous; CO_2 gas was released from under the surface and diffused through the pores into the vacuum surrounding the experiment. To this list, it could be added that nucleus models explain very well comet activity assuming that gas diffusion is taking place in a porous medium (see section 6).

The bulk density is one of the most elusive, yet important, physical properties of a cometary nucleus. The methods that can be used for mass or bulk density estimates are mainly based on tidal or rotational break-up and non-gravitational force modeling. For example, it is possible to utilize the fact that a body needs a certain self-gravity and material strength in order to withstand the centrifugal force due to rotation. Davidsson (2001) derived analytical expressions for the critical break-up period, by assuming oblate or prolate nucleus shapes, and by balancing gravitational, material, and centrifugal forces. If the size, shape, and rotational period of a comet are estimated observationally, these expressions can be used to derive a lower limit on the density. An interesting attempt to obtain hints on the density of real bodies has been made by Davidsson & Gutier-rez (2004) and 2005)by Davidsson & Gutierrez (2004). By requiring that the model body simultaneously reproduces the empirical nucleus rotational lightcurve, the water production rate and in particular the non-gravitational changes induced on the orbital parameters, the authors are able to estimate the density of the nucleus. In the case of 67P/Churyumov-Gerasimenko they obtain a value ranging from 100 to 600 kg/m^3; in the case of 19P/Borrelly they obtain a narrower range, 100-300 kg/m^3. However, an assumption has to be made regarding the tensile strength, e.g., considering it as a function of density or simply applying a constant value. By assuming zero strength, Davidsson (2001) analyzed a sample of 14 comets, and seven objects needed, to remain intact, densities in a range which coincides with the density range considered as likely for comets (Rickman 1989; Asphaug & Benz 1996).

Another method can be applied to comets which have an observable perihelion delay or advance, caused by the non-gravitational force due to nucleus sublimation and outgassing. If the non-gravitational force vector can be calculated with some accuracy as a function of orbital position, the mass of the nucleus can be estimated: then, if the size and shape of the nucleus are known with some accuracy, the density can be estimated.

Observations and subsequent modeling of break-up of cometary nuclei in addition to estimates of the mass and density can also give information about the internal nucleus structure. Cometary disruptions are an interesting opportunity to study unprocessed fragments of the nucleus. The structure and internal strength of the nucleus can be examined by investigations of the disruption process. The most prominent example was comet D/Shoemaker-Levy 9 (e.g., Noll, Weaver & Feldman 1996; Asphaug & Benz 1996) which was disrupted during a close encounter with Jupiter in 1992 and whose fragments finally collided with the planet in 1994. Another example of a comet for which such an analysis has been made is C/1999 S4 (LINEAR) (Weaver *et al.* (2001)). From observations made during the disruption of the comet and the runaway fragmentation that took place between 18 July 2000 and 23 July, Bockelée-Morvan *et al.* (2001) concluded that the relative abundances did not change during the breakup phase: at least in that case the nucleus had a homogeneous composition. Instead, from the observation of the fragmentation of comet C/2001 C2, performed with the SOHO Ultraviolet Coronagraph Spectrometer (UVCS), Bemporad, Poletto, Raymond *et al.* (2005) concluded that the collected evidence suggests that the material of the nucleus tended to break up very easily, and that at least two of the fragments had different composition.

An interesting opportunity to study structure and global composition of comets are sungrazing comets. A large number of them has been discovered by the Large Angle and Spectrometric Coronagraph (LASCO) on the SOlar and Heliospheric Observatory (SOHO) spacecraft. Many of the sungrazing comets observed so far are fragments of a single predecessor, the Kreutz comet (Marsden 1989). Most sungrazers are not observed past perihelion, including all the sungrazing comets detected by SOHO. If the reason is the complete evaporation of the nucleus, then the upper limit for its diameter is a few tens of meters for a nucleus consisting of pure water ice (Weissman 1983). Iseli *et al.* (2002) modeled sungrazing comets to constrain properties like radius, density, and tensile strength starting from the observation that these objects are completely destroyed. The application of the model to sungrazing comets showed that the maximum size of a comet which can be disrupted by sublimation alone is several tens of meters. This value depends on albedo, density, and perihelion distance of the comet, but is nearly independent on its thermal conductivity.

5. Hints from the origin of outer solar system planetesimals

Following Gladman (2005) one can say that our planetary system is embedded in a small-body disk of asteroids and comets, remnants of the original planetesimal population that formed the planets. Once formed, those planets dispersed most of the remaining small bodies: therefore, if we want to understand the internal composition of these bodies we have to understand the mechanisms that were responsible for the formation of this disk of objects.

During the last two decades, our understanding of the conditions under which planetesimals formed in the outer solar system has improved significantly. First of all we have to consider what could have been the composition of these bodies: this depends on the chemical evolution of the protosolar nebula at distances larger than 20 AU. The outer solar system is dominated by ices of different kind, being H_2O ice the most important. The kind of chemistry strongly depends on the reference model of the protosolar nebula. Our understanding of the chemical processes taking place in the primitive solar nebula has increased considerably as more detailed models of the dynamic evolution of such nebulae have become available. Early models (Grossman 1972) assumed that a mixture of hot gases present in the solar nebula cooled slowly maintaining thermodynamic equilibrium. At the beginning the more refractory vapors condensed, followed by the lower melting point materials. The model suggested that the major textural features and mineralogical composition of the Ca, Al-rich inclusions in the C3 chondrites were produced during condensation in the nebula characterized by slight departures from chemical equilibrium due to incomplete reaction of high temperature condensates. Fractionation of such a phase assemblage is sufficient to produce part of the lithophile element depletion of the ordinary chondrites relative to the cosmic abundances. This result is surprisingly good, given the very strong assumption of thermodynamic equilibrium made by the author. Morfill *et al.* (1985), instead, introduced the effect of localized turbulence that should be present when viscous accretion disks are considered. This information is used to develop a transport theory for dust and gas phases. In the paper the possible modifications by intermittent turbulence are discussed, chemical fractionation effects are analyzed, and the heterogeneity on small scales as well as the homogeneity on large scales of the primitive bodies in the solar system is examined within the framework of their theory. It is concluded that a turbulent protoplanetary nebula, if the turbulence is intermittent, may also provide the fastest means of growing planets from the solid dust component. The authors do not discuss about the structure of these aggregations, however it is clear that the suggested mechanism lead to very loose aggregates.

Also other authors, as Fegley & Prinn (1989), challenged the idea that the nebula was quiescent demonstrating that even major gas phase species such as N_2 and NH_3 could fail to achieve equilibrium due to the low temperatures and the concurrent slow chemical reaction rates in the region of the outer planets. At the low temperatures characteristic of the outer solar system, kinetics may mean that carbon remains as CO, since less oxygen is available to form water ice. Predicted rock/ice mass ratio in this case is 70/30, which gives a density of $2000 \ kg/m^3$, similar to that observed for both Triton and Pluto. In the hotter nebula, carbon tends to be incorporated in CH_4 and the oxygen is then available to form water ice; rock/ice mass ratio in this case should be close to 1, giving a density of $1500 \ kg/m^3$. Detection of CO is also consistent with low temperatures during the formation of bodies such as Triton, Pluto and other mainly icy bodies. However Fegley & Prinn (1989) point out that several processes can overlap modifying the original cometary chemistry, as a certain mixing of the protosolar nebula material with material formed in circumplanetary nebulae, as the Jovian and Saturnian ones, homogeneous and heterogeneous thermochemical and photochemical reactions, and disequilibration resulting from fluid transport, condensation, and cooling. Therefore, the interplay between chemical, physical, and dynamical processes should be taken into account if one wants to decipher the origin and evolution of the abundant chemically reactive volatiles (H, O, C, N, S) observed in comets.

This kind of considerations can be the basis to infer the possible composition of KBOs and of comets, in which we expect therefore to find a large amount of volatiles, carbon compounds such as CO being the dominant species, but a small amount of CH_4 of circumplanetary nebulae cannot be excluded. N_2 is also more probable than NH_3. The Halley data on the CO/CH_4 and N_2/NH_3 that are intermediate between those typical of the interstellar medium and those expected in a hotter nebula, seem to support this hypothesis. The original chemical evolution is only responsible for the initial chemistry of icy bodies, the further evolution could have partially altered it.

The process of agglomerate formation by gradual accretion of sub- millimeter solid grains has been studied both experimentally and numerically (e.g., Donn & Duva 1994; Blum *et al.* (2000)), and this investigation is in agreement with the idea that the primordial solar nebula was a suitable environment for the production of ice-rich grain clusters with a highly porous and fractal structure. These objects are accumulation of fluffy aggregates. If so, we have to expect that the present comets are remnants of this very primordial situation. The subsequent growth of these small clusters has been investigated by the use of sophisticated numerical modeling (Weidenschilling 1997), up to the point where 10 km sized planetesimals are formed. The size distribution of cometesimals growing by drag-induced collisions develops a narrow peak in the range tens to hundreds of meters. This occurs because drag-induced velocities decrease with size in this range, while gravitational focusing is negligible. Impact velocities have a minimum at the transition from drag-driven to gravitational accretion at approximately kilometer sizes. Bodies accreted in this manner should have low mechanical strength and macroscopic voids in addition to small-scale porosity. They will be composed of structural elements having a variety of scales, but with some tendency for preferential sizes in the range about 10100 m. In terms of internal structure, these bodies are very similar to the comet nucleus models proposed by Donn (1991) and Weissman (1986), known respectively as fractal and primordial rubble pile models. According to the fractal model, a comet is made of small cometesimals which are bound toghether by the gravitational force that, for these small aggregates, is pretty weak. Macro-porosity with large internal void spaces is therefore superimposed to the micro-porosity foreseen by the Donn (1991) model. The primordial rubble pile model is very similar to the fractal model, except that the cometesimals are more tightly packed, and welded together by collision-induced evaporation and the subsequent freezing of ice.

Recently laboratory simulation experiments were performed using micron-sized dust particles, impacting solid targets at various velocities. Again the result is consistent with the formation of open aggregates (Blum *et al.* (2000)). Slow bombardment of the target generally results in the formation of fluffy dust layers. At higher impact velocities, compact dust-layer growth is observed. Above a certain collision energy, the dust aggregates are disrupted. The Blum *et al.* (2000) experimental results suggest that aggregates restructuring in the solar nebula become an important process when the diameters of the dust agglomerates exceed a few centimeters. Dust aggregates below that size are not subjected to impact compaction. It has also been shown that heating and evaporation during a collision are rather limited even for collisions between large (~ 100 m) cometesimals, eventhough local thermal and possibly chemical alterations cannot be excluded (as in the primordial rubble pile model). Furthermore, bodies with sizes below a few tens of kilometers are not affected by gravitational compression. As a result, comets can be seen as low-density objects, formed slowly at low temperature, but characterized possibly by a complex internal structure which can allow their fragmentation under high-medium velocity impact conditions. Larger bodies, however, shall undergo different histories, due to the contribution of short (as ^{26}Al) and long-life radioactive decay, degassing and impact compaction.

6. Hints from thermal evolution modeling: Kuiper Belt objects

The thermal evolution modeling of Kuiper Belt objects has been dealt with two different kinds of models, corresponding to two different heritages and in turn to two different points of view: models originally developed for comet nuclei and models developed for icy satellites. One could say, following McKinnon (2002), that in one case we are scaling up from the traditional small cometary sizes, while in the other case we are starting from mid-sized icy satellites and moving downward to smaller sizes.

6.1. *Inheritage from comets modeling*

If we think to objects also larger than comets, like KBOs, as porous, ice-rich bodies, it is quite straightforward apply to them the models initially developed to study the thermal evolution of cometary nuclei. In fact it is very difficult to draw a clear line beyond which objects of a certain size begin to be compacted and differentiated: in fact, if for a (almost) homogeneous icy body we can follow the thermal and differentiation history, for a non-compact body characterized by heterogeneities and macro porosity it is more difficult. For this reason the approach commonly used for comets can be applied to a large variety of objects. This has been done by, for example, Capria, De Sanctis, Coradini *et al.* (2000), De Sanctis, Capria, Coradini *et al.* (2000), De Sanctis, Capria & Coradini (2001), Choi, Cohen, Merk *et al.* (2002), Choi, Brosch & Prialnik (2003), Merk, Breuer & Spohn (2002), Merk & Prialnik (2003). The underlying idea is that, if the link between comets and KBOs is real, then the comets observed properties can be used to constrain KBO models, including low formation temperature, low density (high porosity) and high volatile content; this means, in turn, that it is possible to study both kinds of bodies with the same theoretical models.

In order to study the thermal evolution and differentiation of porous, ice-rich bodies many models have been developed in the last years: a very complete discussion on the subject and many references can be found in the book by ISSI Comet Nucleus Team (2005); we will give here only few details.

In the currently used thermal evolution models, heat diffusion and gas diffusion equations are solved in a porous medium, in which sublimating gas can flow through the pores. A mixture of ices and dust is considered, and the flux from surface and subsurface regions is simulated for different gas and dust compositions and properties. The temperature on the surface is obtained by a balance between the solar energy reaching the surface, the energy re-emitted in the infrared, the heat conducted to the interior and the energy used to sublimate surface ices. When the temperature rises, ices can start to sublimate, beginning from the more volatile ones, and the initially homogeneous nucleus can differentiate giving rise to a layered structure in which the boundary between different layers is a sublimation front. Due to the larger, with respect to comet nuclei, sizes of

Kuiper Belt bodies, and to the consequently higher content of refractories, the heating effect of radiogenic elements, both short and long-lived, is usually taken into account. So, these models consider two heating sources of comparable importance, one acting from the surface (solar input) and one present in the whole body: this can give origin to more complex thermal evolution patterns than in the case of comet nuclei.

To give an example, some results from a thermal evolution model that can be applied both to comets and to larger, denser bodies, developed by our group (Capria, De Sanctis, Coradini *et al.* (2000); De Sanctis, Capria, Coradini *et al.* (2000); De Sanctis, Capria & Coradini 2001) will be here briefly described.

The model is one-dimensional. The spherical nucleus is composed by ices (water, CO_2 and CO) and a refractory component. Water ice can be initially amorphous, and in this case more volatile gases can be trapped in the amorphous matrix and released during the transition to crystalline phase. The refractory material is described as spherical grains with given initial size distributions and physical properties. Energy and mass conservation is expressed by a system of coupled differential equations, solved for the whole nucleus:

$$\rho c \frac{\partial T}{\partial t} = \nabla[K \cdot \nabla T] + \sum_{i=1}^{n} Q_i + Q_{am-cr} + Q_{rad} \tag{6.1}$$

$$\frac{1}{RT} \frac{\partial P_i}{\partial t} = \nabla[G_i \cdot \nabla P_i] + Q\prime_i \quad i = 1, n$$

where T is the temperature, t the time, K the heat conduction coefficient, ρ the density of the solid matrix, c the specific heat of the material, Q_i is the energy exchanged by the solid matrix in the sublimation and recondensation of the i_{th} ice, Q_{am-cr} is the heat released during the transition from amorphous to crystalline form, Q_{rad} is the energy released by the decay of radioisotopes, R is the gas constant, P_i the partial pressure of component i, G_i its diffusion coefficient, and $Q\prime_i$ is the gas source term due to sublimation-recondensation processes.

For the radiogenic heating, the effects of ^{40}K, ^{232}Th, ^{235}U,^{238}U radioisotopes, and in some cases of ^{26}Al have been considered. The rate of radioactive energy release, Q_{rad}, is given by

$$Q_{rad} = \rho_{dust} \Sigma \lambda_j X_{oj} exp^{-\lambda_j t} H_j \tag{6.2}$$

where ρ_{dust} is the bulk dust density, λ_j is the decay constant of the $j\prime th$ radioisotope, X_{oj} is its mass fraction within the dust, and H_j is the energy released per unit mass upon decay. The amount of radioisotopes in cometary nuclei is unknown and there is no way to measure it; we are assuming that the abundances of ^{40}K, ^{232}Th, ^{235}U,^{238}U are in the same proportion as in the C1 chondrites (Anders & Grevesse 1989).

This model has been applied to study the evolution of KBOs and Centaurs (Capria, De Sanctis, Coradini *et al.* (2000); De Sanctis, Capria, Coradini *et al.* (2000); De Sanctis, Capria & Coradini 2001). Here we will describe the results of this model applied to two different kinds of body, corresponding to two different hypotheses on the composition and internal structure of KBOs: a body whose composition and density are inherited from the typical ones of comet nuclei, and another one much more dense and rich in refractories.

6.2. *Dense and ice-rich Kuiper Belt objects*

We have applied the model to a typical Kuiper Belt body and have simulated its evolution under two different hypotheses, high and low density, in order to show possible differences in the evolution history. The two models were ran using as much as possible the same set of input parameters, and were followed on the same orbit for the same number of revolutions. The only difference in the two models is the initial density, to study the effect of this parameter in a typical thermal evolution.

Most of the model parameters assumed as reference are the values normally used for cometary nuclei composition. The body has a radius of 450 km, it is made by dust and ices of water and CO (CO/H_2O = 0.01) and the initial temperature is 20 K throughout the whole nucleus. The ice is initially amorphous. A value of 1000 kg/m^3 for the dust density simulates the fact that grains are the result of an accumulation process and are therefore highly porous. Pore radius has been fixed to 10^{-5} m. The orbit has a semimajor axis of 43 AU and an eccentricity value of 0.05. The only difference in the two models is the porosity, 0.8 for the "light" body and 0.3 for the "heavy" body. This gives origin to a density of nearly 400 kg/m^3 in the first case and of nearly 950 kg/m^3 in the second case. Both models were ran with and without the most important short-lived radioisotope, ^{26}Al, in the dust composition.

After several millions of years, depending on the amount and kind of radioisotopes in the models, the CO front reaches a quasi-stationary level in both models. The combined effect of radiogenic and solar heating -the latter coming from outside and the former uniformly distributed through the whole nucleus- leads to an increase of the overall temperature of the nucleus. The sublimation of CO ice has an important role in determining the abrupt slope change present in the thermal profile as soon as the quasi-stationary situation is reached. The thermal evolution of a higher and a lighter density body proceeds in a similar way, however the increase in the internal temperature is higher in the first case: in about 10^4 years the central temperature increases from 20 to 30 K due to the decay of ^{26}Al. When this radioisotope is absent, the increase of temperature is negligible for both kind of bodies, high and low density. The degassing of high density KBOs proceeds faster than for icy bodies. In the figure the internal temperature profile for a ^{26}Al rich (continuous line) and a ^{26}Al poor (dashed line) body are shown in the case of a high density composition: if the ^{26}Al is not contained in the bulk initial composition, the internal temperatures do not reach values sensibly higher than the initial ones. In the case of lower density composition, the figure (and the two curves) would be very similar, but much lower temperatures are reached, even with the presence of ^{26}Al.

It should be taken into account that, while the internal composition can be very different in the two hypotheses, these differences would be in any case hardly recognizable from the observation of the surface.

6.3. *Inheritage from satellite modeling*

Following McKinnon (2002), another analogous that can be used for the modeling of KBOs are the captured satellites such as Triton, the Pluto-Charon system and possibly Phoebe. In this case, the internal density assumed will be much higher than the one needed to explain the cometary activity, also according with recent measurements (9P/Tempel 1 guessed density should be very low, of the order of .5 g/cm^3).

The model by McKinnon (2002) is a spherically symmetric, conductive heat transport model with radiogenic heating and temperature dependent conductivities and heat capacities. Internal transport of volatiles is not considered; accretional heating is considered, along with radioactive heating. The composition is rock and ice, with a rock/rock+ice ratio of 0.7. The porosity is lower than in the traditional comet models, on the basis that it would be impossible for large KBOs have the typical comet porosities, due to the presence of hydrostatic pressure: rearrangement of the internal structure due to hydrostatic pressure is taken into account in the model. The body considered is at the threshold of

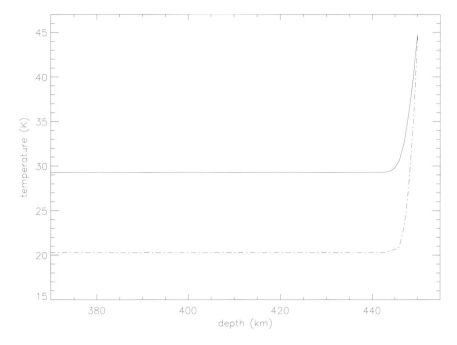

Figure 1. Internal temperature profile for a ^{26}Al rich (continuous line) and a ^{26}Al poor (dashed line) body with high density

pressure-induced densification. As far as regards the results, for a body with a radius of 100 km central temperatures reach 80 K in 50 Myr, then the body is slowly cooling. Irrespective of conductive models, bodies greater than 200 km in radius should have crystalline interiors: the critical size for large scale crystallization lies in the range 75-225 km. Larger KBOs have undergone chemical and structural evolution, whereas smaller ones could be considered cosmochemically primordial.

7. Hints from thermal evolution modeling: comets

The purpose of modeling comet nuclei is not to predict their behavior on the basis of an initial set of parameters, but to reproduce the observed behavior and thereby derive internal properties and processes characteristic of comet nuclei that are inaccessible to observations (see the book by ISSI Comet Nucleus Team (2005)). The general conclusion that emerges from simulations of the evolution of comet nuclei is that, essentially, a nucleus model of porous, grainy material, possibly made of gas-laden amorphous ice and dust, is capable of reproducing activity patterns of comets. From the results of these models, there are some general characteristics that may be expected of comets interiors: loss of the most volatile ices, stratified composition and inhomogeneous structure.

Calculations of the long-term evolution of comets far from the Sun, under the influence of radioactive heating, show that the internal temperatures attained may be quite high, at least several tens kelvin. As a result, comets may have lost volatiles that sublimate below about 40-50 K, that were initially included as ices, and, partially, less volatile species. Observation of such volatiles in comets suggests that they originate from amorphous H_2O ice undergoing crystallization or that radioactive heating was ineffective or did not occur. As far as regards the structure, while the inner part may have been altered by early evolution, the outer layers are altered by recent activity. Thus, the internal composition of comet nuclei is stratified, with increasingly volatile species at increasingly greater depths. Similarly, the internal structure of comets is very likely not uniform: density, porosity, H_2O ice phase and strength vary with depth. Increased porosity arises from volatile depletion, decreased porosity from recondensation.

Particularly in the case of comets, the dynamical history can have a strong influence on the resulting internal structure and composition of the body. Studying the evolution of short period comets, we always tried to simulate their dynamical history (Coradini, Capaccioni, Capria *et al.* (1997)) and we noticed that it has a strong effect. The dynamical evolution that brings a body from the Kuiper Belt region to the inner solar system is very complex (Stagg & Bailey 1989): this process (called multi-capture) involves the reduction of the revolution period, and this can be due to more than a close encounter with a giant planet in succession, beginning with Neptune and ending with Jupiter. The transfer from a Kuiper Belt orbit to a Jupiter family orbit lasts for a very long time, and can be reverted or stopped in any moment. This means that a body presently found on a short period orbit has a long dynamical history and was not injected directly from the Kuiper Belt. For this reason, a body displaced in the inner solar system will be already differentiated, at least in the layers close to the surface (De Sanctis, Capria & Coradini 2001). Differentiation processes due to volatile sublimation begin far from the inner part of the solar system; it is known that even a body on a Centaur-like orbit can be active (Womack & Stern 1999; Capria, Coradini, De Sanctis *et al.* (2000)).

8. The comet paradox

From all the above discussion, it is clear that when we analyze the bodies originated in the outer solar system we are facing a kind of paradox: we have from one side small, fragile, volatile-rich and low-density objects, and from the other side a population of much denser, larger and probably less ice-rich objects; we know that a dynamical link exists between them, but how can we explain this kind of dichotomy? There are, and there were from the beginning two kinds of population, or do they exist physical processes able to transform (or derive from) objects belonging to the Kuiper Belt in the typical comet nucleus? Nor the thermal evolution models nor the formation models can, to the moment, explain this paradox. Obviously there are several effects that depend strongly on the mass of the body, as can be seen by the analysis of planetary satellites, but it is difficult to understand to which extent the internal structure depends on the size and original density of the body. A process not yet discussed and that could shed some light on this dichotomy is the collisional evolution.

Let us imagine a large KBO still, at least partially, volatile-rich: could an impact produce, from this body, volatile rich fragments that could eventually become "real comet nuclei" as we know them? We need to make two assumptions: first, that large, differentiated objects exist (or existed) in the Kuiper Belt, and second, that an impact in this zone is able to shatter a body without completely destroying the volatile content of the fragments. As we saw in the preceding section the first assumption could not be a problem, and maybe impacts themselves contributed to this differentiation; the second assumption must be checked. As far as regards the impact rate, the dissipation time of the heat wave is of the order of millions of years, comparable with the impact rate computed by Farinella, Davis & Stern (2000). Can heat build up from impact to impact so as to trigger some global differentiation process? We already know that the collisions had an important role in shaping the Kuiper Belt, and that those collisions happened at quite a low relative velocity and were for this reason on average less disruptive than in the case of the asteroid main belt.

The role of the impacts in the thermal evolution of the bodies can be better understood using a heat conduction and gas diffusion model for comet nuclei of the kind described above (Orosei, Coradini, De Sanctis *et al.* (2001)): in particular, we could determine if the collisional non-disruptive history of KBOs is able to cause significant alterations in their volatile content. For certain combinations of model parameters, it has been found that the internal temperature of the body can easily reach a temperature of 180 K by the end of the accretion process; the outer layers of the body can be depleted in volatiles. Anyway, the outcome of the computations in the model is strongly dependent on uncertain physical parameters, such as thermal conductivity, radiogenic elements content and impact rate of accreting planetesimals, so it would be difficult to give a final answer on the basis of modeling.

9. A (possible) conclusion

We have seen that, while comet models explain quite well the observed properties of nuclei, KBO models can give very different outcomes depending on the assumptions made on ill constrained parameters such as ice fraction in the initial composition, porosity, density, thermal conductivity, radiogenic elements and dynamical evolution. The upper-size limit for which volatile-rich objects could have been differentiated is still unconstrained; measurements of the OPR (ortho para ratio) in water and other hydrogenated species is suggesting that comets have been preserved at low temperatures (Binzel *et al.* (2003)). If the link between comets and KBOs is real, the results of comets observation should be taken into account when constraining KBO models, in particular when dealing with low formation temperature, low density (high porosity) and high volatile content. Are two populations of KBOs (small, low-density and large, high-density objects) present from the beginning or are comets formed by fragmentation of large bodies only partially de-volatilized and characterized by porosity gradients? We need more data to clarify (or complicate) the problem.

How could we in a near future sample and measure the interiors of these bodies? In the case of masses and densities, it is in some sense simpler: we could use the dynamical properties of binaries systems, for example. To sample the interior properties is much more difficult. An interesting opportunity could be in situ measurements, from drilling to more destructive techniques, but they are probably not feasible in a near future.

Acknowledgements

We would like to acknowledge the useful comments of Julio Fernandez who helped us to improve the manuscript.

References

Anders, E. & Grevesse, N. 1989, *Geochim. Cosmochim. Acta.* 53, 197
Asphaug, E. & Benz, W. 1996, *Icarus* 121, 225
Bemporad, A., Poletto, G., Raymond, J.C., Biesecker, D.A., Marsden, B., Lamy, P., Ko, Y.-K., & Uzzo, M. 2005, *ApJ* 620, 523
Benz, W. & Asphaug, E. 1999, *Icarus* 142, 5
Binzel, R.P., A'Hearn, M., Asphaug, E., Barucci, M. A., Belton, M., Benz, W., Cellino, A., Festou, M.C., Fulchignoni, M., Harris, A.W., Rossi, A., & Zuber, M.T. 2003, *PSS* 51, 443

Blum, J., Wurm, G., Kempf, S., Poppe, T., Klahr, H., Kozasa, T., Rott, M., Henning, T., Dorschner, J., Schrpler, R., Keller, H.U., Markiewicz, W.J., Mann, I., Gustafson, B.A., Giovane, F., Neuhaus, D., Fechtig, H., Grn, E., Feuerbacher, B., Kochan, H., Ratke, L., El Goresy, A., Morfill, G., Weidenschilling, S. J., Schwehm, G., Metzler, K., & Ip, W.-H. 2000, *Phys. Rev. Lett.* 85, 2426

Bockelée-Morvan, D., Biver, N., Moreno, R., Colom, P., Crovisier, J., Gerard, E., Henry, F., Lis, D.C., Matthews, H., Weaver, H.A., Womack, M., & Festou, M.C. 2001, *Science* 292, 1339

Brown, M.E. 2005, *IAU Circ.* 8610, 1

Brown, M.E., Trujillo, C.A., & Rabinowitz, D. 2005, *IAU Circ.* 8577, 1

Capria, M. T., De Sanctis, Coradini, A., & Orosei, R. 2000, *Astron. J* 119, 3112

Choi, Y. J., Cohen, M., Merk, R., & Prialnik, D. 2002, *Icarus* 159, 83

Choi, Y. J., Brosch, N., & Prialnik, D. 2003, *Icarus* 165, 101

Clark, R. N., Brown, R. H., Jaumann, R., Cruikshank, D. P., Nelson, R. M., Buratti, B. J., McCord, T. B., Lunine, J., Baines, K. H., Bellucci, G., Bibring, J.-P., Capaccioni, F., Cerroni, P., Coradini, A., Formisano, V., Langevin, Y., Matson, D. L., Mennella, V., Nicholson, P. D., Sicardy, B., Sotin, C., Hoefen, T. M., Curchin, J. M., Hansen, G., Hibbits, K., & Matz, K.-D. 2004, *Nature* 435, 66

Coradini, A., Capaccioni, F., Capria, M.T., De Sanctis, M.C., Espinasse, S., Orosei, R., & Salomone, M. 1997, *Icarus* 129, 317

Coradini, A., Capaccioni, F., Capria, M.T., De Sanctis, M.C., Espinasse, S., Orosei, R., & Salomone, M. 1997, *Icarus* 129, 337

Cruikshank, D. 2005, *Space Sci. Rev.* 116, 421

Davidsson, B. 2001, *Icarus* 149, 375

Davidsson, B.J.R. & Gutierrez, P.J. 2004, *Icarus* 168, 392

Davidsson, B.J.R. & Gutierrez, P.J. 2005, *Icarus* 176, 453

De Sanctis, M. C., Capria, M. T., Coradini, A., & Orosei, R. 2000, *Astron. J.* 120, 1571

De Sanctis, M. C., Capria, M. T., & Coradini, A. 2001, *Astron. J.* 121, 2792

Donn, B. 1991, in: *Comets in the post-Halley era*, vol. 1, p. 335

Donn, B. & Duva, J.M. 1994, *Astophys. & Space Sci.* 212, 43

Farinella, P. & Davis, D.R. 1996, *Science* 273, 938

Farinella, P., Davis, D.R., & Stern, S.A. 2000, in: V. Mannings, A.P. Boss & S.S. Russell (eds.), *Protostars and Planets IV* (Tucson: Arizona Press), p. 1255

Fegley, B. 1993, in: J.M. Greenberg, C.X. Mendoza-Gomez & V. Pironello (eds.), *The Chemistry of Lifes Origins* (Kluwer), p. 75

Fegley, B. & Prinn, R.G. 1989, in: *The formation and evolution of planetary systems; Proceedings of the Meeting* (Cambridge: Cambridge University Press), p. 171

Fernandez, J.A. 1989, *MNRAS* 192, 481

Funato, Y., Makino, J., Hut, P., Kokubo, E., & Kinoshita, D. 2004, *Nature* 427, 518

Gladman, B. 2005, *Science* 307, 71

Goldreich, P., Murray, N., Longaretti, P.Y., & Banfield, D. 1989, *Science* 1989, 500

Goldreich, P., Lithwick, Y., & Sari, R. 2002, *Nature* 420, 643

Grossman, L. 1972, *Geochim. Cosmochim. Acta.* 36, 597

Hahn, G. & Bailey, M.E. 1990, *Nature* 348, 132

Hartmann, W., Tholen, D.J., Meech, K., & Cruikshank, D.P. 1990, *Icarus* 83, 1

Huebner, W.F. & Benkhoff, J. 1999, *Space Sci. Rev.* 90, 117

Iseli, M., Kueppers, M., Benz, W., & Bochsler, P. 2002, *Icarus* 155, 350

ISSI Comet Nucleus Team 2005, *Heat and gas diffusion in comet nuclei*, International Space Science Institute of Bern, in press

Jewitt, D. 1999, *Annu. Rev. Earth. Planet. Sci.* 27, 287

Jewitt, D. & Sheppard, S.S. 2002, *Astron. J* 123, 2110

Jewitt, D.C. & Luu, J. 2004, *Nature* 432, 731

Kawakita, H., Watanabe, J., Furusho, R., Fuse, T., Capria, M.T., De Sanctis, M.C., & Cremonese, G. 2004, *Astrophys. J.* 601, 1152

Iseli, M., Kueppers, M., Benz, W., & Bochsler, P. 2002, *Icarus* 155, 350

Laemmerzahl, P., Gebhard, J., Gruen, E., & Klees, G. 1995, *PSS* 43, 363

Levison, H.F. & Duncan, M.J. 1994 *Icarus* 108, 18

Licandro, J., Oliva, E., & Di Martino, M. 2001, *Astron. & Astrophys.* 373, L29

Luu, J.X. & Jewitt, D. 2002, *Ann. Rev. Astron. Astrophys.* 40, 63

Marsden, B.G. 1989, *Astron. J.* 98, 2306

McKinnon, W.B. & Leith, C. 1995, *Icarus* 118, 392

McKinnon, W.B. 2002, in: B. Warmbein (ed.), *Proceedings of Asteroids, Comets, Meteors - ACM 2002* (ESA SP-500), p. 29

Meech, K. & Belton, M. 1990, *Astron. J.* 100, 1323

Merk, R., Breuer, D., & Spohn, T. 2002, *Icarus* 159, 183

Merk, R. & Prialnik, D. 2003, *Earth Moon & Planets* 92, 359

Morbidelli, A. 2004, *Science* 306, 1302

Morfill, P., Tscharnuter H., & Volk, H.J. 1985, in: D.C. Black & M.S. Matthews (eds.), *Protostars and Planets II* (Tucson: U. Arizona Press), p. 493

Nazzario, R.C. & Hyde, T.W. 2005, in: *Lunar and Planetary Science XXXVI*

Noll, K.S., Weaver, H.A., & Feldman, P.D. 1996, in: K.S. Noll, H.A. Weaver & P.D. Feldman (eds.), *The Collision of Comet Shoemaker-Levy 9 and Jupiter* (Cambridge: Cambridge University Press), p. 387

Orosei, R., Coradini, A., De Sanctis, M. C., & Federico, C. 2001, *Adv. Space Res.* 28, 1563

Rickman, H. 1989, *Adv. Space Res.* 9, 59

Santos-Sanz, P., Ortiz, J.L., Aceituno, F.J., Brown, M.E., & Rabinowitz, D. 2005 *IAU Circ.* 8577, 2

Stagg, C.R. & Bailey, M.E. 1989, *MNRAS* 241, 507

Tegler, S.C., Romanishin, W., Consolmagno, G.J., Rall, J., Worhatch, R., Nelson, M., & Weidenschilling, S. 2005, *Icarus* 175, 390

Weaver, H.A., Sekanina, Z., Toth, I., Delahodde, C.E., Hainaut, O.R., Lamy, P.L., Bauer, J.M., A'Hearn, M.F., Arpigny, C., Combi, M.R., Davies, J.K., Feldman, P.D., Festou, M.C., Hook, R., Jorda, L., Keesey, M.S.W., Lisse, C.M., Marsden, B.G., Meech, K.J., Tozzi, G.P., & West, R. 2001, *Science* 292, 1329

Weidenschilling, S.J. 1997, *Icarus* 127, 290

Weidenschilling, S.J. 2002, *Icarus* 160, 212

Weissman, P.R. 1983, *Icarus* 55, 448

Weissman, P.R. 1986, *Nature* 320,242

Womack, M. & Stern, S.A. 1999, *Astronomicheskii Vestnik* 33, 187

Asteroids, Comets, Meteors
Proceedings IAU Symposium No. 229, 2005
D. Lazzaro, S. Ferraz-Mello & J.A. Fernández, eds.

© 2006 International Astronomical Union
doi:10.1017/S1743921305006885

Compositional coma investigations: gas and dust production rates in comets

Rita Schulz

ESA Research and Scientific Support Department, ESTEC, Noordwijk, The Netherlands
email: rschulz@rssd.esa.int

Abstract. Although it is presently not possible to extract the true composition of a comet nucleus from its coma composition, the distribution of physical and chemical coma properties among comets may be expedient to establish a comet classification scheme that reflects their origin and/or evolution. Most of the coma species visible in the optical were extensively observed in the past. The analysis of these gas coma constituents, mainly daughter products produced for the most part by photolytic destruction of their parent species, is therefore of major importance, if we want to draw conclusions on diversities and similarities of comets in terms of coma composition on a statistically relevant basis. Hundreds of gas and dust production rates are published, but have never been combined into a single database that would allow identifying whether and how the abundances of coma species differ from comet to comet and how they vary with heliocentric distance and with the number of apparitions. A common database can however only be established if the production rates are re-calculated with a common model and set of parameters.

Keywords. comets: general, astrochemistry

1. Introduction

Comets often show very different phenomenological characteristics, which might indicate the existence of physical and chemical differences of their nuclei. Major efforts have therefore been put over many years into finding criteria for a classification scheme of comets to relate their properties to their place(s) of origin. One that is generally accepted is the classification of comets according to their dynamical characteristics (i.e. dynamically new comets, old long-period comets, young long-period comets, short-period comets of Halley and Jupiter family). Another possibility is the grouping of comets according to their physical and/or chemical properties. Here, the composition of the coma plays an important role, as there is normally no information of the nucleus available. The key to the origin of comets is, however, the composition of their nuclei. Depending on their dynamical characteristics, individual comet nuclei should be in different evolutionary stages. Hence, knowledge of the nucleus composition of comets from different dynamical classes should provide information on how the Solar System was formed, and also lead to clues on the history of the interplanetary environment from the time of planet formation up to the present.

As the composition of a comet nucleus cannot be determined from remote sensing observations, all information has to be derived from coma observations assuming certain physical and chemical conditions in the near-nucleus environment. These conditions vary with heliocentric distance and can a priori be different for different comets. (For instance, the extent of the collision zone is much smaller in weak comets than in very active ones.) Hence, the first step in the compositional analysis of comets is to determine the coma constituents (gas and dust), their abundances, chemistry and dynamics for as many comets as possible. This should be done along the entire part of the orbit for which a comet shows coma activity in order to enable the detection of abundance variations as a function of heliocentric distance.

Several compositional surveys were conducted in the past decades (e.g. A'Hearn and Millis (1980), Newburn and Spinrad (1989), Cochran *et al.* (1992), A'Hearn *et al.* (1995), Fink and Hicks (1996)), which led to conclusions on diversities and similarities of comets in terms of coma composition and gas-to-dust ratio. For the gas coma composition, these surveys focused on the abundances of gas species observable in the optical, mainly daughter products, produced for the most part by photolytic destruction of their parent species. A major breakthrough in the compositional study of comets was achieved when cometary parent molecules became directly detectable by remote sensing observations at sub-mm wavelengths. The investigations of molecules observable at sub-mm and infrared wavelengths are presented and discussed by Crovisier (2005, this issue). This review will give an overview on where we stand in terms of compositional coma analysis of the ensemble of comets concentrating on gaseous daughter products and dust observable in the visible spectral range.

2. Determination of Gas Production Rates

2.1. *The Principles*

The visible spectral range is a good tracer for a number of neutral and ionized gas species as well as the micron sized dust (0.4–0.9 μm) in the cometary coma. The constituents of the gas coma that are observable in the optical, are daughter products, such as OH, CN, NH, C_3, C_2 and NH_2, which are produced, for the most part, by photolytic destruction of their parent species. Many of the above-mentioned species are already known to exist in cometary comae for a very long time. C_2 was already observed in the very first spectrum of a comet which was obtained on 5 August, 1864 by G. Donati. Three emission bands were seen in this spectrum, which W. Huggins later identified as bands of the Swan system of C_2 (d $^3\Pi_g$ – a $^3\Pi_u$). He also took the first photographic record of a cometary spectrum, in which C_3 and CN could be identified as well (Huggins, 1982). The detection of NH_2 was first reported by Swings.

From the radial distribution of the neutral gas species in the coma, direct information about their production and destruction mechanisms can be obtained. For instance, the lifetimes and velocities of parent and daughter species can be inferred by fitting the radial intensity profiles of gaseous species by a photolytic model. These values can be most accurately determined in comets that do not show any spatial or temporal variations in their coma. The coma of such comets can be considered to be in a steady-state, which is the basic assumption of the photolytic models used to derive lifetimes (scale lengths) of individual gas species. Any structure superimposed on a smooth two-dimensional steady-state coma would adulterate the shape of the radial profiles and therefore falsify the values of the lifetimes resulting from fitting a photolytic model.

The most frequently used photolytic models are the Haser Model (Haser, 1957) and the Vectorial Model (Festou, 1981). The Haser Model assumes an isotropic radial outflow of parent and daughter molecules with a constant outflow velocity. The fitting parameters for the shape of the radial profiles are the scale lengths of the parent and daughter species, which are basically a folding between the respective expansion velocities and lifetimes. The Haser Model suffers from its too simplistic assumptions. It can however be used as a mathematical representation of the density in the inner coma ($\leqslant 10^5$ km). The Vectorial Model is more sophisticated in that it takes into account the isotropic emission of the daughter species from their parent molecules and separates the effects of the lifetimes and velocities. Hence it permits the study of velocity dependent phenomena. The variable parameters are the lifetimes and velocities of the parent and daughter species. In addition non-steady-state conditions can be assumed. Neither the Haser nor the Vectorial Model includes any collision effects. Both models are frequently used to determine the production rates of the observed daughter species.

The determination of gas production rates requires a conversion of the observed coma intensity (flux) into column density, for which the excitation mechanism of the observed transition bands needs to be known. For the optical bands of the species discussed here, the excitation mechanism is resonance fluorescence with solar radiation, hence the conversion is conducted by applying the fluorescence efficiencies (g-factors) of the respective transitions to the observed fluxes. The gas production rates are then practically determined by comparing the absolute values of the column densities calculated with the photolytic model (e.g. Haser, Vectorial) to those derived from the observed fluxes.

2.2. *The Reality*

The principles of how to calculate gas production rates from photometric emission band fluxes are well established. However, as no common agreement has been reached on which procedure and which parameters are to be applied, the gas production rates determined by various teams can usually not be compared or combined into a single data set. The problems to be solved are manifold, but can essentially be divided into three categories: (1) models and parameters, (2) available instrumentation, (3) data reduction procedures.

2.2.1. *Models and parameters*

To convert the photometric fluxes, resulting from the basic data reduction, into column densities, the fluorescence efficiency (g-factor) of each transition has to be applied. Although it has been a general notion for a number of years already that the g-factors are well established, most of the teams continue using their preferred values, which in some cases have been determined decades ago. The g-factors used by various authors show discrepancies, which are in some cases as large as a factor of 10. For instance, the g-factors used by A'Hearn *et al.* (1995) for C_2 and C_3 are a factor of 2 and 10 different from those used by Cochran *et al.* (1992). Unfortunately either team keeps using their published values in any subsequent publication, which means that the column densities (representing the abundance of species) calculated from the observed intensities will remain systematically different in publications of these two teams, even if they observe with exactly the same instrumentation. For instance, the OH and NH production rates published for comet 19P/Borrelly for 10 November 1994 (Cochran and Barker, 1999; Schleicher *et al.*, 2003) differ by factors of 9 and 4, respectively, even though both teams used the Haser model with exactly the same scale lengths to derive these values. Interestingly, the CN and C_2 production rates derived for that date only differ by a factor of 2 although not only very different g-factors (for C_2 the g-factors differ by one order of magnitude), but also different scale lengths were used by the two teams. Hence, for CN and C_2 the use of entirely different parameters during all steps of the calculations accidentally led to more similar end-results.

The differences in the production rates derived with the Haser and the Vectorial Model are only minor as long as the cometary coma is in steady-state condition and the chosen values for the Haser scale length equal lifetime × velocity for the Vectorial Model. CN column density profiles in steady-state condition could for instance be well fitted with both, the Haser and the Vectorial Model and resulted in very similar scale lengths and lifetimes, assuming outflow velocities of 1 km/s (Schulz *et al.*, 1993). The analysis of non-steady state column density profiles of comet C/1996 Q1 (Tabur) confirmed that such profiles could only be fitted with sufficient accuracy by the Vectorial Model with time-dependent production rates. Nevertheless, it could also be demonstrated that the production rates derived with the Haser model (and the scale lengths of A'Hearn et al., (1995)) are within 10hence reasonably close to the time-averaged production rates resulting from the Vectorial Model (Lara et al., 2001). Also the comparison of CN and C_2 production rates of comet 46P/Wirtanen derived from photometric fluxes of Farnham and Schleicher (1998) with both, the Haser and the Vectorial Model resulted in differences of only 10-20% (Schulz et al., 1998). Fink and Combi (2004) recalculated published production rates of H_2O, CN, and C_2 for this comet from the original fluxes and demonstrated that, by using a common set of scale lengths, the results of various investigators can be brought into acceptable accord. Originally, the CN and C_2 production rates of comet 46P/Wirtanen derived from the CN (1-0) and the C_2 ($\Delta\nu = -1$) bands (Fink et al., 1998) were systematically higher (up to a factor of 2) than those derived from the CN (0-0) and the C_2 ($\Delta\nu = 0$) bands (Farnham and Schleicher, 1998) although the observations were obtained at the same heliocentric distance.

2.2.2. *Instrumentation*

Gas production rates are usually determined either by imaging or aperture photometry or by spectrophotometry. Details of these techniques can be obtained from recent reviews (Schleicher and Farnham, 2004; Feldman et al., 2004). The direct comparison of column density profiles derived from narrow-band images and spectra of comet C/1995 O1 (Hale-Bopp) demonstrated that the choice of observational method does not affect the resulting column densities (and with that the resulting production rates if the same model and parameters are applied) as long as the same emission bands are used and fully covered (Schulz et al., 2000). Depending on the available equipment, different teams determine coma gas production rates with different observational methods and also from different emission bands. Water production rates are for instance determined from emissions of OH (radio or UV), H (lyman-α) or O ($O(^1D)$). For some comets, this results in large discrepancies between the absolute values of the water production rates, which makes the merging of such data sets into a common database very difficult, if not impossible. For instance, although the water production rate determined for comet 81P/Wild 2 from H lyman-α (Makinen et al., 2001b) is only about 15% higher than that obtained from $O(^1D)$ (Fink et al., 1999), it is about a factor of 2 higher than that determined from OH (Farnham and Schleicher, 2005). For comet 46P/Wirtanen on the other hand, the water production rate from H lyman-α (Makinen et al., 2001b) agrees quite well with that obtained from OH by Farnham and Schleicher (1998) and recalculated by Fink and Combi (2004), whereas it is almost a factor of 2 lower than that determined from $O(^1D)$ (Fink et al., 1998; Fink and Combi, 2004). For comet C/1995 O1 (Hale-Bopp) the water production rates derived around perihelion from H lyman-α (Makinen et al., 2001a) and OH at radio wavelength (e.g. Colom et al., 1997) agree within 10% and are consistent with the expected values extrapolated from OH observations in the UV before perihelion (e.g. Weaver et al., 1997; Schleicher et al., 1997). Hence it appears that the size of the effect depends on the brightness of the observed comet, with weaker comets resulting (as might be expected) in less reliable absolute values for the water production rate. For other species, such as CN and C_2 only few data are available. However, it appears that for those the problem is only minor as long as appropriate g-factor are used. Simultaneous measurements of the integrated band fluxes for the CN $B^2\Sigma^2 - X^2\Sigma^+$ (violet) and $A^2\Pi^2 - X^2\Sigma^+$ (red) systems in comet Austin confirm that the g-factor used for both systems closely describe the radiative properties of CN molecules (Tegler et al., 1992).

2.2.3. *Data Reduction Procedures*

In a few cases discrepancies between published values can be directly related to unconventional procedures in the determination of gas production rates. One example is the difference that resulted in the determination of CN and C_2 production rates in comet C/1996 Q1 (Tabur) by Lara *et al.* (2001) and Turner and Smith (1999). Although both teams used the same observational method (long slit spectroscopy), the same emission bands as well as the same g-factors and Haser scale length, the production rates derived by Turner and Smith (1999) are about a factor of 2 lower than those by Lara *et al.* (2001). A closer look reveals that Turner and Smith (1999) averaged the flux along the entire slit as input for the Haser model, which of course results in decreased (average) flux values, hence smaller production rates. Such unconventional procedures however occur rather rarely and can easily be identified.

3. Determination of Dust Production Rates

In order to be able to compare measurements of cometary dust obtained in different apertures and spectral regions, the quantity $Af\rho$ has been introduced by A'Hearn *et al.* (1984) and is now widely used to measure dust production of dust in comets. It is the product of the Albedo for the scattering angle of the observation ($A(\Theta)$), filling factor of grains in the aperture (f), and effective aperture radius (ρ). The filling factor equals to $f = \frac{N\sigma}{pi\rho^2}$ with σ being the average grain cross section and N being the total number of grains in the aperture. $Af\rho$ is proportional to the observed continuum flux and if the projected density of the dust decreases as ρ^{-1}, it is even independent of the geocentric distance or the aperture size. The relationship between $Af\rho$ and the production of dust varies systematically with the phase angle. There are many more assumptions in using $Af\rho$ as a measure of dust production, which have been summarized by A'Hearn *et al.* (1995). According to Weaver *et al.* (1999), $Af\rho$ can be correlated to the dust mass production rate, Q_{dust} through:

$$Q_{dust} = \frac{0.67 \cdot a \cdot d \cdot v \cdot Af\rho}{A} \qquad (3.1)$$

with: Q_{dust} = dust mass production rate in $kg\,s^{-1}$
  a = average particles radius in μ m
  d = density in $g\,cm^{-3}$
  A_p = geometric albedo
  v = outflow velocity from the nucleus in kms^{-1}
  $Af\rho$ = aperture-independent measure of dust production rate in m.

$Af\rho$ has been used for more than a decade now as a measure of the dust production rate in comets. Almost everybody is using this quantity when publishing dust production rates, which allows comparing and combining the results of various teams. There is of course a number of practical problems, e.g. that some teams do not perform a phase function correction.

The $Af\rho$ system has a very important drawback in that it is not independent of the aperture size any more if the projected density of the dust does not decrease as ρ^{-1}, which is however the reality for most comets observed. Hence the determination of $Af\rho$ will be affected with systematical errors that depend on the slope of the dust coma profiles, if different apertures are used. In summary, with the general use of $Af\rho$ as a measure for the dust production rate of a comet, it was achieved to allow the comparison of values determined by different teams for the same comet in steady-state conditions. Comparison of dust activities of different comets or comets showing short-term variability will however remain very difficult.

3.1. *In-situ dust measurements*

The first in-situ measurements of cometary dust particles were obtained for comet 1P/ Halley in 1986. These measurements made by three spacecraft (Vega 1 & 2 and Giotto) were used to determine the particle size range, the process of dust production, and the composition of the dust. The particles showed a wide distribution of sizes, 0.01–100 μm (McDonnell *et al.*, 1991) and were for the most part composed of silicate refractory and organic molecules (Kissel *et al.*, 1986). In-situ dust measurements were also carried out during the Giotto Extended Mission to comet 26P/Grigg-Skjellerup, however only 3 particles were registered during the fly-by (McDonnell *et al.*, 1993). The data obtained during these fly-by missions indeed remained the only available in-situ measurements of dust particles in comets for a long time.

New in-situ data only became available in 2004, when the Stardust measured the the flux, mass distribution and composition of dust coma particles in comet 81P/Wild 2. The dust coma of this comet was characterized by swarms of particles and bursts of activity, which may be explained by jets and fragmentation (Tuzzolino *et al.*, 2004; Green *et al.*, 2004). The overall mass distribution was similar to that seen in comet 1P/Halley, despite the very large variations detected on small scales. The in-situ compositional analysis of the dust in comet 81P/Wild 2 confirm the predominance of organic matter, which seems to be nitrogen richer and oxygen poorer than interstellar dust (Kissel *et al.*, 2004).

Stardust has collected dust particles from the coma of comet 81P/Wild 2 and will return them to Earth stowed into aerogel in a sample return capsule. The in depth analysis of this dust sample, when returned to the Earth on 15 January 2006, will provide independent information on the physical and compositional properties of the dust in comet 81P/Wild 2, hence provide the ground truth necessary to confirm results and interpretation of in-situ measurements from spacecraft.

4. Coma Evolution along the Orbit

It has been realized over the years that studying the evolution of the activity of a comet as it moves along its orbit is of utmost importance to understand the properties of the comet nucleus. Conclusions on the composition of the nucleus can only be drawn if the coma composition is integrated over the entire orbit of the comet (Prialnik, 2005, this issue). The variation of gas and dust production rates as a function of heliocentric distance has been a subject of investigations for many years. The observations indicate that in general gaseous emissions are a stronger function of heliocentric distance than the dust continuum, e.g. most comets show pure continuum spectra at distance beyond 3 AU (most Jupiter-family comets already beyond 2.5 AU). However, as any gas must drag out the dust, this may well just be the practical manifestation of the sensitivity limits of our observational setups. For example, in bright comet C/1995 O1 (Hale-Bopp) CN could be detected in optical spectra from about 5 AU preperihelion (Schleicher *et al.*, 1997) to 9.8 AU postperihelion (Rauer *et al.*, 2003). It is known for many years that in optical spectra the CN emission usually appears first as a comet approaches the sun, while other emissions, like for instance the C_2 bands, are detected only later (Swings and Haser, 1956). It is common that the dust-to-gas ratio becomes systematically smaller with decreasing heliocentric distance (A'Hearn *et al.*, 1995).

4.1. *Production Rate Ratios*

The gas and dust production rates vary with heliocentric distance, r_h, and the approximate r_h-dependence is most often represented by fitting a power law, r_h^{-k}, to the available measurements. The slopes of these fits may vary significantly from comet to comet and from species to species, e.g. the values for k published in the survey by A'Hearn *et al.* (1995) vary from $0.5 < k < 12$. If the production rates of different species vary with different k, the abundance ratio of these species will change as a function of the heliocentric distance. One of the first studies of abundance variations in cometary comae with heliocentric distance led to indications that the C_2/CN production rate ratio is much smaller at heliocentric distances $> 2\,AU$ than at distances $< 1.5\,AU$ (A'Hearn and Millis, 1980). The result was disputed by Combi and Delsemme (1986) arguing that the drop of the C_2 production rate for larger heliocentric distances is an artifact of inappropriate scale lengths law. However, Newburn and Spinrad (1989) were using the revised scale lengths laws in their spectrophotometric survey and still find that the C_2/CN production rate ratio changed continuously with heliocentric distance in the five comets for which they had measurements at different distances. The C_2/CN ratio decreased with increasing r_h in all cases. Unfortunately, the observations only covered a relatively small range of heliocentric distances and no data beyond 2 AU were available. The change in the C_2/CN ratio described by Newburn and Spinrad (1989) is qualitatively also seen in the data of A'Hearn *et al.* (1995), however the size of the effect was much smaller. A significant increase of the C_2/CN abundance ratio was measured for comet 46P/Wirtanen between 1.8 AU and 1.6 AU during a preperihelion monitoring starting at 2.8 AU (Schulz *et al.*, 1998). The effect remains in the production rates recalculated by Fink and Combi (2004) with a common set of scale lengths.

The survey by Cochran *et al.* (1992) revealed evidence for a heliocentric distance dependence of the NH_2/CN ratio. Later a hint was reported that the C_3/CN and CH/CN production rate ratios in comet 19P/Borrelly may also vary with heliocentric distance (Cochran and Barker, 1999), however, the heliocentric distance covered was too small to permit unambiguous conclusions. The analysis of gas production rates in comet 1P/Halley from 2.6 AU preperihelion to 5.1 AU postperihelion showed that the relative abundances of CN, C_2, and C_3 remained essentially constant with respect to each other, but change markedly with respect to OH (Schleicher *et al.*, 1998).

4.2. *Available Data as $f(r_h)$*

Although the production rates of a large number of comets have been surveyed, the systematic sampling of production rates as a function of heliocentric distance is very poor. Only a hand full of individual comets has been monitored along a sufficiently long part of the orbit, to allow a more detailed analysis of the shape of their activity curves. At least three of these comets clearly show activity curves of complex shape that cannot be fitted by a simple power law and include a sudden preperihelion increase or postperihelion decrease of activity at a certain heliocentric distance. Prominent Comet C/1995 O1 (Hale-Bopp) is one of them, but also the rather faint comets 46P/Wirtanen and 67P/Churyumov-Gerasimenko have been studied extensively, because they were selected as the target for a space mission. To fit the evolution of the OH production rate in Comet C/1995 O1 (Hale-Bopp) the preperihelion data had to be split into three distance regimes and each was fitted separately. The OH production rates increase as $r_h^{-6.8}$ for $r_h > 3$ AU; $r_h^{-1.8}$ for $3\,AU > r_h > 1.3$ AU; and $r_h^{-3.7}$ for $r_h < 1.3$ AU (Colom *et al.*, 1997). The postperihelion data show only two such distinct regimes (Biver *et al.*, 2002). Preperihelion gas and dust production rate curves obtained of comet 46P/Wirtanen during its 1996/97 apparition showed a steep increase between 1.8 AU and 1.6 AU (Schulz *et al.*, 1998;

Schulz & Schwehm, 1999), which was indicated already in the visual light curve during previous apparitions (Morris 1994). Comet 67P/Churyumov-Gerasimenko had a major drop of gas and dust production rates between 2.5 and 2.9 AU postperihelion (Schulz et al., 2004). A comparison of pre- and postperihelion production rate curves of Comet 1P/Halley confirmed that, unlike Hale-Bopp, this comet was significantly more active after perihelion (Schleicher et al., 1998). Fits to the production rate curves showed that for each species the postperihelion curve is flatter that the preperihelion one. However, one has to keep in mind that comet 1P/Halley showed strong short-term variability, which makes any accurate determination of the shape of its production curves, as a function of heliocentric distance very difficult. In summary, the study of those comets for which sufficient data exist reveals a rather complex activity evolution along the orbit. It is therefore absolutely necessary to continue these investigations on a statistically more relevant sample.

5. Classification of Comets

One of the most discussed questions in view to the origin of comets is whether all or at least some of the Jupiter-family comets come from the Edgeworth-Kuiper Belt rather than from the Oort Cloud. Rather convincing evidence for this assumption has already been establish from dynamical investigations (e.g. Levison and Duncan, 1994). A'Hearn et al. (1995) therefore evaluated their production rate survey of 85 comets also in this respect and reported the discovery of significant compositional groupings of comets apparently related to their place of formation. One of their main conclusions is that a significant amount of the short-period comets (most of them Jupiter-family) shows a depletion in carbon-chain molecules, best recognizable from the C_2/CN production rate ratio. On this premises they have introduced a new comet taxonomy, distinguishing comets with *typical* abundance ratios from *carbon-chain depleted* comets and have postulated that the latter comets and only those originate in the Kuiper Belt. The distinction between *typical* and *depleted* abundance ratios in comets has clearly proved to be reasonable and useful, the hypothesis that *depleted* comets might originate in the Edgeworth-Kuiper Belt, however, shows a number of inconsistencies. Firstly, most, but not all *depleted* comets are Jupiter-family comets. About one third of the comets designated as *depleted* were long-period or even dynamically new comets with highly inclined orbits. Secondly, among the Jupiter-family comets, *typical* and *depleted* C_2/CN abundance ratios seem to be more or less evenly distributed (16 typical versus 20 depleted Jupiter-family comets). Thirdly, most of the 85 comets studied by A'Hearn et al. (1995) were observed at a rather limited range of heliocentric distances although the production rates of C_2 and CN are known to vary along the orbit. Therefore, much more work is required to understand why and under which circumstances comets show *typical* or *depleted* abundance ratios and whether this criterion may be connected to their place of formation. For this it is vital to first fully understand the compositional evolution of a cometary coma with heliocentric distance.

6. Summary, Suggestions and Conclusions

A huge amount of gas and dust production rates has been collected over the past three decades. Values exist for about 2/3 of the 168 currently known numbered periodic comets and for more than 60 unnumbered ones. For short-period comets, mostly of the Jupiter-family, data are also available over multiple apparitions. Observations of the same comet obtained by various teams very often excellently complement each other.

Nevertheless, all these data have never been combined into a common database, because production rates are only directly comparable if they have been determined from the measured photometric fluxes using the same photolytic model and parameters. Unfortunately this is not the case. The various teams conducting compositional surveys are not even using the same values for the fluorescence efficiencies (g-factors) for the molecular bands they observe. Sentences like: "the adopted constants for converting from fluxes to column densities may not be identical to the constants adopted by other groups", "comparison with other observers is not straightforward because of differences in models to compute production rates", "considerable different scale lengths causes the production rates to diverge by factors of 3" or "users must be cautious when comparing observations from different groups" can be found in almost every publication made on production rates.

However, if we want to draw conclusions on diversities and similarities of comets in terms of their composition, we need to combine the available data for each comet and the ensemble of comets into a common database. No single data set will be sufficient to reach statistically sound conclusions. By know we have realized that although most of the species visible in the optical were extensively observed in the past, we are still far from a perfect understanding of exactly how these species are produced and destructed, whether and how their abundances differ from comet to comet or how they vary with heliocentric distance and with the number of apparitions of a comet. To ensure that data of the individual comets can be combined into a larger database, or at least to permit a direct comparison of the results of the various groups, we need to re-calculate the gas production rates of all comets from the measured photometric fluxes with a single model and a single set of parameters.

Many teams have been publishing the measured fluxes in addition to production rates that can in principle be used to create a common set of all these data. For instance, most of the teams that have recently published production rates of Comet 81P/Wild 2 have also included the measured photometric fluxes (e.g. Fink *et al.*, 1999; Schulz *et al.*, 2003; Farnham and Schleicher, 2004). To facilitate comparison with other data sets, Lara *et al.* (2004) have started to publish not just the gas production rates they have determined with their favorite model and parameters, but also those that would result with the parameters used in the study by A'Hearn *et al.* (1995), the so far most extensive data base of a single team. This may be the first step into establishing the common database of gas production rates, which is definitely needed if we not just want to collect data, but actually benefit from the huge data pool we have already at hand. In the not too long term, we need to agree on a standard model and a set of parameters for the determination of gas production rates. The least we should do is publishing not just production rates, but also fluxes and when we write a publication, use previously published flux values of other teams to re-calculate for comparison with and complementation of our data set.

References

A'Hearn, M.F. & Millis, R.L. 1980, *A.J.* 85, 1528–1537

A'Hearn, M.F., Schleicher, D.G., Feldman, P.D., Millis, R.L., & Thompson, D.T. 1984, *A. J.* 89, 579–591

A'Hearn, M.F., Schleicher, D.G., Millis, R.L., Feldman, P.D., & Thompson, D.T. 1984, *A. J.* 89, 579–591

A'Hearn, M.F., Millis, R.L., Schleicher, D.G., Osip, D.J., & Birch, P.V. 1995, *Icarus* 118, 223–270

Biver, N., Bockelée-Morvan, D., Colom, P., Crovisier, J., Henry, F., Lellouch, E., Winnberg, A., Johansson, L.E.B., Gunnarson, M., Rickman, H., Rantakyrö, F., Davies, J.K., Dent, W.R.F., Paubert, G., Moreno, R., Wink, J., Despois, D., Benford, D., Gardner, M., Lis, D.C., Mehringer, D., Phillips, T.G., & Rauer, H. 2002, *Earth, Moon & Planets* 90, 5–14

Combi, M.R. & Delsemme, A.H. 1986, *Ap. J.* 308, 472-484

Cochran, A.L., Barker, E.S., Ramseyer, T.F., & Storrs, A.D. 1992, *Icarus* 98, 151–162

Cochran, A.L. & Barker, E.S. 1999, *Icarus* 141, 307–315

Colom, P., Gérard, E., Crovisier, J., & Bockelée-Morvan, D. 1997, *Icarus* 141, 307

Crovisier, J. 2005, *this issue*

Farnham, T.L. & Schleicher, D.G. 1998, *A. & A.* 335, L50

Farnham, T.L. & Schleicher, D.G. 2004, *Icarus* 173, 533

Feldman, P.D., Cochran, A.L., & Combi, M.R. 2004, in: M.C. Festou, H.U. Keller, H.A. Weaver (eds.), *Comets II*, (Tucson: University of Arizona Press), p. 425

Festou, M.C. 1981, *A. & A.* 95, 69

Fink, U. & Hicks, M.D. 1996, *Ap. J.* 459, 729

Fink, U., Hicks, M.D., Fevig, R.A., & Collins, J. 1998, *A. & A.* 335, L37

Fink, U., Hicks, M.D., & Fevig, R.A. 1999, *Icarus* 141, 331

Fink, U. & Combi, M.R. 2004, *Planetary & Space Science* 52, 573

Green, S.F., McDonnell, J.A.M., McBride, N., Colwell, M.T.S.H. Tuzzolino, A.J., Economou, T.E., Tsou, P., Clark, B.C., & Brownlee, D.E. 2004, *JGR* 109, E12S04

Haser, L. 1957, *Bull. Acad. Roy. Belgique, Classe de Sciences* 43, 740

Huggins, W. 1882, *Proc. Roy. Soc.* 33, 1

Kissel, J., Sagdeev, R.Z., Bertaux, J.L., Angarov, V.N., Audouze, J., Blamont, J.E., Büchler, K., Evlanov, E.N., Fechtig, H., Fomenkova, M.N., von Hoerner, H., Inogamov, N.A., Khromov, V.N., Knabe, W., Krueger, F.R., Langevin, Y., Leonas, V.B., Levasseur-Regourd, A.C., Managdze, G.G., Podkolzin, S.N., Shapiro, V.D., Tabaldyev, S.R., & Zubkuv, B.V. 1986, *Nature* 321, 280

Kissel, J., Krueger, F.R., Silen, J., & Clark, B.C. 2004, *Science* 304, 1774

Lara, L.M., Schulz, R., Stüwe, J.A., & Tozzi, G.P. 2001, *Icarus* 150, 124

Lara, L.M., Tozzi, G.P., Boehnhardt, H., DiMartino, M., & Schulz, R. 2004, *A. & A.* 422, 717

Levison, H.F. & Duncan, M.J. 1987, *Icarus* 108, 18-36

Makinen, J.T.T., Bertaux, J.-L., Pulkkinen, T.I., Schmidt, W., Kyrölä, E., Summanen, T., Quèmerais, E., & Lallement, R. 2001a, *A. & A.* 368, 292

Makinen, J.T.T., Silèn, J., Schmidt, W., Kyrölä, E., & Summanen, T. 2001b, *Icarus* 152, 268

McDonnell, J.A.M., Lamy, P.L., & Pankiewicz, G.S. 1991, in: R.L. Newburn Jr., M. Neugebauer & J. Rahe (eds.), *Comets in the Post Halley Era*, (Kluwer Academic Publishers), p. 1043

McDonnell, J.A.M., McBride, N., Beard, R., Bussoletti, E., Colangeli, L., Eberhardt, P., Firth, J. G., Grard, R., Green, S.F., & Greenberg, J.M. 1993, *Nature*, 362, 732

Morris, C. 1994, *International Comet Quarterly* 92, 178

Newburn, R.L., Jr. & Spinrad, H. 1989, *A. J.* 97, 552

Prialnik, D. 2005, *this issue*

Rauer, H., Helbert, J., Arpigny, C., Benkhoff, J., Bockelèe-Morvan, D., Boehnhardt, H., Colas, F., Crovisier, J., Hainaut, O., Jorda, L., Kueppers, M., Manfroid, J., & Thomas, N. 2003, *A. & A.* 397, 1109

Schleicher, D.G., Lederer, S.M., Millis, R.L., & Farnham, T.L. 1997, *Science* 275, 1913

Schleicher, D.G., Millis, R.L., & Birch, P.V. 1998 *Icarus* 132, 397

Schleicher, D.G., Woodney, L.M., & Millis, R.L. 2003, *Icarus* 162, 415

Schleicher, D.G. & Farnham, T.L. 2004, in: M.C. Festou, H.U. Keller, H.A. Weaver (eds.), *Comets II* (Tucson: University of Arizona Press), p. 449

Schulz, R., A'Hearn, M.F., Birch, P.V., Bowers, C., Kempin, M., & Martin, R. 1993b, *Icarus* 104, 206

Schulz, R., Arpigny, C., Manfroid, J., Stüwe, J.A., Tozzi, G.P., Rembor, K., Cremonese, G., & Peschke, S. 1998, *A. & A.* 335, L46

Schulz, R. & Schwehm, G. 1999, *Space Science Review* 90, 321

Schulz, R., Tozzi, G.P., Stüwe, J.A., & Owens, A. 2000, *A. & A.* 361, 359

Schulz, R., Stüwe, J.A., Boehnhardt, H., Gaessler, W., & Tozzi, G.P. 2003, *A. & A.* 398, 345

Schulz, R., Stüwe, J.A., & Boehnhardt, H. 2004, *A. & A.* 422, L19

Swings, P. 1943, *M.N.R.A.S.* 103, 86

Swings, P. & Haser L. 1956, *University of Liège Astrophysical Institute, Tech. Rept. AF61* 514

Tegler, S.C., Campins, H., Larson, S., Kleine, M., Kelley, D., & Rieke, M. 1992, *Ap.J.* 396, 711

Turner, N.J. & Smith, G.H. 1999, *A.J.* 118, 3039

Tuzzolino, A.J., Economou, T.E., Clark, B.C., Tsou, P., Brownlee, D.E., Green, S.F., McDonnell, J.A.M., McBride, N., & Colwell, M.T.S.H. 2004, *Science* 304, 1778

Weaver, H.A., Feldman, P.D., A'Hearn, M.F., Arpigny, C., Brandt, J.C., & Stern, S.A. 1999, *Icarus* 141, 1

Weaver, H.A., Feldman, P.D., A'Hearn, M.F., Arpigny, C., Brandt, J.C., Festou, M.C., Haken, M., McPhate, J.B., Stern, S.A., & Tozzi, G.P. 1997, *Science* 271, 1900

Asteroids, Comets, Meteors
Proceedings IAU Symposium No. 229, 2005
D. Lazzaro, S. Ferraz-Mello & J.A. Fernández, eds.

© 2006 International Astronomical Union
doi:10.1017/S1743921305006897

Active asteroids: mystery in the Main Belt

Henry H. Hsieh and David Jewitt

Institute for Astronomy, University of Hawaii, 2680 Woodlawn Drive,
Honolulu, HI, 96822, USA
Emails: hsieh@ifa.hawaii.edu and jewitt@ifa.hawaii.edu

Abstract. Classically, comets from the outer solar system (beyond the orbit of Neptune), are expected to be icy, and thus active near the Sun, while asteroids in the inner solar system (interior to the orbit of Jupiter) are expected to be relatively ice-deficient, and thus inert. Studies of anomalous objects, most recently 133P/Elst-Pizarro, challenge this classical picture, however, and suggest that either (1) subsurface ice can in fact be preserved over billions of years in small bodies in the inner solar system but still be close enough to the surface to be excavated by an impact by another body, or (2) non-gravitational dynamical evolution (primarily driven by asymmetrical outgassing) of icy bodies from the outer solar system can drive these cometary bodies onto thoroughly asteroid-like orbits, erasing all dynamical signs of their trans-Neptunian origins in the process. The question thus boils down to whether occasionally sublimating icy bodies on stable asteroid-like orbits in the inner solar system, particularly in the main asteroid belt, may in fact be native to the region or whether they must necessarily be recent arrivals.

Keywords. comets: general comets: individual (133P/Elst-Pizarro), minor planets, asteroids

1. Introduction

Classically, comets and asteroids are considered distinct on four levels: observational, physical, dynamical, and evolutionary. The observational distinction is the simplest: comets are "fuzzy", displaying comae and tails, while asteroids are not. This arises from an assumed physical distinction: comets contain significant quantities of volatile material, *i.e.* ices, which sublimate when heated by the Sun, while asteroids do not. This physical distinction is in turn a consequence of dynamics: comets occupy highly eccentric orbits, spending large amounts of time in the cold outer solar system where their primordial volatile supplies are preserved against depletion by solar heating, while most asteroids occupy more circular orbits in close proximity to the Sun and thus receive much more solar heating over their lifetimes, making the survival of primordial ices (if any) correspondingly more difficult. This dynamical distinction is frequently defined by the Tisserand parameter, T_J, an invariant of motion in the restricted three-body problem. Objects with $T_J < 3$ are dynamically coupled to Jupiter and are considered cometary. Objects with $T_J > 3$ are dynamically decoupled from Jupiter and considered asteroidal. Finally, this dynamical distinction arises from different evolutionary histories: the eccentric orbits of comets reflect their origins in the distant, icy Kuiper Belt and Oort Cloud, while the stable, circular orbits of the main-belt asteroids indicate the occupation of their current locations in the inner solar system since their formation.

These classical definitions are of course of declining usefulness in light of recent research. Observationally, comets do not appear fuzzy at all times. They pass through inactive phases, either on the outer portions of their orbits where the temperatures are too low for sublimation to occur, as a result of mantling, or at the end of their active lifetimes when repeated visits near the Sun have exhausted their volatile supplies (*e.g.*, Hartmann *et al.* 1987). Even when comets are active, their activity may be so weak as to escape notice, further complicating the observational distinction between asteroids and comets.

Physically, asteroids are also known to contain volatile materials, or at least to have contained them in the past. Studies of meteorites traced back to the main asteroid belt have revealed the presence of aqueously altered minerals, indications that liquid water and therefore water ice were at least once present in main belt objects (*e.g.*, Hiroi *et al.* 1996; Burbine 1998; Keil 2000). Reflectance spectra of the asteroid 1 Ceres have also been found to possess an absorption feature at 3.1 μm, an indication of possible current surface water ice (Lebofsky *et al.* 1981; Vernazza *et al.* 2005).

From a dynamical standpoint, distinguishing comets and asteroids becomes difficult when one takes non-gravitational orbit perturbations into account. Despite being undisputed observationally as a comet, 2P/Encke occupies an orbit considered by conventional dynamical measures to be non-cometary, with strong gas jet-driven dynamical evolution determined to be the likely reason for this discrepancy (Steel & Asher; Fernández *et al.* 2002; Pittich *et al.* 2004). Likewise, it is possible for main-belt asteroids to be pushed into chaotic resonance zones and subsequently delivered onto Earth-crossing, almost comet-like, orbits via the Yarkovsky effect (Bottke *et al.* 2002 and references within).

In recent years, many objects have been found to possess characteristics of both asteroids and comets, thus forming a broad new category of so-called comet-asteroid transition objects. Transition objects that have been recently studied include the dynamically asteroidal ($T_J = 3.08$) but observationally cometary 107P/(4015) Wilson-Harrington (Osip *et al.* 1995; Campins *et al.* 1995; Chamberlin *et al.* 1996; Fernández *et al.* 2005), near-Earth asteroids with comet-like orbits (*e.g.*, Binzel *et al.* 2004; Fernández *et al.* 2005), the Damocloids (*e.g.*, Jewitt 2005; Fernández *et al.* 2005), and asteroids associated with meteor streams (*e.g.*, Babadzhanov 2001; Meng *et al.* 2004; Williamns *et al.* 2004; Hsieh & Jewitt 2005). In this paper, we review a hitherto less-discussed class of transition objects, that of bodies orbiting in the main asteroid belt yet displaying comet-like behavior: active asteroids. The reader is also referred to a recent review of this topic in *Comets II* (Jewitt 2004).

2. The Strange Case of 133P/(7968) Elst-Pizarro

A discussion of active asteroids necessarily begins with 133P/(7968) Elst-Pizarro (hereafter EP), it being the first and only currently known active main-belt asteroid. An extensive analysis of EP has been presented in Hsieh *et al.* (2004). The following is a summary of the main points presented in that paper.

Previously identified as an inactive asteroid (1979 OW$_7$), EP's active nature became apparent in 1996 when a dust trail was observed by Eric Elst and Guido Pizarro (Elst *et al.* 1996). Numerical modeling by Boehnhardt *et al.* (1996) indicated that the dust trail was formed from dust emitted over several weeks or months, a finding inconsistent with a single instantaneous impact event and strongly suggestive of cometary (sublimation-driven) activity. Given the unexpectedness of such activity, however, considering EP's apparently stable orbit in the main asteroid belt where no other comet-like bodies had ever been observed, controversy remained over the true cause of EP's activity, whether cometary (*e.g.*, Boehnhardt *et al.* 1998) or impact-driven (*e.g.*, Tóth 2000).

Observations in 2002 showing that EP's dust trail had returned, however, eliminated the possibility that impacts alone could be responsible. In Hsieh *et al.* (2004), we reported observations of EP from four occasions in 2002 (August, September, November, and December) during which the trail was visible (Figure 1). The recurrence of the trail is itself a strong argument for a sublimation-driven origin due to the implausibility of two

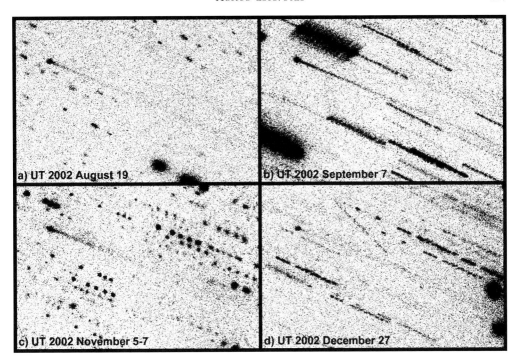

Figure 1. Composite R-band images of 133P/Elst-Pizarro from observations on the University of Hawaii 2.2 m telescope. Images represent 2500, 3900, 4500, and 4200 s in effective exposure time, respectively, and are 2 arcmin by 1.5 arcmin in size, with North at the top and East to the left. The nucleus is located at the same position in each image in the upper left corner with the dust trail extending down and to the right. Shorter, dotted trails and other point-like sources are background stars, galaxies, and field asteroids trailed due to the non-sidereal motion of the telescope in tracking the object. Image from Hsieh *et al.* (2004).

impact events on the same object over the span of 6 years, while no other similar events have been observed on any other main belt object in the entire history of observations of this population. The longevity of EP's dust trail lends further support for a sublimation-driven explanation of EP's activity, since ejecta from a simple impact would not linger for the months that EP's trail was observed to persist in both 1996 and 2002 and then rapidly dissipate in a matter of weeks, as observed in 2002.

In Hsieh *et al.* (2004), we also presented detailed results of Finson-Probstein modeling in which we found that no single impulsive emission model (equivalent to an impact event) could fully account for the behavior of EP's trail (its persistence over months, great length with no observable detachment from the nucleus, and eventual rapid disappearance over weeks). The incompatibility of impulsive emission models and actual observations of EP is exemplified by Figure 2, in which linear surface brightness profiles of modeled dust tails due to impulsive emission are seen to be clearly poor fits to the linear surface brightness profile of EP's dust trail as observed in 2002 September. In contrast, a continuous emission model (approximated as a superposition of multiple, consecutive impulsive emission models like those in Figure 2) can be seen in Figure 3 to be far superior in matching EP's observed surface brightness profile. The much closer fit of this model to the data emphasizes our previous conclusion: EP's trail must have been produced by extended dust emission episode, a scenario consistent with sublimation-driven, *i.e.* cometary, activity.

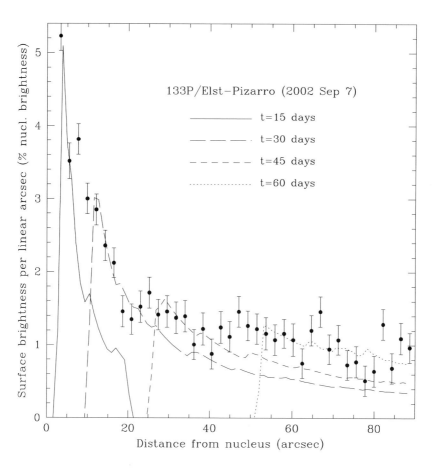

Figure 2. Sample fit of superposed profiles of Finson-Probstein model trails generated from an impulsive emission model to observed data points from 2002 Sep 7, where t is the time of emission in terms of the number of days prior the observation period, and particle sizes and ejection velocities are held constant. For an average nucleus R-band magnitude of 19.7 from this observing period, one trail surface brightness unit is equivalent to 24.7 mag per linear arcsec. Model profiles are arbitrarily scaled to approximate observed data. Image from Hsieh *et al.* (2004).

2.1. *Activity Modulation*

Ordinarily, the onset of sublimation-driven activity in a comet corresponds with its transition from the outer solar system (roughly beyond the orbit of Jupiter) into the inner solar system along its highly eccentric orbit. EP's orbit in the main asteroid belt, however, is entirely confined to the inner solar system (ranging from 3.7 AU at aphelion to 2.6 AU at perihelion), indicating that another modulation mechanism may be required. The 6 year interval between observations of EP's dust trail in 1996 and 2002 corresponds closely to EP's 5.6 year orbit period, suggesting that its activity could be seasonally modulated. The geometry that would give rise to such a scenario is illustrated in Figure 4, where EP's obliquity is non-zero and exposed volatile material is confined to a single isolated patch near one of the nucleus's rotational poles.

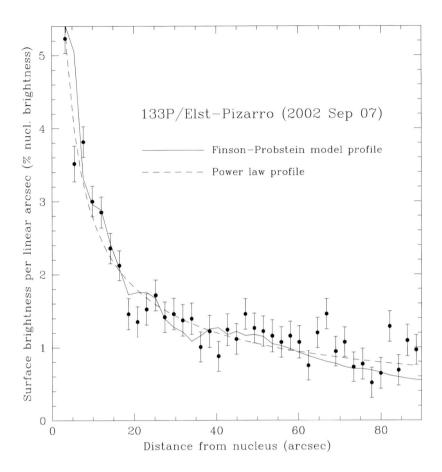

Figure 3. Sample fit of profile of single Finson-Probstein model trail generated from 12 impulsive events of varying intensity over 60 days prior to the observation period, intended to simulate continuous emission, to observed data points from 2002 Sep 7. Particle sizes and ejection velocities are held constant. Also shown for reference is a power-law profile, $\gamma^{-0.6}$, where γ is projected distance from the nucleus in arcsec. For an average nucleus R-band magnitude of 19.7 from this observing period, one trail surface brightness unit is equivalent to 24.7 mag per linear arcsec. The model profile is arbitrarily scaled to approximate observed data. Image from Hsieh *et al.* (2004).

As shown in the diagram, during the volatile hemisphere's "winter," when its pole points away from the Sun, the volatile patch receives little solar heating and thus does not sublimate appreciably, causing EP to appear inactive. During the active hemisphere's "summer," however, the volatile patch receives much greater exposure to the Sun and thus begins to sublimate, ejecting dust particles as it does, producing EP's observed dust trail. This hypothesis is consistent with current observations, in which EP is only seen to be active while in the quadrant of its orbit immediately after its perihelion passage, and will soon permit a simple observational test following EP's next perihelion passage on 2007 July 1. If the seasonal heating hypothesis holds true, a dust trail should develop again near this time and persist until late 2008. Recurrence of dust emission prior to this time, while not ruling out the seasonal heating hypothesis, would imply the presence of multiple active sites, a conclusion which would then have implications for the possible

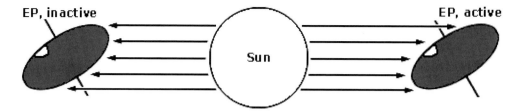

Figure 4. Schematic diagram illustrating seasonal heating of an isolated volatile patch on EP's surface. During EP's inactive phase, its rotational obliquity prevents an isolated patch of volatiles (shown in white) from receiving enough solar radiation to sublimate. During EP's active phase, the volatile patch becomes fully exposed to the Sun (during that hemisphere's "summer"), is heated, and ejects surface material as it sublimates, generating a visible dust trail.

origin of these active sites, as we will discuss in the following section. The non-recurrence of dust emission, however, would effectively exclude the seasonal heating hypothesis and additionally cast doubt on the sublimation hypothesis itself.

3. Explaining Elst-Pizarro

Given this surprising find of a dynamically asteroidal object outgassing like a comet, we must now ask how such a strange object is possible. We consider two hypotheses: either EP is a barely active Jupiter-family comet (JFC) that perhaps has evolved onto its current orbit via the long-term non-gravitational influence of asymmetrical cometary outgassing, or (2) EP is a true native member of the asteroid belt on which preserved, buried ice has been excavated by a recent impact, *i.e.* an activated asteroid. (For clarity, for the remainder of this paper, we will use the term "activated asteroid" to specifically refer to objects described by this latter hypothesis, while "active asteroid" will be used to refer to any currently dynamically asteroidal object displaying cometary behavior without regard to the origin of the body or the activity.)

3.1. *A Lost Comet?*

If EP is an ordinary JFC that has somehow evolved onto an asteroidal orbit, the fact that it contains volatiles and is currently outgassing is easily explained. Its orbit, however, is not. With $T_J = 3.18$, EP is completely decoupled from Jupiter, unlike most other JFCs. Such a decoupling may have come about through a combination of perturbations from close encounters with the terrestrial planets or the influence of other non-gravitational forces, such as asymmetrical cometary outgassing. EP's activity is presently extremely weak, and at its current strength, is unlikely to be able to significantly affect its orbit. It should be noted, however, that earlier in its life, EP probably contained significantly larger quantities of volatile material and thus may have had much stronger outgassing that could then have played a much larger role in EP's dynamical behavior and evolution.

Ipatov & Hahn (1997) found EP's current orbit to be quite stable under purely gravitational influences, meaning that the evolution of a JFC onto such an orbit under those same influences would be unlikely. Fernández *et al.* (2002) also attempted to model the dynamical evolution of a JFC (D/Pigott) onto an EP-like orbit under gravitational influences but could not reproduce EP's low inclination. The inclusion of non-gravitational forces, *i.e.*, cometary outgassing, could significantly change the results of these models, however, and for this reason, we cannot rule out the possibility that EP is a lost comet.

It should also be noted that Comet 2P/Encke also possesses a purportedly asteroidal Tisserand invariant ($T_J = 3.03$) though its identification as a comet is not in question.

Non-gravitational forces are thought to have played and be currently playing a large role in Encke's dynamical evolution (Steel & Asher; Fernández *et al.* 2002; Pittich *et al.* 2004). Clearly, comets are capable of occupying canonically non-cometary orbits and likely move onto such orbits under the influence of non-gravitational cometary outgassing. That being said, EP certainly has the "most" asteroidal orbit of any known comet, sitting precisely amid one of the most populous asteroid families in the main belt. The evolution of a JFC onto such an orbit would certainly be an extraordinary coincidence, but given EP's current observational uniqueness, cannot be completely ruled out.

3.2. *Or an Icy Asteroid?*

The idea that ordinary main belt asteroids might contain ice is not a new one. Spectral features attributed to water of hydration have been observed for main-belt asteroids in the infrared at 3 μm (*e.g.*, Lebofsky 1980; Lebofsky *et al.* 1981; Feierberg *et al.* 1985; Jones *et al.* 1990; Hasegawa *et al.* 2003) and in the visible at 0.7 μm (*e.g.*, Vilas *et al.* 1994; Barucci *et al.* 1998; review by Rivkin *et al.* 2002). The primordial presence of water ice and liquid water in the main belt has likewise been inferred by the presence of hydrated minerals in meteorites, specifically aqueously altered CI and CM carbonaceous chondrites, found here on Earth and determined to have originated from the main belt, specifically from C, G, B, and F-type asteroids (*e.g.*, Hiroi *et al.* 1996; Burbine 1998; Keil 2000). These minerals are generally thought to have formed via the primordial accretion of icy grains into the parent bodies of the carbonaceous chondrites (yr *et al.* 1998; Mousis & Alibert 2005) and the subsequent heating (either by the radioactive decay of ^{26}Al or electromagnetic induction from the solar wind) and liquification of that water ice, leading to aqueous alteration of those parent bodies (Grimm & McSween 1989; Cohen & Coker 2000; Rosenberg *et al.* 2001).

Given the apparent ubiquity of primordial water implied by spectroscopic and meteoritic evidence, Jones *et al.* (1990)] then argued that a decline in detections of hydrated silicates in asteroids of increasing semimajor axis in the main belt could simply be indicative of declining heating effects with increasing distance from the Sun. The absence of hydration features would therefore be because the ice in more distant objects was never heated to the liquid state and thus no hydration reactions were possible. That ice could then still exist today. A similar conclusion was reached by Scott & Krot (2005) who argued that the existence of pristine, unaltered carbonaceous chondrites indicated not that the parent bodies of those chondrites had actually been ice-free, but had instead simply escaped significant heating and could therefore still contain significant quantities of preserved water ice at the present day.

Thermal models appear to support such conclusions. Indeed, even when alteration has occurred, modeling by Grimm & McSween (1989) of CM chondrite production predicts that for small parent bodies, the bulk of each body's ice must remain frozen in order to produce the temperature distribution dictated by the aqueous alteration models. They additionally suggest that ice in large asteroids, *e.g.*, Ceres, at 3 AU that survived primordial heating could be stable against sublimation from ordinary solar heating at depths of only tens of meters, a result corroborated by modeling by Fanale & Salvail (1989).

In fact, water ice has actually been detected for Ceres. Lebofsky *et al.* (1981) reported an absorption feature at 3.1 μm in a near-infrared reflectance spectrum of the body, indicating possible water ice, perhaps in the form of a "surface frost." This detection was later bolstered by A'Hearn & Feldman (1992), who reported detecting OH vapor from Ceres that could have been photodissociated H_2O evaporating from a seasonally-varying northern polar ice cap, and Vernazza *et al.* (2005), who confirmed the existence of absorption at 3.06 μm, finding it to be consistent with a mix of crystalline water ice and residues of ion-irradiated asphaltite. Thus the question of whether water ice can exist in the main belt, at least on a body as large as Ceres, has already been answered. With the discovery and confirmation of EP's comet-like activity, the present-day existence of water ice on much smaller asteroids is now also established.

4. Why Haven't Other Active Asteroids Been Seen?

The previous discussion brings up a natural question: if water ice is thought to be prevalent in outer belt asteroids, why hasn't outgassing been observed in other asteroids besides EP? Perhaps the most straightforward explanation may be that no one has yet looked closely enough at enough asteroids to notice the weak, transient, EP-like activity that may be typical of active asteroids. While many larger asteroids have been studied in detail, EP ($r_{eff} \sim 2.5$ km; Hsieh *et al.* 2004) falls in the far more populous group of small main-belt asteroids that are generally not imaged deeply enough in either discovery or recovery images that are taken for astrometric purposes only, and then subsequently do not attract significant follow-up study due to their small sizes (and thus faint apparent magnitudes) and ordinary dynamical properties. In August 2002, EP's trail was just barely visually noticeable in a 5-minute, R-band image taken with the UH 2.2m telescope on Mauna Kea, indicating that comparable activity from other asteroids may only be detectable by imaging potentially active asteroids to comparable depths from comparably-sized telescopes under comparably good observing conditions.

Observational biases aside, not every small asteroid is expected to be active, of course. Under the activated, icy asteroid hypothesis, a recent collision is also necessary for the creation of an activated asteroid, and presuming the struck asteroid is actually icy (which of course is not guaranteed), this collision must specifically strike an icy portion of that asteroid. If subsurface ice is non-uniformly distributed (not an implausible scenario given likely uneven solar heating effects due to particular rotational axis orientations or local non-uniformities in composition throughout an asteroid), not every impact onto an icy asteroid would necessarily constitute an activation. Even the successful activation of an asteroid, however, does not then guarantee the *detection* of the resulting activity. In addition to possibly being quite weak, as discussed above, the activity of an activated asteroid is expected to be intermittent, requiring observations to be made at the right time. In principle, this means that even inactive objects would need to be continuously monitored in order to ensure that any outbursts are not missed. Furthermore, in analogy to the comets, active sites on activated asteroids likely have finite lifetimes, limited by either actual total local devolatilization of the active site or mantling (Jewitt 1996 and references within). Thus, asteroids that may have been active in the recent past may not necessarily show current activity, much as past activity is inferred for asteroidal parents of meteor streams, such as 3200 Phaethon (parent of the Geminids), despite the absence of any currently observed activity (*e.g.*, Hsieh & Jewitt 2005).

In addition to all the above considerations, the creation and detection of activated asteroids may be even further complicated by a conflict in size preferences that suggests that detectable activated asteroids may only occupy a very narrow range of sizes. On the one hand, the rate of impact excavations of a particular asteroid is dependent on the collisional cross-section of a body. Larger bodies are expected to be struck more often. Thus, it may appear that larger asteroids are more likely to become activated and therefore should form the focus of any observational search for additional activated asteroids. Larger asteroids are also better suited for preserving volatiles from solar heating over large timescales simply because those volatiles can be buried at greater depths than on smaller asteroids, and become exposed later via a particularly deep collisional excavation, collisional disruption of the parent body (either complete fragmentation of the body or the shearing-off of large surface fragments), or gradual "percolation" to the surface.

On the other hand, another key issue may be the extremely low dust ejection velocity ($\sim$1-2 m s^{-1}) we found for EP, comparable to the gravitational escape velocity ($v_{esc} \sim$ 1 m s^{-1}) of the roughly 5-km body. Dust emission on larger bodies may never actually become observable due to the larger escape velocities on those bodies. Volatile material could be sublimating but the gas drag forces generated might simply be too weak to eject dust particles into an observable dust coma or trail. For example, this may be why no dust emission has ever been observed for Ceres, despite the detection of water vaporization by A'Hearn & Feldman (1992). Ejected dust from smaller bodies would be more likely to be able to escape, but those smaller bodies also present smaller collisional cross-sections per object and would not experience as high a rate of activations in the first place as larger bodies. Smaller bodies are also less able to insulate interior ices from long-term solar heating, as discussed above, but are also less effectively heated from within by ^{26}Al (since smaller objects have a greater surface area per unit volume and thus radiate the heat generated by ^{26}Al decay more efficiently than larger objects) and so may therefore be less altered. This conflict in size preferences suggests that activated asteroids may be even rarer and more difficult to discover than initial considerations might indicate.

Finally, we note that at the present time, we still cannot eliminate the possibility that EP is a bona fide comet that has managed to stray into the main belt. If this is the case, the expected rarity of this dynamical transition would suggest that EP may be unique, precluding the existence of any other active asteroids in the main belt, explaining why none other than EP have yet been observed.

5. Finding Other Active Asteroids

If EP is an icy asteroid, other EP-like objects should exist and should be found. The question then is how to go about looking for them. An issue of likely importance is EP's location in the Themis collisional family. This family is roughly defined by orbital element ranges 3.047 AU $< a <$ 3.220 AU, 0.119 $< e <$ 0.191, and 0.688° $< \sin i <$ 2.235°, where a is semimajor axis, e is eccentricity, and i is inclination (Zappalá *et al.* 1990), neatly bracketing the orbital elements of EP ($a = 3.156$ AU, $e = 0.165$, $\sin i = 1.386°$). As the result of the catastrophic disruption of a large parent body, perhaps hundreds of km in size (Marzari *et al.* 1995), the Themis family is thought to consist of related asteroids of similar composition. This conclusion has been supported thus far by studies (*e.g.*, Florczak *et al.* 1999; Ivezić *et al.* 2002) of asteroid families that find approximate spectral homogeneity among members, finding the Themis family in particular to be dominated by C-type asteroids, which, again, are considered to be one of the dominant sources of aqueously altered CM and CI meteorites. Thus, in the search for objects that are behaviorally similar (and therefore probably compositionally similar) to EP, assuming EP is not an interloper of cometary origin or from elsewhere in the main belt, the rest of the Themis family is a logical place to start.

The nature of families as being formed from the catastrophic disruption of larger parent bodies also suggests the prospect that these much larger parent bodies may have protected large quantities of interior ices against early heating episodes, such as a possible early solar wind induction heating episode (*e.g.*, Jones *et al.* 1990), as well as ordinary solar heating over the billions of years since the birth of the solar system. For example, the Themis family is estimated to have been formed perhaps hundreds of Myr to 2 Gyr ago from the disruption of a parent body hundreds of kilometers in size (Marzari *et al.*

1995). Ice inside the Themis parent body could therefore have been preserved for the first few billion years of the life of the solar system and then distributed among smaller bodies, of which EP may be one, upon the destruction of the parent body.

The possible need for collisional activation of EP is also consistent with its location among the Themis family, the family being characterized by a higher than typical collision rate due to the large numbers of asteroids being clustered in orbital element space (Farinella & Davis 1992; Dell'Oro *et al.* 2001). This then suggests that the Koronis family, characterized by even higher collision probabilities than the Themis family (Farinella & Davis 1992), could be a similarly promising region to survey for other activated asteroids. Like the Themis family, the Koronis family appears to be the result of the catastrophic fragmentation of a large parent asteroid about 2 Gyr ago (Marzari *et al.* 1995), and so could likewise contain ice that was deeply buried and preserved for billions of years within the Koronis family parent body, but that now resides in much smaller asteroids following the fragmentation of that parent body. Unlike the Themis family, the Koronis family is dominated by S-type asteroids (Bell 1989; Binzel *et al.* 1993; Mothé-Diniz *et al.* 2005) which along with their analogs, the ordinary chondrites, are not known to show significant aqueous alteration (*e.g.*, Rivkin *et al.* 2002) and thus may be less likely to contain reserves of water ice that might be able to become activated. Controversy exists over the purported anhydrousness of the ordinary chondrites (*e.g.*, Grossman *et al.* 2000, Keil 2000), however, and so for this reason, it is not possible to entirely rule out the Koronis family as a potential enclave of activated asteroids.

The recent identification and age determination of extremely young families, namely the Veritas family (8.3 Myr) and Karin and Iannini clusters (5.8 Myr and <5 Myr, respectively; Nesvorný *et al.* 2002; Nesvorný *et al.* 2003), suggest that these too could be fruitful regions to search for activated asteroids. The extremely young ages of these families suggests that any ices buried deeply within the interiors of the families' parent bodies would only have become exposed upon the disruption of those parent bodies and formation of the families in the last few million years. That ice could then very reasonably still persist on or near the surfaces of the resulting fragments at the present day, and thus be able to drive EP-like, sublimation-driven dust emission. Being young, these families are also still quite tightly clustered in orbital element space, indicating a probable enhanced probability of intrafamily collisional activations in the event that ices were not exposed outright by the initial fragmentation of the families' parent bodies. As before, asteroid types may indicate which of these young families is most likely to exhibit activity – the C-type Veritas family appearing to be a more promising search region than the S-type Karin or Iannini clusters – but uncertainty in the role of taxonomic types in indicating true water content means that none of these families should yet be completely dropped from consideration as possible reservoirs of activated asteroids.

6. Challenges and Future Work

Foremost among the challenges we face in the study of active asteroids is, of course, the fact that only one is known: EP. It is therefore difficult to conclusively identify which properties of EP are most significant in its production of comet-like dust emission and which observational signatures would be most usefully exploited in the search for EP analogs. For example, we discussed above the need for caution in using taxonomic types to definitively declare which asteroids should be expected to display activity and which should not. The possible significance of EP's particular size was also discussed earlier. The unusually rapid rotation of EP ($P_{rot} = 3.471$ hr; Hsieh *et al.* 2004) may also play an important role in its cometary behavior by imparting centrifugal force to gas-ejected dust, aiding its escape from the surface of the body into the dust trail. Interestingly, another comet-asteroid transition object, the observationally inactive 3200 Phaethon that has nonetheless been associated with the Geminid meteor stream (Whipple 1983; Gustafson 1989; Williamns & Wu 1993), implying past comet-like dust emission, also rotates quite rapidly ($P_{rot} = 3.60$ hr; Krugly *et al.* 2002). Whether EP-like dust ejection is equally likely on slower rotators, however, is difficult to say without knowing the range of strengths of asteroidal ice sublimation, information which necessarily requires a larger sample (more than one, at least) of known active asteroids to constrain.

Another major obstacle in assessing EP's nature is the difficulty of confirming or rejecting the possibility that EP is a highly-evolved JFC. The intuitive sentiment that such a situation, if true, should be so rare that EP must be unique is currently supported by the available observational evidence. Without a good statistical understanding of exactly how rare this situation should be, however, this hypothesis is difficult to conclusively observationally assess (particularly since the statistical likelihood of the competing hypothesis that EP is an activated asteroid is also considered low but is currently likewise poorly constrained). More detailed dynamical evolution models of JFCs, with particular attention paid to the role of non-gravitational perturbations exerted by cometary outgassing, to determine whether the transition of a JFC onto a main-belt orbit (and into the midst of the Themis family in particular) is actually impossible, merely nearly impossible, or perhaps not as difficult as currently thought, would therefore obviously be quite valuable in helping to reveal EP's true nature and determining whether we should actually expect to find any other EP-like objects in the main belt.

In any event, it is clear that an effective examination of EP and the nature of its activity is impossible without accompanying campaigns to discover other EP-like objects or ruling out their existence. This can (and is) being done at the current time in pointed surveys of selected small main-belt asteroids such as the Hawaii Trails survey project we have been conducting at the University of Hawaii, but the sheer number of possible targets (numbering in the thousands) that satisfy the size and family-association criteria described above means that such campaigns will eventually be much better served by large-scale synoptic surveys, such as the University of Hawaii's Pan-STARRS, soon to come online. In the meantime, the discovery of even one more active asteroid in the main belt either by the Hawaii Trails project or a similar survey, or by serendipitous discovery, would obviously be quite a significant step forward, permitting the refinement of target selection criteria, thus focusing the scope of future survey campaigns and greatly facilitating the discovery of even more active asteroids and furthering our understanding of these mysterious objects.

Acknowledgements

We thank Humberto Campins for helpful comments on this manuscript. This work was supported in part by a grant to DJ from NASA and a travel grant to HHH from the International Astronomical Union.

References

A'Hearn, M.F. & Feldman, P.D. 1992, *Icarus* 98, 54

Babadzhanov, P.B. 2001, *A&A*, 373, 329

Barucci, M.A., Doressoundiram, A., Fulchignoni, M., Florczak, M., Lazzarin, M., Angeli, C., & Lazzaro, D. 1998, *Icarus* 132, 388

Bell, J.F. 1989, *Icarus*, 78, 426

Binzel, R.P., Xu, S., & Bus, S.J. 1993, *Icarus*, 106, 608

Binzel, R.P., Rivkin, A.S., Stuart, J.S., Harris, A.W., Bus, S.J., & Burbine, T.H. 2004, *Icarus*, 170, 259

Boehnhardt, H., Schulz, R., Tozzi, G.P., Rauer, H., & Sekanina, Z. 1996, *IAUC* 6495

Boehnhardt, H., Sekanina, Z., Fiedler, A., Rauer, H., Schulz, R., & Tozzi, G. 1998, *Highlights in Astronomy* 11, 233

Bottke, W.F., Vokrouhlický, D., Rubincam, D.P., & Broz, M. 2002, in: W.F. Bottke Jr., A. Cellino, P. Paolicchi & R.P. Binzel (eds.), *Asteroids III* (Tucson: University of Arizona Press), p 395

Burbine, T.H. 1998, *Meteoritics & Planetary Science*, 33, 253

Campins, H., Osip, D. J., Rieke, G. H., & Rieke, M. J. 1995, *Planet. & Space Sci.*, 43, 733

Chamberlin, A.B., McFadden, L.-A., Schulz, R., Schleicher, D.G., & Bus, S.J. 1996, *Icarus*, 119, 173

Cohen, B. A. & Coker, R. F. 2000, *Icarus*, 145, 369

Cyr, K.E., Sears, W.D., & Lunine, J.I. 1998, *Icarus* 135, 537

Dell'Oro, A., Paolicchi, P., Cellino, A., Zappalà, V., Tanga, P., & Michel, P. 2001, *Icarus*, 153, 52

Elst, E. W., Pizarro, O., Pollas, C., Ticha, J., Tichy, M., Moravec, Z., Offutt, W., & Marsden, B.G. 1996, *IAUC* 6496

Fanale, F.P. & Salvail, J.R. 1989, *Icarus*, 82, 97

Farinella, P. & Davis, D.R. 1992, *Icarus*, 97, 111

Feierberg, M. A., Lebofsky, L. A., & Tholen, D. J. 1985, *Icarus*, 63, 183

Fernández, Y.R., McFadden, L.A., Lisse, C.M., Helin, E.F., & Chamberlin, A.B. 1997, *Icarus*, 128, 114

Fernández, J.A., Gallardo, T., & Brunini, A. 2002, *Icarus*, 159, 358

Fernández, Y.R., Jewitt, D.C., & Sheppard, S.S. 2005, *AJ*, 130, 308

Florczak, M., Lazzaro, D., Mothé-Diniz, T., Angeli, C.A., & Betzler, A. S. 1999, *A&AS*, 134, 463

Grimm, R.E. & McSween, H.Y., Jr. 1989, *Icarus*, 82, 244

Grossman, J.N., Alexander, C.M.O., Wang, J., & Brearley, A.J. 2000, *Meteoritics and Planetary Science*, 35, 467

Gustafson, B. Å. S. 1989, *A&A*, 225, 533

Hartmann, W.K., Tholen, D.J., & Cruikshank, D.P. 1987, *Icarus*, 69, 33

Hasegawa, S., Murakawa, K., Ishiguro, M., Nonaka, H., Takato, N., Davis, C.J., Ueno, M., & Hiroi, T. 2003, *GeoRL*, 30, 2

Hiroi, T., Zolensky, M.E., Pieters, C.M., & Lipschutz, M.E. 1996, *M&PS*, 31, 321

Hsieh, H.H., Jewitt, D.C., & Fernández, Y.R. 2004, *AJ*, 127, 2997

Hsieh, H. H. & Jewitt, D. 2005, *ApJ*, 624, 1093

Ipatov, S. I. & Hahn, G. J. 1997, in: *Lunar and Planetary Science XXVIII* (Houston: Lunar Planet. Inst.), 619

Ivezić, Ž., Lupton, R.H., Jurić, M., Tabachnik, S., Quinn, T., Gunn, J.E., Knapp, G.R., Rockosi, C.M., & Brinkmann, J. 2002, *AJ*, 124, 2943

Jewitt, D. 1996, *Earth, Moon and Planets*, 72, 185

Jewitt, D. 2004, in: M.C. Festou, H.U. Keller & H.A. Weaver (eds.), *Comets II* (Tucson: University of Arizona Press), p 659

Jewitt, D. 2005, *AJ*, 129, 530

Jones, T.D., Lebofsky, L.A., Lewis, J.S., & Marley M.S. 1990, *Icarus*, 88, 172

Keil, K. 2000, *Planet. & Space Sci.*, 48, 887

Krugly, Yu. N., Belskaya, I.N., Shevchenko, V.G., Chiorny, V.G., Velichko, F.P., Mottola, S., Erikson, A., Hahn, G., Nathues, A., Neukum, G., Gaftonyuk, N.M., & Dotto, E. 2002, *Icarus*, 158, 294

Lebofsky, L. A. 1980, *AJ*, 85, 573

Lebofsky, L.A., Feierberg, M.A., Tokunaga, A.T., Larson, H.P., & Johnson, J.R. 1981, *Icarus*, 48, 453

Marzari, F., Davis, D., & Vanzani, V. 1995, *Icarus*, 113, 168

Meng, H., Zhu, J., Gong, X., Li, Y., Yang, B., Gao, J., Guan, M., Fan, Y., & Xia, D. 2004, *Icarus*, 169, 385

Mothé-Diniz, T., Roig, F., & Carvano, J. M. 2005, *Icarus*, 174, 54

Mousis, O. & Alibert, Y. 2005, *MNRAS* 358, 188

Nesvorný, D., Bottke, W.F., Dones, L., & Levison, H.F. 2002, *Nature*, 417, 720

Nesvorný, D., Bottke, W.F., Levison, H.F., & Dones, L. 2003, *ApJ*, 591, 486

Osip, D., Campins, H., & Schleicher, D.G. 1995, *Icarus*, 114, 423

Pittich, E.M., D'Abramo, G., & Valsecchi, G.B. 2004, *A&A*, 422, 369

Rivkin, A.S., Howell, E.S., Vilas, F., & Lebofsky, L.A. 2002, in: W. F. Bottke Jr., A. Cellino, P. Paolicchi, & R. P. Binzel (eds.), *Asteroids III* (Tucson: University of Arizona Press), p 235

Rosenberg, N.D., Browning, L., & Bourcier, W.L. 2001, *M&PS*, 36, 239

Scott, E.R.D. & Krot, A.N. 2005, *ApJ*, 623, 571

Steel, D.I. & Asher, D.J. 1996, *MNRAS*, 281, 937

Tóth, I. 2000, *A&A*, 360, 375

Vernazza, P., Mothé-Diniz, T., Barucci, M.A., Birlan, M., Carvano, J.M., Strazzulla, G., Fulchignoni, M., & Migliorini, A. 2005, *A&A*, 436, 1113

Vilas, F., Jarvis, K.S., & Gaffey, M.J. 1994, *Icarus* 109, 274

Whipple, F. L. 1983, *IAUC* 3881

Williams, I.P. & Wu, Z. 1993, *MNRAS*, 262, 231

Williams, I.P., Ryabova, G.O., Baturin, A.P., & Chernitsov, A.M. 2004, *MNRAS*, 355, 1171

Zappalá, V., Cellino, A., Farinella, P., & Knežević, Z. 1990, *AJ*, 100, 2030

Asteroids, Comets, Meteors
Proceedings IAU Symposium No. 229, 2005
D. Lazzaro, S. Ferraz-Mello & J.A. Fernández, eds.

© 2006 International Astronomical Union
doi:10.1017/S1743921305006903

Rotational properties of asteroids, comets and TNOs

Alan W. Harris[1] and Petr Pravec[2]

[1]Space Science Institute, Boulder, CO 80301, USA
email: awharris@spacescience.org

[2]Astronomical Institute, Academy of Sciences of the Czech Republic,
Fričova 1, CZ-25165 Ondřejov, Czech Republic
email: ppravec@asu.cas.cz

Abstract. Over the past 25 years the number of reliably determined rotation rates of asteroids has increased by an order of magnitude, from 157 in 1979 to 1686 in 2005. As the numbers have increased, various special classes and features have emerged. Asteroids larger than $\sim 50\,\mathrm{km}$ diameter have a dispersion of spin rates that is well represented by a single Maxwellian distribution. Smaller asteroids have a more dispersed distribution, with both slow and fast spinning populations. We see a "spin rate barrier" in the size range of 1–10 km diameter that suggests that even rather small asteroids are "rubble piles". Among the very slow rotators are some (but not all) that are "tumbling" in non-principal axis rotation states. Among the smallest asteroids (less than a few hundred m diameter) are some that spin dramatically faster than the "spin barrier", indicating that they must have some tensile strength rather than consisting of loose regolith. In the last few years it has been recognized that the spins of asteroids smaller than a few tens of km diameter are affected by radiation pressure torques that tend to either speed up or slow down asteroid spin rates, thus providing an explanation for the dispersion of small asteroid spins, and also their non-random axis orientations. Lightcurves have also revealed the presence of binary asteroids among both Near-Earth and Main-Belt populations. Automated robotic observatories and next-generation survey instruments promise to increase the rate of production of asteroid lightcurves so that we may soon have tens of thousands of lightcurve results, extending down to even smaller sizes. In contrast, there are only about 20 rotation rates known for comets, and 15 for TNOs. Very little can be said from such meager statistics; the mean spin rate of TNOs appears to be comparable to that of asteroids, without extremes of fast or slow rotation; the mean spin rate of comets appears to be a bit slower than asteroids, perhaps due to lower mean density, and there may be an excess of slow rotators, probably due to gas jetting effects. The future is promising for studies of these objects as larger telescopes become available to do photometry to fainter magnitudes, so that comet nuclei can be studied at greater heliocentric distance with less coma interference, and more TNOs can be observed.

Keywords. minor planets, asteroids, Kuiper Belt

1. Introduction

At the time of the first *Asteroids* conference and book in 1979, the total number of reliably determined rotation periods was about 157. In the 26 years since that time, the number has grown steadily to the present number of 1686 (Figure 1). In 1979, only a few general trends and properties could be discerned (Figure 2): that the mean rotation rate was nearly constant over size with possibly just a slight dip in the mid-size range, and that there might be differences in mean rotation rate among taxonomic classes, most notably the M-class, which seemed to have shorter rotation periods on average.

The current data set consists of 1686 reliably determined rotation rates of minor planets, including NEAs, Trojans, Centaurs, TNOs, and a few "transition objects" such as

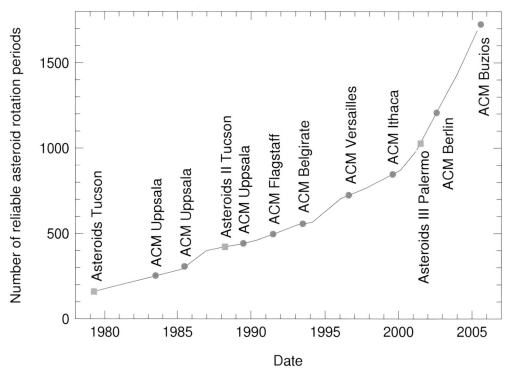

Figure 1. Number of reliably determined rotation rates of asteroids from 1979 to 2005.

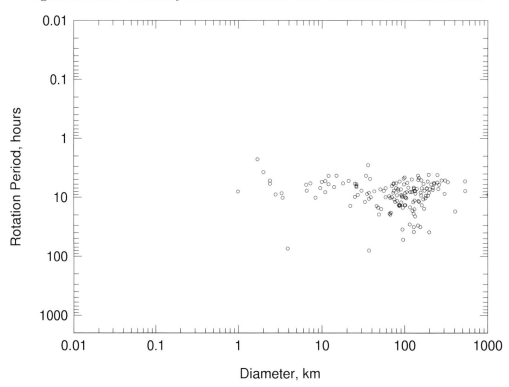

Figure 2. Rotation rate versus diameter for 157 asteroid rotations known in 1979.

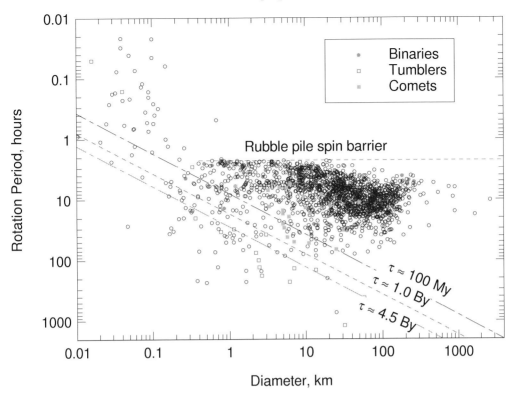

Figure 3. Rotation rate versus size for asteroids, comets and TNOs as of 2005. See text for details.

(2060) Chiron, which are also designated as "comets". Figure 3 is a plot of those rotations versus diameter of objects. The scale has been expanded a bit to include recent additions of (50000) Quaoar and (90377) Sedna, both likely larger than (1) Ceres, and we include the 20 or so comet nuclei with reasonably reliably determined rotation periods. A considerable amount of structure is apparent, which we will describe in the sections that follow. For references to the results, see Pravec *et al.* (2002) and references cited therein.

2. Rotations of large asteroids

Asteroids larger than a few tens of kilometers in diameter spin with a mean rotation period around 10 hours, with some minor variation with size. We have done a running-box calculation of the mean rate versus diameter, and then normalized individual rotation rates to the mean for that size. Thus, if an asteroid of a size where the mean spin period is 10 hours has a rotation period of 8 hours, it's normalized spin frequency is 1.25 (inverse period, compared to the mean). But in a size range where the mean spin period is 8 hours, the same 8-hour period would correspond to a normalized spin frequency of 1.0. Figure 4 is a histogram of normalized spin frequencies of large asteroids, with a 3-dimensional Maxwellian distribution appropriately normalized for comparison. The quality of this fit suggests that for asteroids larger than $\sim 50\,\mathrm{km}$ diameter, the dispersion of spin rates is as one would expect for a collisionally relaxed distribution, and in particular that the dispersion is close to isotropic in three dimensions, that is, there is not a significant

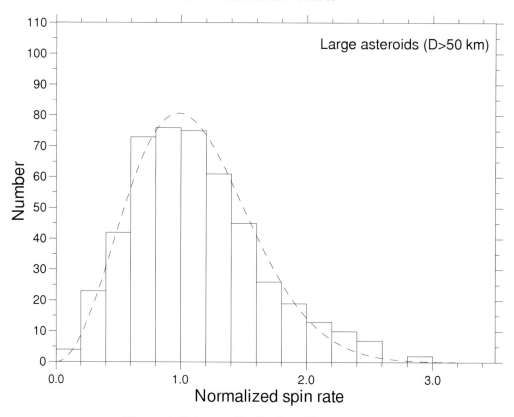

Figure 4. Spin rate distribution of large asteroids.

preferred axis alignment among larger asteroids. The small excess at $f/\langle f \rangle$ greater than 2.2 is due to faster rotating M-type asteroids.

3. Rotations of smaller asteroids – YORP alteration

Beginning in the range of about 50 km diameter, the distribution of spins deviate increasingly with decreasing size from a simple Maxwellian, becoming almost bimodal with fast and slow spins among asteroids under 10 km diameter. Among fast rotators, in the size range from $\sim 1 - 10$ km diameter there appears to be a "barrier" to spins faster than ~ 11 cycles/day (~ 2.2 hours period), as is readily apparent in Figure 3. This is about the spin period at which centrifugal force at the equator equals the acceleration of gravity for a near-spherical body of expected asteroidal density in the range of $2 - 3$ g/cm^3. Thus we infer that the appearance of this "barrier" suggests that even such small asteroids are in some sense "rubble piles" with no substantial tensile strength. The stability limits and equilibrium shapes of rubble piles have been further studied by Holsapple (2001, 2004), and Richardson *et al.* (2005). As Holsapple (2004) correctly points out, the evidence is circumstantial: asteroids spinning slower than the limit *may* have tensile strength, we can only show that they don't need it to maintain their shapes with their given spin rates.

Below a few hundred meters in diameter, asteroids no longer "obey the speed limit", with indeed a majority spinning faster. These asteroids must be "monolithic" in the sense of possessing some material strength. It has been amply pointed out that, due to their very small size, only very modest material strength is needed to resist the centrifugal force of their spins, even for periods as short as a minute or so. Even the softest rock or soft clay, if not fractured, would have enough strength to retain coherence as a single body.

In the size range from a few tens of km diameter and extending down to the smallest observed asteroids, we see a small population, of order 10% of the total, that are very slow rotators, lying statistically well outside of a Maxwellian distribution of spins. The distribution of these spin rates is approximately uniform with spin rate, that is, the cumulative number, $N(<f)$, spinning slower than a frequency f is proportional to f (Harris 2002). This is essentially the distribution one might expect from a retarding force that acts like sliding friction, or for that matter tidal friction, where the magnitude of the retarding force does not depend on velocity, in this case spin frequency f. Rubincam (2000) proposed that thermal re-radiation from irregularly shaped bodies such as asteroids could provide the needed mechanism for spin-up as well as spin-down. Building on Rubincam's formalism, Vokrouhlický *et al.* (2003) showed that spin rate and axial alignments of Koronis family members (Slivan *et al.* 2003) provide an unmistakable "fingerprint" of so-called YORP (Yarkovsky-O'Keefe-Radzievskii-Paddack) spin evolution. While the Koronis family members provided the diagnostic evidence to prove the effect of YORP evolution, it is by no means confined to asteroids in that family, but is ubiquitous among asteroids smaller than a few tens of km in diameter and provides a new paradigm to explain both fast and slow spins as well as non-random spin axis alignments and possibly even shapes and binary configurations of small asteroids.

While YORP provides a mechanism for slowing down spins of some asteroids it leaves a couple questions unanswered. Many of the slowest-spinning asteroids are "tumbling", that is, in a state of non-principal axis rotation. This is in some sense not unexpected, as the damping time scale to principal axis rotation becomes long compared to asteroid ages for small and slow spinning asteroids (Burns & Safronov 1973; Harris 1994). We plot lines of constant damping time scale in Figure 3. Confirmed tumblers lie mostly below lines of damping time scale commensurate with their expected ages (smaller asteroids are expected to be younger due to collisional disruptions). However, if YORP slows an asteroid gently over a long time from a previously faster spin, one would expect the wobble from the last disruptive collision to have damped while the asteroid was spinning fast, leaving the slowed asteroid in a state of principal-axis rotation. We see some evidence that this is so in that not all very slowly spinning asteroids are tumbling. Most of those below the long-age lines have insufficient data to say if they are tumbling or not, but a few are rather definitely determined to have little if any non-principal axis spin. Perhaps these are asteroids that were damped to principal axis spins before YORP slowed the spins, and have suffered no further excitation. Among the largest slow-spinning asteroids (e.g. (288) Glauke, (253) Mathilde), we can calculate the expected excitation of angular momentum from sub-catastrophic collisions in the main belt and show that this is sufficient to excite "tumbling." However, among smaller asteroids, YORP is so powerful that collisional excitation does not appear sufficient to excite tumbling faster than spin rate is damped. Perhaps in these cases YORP de-spins the asteroid before the tumbling component of spin has had time to damp.

Another puzzle is why YORP de-spinning does not progress all the way to spin-orbit synchronicity, as does tidal friction in some cases. Collisional excitation may provide the answer to that as well for the larger de-spun asteroids, but as in the case of tumbling, YORP appears to be too powerful for collisions to make much difference among asteroids smaller than a few km in diameter. Further study of these competing processes is needed.

Returning briefly to the other end of the spin distribution, YORP is fully capable of spinning up small asteroids of diameters less than 10 km to rates up against the "spin barrier", and among sub-km sized asteroids, even to the extremely fast spins observed in this size range. This raises the question of what happens when YORP spins a rubble pile asteroid up to speeds where it can no longer maintain coherence as a single body.

4. Binary Asteroids

After decades, even centuries, of speculation and searching, satellites of asteroids have been found, and are now being discovered at a rate of about one a month by a variety of techniques, including radar, space based imaging (HST), ground-based (mostly adaptive optics) imaging, and lightcurve monitoring (Merline *et al.* 2002; Pravec *et al.* 2006; Richardson & Walsh 2006).

Among the earliest discovered binaries were near-Earth asteroids, generally less than one or two km in diameter, with satellites typically half the diameter of the primary or smaller, and in orbits not synchronously locked to the primary spin (with one exception, (69230) Hermes, which has two nearly equal components and a tidally locked spin-orbit period of 13.89 hours). The prevalence of binaries among NEAs (estimated at $\sim 15\%$ of the total population by Merline *et al.* (2002), and Pravec *et al.* (2006)) and the apparent lack of similar numbers in the main belt suggests that the mechanism of formation might be tied to planet-crossing. The tidal disruption of comet Shoemaker-Levy 9 by Jupiter provided an obvious visual analogy at about the same time as the discovery of the first binaries (see Merline *et al.* (2002) for a discussion of origin scenarios and references). In the last year or two, as we obtain lightcurves of smaller and smaller main-belt asteroids, we are discovering binaries among the smaller members of the main belt (e.g., Ryan *et al.* 2004; Warner *et al.* 2005). Indeed, preliminary estimates of the frequency of binaries among MBAs smaller than $\sim 10\,\mathrm{km}$ diameter may be nearly as great as the frequency of binaries among somewhat smaller NEAs. This calls into question the idea that the formation of binaries of this sort is dominantly by tidal interaction with planets.

In Table 1 we list the properties of selected binaries (we have not included TNO binaries, which seem to have formed by other means and have little to do with rotational properties). The four columns following the object identifications list the diameter of the primary, the ratio of diameter of secondary to diameter of primary, the rotation period of the primary, and the orbit period of the secondary. The last two columns are a measure of the total angular momentum of the system, including spin and orbital angular momenta, which we call α, in dimensionless units. The unit used is equal to the angular momentum of a homogeneous sphere of mass equal to that of the binary (primary plus secondary) if it were spinning at a rate equal to the surface orbit frequency about the sphere - that is, a rate such that a test particle on the equator would just levitate off the surface with gravity balanced by centrifugal force. This necessarily involves the density of the bodies, which is generally not known, so we tabulate values of angular momentum for densities of 2.0 and $3.0\,\mathrm{g/cm^3}$. A spinning, homogeneous fluid takes the form of an oblate (so-called Maclaurin) spheroid for values of $\alpha < 0.76$. In the range $0.76 < \alpha < 0.975$, the fluid takes on the form of a triaxial prolate ellipsoid, known as a Jacobi ellipsoid. At a value of α greater than this, the fluid fissions into two orbiting bodies (see Weidenschilling 1981 for a discussion relating to asteroid spins). Asteroids, however, are not fluids. In defining the "rotational speed limit", Harris (1996) in effect took $\alpha = 1.0$ as the limit for stability of a "rubble pile". More recently, Holsapple (2001, 2004) has refined this limit for realistic unconsolidated materials, and finds that stability can be maintained up to values of α of about 1.3. If one imagines starting from a spherical rubble pile and adding angular momentum, at a value of α of about 0.9, the figure will start to "landslide" into a Jacobi-like form, becoming more and more elongate until it reaches an elongation

Table 1. Properties of some Asteroid binaries

	D_P	D_S/D_P	P_{rot}	P_{orb}	α $\rho = 2$	$\rho = 3$
NEA binaries						
3671 Dionysus	1.5	0.20	2.71	27.74	0.90	0.74
35107 1991 VH	1.2	0.38	2.62	32.66	1.11	0.98
65803 Didymos	0.8	0.22	2.26	11.91	1.06	0.88
66063 1998 RO1	0.8	0.48	2.49	14.54	1.20	1.09
66391 1999 KW4	1.2	0.35	2.77	17.44	0.98	0.85
69230 Hermes	0.6	0.90	13.89	13.89	1.21	1.27
85938 1999 DJ4	0.4	0.50	2.51	17.73	1.25	1.15
88710 2001 SL9	0.8	0.28	2.40	16.40	1.04	0.87
1994 AW1	1.0	0.49	2.52	22.30	1.27	1.17
1996 FG3	1.5	0.31	3.59	16.14	0.75	0.65
1999 HF1	3.5	0.23	2.32	14.03	1.04	0.86
2000 DP107	0.8	0.41	2.77	42.20	1.15	1.04
2000 UG11	0.2	0.50	4.44	18.40	0.93	0.88
2002 CE26	3.0	0.07	3.29	16.00	0.71	0.58
2003 YT1	1.0	0.18	2.34	30.00	1.02	0.84
5407 1992 AX	3.9	0.20	2.55	13.52	0.94	0.78
2005 AB	1.1	0.24	3.34	17.93	0.75	0.63
5381 Sekhmet	1.0	0.30	2.7	12.00	0.94	0.79
1990 OS	0.3	0.15	3.	21.00	0.79	0.65
1998 ST27	0.6	0.13	3.0	154.36	0.80	0.66
MB binaries						
22 Kalliope	181.0	0.23	4.14	85.92	0.65	0.56
90 Antiope	100.0	1.00	16.53	16.53	1.29	1.36
617 Patroclus	105.0	0.90	81.84	81.84	2.01	2.15
854 Frostia	9.0	0.86	37.56	37.56	1.54	1.64
1089 Tama	13.0	0.70	16.44	16.44	1.01	1.06
1313 Berna	10.0	0.90	25.46	25.46	1.41	1.49
1509 Esclangona	8.0	0.33	2.64	474.93	1.33	1.21
3782 Celle	6.0	0.43	3.84	36.57	0.96	0.89
3703 Volkonskaya	3.0	0.40	3.24	24.00	0.96	0.86
4492 Debussy	11.0	0.80	26.59	26.59	1.32	1.40
5905 Johnson	3.6	0.40	3.78	21.79	0.86	0.77
9069 Hovland	2.8	0.5	4.22	30.35	1.04	0.99
Planets						
Pluto-Charon	2302	0.52	153.29	153.29	1.10	1.18
Earth-Moon	12742	0.27	23.93	655.72	0.40	0.41
Non-binaries						
433 Eros	17.	–	5.27	–		1.01
1566 Icarus	1.3	–	2.27	–	1.10	0.90
1620 Geographos	3.	–	5.22	–	0.98	0.80

axis ratio of two or three to one and a value of α of about 1.3. At still higher angular momentum, it should bifurcate, or perhaps shed mass from the equator in a nearly "fluid" way. Richardson *et al.* (2005) suggest the latter, although probably that depends on the detailed structure of the "rubble pile".

The slow spin-up of small asteroids by the YORP effect suggests a mechanism for making binaries among small asteroids, and indeed a mechanism that should be nearly as effective among small main-belt asteroids as among planet-crossing asteroids. In this scenario, as a rubble pile asteroid is spun up by the YORP effect, it might "landslide" into increasingly elongate shapes until it either sheds matter in bits and pieces to re-form into a satellite, or it might spontaneously bifurcate into two co-orbiting bodies. To investigate this further, we have attempted to calculate the angular momentum of observed binary systems, as listed in the final two columns of Table 1. The fact that most of the primaries are spinning very fast suggests that rotation near the critical rate is at least a requirement, if not the cause, of binary formation. It is particularly noteworthy that many of the binaries contain just about exactly the critical angular momentum for splitting into binaries. However, it is also noteworthy that quite a few binaries contain less than the critical angular momentum, so that they must have either lost some angular momentum after fissioning, or were created in some other way. It also seems remarkable that almost no binaries have significantly more than the critical angular momentum. Thus, if formed by spin-up to fission, the spin-up torque must be shut off as the binary is formed.

We call attention to the few entries at the end of the table. The Pluto/Charon binary system fits nicely into the size ratio and angular momentum profile to have been formed by fission followed by angular momentum conserving tidal evolution to its present state. But certainly it was not spun up by YORP! The Earth-Moon system is a factor of two deficient in angular momentum for the moon to have formed by fission, a fact known for more than a century. And finally, we list a few non-binaries, very fast-spinning single asteroids. These single bodies contain about the same critical angular momentum as many of the binary systems, and more than some of them. Why do these asteroids remain single bodies while others with the same angular momentum are binaries?

5. Concluding remarks

So far we have said very little about the spins of comets, Centaurs, and TNOs. The number of reliably measured spins of these classes of objects is too few to make detailed comparisons or to draw conclusions. Comet nuclei spins (plotted in Figure 3) appear more dispersed than "Maxwellian", and maybe a bit slower on average than asteroids. This is consistent with presumed lower density and with outgassing torques which could tend to spin them up or down, in the same way YORP affects small asteroid spins. Centaurs and TNOs are plotted in Figure 3 but not with special symbols, so they cannot be discerned from the asteroids. However, they appear to have a somewhat longer mean rotation period than asteroids, but otherwise similar statistics, consistent with "Maxwellian", but with too few measured values to critically test that hypothesis.

We conclude by mentioning some avenues of investigation that seem ripe for pursuit in the study of rotations of small bodies:

• Robotic and/or remote controlled telescopes with CCD cameras provide an ideal means to obtain vastly greater volumes of data than was possible using "hands on" observations with photoelectric or even CCD systems. This opens up the possibility of conducting surveys of many more objects with more densely sampled lightcurves, going to smaller sizes and looking for complex lightcurves revealing binary systems and "tumbling" asteroids.

• The authentication of young asteroid families by tracing back orbits to a common nodal alignment (Nesvorný et al. 2002; Nesvorný & Bottke 2004) provides us with the opportunity to study asteroid spins presumably unevolved since the time of the parent asteroid break-up. Small members of such families (Karin, for example) should not show a bimodal spin distribution nor anisotropic spin axis alignments that have been demonstrated among the rest of the Koronis family. One might even be able to age date some of the medium age families (those too old to date by nodal alignment) by determining the degree of YORP alteration among members of the family.

• We are on the verge of directly measuring the slow-down of at least one asteroid (25143 Itokawa; Vokrouhlický et al. 2004). Because a change in rotation rate results in an accumulating shift in rotation phase, an exceedingly small deceleration can be detected over years of observation, so directly measuring the YORP change in rotation rates of small asteroids is entirely feasible and we can expect to obtain such results from detailed observations over the coming few years or decade.

- Kaasalainen (2004) has recently presented a method for extracting shape and pole information from sparse photometry such as will be obtained for thousands of asteroids by the next generation of sky surveys (Pan-STARRS, LSST, GAIA). Applying this method to data from next-generation surveys holds the promise of increasing the number of asteroids for which we have rotation and shape information by two orders of magnitude or more. One can only guess what "fine structure" might emerge from a plot like Figure 3 with 100 times more "resolution".

- The present and ever-increasing flood of lightcurve data is leading to binaries galore. Within a year or two there will likely be more than 100 known, so serious theoretical and modeling efforts to understand their origin and evolution can get underway. The study is just beginning.

Acknowledgements

The work at Space Science Institute (A.W.H.) was supported by grant NAG5-13244 from the NASA Planetary Geology-Geophysics Program. The work at Ondřejov (P.P.) was supported by the Grant Agency of the Czech Republic, Grant 205/05/0604.

References

Burns, J.A. & Safronov, V.S. 1973, *Mon. Not. Roy. Astron. Soc.* 165, 403

Harris, A.W. 1994, *Icarus* 107, 209

Harris, A.W. 1996, *Lunar & Planetary Sci.* XXVII, 493

Harris, A.W. 2002, *Icarus* 156, 184

Holsapple, K.A. 2001, *Icarus* 154, 432

Holsapple, K.A. 2004, *Icarus* 172, 272

Kaasalainen, M. 2004, *Astron. & Astrophys.* 422, L39

Merline, W.J., Weidenschilling, S.J., Durda, D.D., Margot, J.-L., Pravec, P., & Storrs, A.D. 2002, in: W.F. Bottke Jr., A. Cellino, P. Paolicchi, & R.P. Binzel (eds.), *Asteroids III* (Tucson: University of Arizona Press), p. 289

Nesvorný, D., Bottke, W.F., Levison, H., & Dones, L. 2002, *Nature* 417, 720

Nesvorný, D. & Bottke, W.F. 2004, *Icarus* 170, 324

Pravec, P., Harris, A.W., & Michałowski, T. 2002, in: W.F. Bottke Jr., A. Cellino, P. Paolicchi, & R.P. Binzel (eds.), *Asteroids III*, (Tucson: University of Arizona Press), p. 113

Pravec, P., & 56 co-authors 2005, *Icarus* in press

Richardson, D.C., Elankumaran, P., & Sanderson, R.E. 2005, *Icarus* 173, 349

Richardson, D.C. & Walsh, K.J. 2006, *Ann. Rev. Earth & Planetary Sci.*, in press

Rubincam, D.P. 2000, *Icarus* 148, 2

Ryan, W.H., Ryan, E.V., & Martinez, C.T. 2004, *Planetary and Space Sci.* 52, 1093

Slivan, S.M., Binzel, R.P., Crespo da Silva, L.D., Kaasalainen, M., Lyndaker, M.M., & Krčo, M. 2003, *Icarus* 162, 285

Vokrouhlický, D., Čapek, D., Kaasalainen, M., & Ostro, S.J. 2004, *Astron. & Astrophys.* 414, L21

Vokrouhlický, D., Nesvorný, D., & Bottke, W.F. 2003, *Nature* 425, 147

Warner, B.D., Pravec, P., Harris, A.W., Galad, A., Kusnirak, P., Pray, D.P., Brown, P., Krzeminski, Z., Cooney Jr., W.R. & 10 co-authors 2005, *IAU Symp. 229* Abstracts, ∂. 89

Weidenschilling, S.J. 1981, *Icarus* 46, 124

Asteroids, Comets, Meteors
Proceedings IAU Symposium No. 229, 2005
D. Lazzaro, S. Ferraz-Mello & J.A. Fernández, eds.

© 2006 International Astronomical Union
doi:10.1017/S1743921305006915

The surface properties of small asteroids from thermal-infrared observations

Alan W. Harris

DLR Institute of Planetary Research, Rutherfordstrasse 2, 12489 Berlin, Germany
email: alan.harris@dlr.de

Abstract. While the physical characterization of near-Earth objects (NEOs) is progressing at a much slower rate than that of discovery, a substantial body of thermal-infrared data has been gathered over the past few years. A wide variety of taxonomic classes in the NEO population have now been sampled by means of thermal-infrared spectrophotometric observations. The resulting albedo information, together with the distribution of taxonomic types from spectroscopic investigations and the rapidly increasing catalog of orbits and absolute magnitudes derived from NEO search programs, such as LINEAR, facilitates more accurate estimates of the size distribution of the NEO population and the magnitude of the impact hazard. Despite our rapidly increasing knowledge of the NEO population, many questions and uncertainties remain, such as: How does the albedo distribution of NEOs compare with that of main-belt asteroids, and does space weathering play a role? How does the surface structure and regolith coverage of NEOs vary with size and taxonomic type? What fraction of NEOs are extinct comets? A property of particular interest is the surface thermal inertia of small asteroids, which is an indicator of the presence or lack of a thermally-insulating surface layer. Large asteroids can accumulate regolith, but can very small asteroids retain thermally-insulating collisional debris or at least a dust layer? Knowledge of thermal inertia is important for accurate calculations of the Yarkovsky effect, which can significantly influence the orbital evolution of potentially hazardous NEOs, and for the design of instruments for lander missions. Contrary to earlier expectations, evidence appears to be accumulating that even sub-kilometer asteroids often have a significant thermally-insulating surface layer. Recent results from thermal-infrared investigations of NEOs are reviewed and implications for the surface properties of small asteroids discussed.

Keywords. Asteroids, near-Earth objects, physical properties

---

1. Introduction

Whatever method of investigation is chosen, whether it be in-situ measurements from a spacecraft or remote astronomical observations, the primary physical information we obtain about a target asteroid normally relates to its surface properties and shape. While we can deduce some information on bulk properties and internal structure from in-situ techniques such as radio science, radar and seismic tomography, and the study of binary systems, most observational data, whether optical, infrared, or radar, provide direct information only on external characteristics. It is therefore important to study the range of surface characteristics present among asteroids and determine how these relate to properties such as size, internal structure and mineralogical composition, and the history of the object, in order to maximize the amount of information that can be extracted about a body as a whole from the observational data.

Even basic parameters, such as size and albedo, are notoriously difficult to determine accurately, especially in the case of small asteroids, in particular near-Earth objects (NEOs), which often have very irregular shapes and surface properties that differ widely from one object to another. Most estimates of the sizes of asteroids are based on absolute visual magnitudes determined from photometric data and orbital information, together with an assumed geometric albedo, pv. Since the observed range of asteroid albedos is about 2% – 60%, estimates of diameter derived on the assumption of a typical albedo, e.g. 15%, can be in error by a factor of 2 or more (for a given H-value diameter is proportional to $p_v^{-0.5}$). Clearly, for many applications, including studies of the size distribution of asteroid populations, more accurate determinations of size are required.

In some cases it has been possible to determine the sizes of asteroids directly, e.g. from rendezvous or fly-by missions, occultation measurements, or direct imaging by astronomical telescopes. In these cases the known size and the H-value can be combined to derive a relatively reliable value of albedo. However, the vast majority of cataloged albedos have been derived indirectly by means of optical and thermal-infrared observations and the use of a thermal model. Results to date have shown that the albedo distribution of the NEO population appears to be broadly similar to that of the observed main-belt population but there may be some differences in detail. There is evidence that the albedo ranges associated with some spectral classes or taxonomic types in the NEO population are broader than the albedo ranges associated with the same types in the main-belt population. Observations of NEOs sample smaller and younger objects, in general, than observations made to date of main-belt objects, so size- and/or age-dependent phenomena such as the presence, or lack of, impact debris (regolith) and space weathering may offer explanations for apparent differences in the albedo distributions of the NEO and main-belt populations. Furthermore, while most NEOs are thought to be collisional fragments from the main belt, recent studies have shown that at least several percent of the NEO population may have a cometary origin.

Reliable modeling of scattered solar and thermally emitted radiation from small asteroids is very difficult and the interpretation of observational data is prone to error due to the use of techniques that over-simplify the physical description of the objects. Models based on a simplified surface temperature distribution and spherical geometry are often unavoidable due to the lack of information on the physical characteristics of the object. However, the rapidly increasing body of observational data has encouraged the development of sophisticated thermophysical models that take account of shape, spin-axis orientation, thermal inertia and surface roughness. Shape models derived from lightcurve inversion techniques or extensive radar observations can be used as the basis of a thermophysical model. Given the rotation rate and spin-axis orientation, model parameters such as thermal inertia, crater geometry and the surface density of craters can be varied to give the best fit of the model thermal-infrared fluxes to observational data taken at different wavelengths and solar phase angles. Recent results for a number of well-studied asteroids are indicative of marked differences between the thermal characteristics of large main-belt asteroids and NEOs. One area in which such results are of crucial importance is the study of the orbital evolution of NEOs, especially potentially hazardous objects (PHOs). Due to the momentum they carry, the asymmetric emission of thermal photons can cause significant drift in the orbit of a PHO over long time scales and therefore influence the long-term evolution of impact probabilities. For accurate calculations of the orbital drift, resulting from what has become known as the Yarkovsky effect, knowledge of the surface thermal properties of the asteroid is required.

2. Thermal modeling

2.1. *Models based on spherical geometry*

The optical brightness of an asteroid depends on the product of its geometric albedo and projected area. While these parameters cannot be individually determined from optical photometry alone, if observations of the optical brightness can be combined with measurements of the objects thermal emission, both its albedo and size can be individually derived.

Thermal-infrared measurements have provided the vast majority of asteroid albedo determinations to date. However, in the case of small asteroids, in particular NEOs, complications can arise due to their irregular shapes, the apparent wide range of surface properties, presumably reflecting the presence or absence of a dusty, insulating regolith (small objects may have insufficient gravity to retain collisional debris), and the fact that they are often observed at large solar phase angles. The problem of irregular shape can be largely overcome by combining thermal-infrared observations with lightcurve-tracing optical photometry obtained at about the same time. The optical photometry allows the infrared fluxes to be corrected for rotational variability.

If no physical data are available for an observed asteroid, one has little choice but to adopt a simple thermal model based on spherical geometry. The most common example is the standard thermal model (STM) described by Lebofsky *et al.* (1986). The STM was designed for use with large main-belt asteroids and incorporates parameters that apply to asteroids having low thermal inertia and/or slow rotation observed at low solar phase angles. In order to extend the applicability of the STM to objects that may have significant thermal inertia and rapid rotation, such as NEOs, Harris (1998) considered a modified approach, the near-Earth asteroid thermal model (NEATM), in which the model temperature distribution is adjusted via the parameter η to force consistency with the observed apparent color temperature of the asteroid, which depends on thermal inertia, surface roughness and spin vector (the value of η is kept constant in the STM, in which it was originally introduced to take account of the beaming effect at low phase angles due to surface roughness).

In the NEATM and the STM the surface temperature distribution is given by

$$T(\theta) = T(0) \cos^{1/4} \theta \qquad (2.1)$$

where θ is the angular distance from the sub-solar point. For surface elements not visible to the observer at the given phase angle, or on the night side, $T(\theta)$ is set to zero. $T(0)$, the sub-solar temperature, is given by

$$T(0) = [1 - A)S/(\eta\varepsilon\sigma)]^{1/4} \qquad (2.2)$$

where A is the bolometric Bond albedo, S the solar flux at the asteroid, η the beaming parameter, ε the emissivity, and σ the Stefan-Boltzmann constant. In the NEATM the beaming parameter is treated as a variable for the purposes of fitting the model continuum to the observed thermal-IR fluxes.

The NEATM requires flux measurements at a minimum of two wavelengths (preferably several, widely spaced around the thermal continuum peak in the range $5 - 20\mu$m), whereas for use of the STM only one is required. A further difference between the NEATM and the STM is the treatment of the thermal-infrared phase effect: the STM uses an empirical phase coefficient, typically 0.01 mag/deg, derived from observations of main-belt asteroids at solar phase angles below 30°. Since NEOs are often observed at larger phase angles, the NEATM takes account of the phase angle by means of a numerical integration of the observable thermal emission from that part of the surface of the spherical model illuminated by the Sun. For asteroids with STM-type characteristics observed at low phase angles the results from the STM and the NEATM are very similar (as can be demonstrated by application of both models to the IRAS flux data from which the IRAS Minor Planet Survey results (Tedesco 1992) are derived - see Walker 2003, and http://www.mira.org/research/irasIntro.htm).

The NEATM is a relatively simple and practical approach to the derivation of diameters and albedos of asteroids in general (not just NEOs!) applicable to the majority of observational circumstances. However, as in the case of the STM, thermal emission from the night side of the object is not accounted for in the model, which is therefore expected to lead to overestimation of diameters and underestimation of albedos for objects with large thermal inertias observed at large solar phase angles. In such cases, use of a simple model with a longitude-independent temperature distribution, often called the fast-rotating or isothermal-latitude model (FRM/ILM, Lebofsky & Spencer 1989), may be preferable (see §6).

The above discussion of the commonly used thermal models is a very brief summary for the purposes of this review. For more detailed discussions of the NEATM and other thermal models outlined here see Harris (1998), Delbó & Harris (2002), Harris & Lagerros (2002), and Delbó *et al.* (2003), and references therein. Delbó & Harris (2002) give the mathematical expressions for calculating the wavelength-dependent observable thermal-infrared fluxes for all three models.

2.2. *Thermophysical models*

Improvements in computing power in recent years have paved the way for the development of sophisticated thermophysical models that incorporate physical descriptions of thermal inertia and surface roughness. Clearly such models should provide more reliable results than the simpler models described above but they require some prior information about the asteroid. For example, information on shape and spin-axis orientation may be available from lightcurve or radar observations. If extensive lightcurve data are available then it may be possible to construct a shape model of an asteroid from lightcurve inversion techniques (see, for example, Kaasalainen *et al.* 2004, and references therein) that can be used as the basis of the thermophysical model. Typically, the surface of the asteroid is modeled by means of a convex mesh of several thousand facets. Hemispheres, or sections thereof, can be added to the facets to model the effects of shadowing and multiple reflections of solar and thermally emitted radiation in craters. Unfortunately most shape models currently available do not include intermediate-scale topographical structure, so a more realistic representation of surface roughness is normally not possible. The lack of information on surface structure is a serious problem in the case of NEOs, which are often observed at large phase angles at which shadowing effects are more significant and measured fluxes are presumably strongly dependent on surface roughness.

The temperature of each facet can be determined by applying a simple one-dimensional vertical heat-transfer algorithm (e.g. that of Spencer, Lebofsky & Sykes 1989) assuming values of albedo and thermal inertia. By summing the contributions from each facet visible to the observer the total model thermal-infrared flux can be compared to flux measurements made at the telescope, and model parameters adjusted to give the best overall fit to the observational data, thereby constraining the physical properties of the observed asteroid.

The detailed thermophysical modeling of main-belt asteroids has been pioneered by Johan Lagerros, who has published a series of detailed papers on various aspects of the field. For an overview and list of references see Harris & Lagerros (2002). Sophisticated thermophysical modeling based on the work of Lagerros has been applied to the study of NEOs by Delbó (2004), Müller *et al.* (2004), Mueller *et al.* (2005), and Harris *et al.* (2005).

3. The albedo distribution of NEOs

Reliable albedo data for NEOs are available for only a very small sample of the known population. Harris & Lagerros (2002) list results for 20 NEOs and compare results obtained on the basis of the three commonly used thermal models described in §2 with data from radar observations, where available, and taxonomic types. There are serious discrepancies in many cases between diameters and albedos derived from the STM and FRM and values deduced from radar observations or expected on the basis of taxonomic type. The NEATM appears to give more consistent results but also fails in at least two cases in which a default value of the model parameter, η, was used due to lack of spectral data for model fitting.

The NEATM has been used by Delbó *et al.* (2003) to determine the sizes and albedos of 20 NEOs observed with the Keck-I telescope on Mauna Kea. The spread of albedos was found to be very large ($p_v = 0.02$ - 0.55) although for the most part the values are consistent with those expected for the spectral types observed. A serious problem in surveys of NEO albedos is the under abundance of the darker taxonomic types (C, D, P, etc.) in the known population due to detection bias against optically weak objects. Detection and observational biases have to be taken into account in attempts to relate surveys of samples of the NEO population to the population as a whole. In order to apply the albedo data of Delbó *et al.* (2003) to a study of the size distribution of NEOs, Stuart & Binzel (2004) applied a magnitude limited de-biasing technique to produce de-biased average albedos for the taxonomic types sampled by Delbó *et al.* The resulting de-biased albedos were combined with a large body of taxonomic data to convert the absolute magnitude distribution of NEOs to a diameter distribution, which was used as the basis of estimates of impact rates of NEOs of different sizes. For NEOs with diameters of 1 km or more, Stuart & Binzel (2004) determined an abundance of 1090 ± 180 and an average impact frequency on the Earth of 1 per 0.6 ± 0.1Myr. They obtained a value of 0.14 ± 0.02 for the de-biased mean albedo of the NEO population, which is higher than would be expected from comparison with the albedos in the IRAS Minor Planet Survey (~ 0.1, Tedesco, 1992). In particular, the albedos of C-type NEOs appear to be some 70% higher than known C-types in the main belt, although caution is due given that this conclusion is based on a sample of only 6 NEO C-types. In any case, it is not clear why NEO C-types should have higher albedos than those observed in the main belt. Possible explanations include erroneous absolute magnitudes (could the absolute brightnesses of dark NEOs be consistently overestimated?), and the emissivity in the thermal infrared for C-type NEOs being much lower than the commonly assumed value for all asteroids of 0.9. If the latter turned out to be the case it would offer valuable insight into the nature of C-type NEO surfaces. The uncertainty associated with the emissivity of asteroid surfaces is illustrated by the work of Walker (2003), who applied the NEATM to the IRAS Minor Planet Survey (IMPS) 12μm and 25μm fluxes for 54 asteroids with measured occultation diameters and compared those diameters with the resulting NEATM diameters. His results led Walker (2003) to conclude that the average infrared emissivity, ε, of the 54 IMPS asteroids is 0.79, rather than the commonly adopted value of 0.9. However, Walker reports that his results appear to be independent of taxonomic class. The questions of whether Walkers results also hold for NEOs, and whether emissivity depends on taxonomic class in the case of NEOs, constitute an interesting subject for future study.

Delbó *et al.* (2003) noted a possible trend of increasing albedo with decreasing diameter in the case of S-type NEOs with diameters below 10km. While the significance of this trend is not clear due to the small sample size and the possible influence of a detection bias against small, dark objects, it is consistent with the idea that collisional processing leads to reduced lifetimes for small NEOs and therefore reduced exposure to space weathering, a process that may darken the surfaces of the olivine-rich S and Q taxonomic types (Clark, Hapke, Pieters, *et al.* 2002). I have reviewed this apparent correlation in the light of further albedo data that have become available since the study of Delbó *et al..*

Table 1. Linear correlation coefficients for the data sets plotted in figure 1

Data set	No. points	Linear correlation coefft., R	% probability that R = 0
Excluding S & Q types	21	-0.57	0.66
S & Q only	17	-0.76	0.04
Excluding S & Q ($D > 0.5$ km only)	17	-0.24	36.0
S & Q only ($D > 0.5$ km only)	13	-0.71	0.66

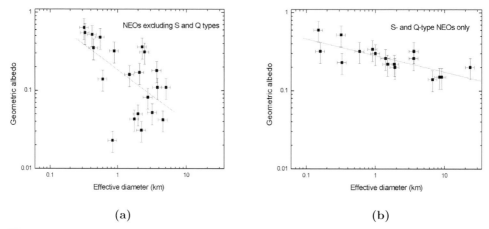

(a)   (b)

Figure 1. Plots of geometric albedo against diameter for a set of 21 NEOs excluding S and Q typers (a - left-hand frame), and for a set of 17 S- and Q-type NEOs (b - right-hand frame). A best-fit line is superimposed in both cases: the corresponding linear correlation coefficients are listed in table 1. The sources of the (NEATM-derived) diameter and albedo data are: Delbó *et al.*. (2003) and references therein, Delbó (2004), Harris *et al.* (2005), and Wolters *et al.* (2005).

In figure 1a albedo values derived on the basis of the NEATM for NEOs for which taxonomic information is available, excluding S- and Q-types, are plotted against diameter. In figure 1b albedo data for S- and Q-types only are plotted. There appears to be a significant correlation in the first case (linear correlation coefficient in figure 1a $= -0.57$), which excludes S- and Q-type NEOs, indicating that detection bias may indeed play a role (note that the correlation is much weaker if the four objects with diameters less than 0.5km are excluded see table 1). However, the correlation in the second case (S and Q only) is more significant (correlation coefficient $= -0.76$), suggesting that another effect in addition to detection bias is strengthening the correlation of albedo with diameter in the case of the S- and Q-type asteroids (note that in this case the correlation is only marginally weaker if the four objects with diameters less than 0.5km are excluded - see table 1). Space weathering may be responsible for this effect, in addition to the trend to ordinary-chondrite-type reflection spectra with decreasing size observed in the NEO population (Binzel, Lupishko, Di Martino, *et al.* 2002). It is also possible that size-dependent regolith properties, such as the size distribution of grains, may give rise to a correlation of NEO diameter with albedo. Turning to the low end of the albedo scale, a long-running debate concerns the fraction of the NEO population that has a cometary origin. Since the dynamical lifetimes of comets generally exceed their active lifetimes, there is expected to be a large number of dormant or extinct comets that are catalogued as asteroids. Fernández, Jewitt & Sheppard (2005) have performed a thermal-infrared survey of 26 asteroids having comet-like orbits, including six NEOs. Diameters and geometric albedos for the target asteroids were derived on the basis of the NEATM. Fernández *et al.* found that some 64% of the objects had low ($p_v < 0.075$), comet-like albedos and considered these to be candidate dormant or extinct comets. On the basis of their results together with albedos in the literature for a further 4 NEOs, Fernández, Jewitt & Sheppard suggest that some 53% of NEOs on comet-like orbits have comet-like albedos; assuming 7% of all NEOs are on comet-like orbits they conclude that some 4% of all NEOs are dormant or extinct comets. However, Binzel *et al.* (2004) estimate that 10 - 18% of the NEO population may be extinct comets, after correcting for observational bias against the detection of low-albedo objects in the known population of NEOs.

4. Thermal inertia and the Yarkovsky effect

Thermal inertia, i.e. the resistance of a material to temperature change, is defined as $(\kappa\rho c)^{0.5}$, where κ is the thermal conductivity, ρ the density and c the specific heat. The surface temperature distribution of an asteroid is determined by a number of factors, including the rate of rotation, the spin-axis orientation with respect to the Sun, the surface roughness and the surface thermal inertia. An asteroid covered in a thermally-insulating dusty regolith would be expected to have a low thermal inertia. Values of thermal inertia derived for some main-belt asteroids are around 10–20 $\mathrm{Jm^{-2}s^{-0.5}K^{-1}}$ (Müller & Lagerros 1998). The thermal inertia of the lunar surface is some 50 $\mathrm{Jm^{-2}s^{-0.5}K^{-1}}$ (see for example Spencer, Lebofsky & Sykes 1989) and that of bare rock about 2500 $\mathrm{Jm^{-2}s^{-0.5}K^{-1}}$ (Jakosky 1986). The surface temperature distribution of a body with low thermal inertia rotating slowly, with the Sun above its equator, has a prominent peak near the sub-solar point. The amplitude of the equatorial temperature distribution decreases with increasing surface thermal inertia or rotation rate.

Areas in which knowledge of thermal inertia is of crucial importance are the design of asteroid lander missions, for which diurnal thermal cycling can affect the lifetime and performance of scientific instruments, and the calculation of the gradual orbital drift of small asteroids due to the anisotropic emission of thermal-infrared photons, known as the Yarkovsky effect. The very small force that results from the anisotropic radiation of absorbed solar energy has been invoked to explain the apparent long time scale for the delivery of collisional fragments into orbital resonances in the main belt, from which they are injected into near-Earth orbits. The Yarkovsky effect has a significant influence on the orbital evolution of potentially hazardous near-Earth objects (PHOs). For example, considering two extreme but plausible values of thermal inertia, Giorgini *et al.* (2002) showed that the uncertainty in the Yarkovsky effect alone leads to an uncertainty of 83×10^6 km in the along-track position of the PHO 1950 DA just before the time of nominal orbit intersection with the Earth in the year 2880. This example highlights the importance of investigations of the thermal properties of NEO surfaces for calculations of collision probabilities of potentially hazardous objects.

The derivation of a reliable value for the surface thermal inertia of an asteroid, especially a small object or NEO, is a challenging task that requires accurate photometry of the thermal continuum emission and a detailed thermal model. However, some information on thermal inertia can be obtained from the apparent color temperature of an asteroid observed at a low solar phase angle. The model parameter, η, in the NEATM can be used as a measure of the deviation of the observed colour temperature from that expected on the basis of the STM (in which $\eta = 1$ for a smooth sphere, or less than 1 in the case of beaming due to roughness). As explained above, the sub-solar temperature decreases with increasing thermal inertia for an object rotating with the Sun on its equator. A low sub-solar temperature observed at a low phase angle will be reflected in a relatively high best-fit value of η (cf. 2.2), so high values of η at low phase angles are indicative of high thermal inertia. In a set of 17 NEOs in the size range 0.6–25 km for which η has been determined by thermal continuum fitting, Delbó *et al.* (2003) noted that η appears to increase with phase angle and that the only objects displaying values of η significantly higher than unity were observed at solar phase angles exceeding 35°. An updated plot of η versus solar phase angle for a set of NEOs with adequate multi-filter photometric data to enable η to be derived via NEATM fitting is shown in figure 2; values of η resulting from applying the NEATM to synthetic fluxes from perfect STM (zero thermal inertia and no beaming) and FRM (fast rotation/high thermal inertia, Sun on equator) asteroids are traced by the dashed straight line and dotted curve, respectively. Note that the η

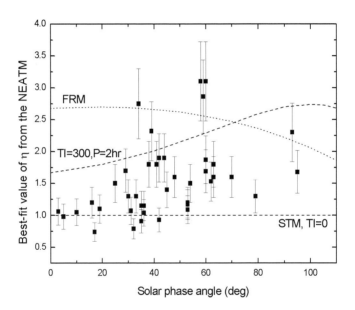

Figure 2. Plot of the best-fit η value from the NEATM against solar phase angle for a set of NEOs with adequate multi-filter photometric data to enable η to be derived via spectral fitting (the data set includes multiple values of η for some objects observed at more than one phase angle). The error bars represent a conservative 20% uncertainty, which is based on the reproducibility of η for those objects for which more than one measurement from independent data sets is available. The dashed line at $\eta = 1$ represents a lower limit given by an STM-type asteroid (zero thermal inertia or spin axis pointing at the Sun) with a smooth surface. If beaming due to surface roughness is present, η may be lower than unity at low phase angles. The dotted curve labeled FRM represents an upper limit for a population of FRM-type asteroids (fast rotation/high thermal inertia surface, e.g. bare rock). The dashed curve represents an upper limit for a population of spherical, *smooth* asteroids at 1 AU from the Sun with thermal inertia $= Jm^{-2}s^{-0.5}K^{-1}$ and rotation periods longer than 2 hr (note: the presence of surface roughness could increase η above this lever at high phase angles). In particulr, the fact that the points cluster around the low-thermal-inertia limit at low phase angles (i.e. less than 30°) suggests there is a lack of NEOs with high thermal inertia in the set of objects observed to data. For data sources see captions to figure 1 (some additional unpublished data have been included).

value of a fast-rotating asteroid with high thermal inertia decreases as the angle between the spin axis and the solar direction decreases. The temperature distribution of an FRM asteroid with its spin axis pointing towards the Sun is identical to that of the STM. So for a population of FRM-type asteroids (fast rotation/high thermal inertia surface, e.g. bare rock) with randomly oriented spin axes, the dashed curve labeled FRM represents an upper limit for η. Likewise, for a population of spherical, smooth asteroids with thermal inertia $= 300$ Jm^{-2}s$^{-0.5}$K^{-1}, rotation periods longer than 2 hr (true for nearly all NEOs with known periods) and random spin-axis orientations, the η value distribution in figure 2 would be bounded by the dashed STM line and the dashed curve labeled $TI = 300$, $P = 2$ hr. While the data set is small, the distribution of η values in figure 2 appears to favor the latter case, rather than a population consisting largely of high thermal inertia objects. In particular, the absence of high values of η at low phase angles suggests that objects with high thermal inertia, in the size range investigated, may be relatively rare (see §5). The presence of a significant beaming effect due to surface roughness would reduce η below unity at low phase angles. At high phase angles the beaming effect, which enhances thermal emission at low phase angles, tends to reduce the amount of thermal emission observed and the apparent color temperature. Therefore at high phase angles both surface roughness and thermal inertia lead to increased values of η and unraveling their contributions requires sophisticated thermal modeling.

Note that for fixed η thermal inertia is proportional to $P^{0.5}$, where P is the rotation period (see, for example, Spencer, Lebofsky & Sykes 1989). Therefore the dashed upper limit curve in figure 2 labeled $TI = 300$, $P = 2$ hr also holds for a population of NEOs in which each object has a typical rotation period of $P = 6hr$ and thermal inertia not exceeding 520 Jm^{-2}s$^{-0.5}$K^{-1}. Delbó (2004) has analyzed the η-value distribution shown in figure 2 with the aid of a thermophysical model that includes physical descriptions of both thermal inertia and beaming resulting from a uniform distribution of craters, and takes account of the assumed random distribution of spin axes. Assuming all objects have the same thermal inertia, and taking account of their known rotation rates, Delbó varied the thermal inertia and the degree of surface roughness until he obtained the best match to the η versus solar phase angle distribution; his resulting best-fit value of thermal inertia is 500 ± 100 Jm^{-2}s$^{-0.5}$K^{-1}.

Three estimates of thermal inertia obtained via thermophysical modeling for individual NEOs for which shape models are available are given in table 2, together with some values derived for main-belt asteroids by Müller & Lagerros (1998).

In summary, the results reviewed here suggest that the appropriate values of thermal inertia to adopt in calculations of the Yarkovsky effect on NEOs are an order of magnitude

Table 2. Comparison of asteroid thermal inertia determinations from thermophysical modeling

Object	D_{eff} (km)	Thermal inertia (Jm^{-2}s$^{-0.5}$K^{-1})	Source
(1) Ceres	923	10	Müller & Lagerros (1998)
(4) Vesta	530	25	Müller & Lagerros (1998)
(2) Pallas	500	10	Müller & Lagerros (1998)
(3) Juno	240	5	Müller & Lagerros (1998)
(532) Herculina	218	15	Müller & Lagerros (1998)
(433) Eros (NEO)	17.5	~ 150	Mueller *et al.* (2005)
(1580) Betulia (NEO)	4.57 ± 0.46	~ 180	Harris *et al.* (2005)
(25143) Itokawa (NEO)	~ 0.3	~ 350	Mueller *et al.* (2005)
Bare rock		~ 2500	Jakosky (1986)

or more higher than those derived for main-belt asteroids, but much less than that of bare rock.

5. Where are the asteroids with bare rock surfaces?

The C-type asteroid Betulia is significant in being the first NEO for which thermal-infrared observations (Lebofsky, Veeder, Lebofsky, *et al.* 1978) indicated a surface of high thermal inertia. In contrast to all asteroids previously observed in the thermal infrared, the measured infrared fluxes of Betulia, together with the size estimates from radar and polarimetric observations, appeared to be inconsistent with the standard thermal model, which assumes no or very little thermal inertia. Lebofsky *et al.* (1978) found that a thermal model having the characteristics of bare rock, similar to the FRM, was required to give a size consistent with that derived from polarimetry and radar observations. The diameter resulting from their model is 7.5 ± 0.34 km. Harris *et al.* (2005) applied a thermophysical model developed by M. Mueller (paper in preparation) to recent thermal-infrared flux measurements of Betulia obtained at the NASA Infrared Telescope Facility (IRTF) and derived a diameter of 4.57 ± 0.46 km and an albedo of $p_v = 0.077 \pm 0.015$ (figure 3). The resulting thermal inertia is around 180 $\mathrm{Jm^{-2}s^{-0.5}K^{-1}}$, or some three times the lunar value. This value of thermal inertia is less than 10% of that expected for a bare-rock surface and implies that the surface of Betulia has a significant thermally-insulating regolith, in contrast to the conclusions of earlier work. The reason why the more recent results for the size and thermal inertia of Betulia differ markedly from the earlier results discussed by Lebofsky *et al.* (1978) is not clear. Betulia is well known for its unusual lightcurve, the amplitude and form of which changes dramatically with

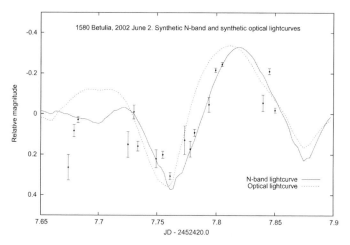

Figure 3. Synthetic thermal infrared (N-band) and synthetic optical lightcurves for (1580) Betulia generated using a thermophysical model (incorporating a shape model) for the observing geometry of 2002 June 2. Points with 1-σ statistical error bars corresponding to the measured N-band fluxes are superimposed. The zero-magnitude level of the N-band lightcurve refers to the mean flux level; the zero-magnitude level in the case of the N-band measurements refers to the NEATM thermal contiuum fit to the lightcurve-corrected data. No relative manual adjustment to the N-band synthetic lightcurve or measurement points were made. The shape model and thermophysical model generate optical and thermal-infrared lightcurves that are in good agreement with the observations. The value of thermal inertia resulting from the thermophysical model fit to the data is 180 $\mathrm{Jm^{-2}s^{-0.5}K^{-1}}$, which is much less than the expected for a surface of bare rock. For further deatils see Harris *et al.* (2005).

changing solar phase angle, probably as a result of a highly irregular shape and/or unusual topographic features (e.g. Tedesco, Drummond, Candy, *et al.* 1978). It is important to bear the unusual nature of Betulia in mind when interpreting observational data with the aid of standard optical, thermal-infrared, and other techniques. It is interesting to note that if the data of Lebofsky *et al.* (1978) are analyzed on the assumption of *low* thermal inertia, a similar (smaller) diameter is obtained to that derived by Harris *et al.* (2005). So the two sets of infrared data appear to be consistent if a relatively low thermal inertia model is used, but not if a model with the thermal characteristics of bare rock is used. On the other hand, the earlier polarimetry and radar observations appear to support a larger size. Further observations of Betulia will be required to resolve these issues. In any case, observations of unusual objects such as Betulia are particularly valuable for exploring the limits of applicability and reliability of the various analysis techniques applied to asteroid data.

While the absence of high η values at low solar phase angles in figure 2 suggests that in general NEOs with high thermal inertia are lacking in the population observed to date, it should be noted that some of the relatively high η values evident in figure 2 might be at least partly due to unusually high thermal inertia. However, as mentioned in §4, surface roughness confuses the issue at large phase angles where it can also lead to increased η values. Judging from their high η values measured at moderate phase angles, the best candidates for unusually high thermal inertia surfaces amongst the objects studied to date appear to be 2002 HK_{12} (Wolters, Green, McBride, *et al.* 2005), with $\eta = 2.75$ at a phase angle of 33° and (2100) Ra-Shalom (Delbó, Harris, Binzel, *et al.* 2003), with $\eta = 2.32$ at a phase angle of 39° (see figure 2).

There is good reason to expect that regolith-free surfaces will be found amongst very small, rapidly rotating NEOs. While confirmation of a size-dependence of thermal inertia must await the collection of data for many more objects in the sub-km size range, it seems likely that the very weak surface gravities of such objects, especially if they rotate rapidly, would be unable to retain significant thermally-insulating surface debris from impacts, unless some other effect is causing material such as dust to adhere to their surfaces. One object that promises to throw some light on this question is the NEO (54509) 2000 PH_5, which is in a dynamically interesting Sun-Earth horseshoe orbit. The size of 2000 PH_5 is 100 – 200m and its rotation period is just 0.2 hr (from P. Pravec: http://sunkl.asu.cas.cz/p̃pravec/). Current observational programs should provide information on the thermal inertia of 2000 PH_5 and whether anisotropic thermal emission is gradually modifying its rotation rate (the YORP effect).

6. How reliable are the results from thermal models?

Unfortunately at the present time very few ground-truth data are available on the sizes, albedos, thermal inertia and surface roughness of asteroids to compare with results from astronomical observations and thermal modeling. It is therefore very difficult to calibrate the models against the characteristics of known objects. Even the sophisticated thermophysical models can do no better than adopt simplified and idealized descriptions of thermal inertia and surface roughness. For example, it is not clear that the effects of surface roughness on very small, irregular bodies can be adequately described by a uniform distribution of hemispherical craters (cf. §2). A more sophisticated treatment may be required that takes account of intermediate-scale topographical structure below the resolution of the shape models. While sizes and albedos derived from thermophysical models should be more reliable than those from the simpler models outlined in §2, the

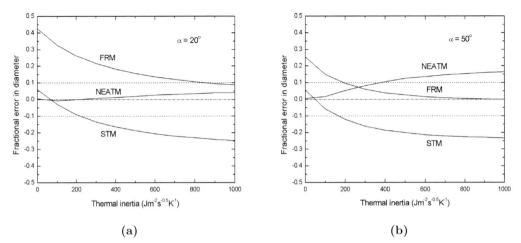

Figure 4. Plots of the performances of the STM, NEATM, and FRM against thermal inertia for solar phase angles of $\alpha = 20°$ and $\alpha = 50°$. Test data were generated using a smooth spherical model incorporating the effects of thermal inertia. In the STM the beaming parameter, η, was set to unity. The model asteroid had a rotation period of 6 hr and the sub-solar and sub-Earth latitudes were zero; other parameters were chosen to be typical of NEOs. The sense of the phase angle is such that the cooler, morning side of the asteroid was viewed. The small positive error at thermial inertia $= 0$ in the case of the STM results from the use of the phase coefficient of 0.01 mag/deg (see § 2)

corresponding values of thermal inertia and surface roughness should be treated with caution until more ground-truth data are available from space missions.

The simpler models based on spherical geometry serve the purpose of giving reasonable estimates of size and albedo in the absence of knowledge of an asteroids shape and spin vector. The accuracy obtainable with the simpler models depends on the circumstances in which they are applied. Given the range of values of thermal inertia for main-belt asteroids and NEOs listed in table 2, it is possible to gain insight into the limits of applicability of the simpler models.

To provide an illustration, synthetic flux values were generated using a smooth spherical model that incorporates the effects of thermal inertia by means of a one-dimensional vertical heat-transfer algorithm. The model asteroid had a rotation period of 6 hr and the sub-solar and sub-Earth latitudes were zero; other parameters were chosen to be typical of NEOs. Values of diameter were derived from the synthetic fluxes using the STM (with $\eta = 1.0$, i.e. smooth surface), NEATM and FRM outlined in § 2 for various values of thermal inertia and phase angle, and the fractional error in the diameter calculated in each case. Figure 4 shows the dependence of the fractional error in diameter on thermal inertia for solar phase angles, α, of 20° and 50° (n.b. the sense of the phase angle is such that the cooler, morning side of the asteroid was viewed). For low values of thermal inertia, typical of main-belt asteroids (see table 2) and the Moon, the STM and the NEATM both perform well at both values of phase angle, whereas the FRM grossly overestimates the diameter. At $\alpha = 20°$ the NEATM performs best over the full range of thermal inertia, while for thermal inertia exceeding 100 $\mathrm{Jm^{-2}s^{-0.5}K^{-1}}$ the STM underestimates the diameter to an extent that worsens with increasing thermal inertia. On the other hand the performance of the FRM improves with increasing thermal inertia, as expected. At $\alpha = 50°$ the performance of the FRM is better than that of the NEATM for

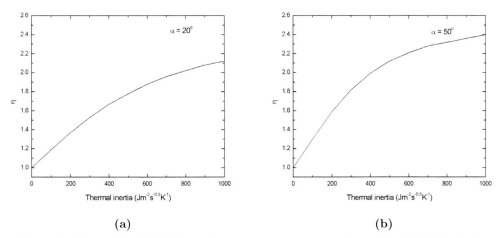

(a)   (b)

Figure 5. Plots of the NEATM model parameter η corresponding to the NEATM solutions plotted in figure 4.

values of thermal inertia above 250 $\mathrm{Jm^{-2}s^{-0.5}K^{-1}}$, although the NEATM error remains below 15% over most of the thermal inertia range. The NEATM overestimates the diameter (and therefore *underestimates* the albedo) due to the fact that it ignores thermal emission from the night side of the object, a problem that worsens with increasing phase angle and thermal inertia.

It should be noted that the plots of figure 4 represent worst cases for the NEATM and the STM in the sense that the errors were calculated for zero sub-solar latitude. For other values of sub-solar latitude (i.e. non-perpendicular inclinations of the spin-axis to the solar direction) the errors due to thermal inertia will be lower: as mentioned above, in the limit of sub-solar latitude $= 90°$, i.e. spin axis pointing towards the Sun, the temperature distribution of an FRM-type object is identical to that of the STM, i.e. an object with zero thermal inertia.

Values of η corresponding to the NEATM solutions plotted in figure 4 are shown in figure 5. There is a strong dependence of η on thermal inertia, which demonstrates why the STM, in which η is held constant, is not applicable to asteroids with significant thermal inertia.

7. Summary and discussion

Observations of asteroids in the thermal infrared are the main source of information on their sizes, albedos, and thermal inertia. Knowledge of these parameters is crucial for many aspects of asteroid research, including studies of the size distribution and impact hazard of NEOs, the taxonomic composition of populations of asteroids, the physical properties of asteroid surfaces and regolith, and calculations of the magnitude of the Yarkovsky effect. To derive physical properties from thermal-infrared data a thermal model of the asteroid is required. Commonly used models based on spherical geometry can provide acceptable results, especially when combined with knowledge of an objects optical lightcurve, but have to be applied with care, especially in the case of small objects such as NEOs. Modern computers enable the use of sophisticated thermophysical models based on realistic shapes that incorporate physical descriptions of thermal inertia and surface roughness. However, some prior knowledge of an objects rotation vector and shape is a prerequisite for the use of thermophysical models.

A very broad range of albedos is observed amongst the NEO population. There is some evidence for the effects of space weathering, or some other surface phenomenon, leading to smaller S- and Q-type objects having higher albedos. However, detection bias against small, dark objects can also give rise to such a correlation. While the present data set suggests that detection bias alone cannot explain the correlation, a final resolution of this question must await the availability of a much larger data set.

Insight into the thermal inertia of asteroids can be gained from a study of the variation of the model parameter η with solar phase angle. The parameter η in the NEATM is a measure of the deviation of the observed color temperature from that expected for a smooth sphere with zero thermal inertia. Results suggest that the thermal inertia values of most NEOs observed to date are well below the value for bare rock (~ 2500 $Jm^{-2}s^{-0.5}K^{-1}$). Values of thermal inertia for a few individual NEOs derived on the basis of thermophysical modeling are in the range $100 - 400$ $Jm^{-2}s^{-0.5}K^{-1}$, consistent with surfaces largely covered in thermally-insulating regolith or dust but an order of magnitude higher than values derived for large main-belt asteroids. Thermal inertia is an important parameter in the calculation of the orbital drift of potentially hazardous asteroids due to the Yarkovsky effect. Further observations are required to probe smaller objects and to determine the extent to which the presence of thermally-insulating surface material, and therefore thermal inertia, is size dependent.

The application of sophisticated thermophysical models incorporating information on shape and spin vector to observational data has the potential to teach us a great deal about asteroid surface properties. Unfortunately information on topographical structure analogous to hills, valleys, ridges, etc., is rarely available, so the modeling of surface struc-ture is normally limited to an idealized uniform distribution of craters represented by hemispheres, or sections thereof. The over-simplification of surface structure in thermo-physical models may seriously limit their accuracy in the case of NEOs, which are often observed at large phase angles at which shadowing effects are more significant and mea-sured fluxes are presumably strongly dependent on topographical structure, in addition to cratering. The influence of intermediate-scale surface structure on the thermophysical modeling of asteroids is an important area for future study.

References

Binzel, R.P., Lupishko, D., Di Martino, M., Whiteley, R.J., & Hahn, G.J. 2002, in: W. Bottke, A. Cellino, P. Paolicchi & R.P. Binzel (eds.), *Asteroids III*, (Tucson: Univ. Arizona Press), p. 255

Binzel, R.P., Rivkin, A.S., Stuart, J.S., Harris, A.W., Bus, S.J., & Burbine, T.H. 2004, *Icarus* 170, 259

Clark, B.E., Hapke, B., Pieters, C., & Britt, D. 2002, in: W. Bottke, A. Cellino, P. Paolicchi & R.P. Binzel (eds.), *Asteroids III*, (Tucson: Univ. Arizona Press), p. 585

Delb, M. 2004, Doctoral thesis, Freie Universität Berlin.

Delbó, M. & Harris, A.W. 2002, *Meteoritics and Planet. Sci.* 37, 1929

Delbó, M., Harris, A.W., Binzel, R.P, Pravec, P., & Davies, J.K. 2003, *Icarus* 166, 116

Fernández, Y.R., Jewitt, D.C., & Sheppard, S.S. 2005, *A.J.* in press.

Giorgini, J.D., Ostro, S.J., Benner, L.A.M., Chodas, P.W., Chesley, S.R., Hudson, R.S., Nolan, M.C., Klemola, A.R., Standish, E.M., Jurgens, R.F., Rose, R., Chamberlin, A.B., Yeomans, D.K., & Margot, J.-L. 2002, *Science* 296, 132

Harris, A.W. 1998, *Icarus* 131, 291

Harris, A.W. & Lagerros, J.S.V. 2002, in: W. Bottke, A. Cellino, P. Paolicchi & R.P. Binzel (eds.), *Asteroids III*, (Tucson: Univ. Arizona Press), p. 205

Harris A.W., Mueller, M., Delbó, M., & Bus, S.J. 2005, *Icarus* in press.

Jakosky B.M. 1986, *Icarus* 66, 117

Kaasalainen, M., and 21 colleagues 2004, *Icarus* 167, 178

Lebofsky, L.A. & Spencer, J.R. 1989, in: R. P. Binzel, T. Gehrels & M. S. Matthews (eds.), *Asteroids II*, (Tucson: Univ. Arizona Press), p. 128

Lebofsky, L.A., Sykes, M.V., Tedesco, E.F., Veeder, G.J., Matson, D.L., Brown, R.H., Gradie, J.C., Feierberg, M.A., & Rudy, R.J. 1986, *Icarus* 68, 239

Lebofsky, L.A., Veeder, G.J., Lebofsky, M.J., & Matson, D.L. 1978, *Icarus* 35, 336

Mueller, M., Delbó, M., Di Martino, M., Harris, A.W., Kaasalainen, M., & Bus, S.J. 2005, *ASP Conference Series*, (submitted Dec. 2004)

Müller, T.G. & Lagerros, J.S.V. 1998, *Astron. Astrophys.* 338, 340

Müller, T.G., Sterzik, M.F., Schütz, O., Pravec, P., & Siebenmorgen, R. 2004, *Astron. Astrophys.* 424, 1075

Spencer, J.R., Lebofsky, L.A., & Sykes, M.V. 1989, *Icarus* 78, 337

Stuart, J.S. & Binzel, R.P. 2004, *Icarus* 170, 295

Tedesco, E.F. (ed. 1992, *The IRAS minor planet survey* Tech. Rep. PL-TR-92-2049. Phillips Lab., Hanscom Air Force Base, MA.

Tedesco, E., Drummond, J., Candy, M., Birch, P., Nikoloff, I., & Zellner, B. 1978, *Icarus* 35, 344

Walker, R.G. 2003, *Bulletin of the American Astronomical Society* 35, abstract 34.19

Wolters, S.D., Green, S.F., McBride, N., & Davies, J.K. 2005, *Icarus* 175, 92

Author Index